FOURTH EDITION

Fundamental MATHEMATICS

MARVIN L. BITTINGER

Fundamental Mathematics

Fundamental Mathematics

4th EDITION

Marvin L. Bittinger

Indiana University Purdue University Indianapolis

Boston San Francisco New York
London Toronto Sydney Tokyo Singapore Madrid
Mexico City Munich Paris Cape Town Hong Kong Montreal

Publisher	Greg Tobin
Editor in Chief	Maureen O'Connor
Executive Editor	Jennifer Crum
Executive Project Manager	Kari Heen
Project Editor	Katie Nopper
Editorial Assistants	Elizabeth Bernardi, Alison Macdonald
Managing Editor	Ron Hampton
Senior Designer	Dennis Schaefer
Cover Designer	Dennis Schaefer
Supplements Supervisor	Emily Portwood
Media Producers	Sharon Smith, Ceci Fleming
Software Development	Mary Durnwald, TestGen; Jozef Kubit, MathXL
Marketing Manager	Jay Jenkins
Marketing Coordinator	Tracy Rabinowitz
Senior Author Support/ Technology Specialist	Joseph K. Vetere
Senior Prepress Supervisor	Caroline Fell
Rights and Permissions	Dana Weightman, Ellen Keohane
Senior Manufacturing Buyer	Evelyn M. Beaton
Art and Design Services	Geri Davis/The Davis Group, Inc.
Production Coordination	Kathy Diamond
Composition	Beacon Publishing Services
Illustrations	William Melvin, Network Graphics
Cover Photo	© Lindsey P. Martin/Corbis

About the cover: The aurora borealis, or northern lights, as photographed in the skies over Alaska.

Photo credits appear on page P-1.

Library of Congress Cataloging-in-Publication Data

Bittinger, Marvin L.
Fundamental mathematics.—4th ed./Marvin L. Bittinger.
p. cm.
Includes indexes.
ISBN 0-321-31907-9 (Student's Edition)
1. Mathematics. I. Title.

QA39.3.B57 2005
513—dc22 2005050999

1 2 3 4 5 6 7 8 9 10–VHP–09 08 07 06 05

Contents

3 FRACTION NOTATION AND MIXED NUMERALS

DECIMAL NOTATION

5 RATIO AND PROPORTION

6 PERCENT NOTATION

Index of Applications

Consumer

Domestic

Economics

Education

Engineering

Environment

Finance

Food

Geometry

Government

Health/Medicine

Labor

Miscellaneous

Physics

Social Sciences

Sports/Entertainment

Statistics/Demographics

Technology

Transportation

Index of Study Tips

CHAPTER 1

CHAPTER 2

CHAPTER 3

CHAPTER 4

CHAPTER 5

CHAPTER 6

Preface

It is with great pride and excitement that we present to you the fourth edition of *Fundamental Mathematics,* a brief version of *Basic Mathematics,* Tenth Edition, by Marvin L. Bittinger. *Fundamental Mathematics* is appropriate for a course in prealgebra or arithmetic and is part of a series of texts that has evolved dramatically over the past 36 years through your comments, responses, and opinions. This feedback, combined with our overall objective of presenting the material in a clear and accurate manner, drives each revision. It is our hope that *Fundamental Mathematics,* Fourth Edition, and the supporting supplements will help provide an improved teaching and learning experience by matching the needs of instructors and successfully preparing students for their future.

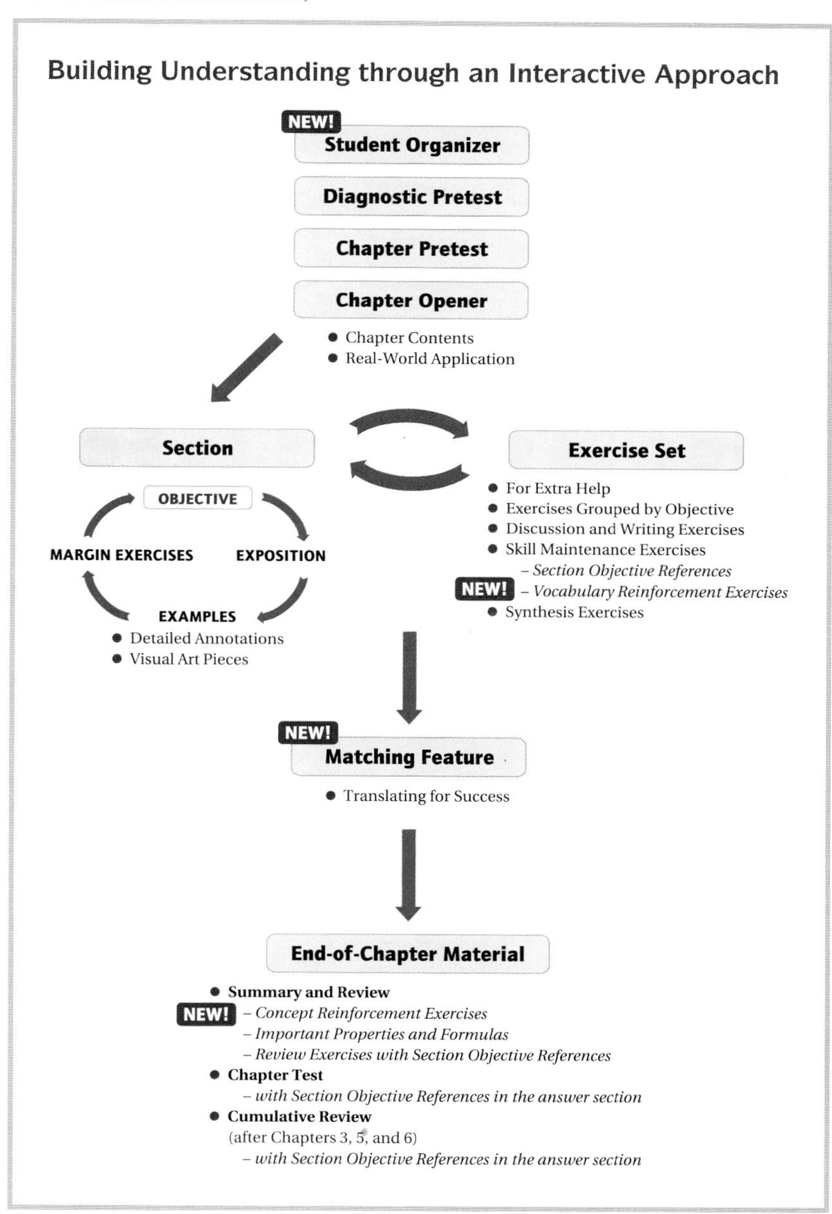

This text is the second in a series that includes the following:

Bittinger: *Basic Mathematics,* Tenth Edition

Bittinger: *Fundamental Mathematics,* Fourth Edition

Bittinger: *Introductory Algebra,* Tenth Edition

Bittinger: *Intermediate Algebra,* Tenth Edition

Bittinger/Beecher: *Introductory and Intermediate Algebra,* Third Edition

Building Understanding through an Interactive Approach

The pedagogy of this series is designed to provide an interactive learning experience between the student and the exposition, annotated examples, art, margin exercises, and exercise sets. This unique approach, which has been developed and refined over ten editions and is illustrated at the right, provides students with a clear set of learning objectives, involves them with the development of the material, and provides immediate and continual reinforcement and assessment through the margin exercises.

Let's Visit the Fourth Edition

The style, format, and approach of the third edition have been strengthened in this new edition in a number of ways. However, the accuracy that the Bittinger books are known for has not changed. This edition, as with all editions, has gone through an exhaustive checking process to ensure accuracy in the problem sets, mathematical art, and accompanying supplements. We know what a critical role the accuracy of a book plays in student learning and we value the reputation for accuracy that we have earned.

NEW!

IN THE FOURTH EDITION

Each revision gives us the opportunity to incorporate new elements and refine existing elements to provide a better experience for students and teachers alike. Below are four new features designed to help students succeed.

- Student Organizer
- Translating for Success matching exercises
- Vocabulary Reinforcement exercises
- Concept Reinforcement exercises

These features, along with the hallmark features of this book, are discussed in the pages that follow.

In addition, the fourth edition has been designed to be open and flexible, helping students focus their attention on details that are critical at this level through prominent headings, boxed definitions and rules, and clearly labeled objectives. *Chapter Pretests*, now located along with the Diagnostic Pretest in the *Printed Test Bank* and in MyMathLab, diagnose at the section and objective level and can be used to place students in a specific section, or objective, of the chapter, allowing them to concentrate on topics with which they have particular difficulty. Answers to these pretests are available in the *Printed Test Bank* and in MyMathLab.

NEW! STUDENT ORGANIZER

Along with the study tips found throughout the text, we have provided a pull-out card that will help students stay organized and increase their ability to be successful in this course. Students can use this card to keep track of important dates and useful contact information and to access information for technology and to plan class time, study time, work time, travel time, family time, and, sometimes most importantly, relaxation time.

PLANNING FOR SUCCESS

Bittinger Student Organizer COURSE INFORMATION

Study Tips
Throughout this text, you will find a feature called *Study Tips.* We discuss these in the Preface of this text. They are intended to help improve your math study skills. An index of all t... can ... fron...

On the first day of class, complete this chart and the weekly planner that follows on the reverse page.

Instructor Information:
Name
Office Hours and Location
Phone Number

SCHEDULING FOR SUCCESS

Bittinger Student Organizer WEEKLY PLANNER

Success is planned. On this page, plan a typical week. Consider time allotments for class, study, work, travel, family, and relaxation.

Important Dates
Midterm Exam
Final Exam
Holidays
Other

TIME	Sun.	Mon.	Tues.	Wed.	Thurs.	Fri.	Sat.
6:00 A.M.							
6:30							
7:00							
7:30							
8:00							
8:30							
9:00							
9:30							
10:00							
10:30							
11:00							
11:30							
12:00 P.M.							
12:30							
1:00							
1:30							
2:00							
2:30							
3:00							
3:30							
4:00							
4:30							
5:00							
5:30							
6:00							
6:30							
7:00							
7:30							
8:00							
8:30							
9:00							
9:30							
10:00							
10:30							
11:00							
11:30							
12:00 A.M.							

For more information about this text or any of its supplements, visit www.aw-bc.com/math

CHAPTER OPENERS

To engage students and prepare them for the upcoming chapter material, gateway chapter openers are designed with exceptional artwork that is tied to a motivating real-world application. (See pages 1, 89, and 149.)

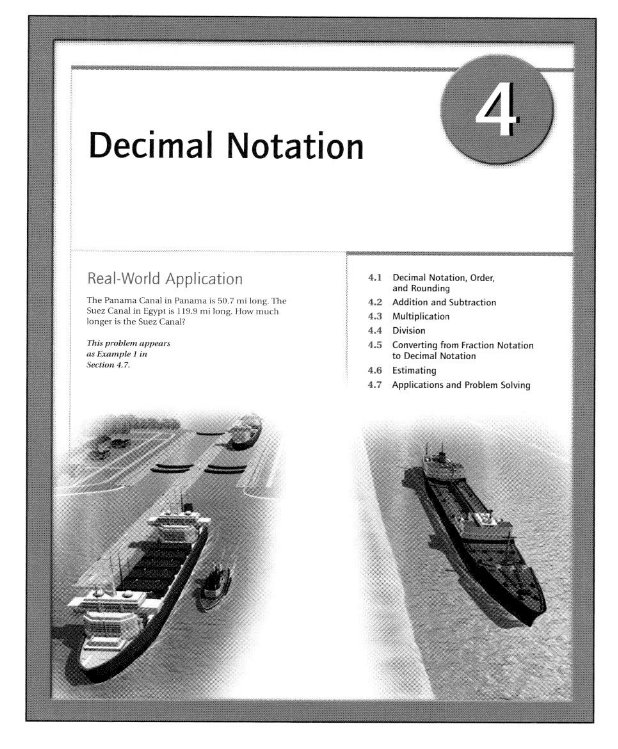

4

Decimal Notation

Real-World Application

The Panama Canal in Panama is 50.7 mi long. The Suez Canal in Egypt is 119.9 mi long. How much longer is the Suez Canal?

This problem appears as Example 1 in Section 4.7.

4.1 Decimal Notation, Order, and Rounding
4.2 Addition and Subtraction
4.3 Multiplication
4.4 Division
4.5 Converting from Fraction Notation to Decimal Notation
4.6 Estimating
4.7 Applications and Problem Solving

OBJECTIVE BOXES

At the beginning of each section, a boxed list of objectives is keyed by letter not only to section subheadings, but also to the exercises in the Pretest (located in the *Printed Test Bank* and MyMathLab), the section exercise sets, and the Summary and Review exercises, as well as to the answers to the questions in the Chapter Tests and Cumulative Reviews. This correlation enables students to easily find appropriate review material if they need help with a particular exercise or skill at the objective level. (See pages 90, 165, and 218.)

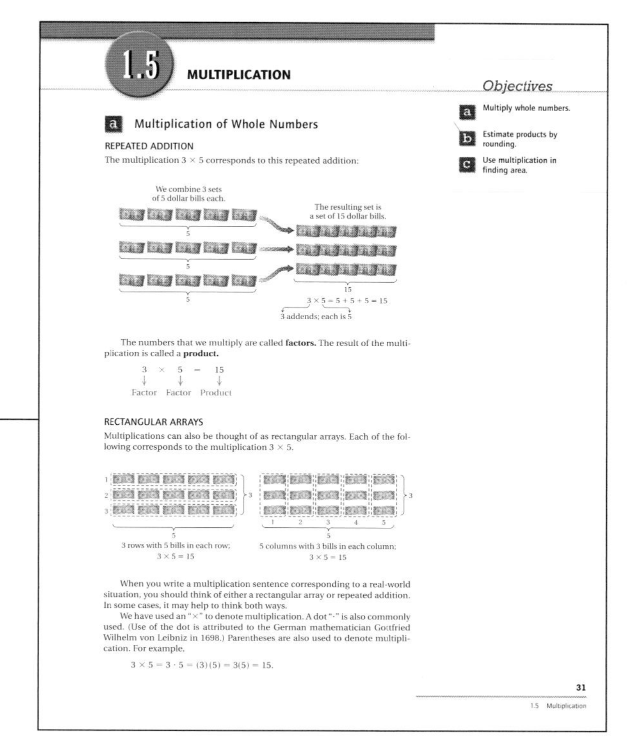

1.5 MULTIPLICATION

Objectives

a Multiply whole numbers.
b Estimate products by rounding.
c Use multiplication in finding area.

a Multiplication of Whole Numbers

REPEATED ADDITION

The multiplication 3×5 corresponds to this repeated addition:

We combine 3 sets of 5 dollar bills each.
The resulting set is a set of 15 dollar bills.
$3 \times 5 = 5 + 5 + 5 = 15$
3 addends; each is 5

The numbers that we multiply are called **factors.** The result of the multiplication is called a **product.**

$3 \times 5 = 15$
Factor Factor Product

RECTANGULAR ARRAYS

Multiplications can also be thought of as rectangular arrays. Each of the following corresponds to the multiplication 3×5.

3 rows with 5 bills in each row; $3 \times 5 = 15$
5 columns with 3 bills in each column; $3 \times 5 = 15$

When you write a multiplication sentence corresponding to a real-world situation, you should think of either a rectangular array or repeated addition. In some cases, it may help to think both ways.

We have used an "×" to denote multiplication. A dot "·" is also commonly used. (Use of the dot is attributed to the German mathematician Gottfried Wilhelm von Leibniz in 1698.) Parentheses are also used to denote multiplication. For example,

$3 \times 5 = 3 \cdot 5 = (3)(5) = 3(5) = 15.$

31
1.5 Multiplication

ANNOTATED EXAMPLES

Detailed annotations and color highlights lead the student through the structured steps of the examples. The level of detail in these annotations is a significant reason for students' success with this book. (See pages 168, 192, and 314.)

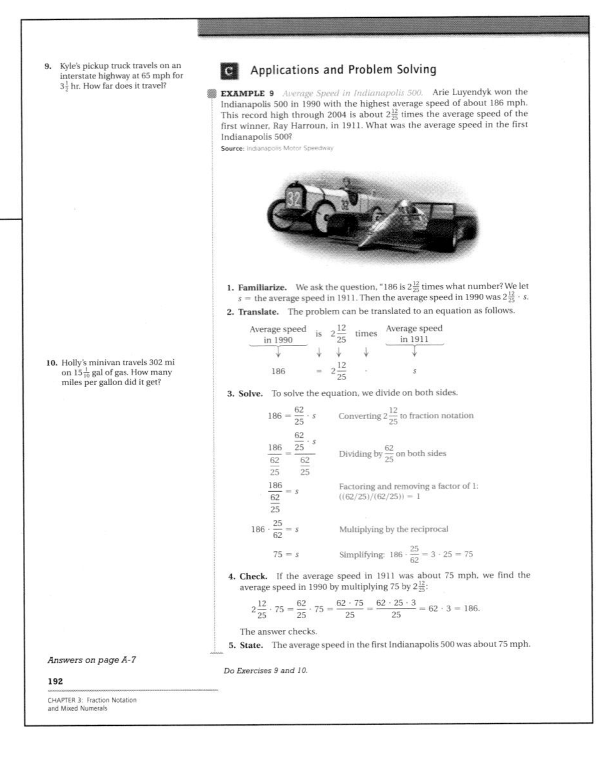

9. Kyle's pickup truck travels on an interstate highway at 65 mph for $3\frac{1}{2}$ hr. How far does it travel?

10. Holly's minivan travels 302 mi on $15\frac{1}{10}$ gal of gas. How many miles per gallon did it get?

Answers on page A-7

C Applications and Problem Solving

EXAMPLE 9 *Average Speed in Indianapolis 500.* Arie Luyendyk won the Indianapolis 500 in 1990 with the highest average speed of about 186 mph. This record high through 2004 is about $2\frac{12}{25}$ times the average speed of the first winner, Ray Harroun, in 1911. What was the average speed in the first Indianapolis 500?

Source: Indianapolis Motor Speedway

1. **Familiarize.** We ask the question, "186 is $2\frac{12}{25}$ times what number? We let s = the average speed in 1911. Then the average speed in 1990 was $2\frac{12}{25}\cdot s$.
2. **Translate.** The problem can be translated to an equation as follows.

Average speed in 1990 is $2\frac{12}{25}$ times Average speed in 1911

$186 = 2\frac{12}{25}\cdot s$

3. **Solve.** To solve the equation, we divide on both sides.

$186 = \frac{62}{25}\cdot s$ Converting $2\frac{12}{25}$ to fraction notation

$\frac{186}{\frac{62}{25}} = \frac{\frac{62}{25}\cdot s}{\frac{62}{25}}$ Dividing by $\frac{62}{25}$ on both sides

$\frac{186}{\frac{62}{25}} = s$ Factoring and removing a factor of 1: $((62/25)/(62/25)) = 1$

$186\cdot\frac{25}{62} = s$ Multiplying by the reciprocal

$75 = s$ Simplifying: $186\cdot\frac{25}{62} = 3\cdot 25 = 75$

4. **Check.** If the average speed in 1911 was about 75 mph, we find the average speed in 1990 by multiplying 75 by $2\frac{12}{25}$:

$2\frac{12}{25}\cdot 75 = \frac{62}{25}\cdot 75 = \frac{62\cdot 75}{25} = \frac{62\cdot 25\cdot 3}{25} = 62\cdot 3 = 186.$

The answer checks.

5. **State.** The average speed in the first Indianapolis 500 was about 75 mph.

Do Exercises 9 and 10.

192

CHAPTER 3: Fraction Notation and Mixed Numerals

MARGIN EXERCISES

Throughout the text, students are directed to numerous margin exercises that provide immediate practice and reinforcement of the concepts covered in each section. Answers are provided at the back of the book so students can immediately self-assess their understanding of the skill or concept at hand. (See pages 91, 250, and 292.)

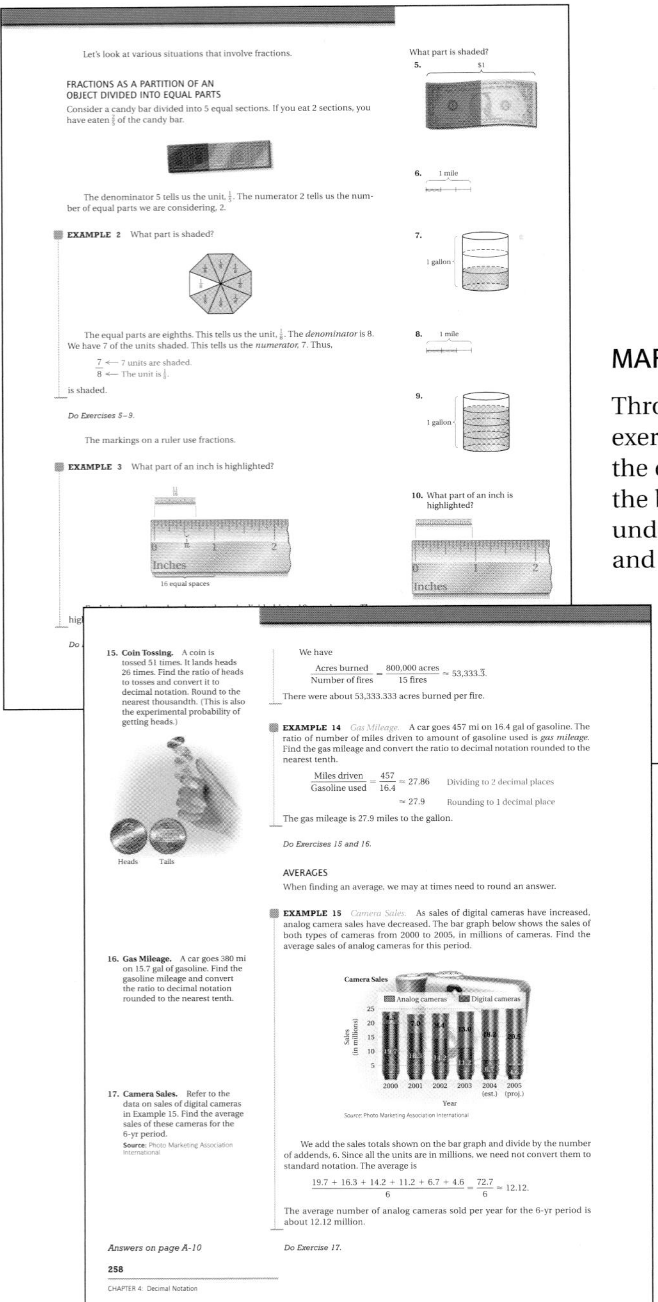

Let's look at various situations that involve fractions.

FRACTIONS AS A PARTITION OF AN OBJECT DIVIDED INTO EQUAL PARTS

Consider a candy bar divided into 5 equal sections. If you eat 2 sections, you have eaten $\frac{2}{5}$ of the candy bar.

The denominator 5 tells us the unit, $\frac{1}{5}$. The numerator 2 tells us the number of equal parts we are considering, 2.

EXAMPLE 2 What part is shaded?

The equal parts are eighths. This tells us the unit, $\frac{1}{8}$. The *denominator* is 8. We have 7 of the units shaded. This tells us the *numerator*, 7. Thus,

$\frac{7}{8}$ ← 7 units are shaded. ← The unit is $\frac{1}{8}$.

is shaded.

Do Exercises 5–9.

The markings on a ruler use fractions.

EXAMPLE 3 What part of an inch is highlighted?

What part is shaded?

5. $1

6. 1 mile

7. 1 gallon

8. 1 mile

9. 1 gallon

10. What part of an inch is highlighted?

15. **Coin Tossing.** A coin is tossed 51 times. It lands heads 26 times. Find the ratio of heads to tosses and convert it to decimal notation. Round to the nearest thousandth. (This is also the experimental probability of getting heads.)

Heads Tails

16. **Gas Mileage.** A car goes 380 mi on 15.7 gal of gasoline. Find the gasoline mileage and convert the ratio to decimal notation rounded to the nearest tenth.

17. **Camera Sales.** Refer to the data on sales of digital cameras in Example 15. Find the average sales of these cameras for the 6-yr period.

Source: Photo Marketing Association International

Answers on page A-10

We have

$\frac{\text{Acres burned}}{\text{Number of fires}} = \frac{800{,}000 \text{ acres}}{15 \text{ fires}} \approx 53{,}333.\overline{3}.$

There were about 53,333.333 acres burned per fire.

EXAMPLE 14 *Gas Mileage.* A car goes 457 mi on 16.4 gal of gasoline. The ratio of number of miles driven to amount of gasoline used is *gas mileage*. Find the gas mileage and convert the ratio to decimal notation rounded to the nearest tenth.

$\frac{\text{Miles driven}}{\text{Gasoline used}} = \frac{457}{16.4} \approx 27.86$ Dividing to 2 decimal places

≈ 27.9 Rounding to 1 decimal place

The gas mileage is 27.9 miles to the gallon.

Do Exercises 15 and 16.

AVERAGES

When finding an average, we may at times need to round an answer.

EXAMPLE 15 *Camera Sales.* As sales of digital cameras have increased, analog camera sales have decreased. The bar graph below shows the sales of both types of cameras from 2000 to 2005, in millions of cameras. Find the average sales of analog cameras for this period.

Source: Photo Marketing Association International

We add the sales totals shown on the bar graph and divide by the number of addends, 6. Since all the units are in millions, we need not convert them to standard notation. The average is

$\frac{19.7 + 16.3 + 14.2 + 11.2 + 6.7 + 4.6}{6} = \frac{72.7}{6} \approx 12.12.$

The average number of analog cameras sold per year for the 6-yr period is about 12.12 million.

Do Exercise 17.

258

CHAPTER 4: Decimal Notation

REAL-DATA APPLICATIONS

This text encourages students to see and interpret the mathematics that appears every day in the world around them. Throughout the writing process, an extensive and energetic search for real-data applications was conducted, and the result is a variety of examples and exercises that connect the mathematical content with the real world. A large number of the applications are new to this edition, and many are drawn from the fields of business and economics, life and physical sciences, social sciences, and areas of general interest such as sports and daily life. To further encourage students to understand the relevance of mathematics, many applications are enhanced by graphs and drawings similar to those found in today's newspapers and magazines, and feature source lines as well. (See pages 164, 239, and 301.)

NEW! TRANSLATING FOR SUCCESS

Translating for Success The goal of the matching exercises in this new feature is to practice step two, *Translate,* of the five-step problem-solving process. Students translate each of ten problems to an equation and select the correct translation from fifteen given equations. This feature appears once in each chapter and reviews skills and concepts with problems from all preceding chapters. (See pages 195, 277, and 319.)

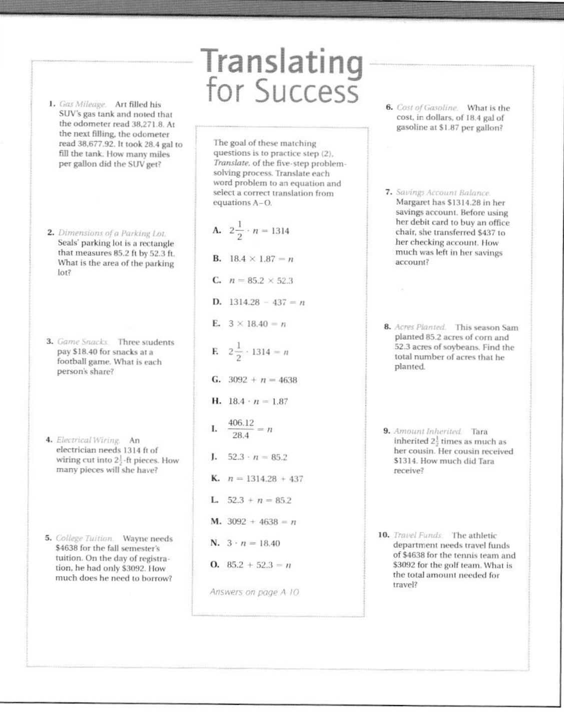

Translating for Success

The goal of these matching questions is to practice step (2), *Translate,* of the five-step problem-solving process. Translate each word problem to an equation and select a correct translation from equations A–O.

1. *Gas Mileage.* Art filled his SUV's gas tank and noted that the odometer read 38,271.8. At the next filling, the odometer read 38,677.92. It took 28.4 gal to fill the tank. How many miles per gallon did the SUV get?
2. *Dimensions of a Parking Lot.* Seals' parking lot is a rectangle that measures 85.2 ft by 52.3 ft. What is the area of the parking lot?
3. *Game Snacks.* Three students pay $18.40 for snacks at a football game. What is each person's share?
4. *Electrical Wiring.* An electrician needs 1314 ft of wiring cut into $2\frac{1}{2}$-ft pieces. How many pieces will she have?
5. *College Tuition.* Wayne needs $4638 for the fall semester's tuition. On the day of registration, he had only $3092. How much does he need to borrow?
6. *Cost of Gasoline.* What is the cost, in dollars, of 18.4 gal of gasoline at $1.87 per gallon?
7. *Savings Account Balance.* Margaret has $1314.28 in her savings account. Before using her debit card to buy an office chair, she transferred $437 to her checking account. How much was left in her savings account?
8. *Acres Planted.* This season Sam planted 85.2 acres of corn and 52.3 acres of soybeans. Find the total number of acres that he planted.
9. *Amount Inherited.* Tara inherited $2\frac{1}{2}$ times as much as her cousin. Her cousin received $1314. How much did Tara receive?
10. *Travel Funds.* The athletic department needs travel funds of $4638 for the tennis team and $3092 for the golf team. What is the total amount needed for travel?

A. $2\frac{1}{2} \cdot n = 1314$
B. $18.4 \times 1.87 = n$
C. $n = 85.2 \times 52.3$
D. $1314.28 - 437 = n$
E. $3 \times 18.40 = n$
F. $2\frac{1}{2} \cdot 1314 = n$
G. $3092 + n = 4638$
H. $18.4 \cdot n = 1.87$
I. $\frac{406.12}{28.4} = n$
J. $52.3 \cdot n = 85.2$
K. $n = 1314.28 + 437$
L. $52.3 + n = 85.2$
M. $3092 + 4638 = n$
N. $3 \cdot n = 18.40$
O. $85.2 + 52.3 = n$

Answers on page A-10

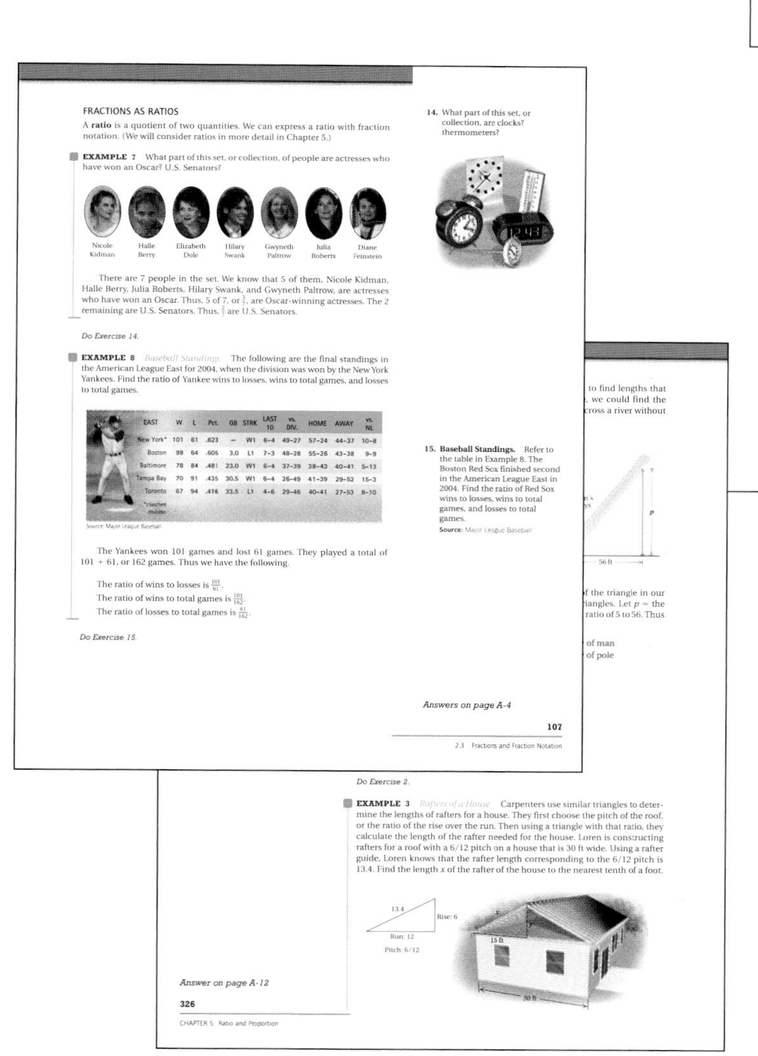

FRACTIONS AS RATIOS

A **ratio** is a quotient of two quantities. We can express a ratio with fraction notation. (We will consider ratios in more detail in Chapter 5.)

EXAMPLE 7 What part of this set, or collection, of people are actresses who have won an Oscar? U.S. Senators?

Nicole Kidman, Halle Berry, Elizabeth Dole, Hilary Swank, Gwyneth Paltrow, Julia Roberts, Diane Feinstein

There are 7 people in the set. We know that 5 of them, Nicole Kidman, Halle Berry, Julia Roberts, Hilary Swank, and Gwyneth Paltrow, are actresses who have won an Oscar. Thus, 5 of 7, or $\frac{5}{7}$, are Oscar-winning actresses. The 2 remaining are U.S. Senators. Thus, $\frac{2}{7}$ are U.S. Senators.

Do Exercise 14.

EXAMPLE 8 *Baseball Standings.* The following are the final standings in the American League East for 2004, when the division was won by the New York Yankees. Find the ratio of Yankee wins to losses, wins to total games, and losses to total games.

EAST	W	L	Pct.	GB	STRK	LAST 10	vs. DIV.	HOME	AWAY	vs. NL
New York*	101	61	.623	–	W1	6-4	49-27	57-24	44-37	10-8
Boston	98	64	.605	3.0	L1	7-3	48-28	55-26	43-38	9-9
Baltimore	78	84	.481	23.0	W1	6-4	37-39	38-43	40-41	5-13
Tampa Bay	70	91	.435	30.5	W1	6-4	26-49	41-39	29-52	15-3
Toronto	67	94	.416	33.5	L1	4-6	29-46	40-41	27-53	8-10

*clinched division

Source: Major League Baseball

The Yankees won 101 games and lost 61 games. They played a total of 101 + 61, or 162 games. Thus we have the following.

The ratio of wins to losses is $\frac{101}{61}$.
The ratio of wins to total games is $\frac{101}{162}$.
The ratio of losses to total games is $\frac{61}{162}$.

Do Exercise 15.

14. What part of this set, or collection, are clocks? thermometers?

15. **Baseball Standings.** Refer to the table in Example 8. The Boston Red Sox finished second in the American League East in 2004. Find the ratio of losses, wins to total games, and losses to total games.
Source: Major League Baseball

Answers on page A-4

107

2.3 Fractions and Fraction Notation

Do Exercise 2.

EXAMPLE 3 *Rafters of a House.* Carpenters use similar triangles to determine the lengths of rafters for a house. They first choose the pitch of the roof, or the ratio of the rise over the run. Then using a triangle with that ratio, they calculate the length of the rafter needed for the house. Loren is constructing rafters for a roof with a 6/12 pitch on a house that is 30 ft wide. Using a rafter guide, Loren knows that the rafter length corresponding to the 6/12 pitch is 13.4. Find the length x of the rafter of the house to the nearest tenth of a foot.

Answer on page A-12

326

CHAPTER 5: Ratio and Proportion

ART PROGRAM

Today's students are often visually oriented and their approach to a printed page is no exception. The art program is designed to improve the visualization of the mathematical concepts and to enhance the real-data applications. (See pages 107, 270, and 320.)

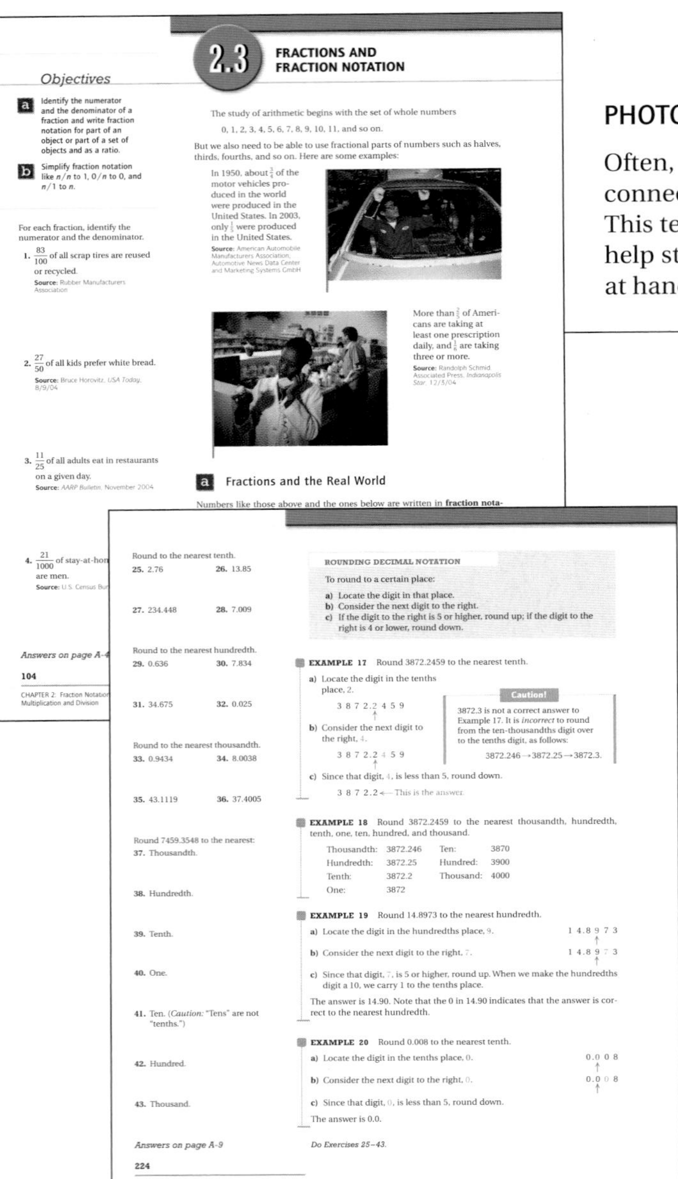

2.3 FRACTIONS AND FRACTION NOTATION

Objectives

a Identify the numerator and the denominator of a fraction and write fraction notation for part of an object or part of a set of objects and as a ratio.

b Simplify fraction notation like n/n to 1, $0/n$ to 0, and $n/1$ to n.

The study of arithmetic begins with the set of whole numbers

0, 1, 2, 3, 4, 5, 6, 7, 8, 9, 10, 11, and so on.

But we also need to be able to use fractional parts of numbers such as halves, thirds, fourths, and so on. Here are some examples:

In 1950, about $\frac{3}{4}$ of the motor vehicles produced in the world were produced in the United States. In 2003, only $\frac{1}{6}$ were produced in the United States.
Source: American Automobile Manufacturers Association, Automotive News Data Center and Marketing Systems GmbH

More than $\frac{2}{5}$ of Americans are taking at least one prescription daily, and $\frac{1}{6}$ are taking three or more.
Source: Randolph Schmid, Associated Press, Indianapolis Star, 12/3/04

For each fraction, identify the numerator and the denominator.

1. $\frac{83}{100}$ of all scrap tires are reused or recycled.
Source: Rubber Manufacturers Association

2. $\frac{27}{50}$ of all kids prefer white bread.
Source: Bruce Horovitz, USA Today, 8/9/04

3. $\frac{11}{25}$ of all adults eat in restaurants on a given day.
Source: AARP Bulletin, November 2004

4. $\frac{21}{1000}$ of stay-at-ho[me] ... are men.
Source: U.S. Census Bu[reau]

a Fractions and the Real World

Numbers like those above and the ones below are written in **fraction nota-**

Answers on page A-4

104

CHAPTER 2: Fraction Notation: Multiplication and Division

Round to the nearest tenth.

25. 2.76
26. 13.85
27. 234.448
28. 7.009

Round to the nearest hundredth.

29. 0.636
30. 7.834
31. 34.675
32. 0.025

Round to the nearest thousandth.

33. 0.9434
34. 8.0038
35. 43.1119
36. 37.4005

Round 7459.3548 to the nearest:

37. Thousandth.
38. Hundredth.
39. Tenth.
40. One.
41. Ten. (*Caution:* "Tens" are not "tenths.")
42. Hundred.
43. Thousand.

ROUNDING DECIMAL NOTATION

To round to a certain place:

a) Locate the digit in that place.
b) Consider the next digit to the right.
c) If the digit to the right is 5 or higher, round up; if the digit to the right is 4 or lower, round down.

EXAMPLE 17 Round 3872.2459 to the nearest tenth.

a) Locate the digit in the tenths place, 2.

3 8 7 2.2 4 5 9

b) Consider the next digit to the right, 4.

3 8 7 2.2 4 5 9

c) Since that digit, 4, is less than 5, round down.

3 8 7 2.2 ← This is the answer.

Caution!

3872.3 is not a correct answer to Example 17. It is *incorrect* to round from the ten-thousandths digit over to the tenths digit, as follows:

3872.246 → 3872.25 → 3872.3.

EXAMPLE 18 Round 3872.2459 to the nearest thousandth, hundredth, tenth, one, ten, hundred, and thousand.

Thousandth:	3872.246	Ten:	3870
Hundredth:	3872.25	Hundred:	3900
Tenth:	3872.2	Thousand:	4000
One:	3872		

EXAMPLE 19 Round 14.8973 to the nearest hundredth.

a) Locate the digit in the hundredths place, 9. 1 4.8 9 7 3

b) Consider the next digit to the right, 7. 1 4.8 9 7 3

c) Since that digit, 7, is 5 or higher, round up. When we make the hundredths digit a 10, we carry 1 to the tenths place.

The answer is 14.90. Note that the 0 in 14.90 indicates that the answer is correct to the nearest hundredth.

EXAMPLE 20 Round 0.008 to the nearest tenth.

a) Locate the digit in the tenths place, 0. 0.0 0 8

b) Consider the next digit to the right, 0. 0.0 0 8

c) Since that digit, 0, is less than 5, round down.

The answer is 0.0.

Answers on page A-9

Do Exercises 25–43.

224

CHAPTER 4: Decimal Notation

PHOTOGRAPHS

Often, an application becomes relevant to students when the connection to the real world is illustrated with a photograph. This text has numerous photographs throughout in order to help students see the relevance and visualize the application at hand. (See pages 104, 219, and 303.)

CAUTION BOXES

Found at relevant points throughout the text, boxes with the "Caution!" heading warn students of common misconceptions or errors made in performing a particular mathematics operation or skill. (See pages 125, 190, and 224.)

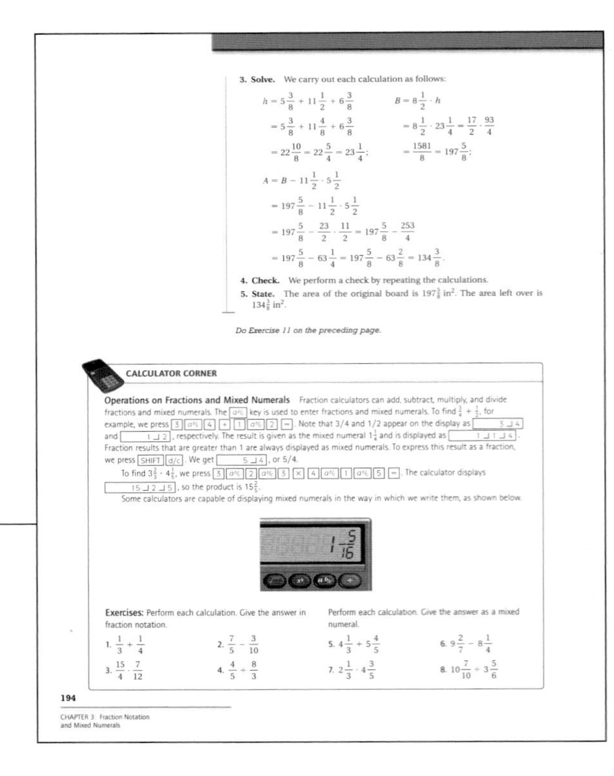

3. **Solve.** We carry out each calculation as follows:

$h = 5\frac{3}{8} + 11\frac{1}{2} + 6\frac{3}{8} = 5\frac{3}{8} + 11\frac{4}{8} + 6\frac{3}{8} = 22\frac{10}{8} = 22\frac{5}{4} = 23\frac{1}{4}$;

$B = 8\frac{1}{2} \cdot h = 8\frac{1}{2} \cdot 23\frac{1}{4} = \frac{17}{2} \cdot \frac{93}{4} = \frac{1581}{8} = 197\frac{5}{8}$;

$A = B - 11\frac{1}{2} \cdot 5\frac{1}{2} = 197\frac{5}{8} - 11\frac{1}{2} \cdot 5\frac{1}{2} = 197\frac{5}{8} - \frac{23}{2} \cdot \frac{11}{2} = 197\frac{5}{8} - \frac{253}{4} = 197\frac{5}{8} - 63\frac{1}{4} = 197\frac{5}{8} - 63\frac{2}{8} = 134\frac{3}{8}$.

4. **Check.** We perform a check by repeating the calculations.

5. **State.** The area of the original board is $197\frac{5}{8}$ in². The area left over is $134\frac{3}{8}$ in².

Do Exercise 11 on the preceding page.

CALCULATOR CORNER

Operations on Fractions and Mixed Numerals Fraction calculators can add, subtract, multiply, and divide fractions and mixed numerals. The a b/c key is used to enter fractions and mixed numerals. To find $\frac{3}{4} + \frac{1}{2}$, for example, we press 3 a b/c 4 + 1 a b/c 2 =. Note that 3/4 and 1/2 appear on the display as 3 ⌟ 4 and 1 ⌟ 2, respectively. The result is given as the mixed numeral $1\frac{1}{4}$ and is displayed as 1 ⌟ 1 ⌟ 4. Fraction results that are greater than 1 are always displayed as mixed numerals. To express this result as a fraction, we press SHIFT d/c. We get 5 ⌟ 4, or 5/4.

To find $3\frac{2}{5} \cdot 4\frac{1}{5}$, we press 3 a b/c 2 a b/c 5 × 4 a b/c 1 a b/c 5 =. The calculator displays 15 ⌟ 2 ⌟ 5, so the product is $15\frac{2}{5}$.

Some calculators are capable of displaying mixed numerals in the way in which we write them, as shown below.

Exercises: Perform each calculation. Give the answer in fraction notation.

1. $\frac{1}{3} + \frac{1}{4}$
2. $\frac{7}{5} - \frac{3}{10}$
3. $\frac{15}{4} \cdot \frac{7}{12}$
4. $\frac{4}{5} \div \frac{8}{3}$

Perform each calculation. Give the answer as a mixed numeral.

5. $4\frac{1}{3} + 5\frac{4}{5}$
6. $9\frac{2}{7} - 8\frac{1}{4}$
7. $2\frac{1}{3} \cdot 4\frac{3}{5}$
8. $10\frac{7}{10} \div 3\frac{5}{6}$

194

CHAPTER 3: Fraction Notation and Mixed Numerals

CALCULATOR CORNERS

Where appropriate throughout the text, students will find optional Calculator Corners. Popular in the third edition, these Calculator Corners have been revised to be more accessible to students and to represent current calculators. (See pages 194, 230, and 310.)

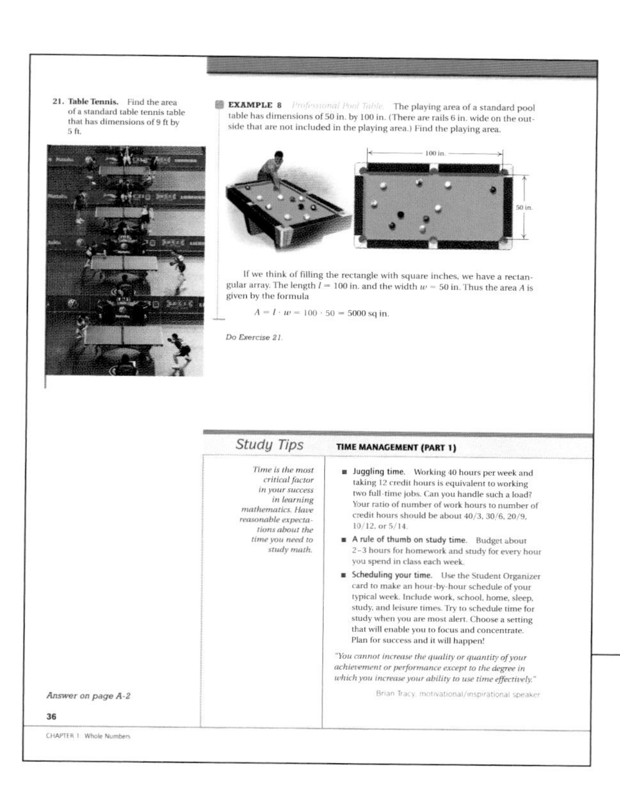

21. **Table Tennis.** Find the area of a standard table tennis table that has dimensions of 9 ft by 5 ft.

EXAMPLE 8 *Professional Pool Table.* The playing area of a standard pool table has dimensions of 50 in. by 100 in. (There are rails 6 in. wide on the outside that are not included in the playing area.) Find the playing area.

If we think of filling the rectangle with square inches, we have a rectangular array. The length $l = 100$ in. and the width $w = 50$ in. Thus the area A is given by the formula

$$A = l \cdot w = 100 \cdot 50 = 5000 \text{ sq in.}$$

Do Exercise 21.

Study Tips **TIME MANAGEMENT (PART 1)**

Time is the most critical factor in your success in learning mathematics. Have reasonable expectations about the time you need to study math.

- **Juggling time.** Working 40 hours per week and taking 12 credit hours is equivalent to working two full-time jobs. Can you handle such a load? Your ratio of number of work hours to number of credit hours should be about 40/3, 30/6, 20/9, 10/12, or 5/14.
- **A rule of thumb on study time.** Budget about 2–3 hours for homework and study for every hour you spend in class each week.
- **Scheduling your time.** Use the Student Organizer card to make an hour-by-hour schedule of your typical week. Include work, school, home, sleep, study, and leisure times. Try to schedule time for study when you are most alert. Choose a setting that will enable you to focus and concentrate. Plan for success and it will happen!

"You cannot increase the quality or quantity of your achievement or performance except to the degree in which you increase your ability to use time effectively."

Brian Tracy, motivational/inspirational speaker

Answer on page A-2

36

CHAPTER 1 Whole Numbers

STUDY TIPS

A variety of Study Tips throughout the text give students pointers on how to develop good study habits as they progress through the course. At times short snippets and at other times more lengthy discussions, these Study Tips encourage students to get involved in the learning process. (See pages 155, 295, and 308.)

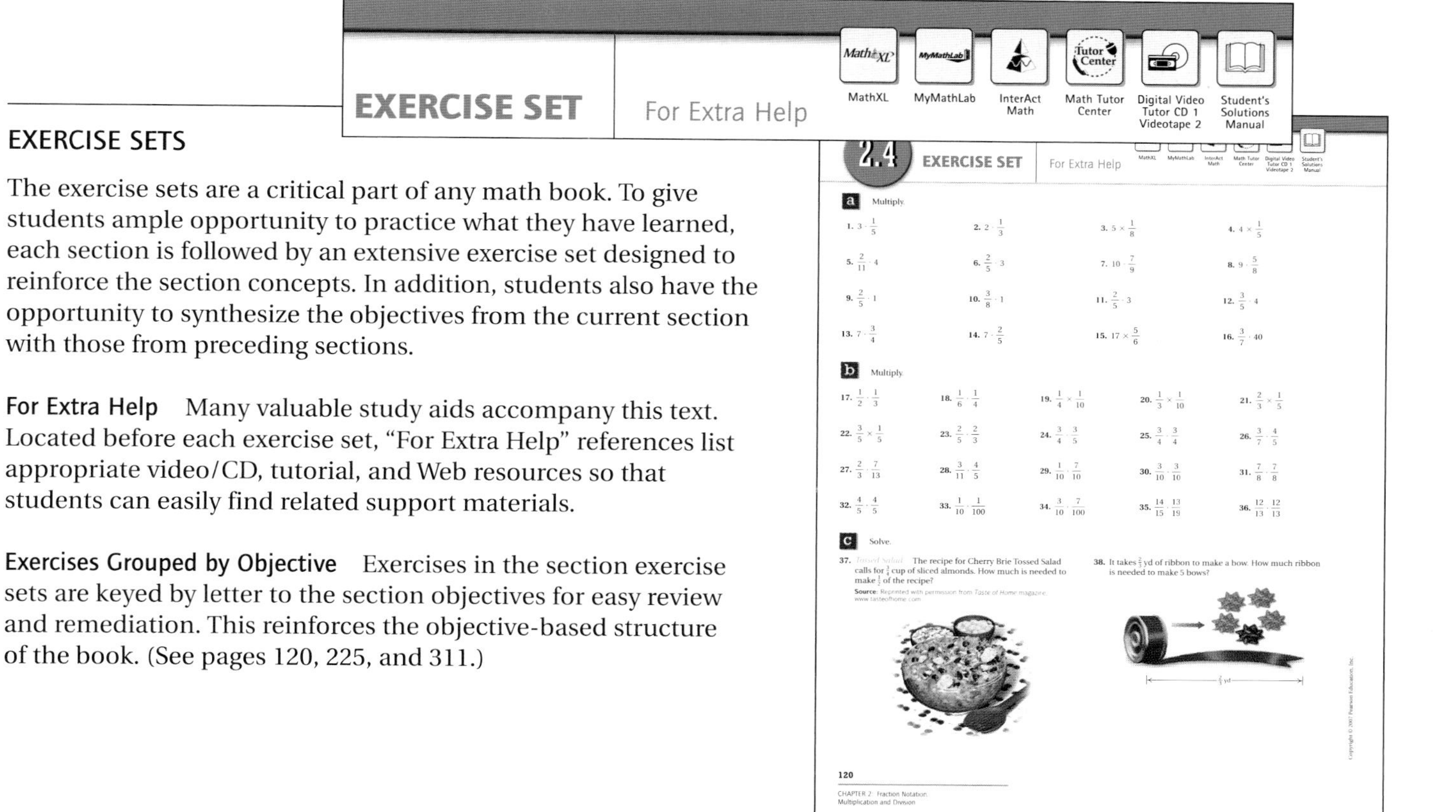

EXERCISE SET For Extra Help

MathXL | MyMathLab | InterAct Math | Math Tutor Center | Digital Video Tutor CD 1 Videotape 2 | Student's Solutions Manual

2.4 EXERCISE SET For Extra Help

a Multiply.

1. $3 \cdot \frac{1}{5}$ 2. $2 \cdot \frac{1}{3}$ 3. $5 \times \frac{1}{8}$ 4. $4 \times \frac{1}{5}$
5. $\frac{2}{11} \cdot 4$ 6. $\frac{2}{5} \cdot 3$ 7. $10 \cdot \frac{7}{9}$ 8. $9 \cdot \frac{5}{8}$
9. $\frac{2}{5} \cdot 1$ 10. $\frac{3}{8} \cdot 1$ 11. $\frac{2}{5} \cdot 3$ 12. $\frac{3}{5} \cdot 4$
13. $7 \cdot \frac{3}{4}$ 14. $7 \cdot \frac{2}{5}$ 15. $17 \times \frac{5}{6}$ 16. $\frac{3}{7} \cdot 40$

b Multiply.

17. $\frac{1}{2} \cdot \frac{1}{3}$ 18. $\frac{1}{6} \cdot \frac{1}{4}$ 19. $\frac{1}{4} \times \frac{1}{10}$ 20. $\frac{1}{3} \times \frac{1}{10}$ 21. $\frac{2}{3} \times \frac{1}{5}$
22. $\frac{3}{5} \times \frac{1}{5}$ 23. $\frac{2}{5} \cdot \frac{2}{3}$ 24. $\frac{3}{4} \cdot \frac{3}{5}$ 25. $\frac{3}{4} \cdot \frac{3}{4}$ 26. $\frac{3}{7} \cdot \frac{4}{5}$
27. $\frac{2}{3} \cdot \frac{7}{13}$ 28. $\frac{3}{11} \cdot \frac{4}{5}$ 29. $\frac{1}{10} \cdot \frac{7}{10}$ 30. $\frac{3}{10} \cdot \frac{3}{10}$ 31. $\frac{7}{8} \cdot \frac{7}{8}$
32. $\frac{4}{5} \cdot \frac{4}{5}$ 33. $\frac{1}{10} \cdot \frac{1}{100}$ 34. $\frac{3}{10} \cdot \frac{7}{100}$ 35. $\frac{14}{15} \cdot \frac{13}{19}$ 36. $\frac{12}{13} \cdot \frac{12}{13}$

c Solve.

37. *Tossed Salad.* The recipe for Cherry Brie Tossed Salad calls for $\frac{3}{4}$ cup of sliced almonds. How much is needed to make $\frac{1}{2}$ of the recipe?

Source: Reprinted with permission from *Taste of Home* magazine, www.tasteofhome.com

38. It takes $\frac{2}{3}$ yd of ribbon to make a bow. How much ribbon is needed to make 5 bows?

$\frac{2}{3}$ yd

120

CHAPTER 2 Fraction Notation: Multiplication and Division

Copyright © 2007 Pearson Education, Inc.

EXERCISE SETS

The exercise sets are a critical part of any math book. To give students ample opportunity to practice what they have learned, each section is followed by an extensive exercise set designed to reinforce the section concepts. In addition, students also have the opportunity to synthesize the objectives from the current section with those from preceding sections.

For Extra Help Many valuable study aids accompany this text. Located before each exercise set, "For Extra Help" references list appropriate video/CD, tutorial, and Web resources so that students can easily find related support materials.

Exercises Grouped by Objective Exercises in the section exercise sets are keyed by letter to the section objectives for easy review and remediation. This reinforces the objective-based structure of the book. (See pages 120, 225, and 311.)

Discussion and Writing Exercises Designed to help students develop a deeper comprehension of critical concepts, Discussion and Writing exercises (indicated by D_W) are suitable for individual or group work. These exercises encourage students to both think and write about key mathematical ideas in the chapter. (See pages 121, 189, and 283.)

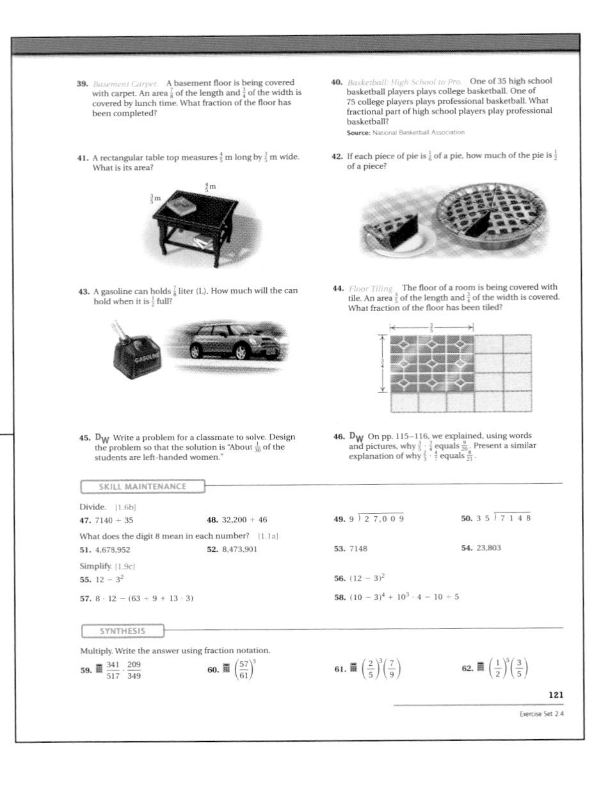

39. *Basement Carpet.* A basement floor is being covered with carpet. An area $\frac{7}{8}$ of the length and $\frac{3}{4}$ of the width is covered by lunch time. What fraction of the floor has been completed?

40. *Basketball: High School to Pro.* One of 35 high school basketball players plays college basketball. One of 75 college players plays professional basketball. What fractional part of high school players play professional basketball?

Source: National Basketball Association

41. A rectangular table top measures $\frac{4}{5}$ m long by $\frac{3}{5}$ m wide. What is its area?

42. If each piece of pie is $\frac{1}{6}$ of a pie, how much of the pie is $\frac{1}{2}$ of a piece?

43. A gasoline can holds $\frac{7}{8}$ liter (L). How much will the can hold when it is $\frac{1}{2}$ full?

44. *Floor Tiling.* The floor of a room is being covered with tile. An area $\frac{3}{5}$ of the length and $\frac{3}{4}$ of the width is covered. What fraction of the floor has been tiled?

45. D_W Write a problem for a classmate to solve. Design the problem so that the solution is "About $\frac{1}{30}$ of the students are left-handed women."

46. D_W On pp. 115–116, we explained, using words and pictures, why $\frac{3}{5} \cdot \frac{3}{4}$ equals $\frac{9}{20}$. Present a similar explanation of why $\frac{2}{3} \cdot \frac{4}{7}$ equals $\frac{8}{21}$.

SKILL MAINTENANCE

Divide. [1.6b]

47. $7140 \div 35$ 48. $32{,}200 \div 46$ 49. $9\overline{)27{,}009}$ 50. $35\overline{)7148}$

What does the digit 8 mean in each number? [1.1a]

51. 4,678,952 52. 8,473,901 53. 7148 54. 23,803

Simplify. [1.9c]

55. $12 - 3^2$ 56. $(12 - 3)^2$

57. $8 \cdot 12 - (63 \div 9 + 13 \cdot 3)$ 58. $(10 - 3)^4 + 10^3 \cdot 4 - 10 \div 5$

SYNTHESIS

Multiply. Write the answer using fraction notation.

59. $\frac{341}{517} \cdot \frac{209}{349}$ 60. $\left(\frac{57}{61}\right)^3$ 61. $\left(\frac{2}{5}\right)^3\left(\frac{7}{9}\right)$ 62. $\left(\frac{1}{2}\right)^5\left(\frac{3}{5}\right)$

121

Exercise Set 2.4

Skill Maintenance Exercises Found in each exercise set, these exercises review concepts from other sections in the text to prepare students for their final examination. Section and objective codes appear next to each Skill Maintenance exercise for easy reference. (See pages 164, 172, and 284.)

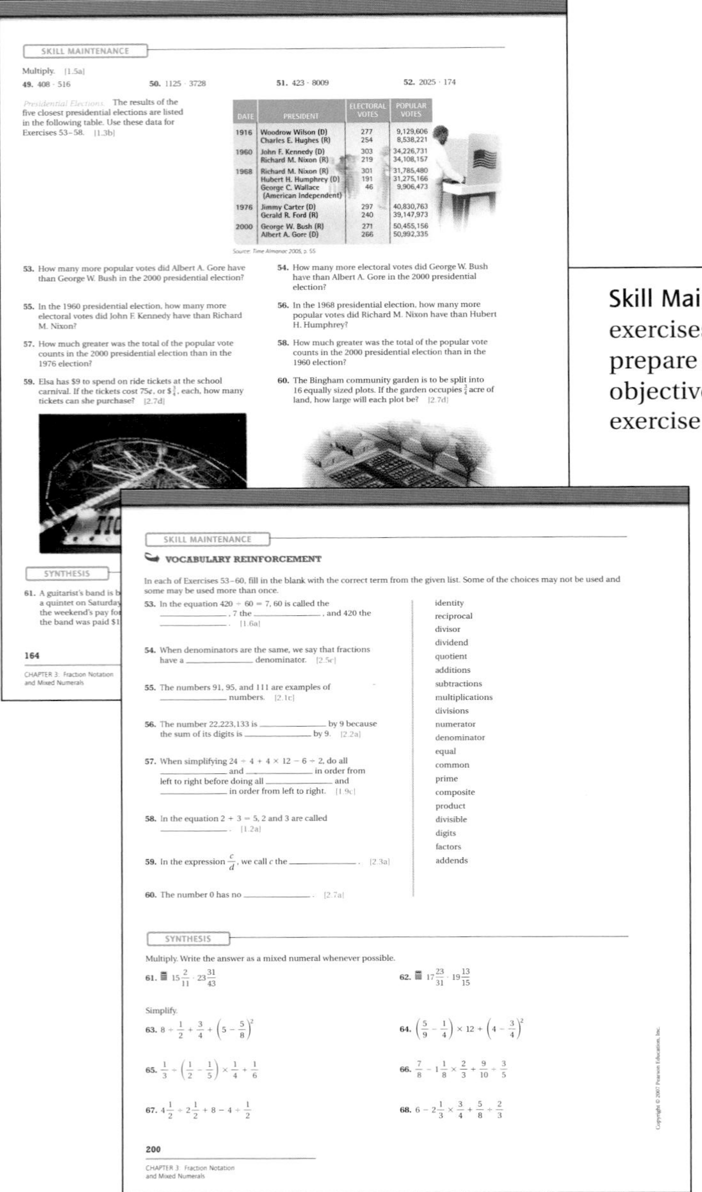

SKILL MAINTENANCE

Multiply. [1.5a]

49. $408 \cdot 516$ 50. $1125 \cdot 3728$ 51. $423 \cdot 8009$ 52. $2025 \cdot 174$

Presidential Elections. The results of the five closest presidential elections are listed in the following table. Use these data for Exercises 53–58. [1.3b]

DATE	PRESIDENT	ELECTORAL VOTES	POPULAR VOTES
1916	Woodrow Wilson (D)	277	9,129,606
	Charles E. Hughes (R)	254	8,538,221
1960	John F. Kennedy (D)	303	34,226,731
	Richard M. Nixon (R)	219	34,108,157
1968	Richard M. Nixon (R)	301	31,785,480
	Hubert H. Humphrey (D)	191	31,275,166
	George C. Wallace (American Independent)	46	9,906,473
1976	Jimmy Carter (D)	297	40,830,763
	Gerald R. Ford (R)	240	39,147,973
2000	George W. Bush (R)	271	50,455,156
	Albert A. Gore (D)	266	50,992,335

Source: Time Almanac 2005, p. 55

53. How many more popular votes did Albert A. Gore have than George W. Bush in the 2000 presidential election?

54. How many more electoral votes did George W. Bush have than Albert A. Gore in the 2000 presidential election?

55. In the 1960 presidential election, how many more electoral votes did John F. Kennedy have than Richard M. Nixon?

56. In the 1968 presidential election, how many more popular votes did Richard M. Nixon have than Hubert H. Humphrey?

57. How much greater was the total of the popular vote counts in the 2000 presidential election than in the 1976 election?

58. How much greater was the total of the popular vote counts in the 2000 presidential election than in the 1960 election?

59. Elsa has \$9 to spend on ride tickets at the school carnival. If the tickets cost 75¢, or \$$\frac{3}{4}$, each, how many tickets can she purchase? [2.7d]

60. The Bingham community garden is to be split into 16 equally sized plots. If the garden occupies $\frac{3}{4}$ acre of land, how large will each plot be? [2.7d]

SYNTHESIS

61. A guitarist's band is b... a quintet on Saturday... the weekend's pay fo... the band was paid \$1...

164

CHAPTER 3 Fraction Notation and Mixed Numerals

SKILL MAINTENANCE

VOCABULARY REINFORCEMENT

In each of Exercises 53–60, fill in the blank with the correct term from the given list. Some of the choices may not be used and some may be used more than once.

53. In the equation $420 \div 60 = 7$, 60 is called the ________, 7 the ________, and 420 the ________. [1.6a]

54. When denominators are the same, we say that fractions have a ________ denominator. [2.5c]

55. The numbers 91, 95, and 111 are examples of ________ numbers. [2.1c]

56. The number 22,223,133 is ________ by 9 because the sum of its digits is ________ by 9. [2.2a]

57. When simplifying $24 \div 4 + 4 \times 12 - 6 \div 2$, do all ________ and ________ in order from left to right before doing all ________ and ________ in order from left to right. [1.9c]

58. In the equation $2 + 3 = 5$, 2 and 3 are called ________. [1.2a]

59. In the expression $\frac{c}{d}$, we call c the ________. [2.3a]

60. The number 0 has no ________. [2.7a]

identity
reciprocal
divisor
dividend
quotient
additions
subtractions
multiplications
divisions
numerator
denominator
equal
common
prime
composite
product
divisible
digits
factors
addends

SYNTHESIS

Multiply. Write the answer as a mixed numeral whenever possible.

61. $15\frac{2}{11} \cdot 23\frac{31}{43}$ 62. $17\frac{23}{31} \cdot 19\frac{13}{15}$

Simplify.

63. $8 \div \frac{1}{2} + \frac{3}{4} + \left(5 - \frac{5}{8}\right)^2$ 64. $\left(\frac{5}{9} - \frac{1}{4}\right) \times 12 + \left(4 - \frac{3}{4}\right)^2$

65. $\frac{1}{3} \div \left(\frac{1}{2} - \frac{1}{5}\right) \times \frac{1}{4} + \frac{1}{6}$ 66. $\frac{7}{8} - 1\frac{1}{8} \times \frac{2}{3} + \frac{9}{10} \div \frac{3}{5}$

67. $4\frac{1}{2} \div 2\frac{1}{2} + 8 - 4 \div \frac{1}{2}$ 68. $6 - 2\frac{1}{3} \times \frac{3}{4} + \frac{5}{8} \div \frac{2}{3}$

200

CHAPTER 3 Fraction Notation and Mixed Numerals

NEW! **Vocabulary Reinforcement Exercises** This new feature checks and reviews students' understanding of the vocabulary introduced throughout the text. It appears once in every chapter, in the Skill Maintenance portion of an exercise set, and is intended to provide a continuing review of the terms that students must know in order to be able to communicate effectively in the language of mathematics. (See pages 143, 200, and 269.)

Synthesis Exercises In most exercise sets, Synthesis exercises help build critical-thinking skills by requiring students to synthesize or combine learning objectives from the current section as well as from preceding text sections. (See pages 114, 208, and 332.)

END-OF-CHAPTER MATERIAL

At the end of each chapter, students can practice all they have learned as well as tie the current chapter content to material covered in earlier chapters.

SUMMARY AND REVIEW

A three-part *Summary and Review* appears at the end of each chapter. The first part includes the Concept Reinforcement Exercises described below. The second part is a list of important properties and formulas, when applicable, and the third part provides an extensive set of review exercises.

NEW! **Concept Reinforcement Exercises** Found in the Summary and Review of every chapter, these true/false exercises are designed to increase understanding of the concepts rather than merely assess students' skill at memorizing procedures. (See pages 209, 285, and 333.)

Important Properties and Formulas A list of the important properties and formulas discussed in the chapter is provided for students in an organized manner to help them prioritize topics learned and prepare for chapter tests. This list is only provided in those chapters in which new properties or formulas are presented. (See page 410.)

Review Exercises At the end of each chapter, students are provided with an extensive set of review exercises. Reference codes beside each exercise or direction line allow students to easily refer back to specific, objective-level content for remediation. (See pages 144, 209, and 285.)

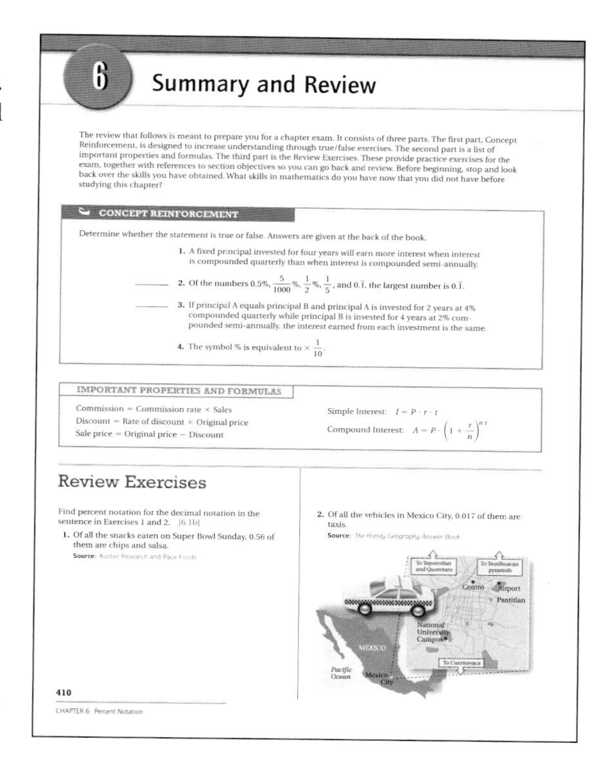

6 Summary and Review

The review that follows is meant to prepare you for a chapter exam. It consists of three parts. The first part, Concept Reinforcement, is designed to increase understanding through true/false exercises. The second part is a list of important properties and formulas. The third part is the Review Exercises. These provide practice exercises for the exam, together with references to section objectives so you can go back and review. Before beginning, stop and look back over the skills you have obtained. What skills in mathematics do you have now that you did not have before studying this chapter?

CONCEPT REINFORCEMENT

Determine whether the statement is true or false. Answers are given at the back of the book.

1. A fixed principal invested for four years will earn more interest when interest is compounded quarterly than when interest is compounded semi-annually.
2. Of the numbers 0.5%, $\frac{5}{1000}\%$, $\frac{1}{2}\%$, $\frac{1}{5}$, and $0.\overline{1}$, the largest number is $0.\overline{1}$.
3. If principal A equals principal B and principal A is invested for 2 years at 4% compounded quarterly while principal B is invested for 4 years at 2% compounded semi-annually, the interest earned from each investment is the same.
4. The symbol % is equivalent to $\times \frac{1}{10}$.

IMPORTANT PROPERTIES AND FORMULAS

Commission = Commission rate × Sales
Discount = Rate of discount × Original price
Sale price = Original price − Discount
Simple Interest: $I = P \cdot r \cdot t$
Compound Interest: $A = P \cdot \left(1 + \frac{r}{n}\right)^{n \cdot t}$

Review Exercises

Find percent notation for the decimal notation in the sentence in Exercises 1 and 2. [6.1b]

1. Of all the snacks eaten on Super Bowl Sunday, 0.56 of them are chips and salsa.
Source: Korbel Research and Pace Foods

2. Of all the vehicles in Mexico City, 0.017 of them are taxis.
Source: The Handy Geography Answer Book

410
CHAPTER 6 Percent Notation

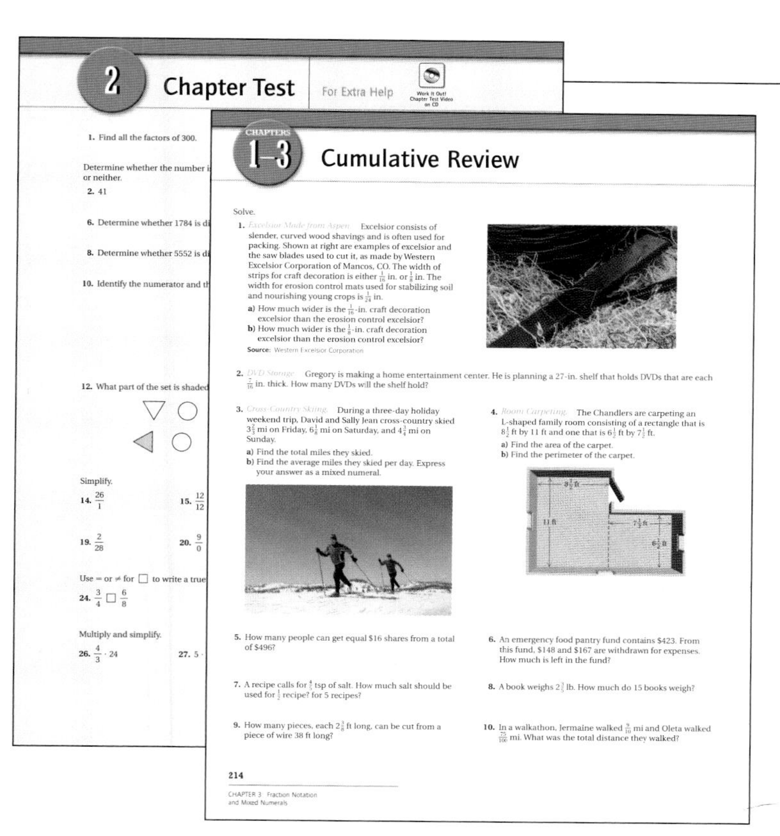

2 Chapter Test For Extra Help Work It Out! Chapter Test Video on CD

1. Find all the factors of 300.

Determine whether the number i… or neither.
2. 41
6. Determine whether 1784 is di…
8. Determine whether 5552 is di…
10. Identify the numerator and th…
12. What part of the set is shaded…

Simplify.
14. $\frac{26}{1}$ 15. $\frac{12}{12}$
19. $\frac{2}{28}$ 20. $\frac{9}{0}$

Use = or ≠ for □ to write a true…
24. $\frac{3}{4} \square \frac{6}{8}$

Multiply and simplify.
26. $\frac{4}{3} \cdot 24$ 27. 5 ·

CHAPTERS 1–3 Cumulative Review

Solve.

1. *Excelsior Made from Aspen* Excelsior consists of slender, curved wood shavings and is often used for packing. Shown at right are examples of excelsior and the saw blades used to cut it, as made by Western Excelsior Corporation of Mancos, CO. The width of strips for craft decoration is either $\frac{1}{16}$ in. or $\frac{1}{8}$ in. The width for erosion control mats used for stabilizing soil and nourishing young crops is $\frac{1}{24}$ in.
a) How much wider is the $\frac{1}{16}$-in. craft decoration excelsior than the erosion control excelsior?
b) How much wider is the $\frac{1}{8}$-in. craft decoration excelsior than the erosion control excelsior?
Source: Western Excelsior Corporation

2. *DVD Storage* Gregory is making a home entertainment center. He is planning a 27-in. shelf that holds DVDs that are each $\frac{7}{16}$ in. thick. How many DVDs will the shelf hold?

3. *Cross-Country Skiing* During a three-day holiday weekend trip, David and Sally Jean cross-country skied $3\frac{2}{3}$ mi on Friday, $6\frac{1}{8}$ mi on Saturday, and $4\frac{3}{4}$ mi on Sunday.
a) Find the total miles they skied.
b) Find the average miles they skied per day. Express your answer as a mixed numeral.

4. *Room Carpeting* The Chandlers are carpeting an L-shaped family room consisting of a rectangle that is $8\frac{1}{2}$ ft by 11 ft and one that is $6\frac{1}{2}$ ft by $7\frac{1}{2}$ ft.
a) Find the area of the carpet.
b) Find the perimeter of the carpet.

5. How many people can get equal \$16 shares from a total of \$496?

6. An emergency food pantry fund contains \$423. From this fund, \$148 and \$167 are withdrawn for expenses. How much is left in the fund?

7. A recipe calls for $\frac{4}{5}$ tsp of salt. How much salt should be used for $\frac{1}{2}$ recipe? for 5 recipes?

8. A book weighs $2\frac{3}{5}$ lb. How much do 15 books weigh?

9. How many pieces, each $2\frac{3}{8}$ ft long, can be cut from a piece of wire 38 ft long?

10. In a walkathon, Jermaine walked $\frac{9}{10}$ mi and Oleta walked $\frac{75}{100}$ mi. What was the total distance they walked?

214
CHAPTER 3 Fraction Notation and Mixed Numerals

CHAPTER TEST

Following the Review Exercises, a sample Chapter Test allows students to review and test comprehension of chapter skills prior to taking an instructor's exam. Answers to all questions in the Chapter Test are given at the back of the book. Section and objective references for each question are included with the answers. (See pages 147, 212, and 288.)

CUMULATIVE REVIEW

Following Chapters 3, 5, and 6, students encounter a Cumulative Review. This exercise set reviews skills and concepts from all preceding chapters to help students recall previously learned material and prepare for a final exam. At the back of the book are answers to all Cumulative Review exercises, together with section and objective references, so that students know exactly what material to study if they miss a review exercise. Additional Cumulative Review Tests for every chapter are available in the *Printed Test Bank*. (See pages 214, 338, and 415.)

Ancillaries

The following ancillaries are available to help both instructors and students use this text more effectively.

Student Supplements

Student's Solutions Manual (ISBN 0-321-29607-9)

- By Judith A. Penna, *Indiana University Purdue University Indianapolis*
- Contains completely worked-out solutions with step-by-step annotations for all the odd-numbered exercises in the text, with the exception of the discussion and writing exercises, as well as completely worked-out solutions to all the exercises in the Chapter Reviews, Chapter Tests, and Cumulative Reviews.

Collaborative Learning Activities Manual
(ISBN 0-321-29604-4)

- Features group activities tied to text sections and includes the focus, time estimate, suggested group size and materials, and background notes for each activity.
- Available as a stand-alone supplement sold in the bookstore, as a textbook bundle component for students, or as a classroom activity resource for instructors.

Videotapes (ISBN 0-321-30569-8)

- Include new chapter openers presented by author Marvin Bittinger.
- Present a series of lectures correlated directly to the content of each section of the text.
- Feature an engaging team of instructors who present material in a format that stresses student interaction, often using examples and exercises from the text.

Digital Video Tutor (ISBN 0-321-29605-2)

- Complete set of digitized videos (as described above) on CD-ROMs for student use at home or on campus.
- Ideal for distance learning or supplemental instruction.
- Are available with captioning on request. Contact your local Addison-Wesley representative for details.

NEW! ***Work It Out! Chapter Test Video on CD***
(ISBN 0-321-41985-5)

- Presented by Judith A. Penna and Barbara Johnson
- Provides step-by-step solutions to every exercise in each Chapter Test from the text.
- Helps students prepare for chapter tests and synthesize content.

Math Study Skills for Students Video on CD
(ISBN 0-321-29745-8)

- Presented by author Marvin Bittinger
- Designed to help students make better use of their math study time and improve their retention of concepts and procedures taught in classes from basic mathematics through intermediate algebra.
- Through carefully crafted graphics and comprehensive on-camera explanation, focuses on study skills that are commonly overlooked.

Instructor Supplements

Annotated Instructor's Edition (ISBN 0-321-30553-1)

- Includes answers to all exercises printed in blue on the same page as those exercises.

Instructor's Solutions Manual (ISBN 0-321-30556-6)

- By Judith A. Penna, *Indiana University Purdue University Indianapolis*
- Contains brief solutions to the even-numbered exercises in the exercise sets, answers to all of the Discussion and Writing exercises, and the completely worked-out solutions to all the exercises in the Chapter Reviews, Chapter Tests, and Cumulative Reviews.

Online Answer Book

- By Judith A. Penna, *Indiana University Purdue University Indianapolis*
- Available in electronic form from the instructor resource center. Contact your local Addison-Wesley representative for details.
- Contains answers to all the section exercises in the text.

Printed Test Bank (ISBN 0-321-30557-4)

- By Laurie Hurley
- Contains one diagnostic test.
- Contains one pretest for each chapter.
- Provides 13 new test forms for every chapter and 8 new test forms for the final exam.
- For the chapter tests, 5 test forms are modeled after the chapter tests in the text, 3 test forms are organized by topic order following the text objectives, 3 test forms are designed for 50-minute class periods and organized so that each objective in the chapter is covered on one of the tests, and 2 test forms are multiple-choice. Chapter tests also include more challenging synthesis questions.
- Contains 2 cumulative tests per chapter beginning with Chapter 2.
- For the final exam, 3 test forms are organized by chapter, 3 test forms are organized by question type, and 2 test forms are multiple-choice.

NEW! ***Instructor and Adjunct Support Manual***
(ISBN 0-321-30554-X)

- Includes Adjunct Support Manual material.
- Features resources and teaching tips designed to help both new and adjunct faculty with course preparation and classroom management.
- Resources include chapter reviews, extra practice sheets, conversion guide, video index, audio index, and transparency masters.
- Also available electronically so course/adjunct coordinators can customize material specific to their schools.

Student Supplements

Audio Recordings

- By Bill Saler
- Lead students through the material in each section of the text, explaining solution steps to examples, pointing out common errors, and focusing on margin exercises and solutions.
- Audio files are available to download in MP3 format. Contact your local Addison-Wesley representative for details.

Addison-Wesley Math Tutor Center

www.aw-bc.com/tutorcenter

- The Addison-Wesley Math Tutor Center is staffed by qualified mathematics instructors who provide students with tutoring on examples and odd-numbered exercises from the textbook. Tutoring is available via toll-free telephone, toll-free fax, e-mail, or the Internet. White Board technology allows tutors and students to actually see problems worked while they "talk" in real time over the Internet during tutoring sessions.

MathXL® Tutorials on CD (ISBN 0-321-29747-4)

- Provides algorithmically generated practice exercises that correlate at the objective level to the content of the text.
- Includes an example and a guided solution to accompany every exercise and video clips for selected exercises.
- Recognizes student errors and provides feedback; generates printed summaries of students' progress.

Instructor Supplements

TestGen with Quizmaster (ISBN 0-321-30567-1)

- Enables instructors to build, edit, print, and administer tests.
- Features a computerized bank of questions developed to cover all text objectives.
- Algorithmically based content allows instructors to create multiple but equivalent versions of the same question or test with a click of a button.
- Instructors can also modify test-bank questions or add new questions by using the built-in question editor, which allows users to create graphs, input graphics, and insert math notation, variable numbers, or text.
- Tests can be printed or administered online via the Internet or another network. Quizmaster allows students to take tests on a local area network.
- Available on a dual-platform Windows/Macintosh CD-ROM.

MathXL® www.mathxl.com

MathXL is a powerful online homework, tutorial, and assessment system that accompanies Addison-Wesley textbooks in mathematics or statistics. With MathXL, instructors can create, edit, and assign online homework and tests using algorithmically generated exercises correlated at the objective level to the textbook. They can also create and assign their own online exercises and import TestGen tests for added flexibility. All student work is tracked in MathXL's online gradebook. Students can take chapter tests in MathXL and receive personalized study plans based on their test results. The study plan diagnoses weaknesses and links students directly to tutorial exercises for the objectives they need to study and retest. Students can also access supplemental animations and video clips directly from selected exercises. MathXL is available to qualified adopters. For more information, visit our Web site at www.mathxl.com or contact your Addison-Wesley sales representative.

MyMathLab www.mymathlab.com

MyMathLab is a series of text-specific, easily customizable online courses for Addison-Wesley textbooks in mathematics and statistics. Powered by CourseCompass™ (Pearson Education's online teaching and learning environment) and MathXL® (our online homework, tutorial, and assessment system), MyMathLab gives instructors the tools they need to deliver all or a portion of their course online, whether students are in a lab setting or working from home. MyMathLab provides a rich and flexible set of course materials, featuring free-response exercises that are algorithmically generated for unlimited practice and mastery. Students can also use online tools, such as video lectures, animations, and a multimedia textbook, to independently improve their understanding and performance. Instructors can use MyMathLab's homework and test managers to select and assign online exercises correlated directly to the textbook, and they can also create and assign their own online exercises and import TestGen tests for added flexibility. MyMathLab's online gradebook—designed specifically for mathematics and statistics—automatically tracks students' homework and test results and gives the instructor control over how to calculate final grades. Instructors can also add offline (paper-and-pencil) grades to the gradebook. MyMathLab is available to qualified adopters. For more information, visit our Web site at www.mymathlab.com or contact your Addison-Wesley sales representative.

InterAct Math® Tutorial Web site www.interactmath.com

Get practice and tutorial help online! This interactive tutorial Web site provides algorithmically generated practice exercises that correlate directly to the exercises in the textbook. Students can retry an exercise as many times as they like with new values each time for unlimited practice and mastery. Every exercise is accompanied by an interactive guided solution that provides helpful feedback for incorrect answers, and students can also view a worked-out sample problem that steps them through an exercise similar to the one they're working on.

ADDISON-WESLEY MATH ADJUNCT SUPPORT CENTER

The Addison-Wesley Math Adjunct Support Center is staffed by qualified mathematics instructors with over 50 years of combined experience at both the community college and university level. Assistance is provided for faculty in the following areas:

- Suggested syllabus consultation
- Tips on using materials packaged with your book
- Book-specific content assistance
- Teaching suggestions including advice on classroom strategies

For more information, visit www.aw-bc.com/tutorcenter/math-adjunct.html

Acknowledgments and Reviewers

Many of you helped to shape the fourth edition by reviewing and spending time with us on your campuses. Our deepest appreciation to all of you and in particular to the following:

Sally Clark, *Ozarks Technical Community College*
Penny Deggelman, *Lane Community College*
Molly Misko, *Gadsden State Community College*
Cassonda Thompson, *York Technical College*
Mary Woestman, *Broome Community College*

We wish to express our heartfelt appreciation to a number of people who have contributed in special ways to the development of this textbook. Our editor, Jennifer Crum, and marketing manager, Jay Jenkins, encouraged our vision and provided marketing insight. Kari Heen, the project manager, deserves special recognition for overseeing every phase of the project and keeping it moving. The unwavering support of the Developmental Math group, including Katie Nopper, project editor, Elizabeth Bernardi and Alison Macdonald, editorial assistants, Ron Hampton, managing editor, Dennis Schaefer, cover designer, and Sharon Smith and Ceci Fleming, media producers, and the endless hours of hard work by Kathy Diamond and Geri Davis have led to products of which we are immensely proud.

We also want to thank Judy Penna for writing the *Student's* and *Instructor's Solutions Manuals* and for her strong leadership in the preparation of the printed supplements, videotapes, and MyMathLab. Other strong support has come from Laurie Hurley for the *Printed Test Bank,* Bill Saler for the audio recordings, and Barbara Johnson, Michelle Lanosga, and Jennifer Rosenberg for their accuracy checking of the manuscript. We also wish to recognize those who wrote scripts, presented lessons on camera, and checked the accuracy of the videotapes.

To the Student

As your author, I would like to welcome you to this study of *Fundamental Mathematics.* Whatever your past experiences, I encourage you to look at this mathematics course as a fresh start. Approach this course with a positive attitude about mathematics. Mathematics is a base for life, for many majors, for personal finances, for most careers, or just for pleasure.

You are the most important factor in the success of your learning. In earlier experiences, you may have allowed yourself to sit back and let the instructor "pour in" the learning, with little or no follow-up on your part. But now you must take a more assertive and proactive stance. This may be the first adjustment you make in college. As soon as possible after class, you should thoroughly read the textbook and the supplements and do all you can to learn on your own. In other words, rid yourself of former habits and take responsibility for your own learning. Then, with all the help you have around you, your hard work will lead to success.

One of the most important suggestions I can make is to allow yourself enough *time* to learn. You can have the best book, the best instructor, and the best supplements, but if you do not give yourself time to learn, how can they be of benefit? Many other helpful suggestions are presented in the Study Tips that you will find throughout the book. You may want to read through all the Study Tips before you begin the text. An Index of Study Tips can be found at the front of the book.

M.L.B.

Fundamental Mathematics

Bittinger Student Organizer

COURSE INFORMATION

Study Tips

Throughout this text, you will find a feature called *Study Tips*. We discuss these in the Preface of this text. They are intended to help improve your math study skills. An index of all the *Study Tips* can be found at the front of the book.

For Extra Help

MyMathLab

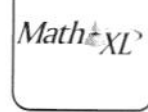

MathXL

Student's Solutions Manual

Digital Video Tutor CD Videotape

Math Tutor Center

InterAct Math

Additional Resources

Basic Math Review Card
(ISBN 0-321-39476-3)

Algebra Review Card
(ISBN 0-321-39473-9)

Math for Allied Health Reference Card
(ISBN 0-321-39474-7)

Graphing Calculator Reference Card
(ISBN 0-321-39475-5)

Math Study Skills for Students Video on CD
(ISBN 0-321-29745-8)

Go to www.aw-bc.com/math for more information.

On the first day of class, complete this chart and the weekly planner that follows on the reverse page.

Instructor Information:

Name ____________________

Office Hours and Location ____________________

____

Phone Number ____________________

Fax Number ____________________

E-mail Address ____________________

Find the names of two students whom you could contact for class information or study questions:

1. Name ____________________

Phone Number ____________________

Fax Number ____________________

E-mail Address ____________________

2. Name ____________________

Phone Number ____________________

Fax Number ____________________

E-mail Address ____________________

Math Lab on Campus:

Location ____________________

Hours ____________________

Phone Number ____________________

Tutoring:

Campus Location ____________________

Hours ____________________

To order the Addison-Wesley Math Tutor Center, call 1-888-777-0463.

(*See the Preface for important information concerning this tutoring.*)

Important Supplements: (*See the Preface for a complete list of available supplements.*)

Supplements recommended by the instructor ____________________

____

Online Log-in Information (*include access code, password, Web address, etc.*)

____

____

Bittinger Student Organizer

WEEKLY PLANNE

Success is planned. On this page, plan a typical week. Consider time allotments for class, study, work, travel, family, and relaxation.

Important Dates

Midterm Exam

____

Final Exam

____

Holidays

____

____

____

____

____

Other

____

____

____

____

____

TIME	Sun.	Mon.	Tues.	Wed.	Thurs.	Fri.	Sat.
6:00 A.M.							
6:30							
7:00							
7:30							
8:00							
8:30							
9:00							
9:30							
10:00							
10:30							
11:00							
11:30							
12:00 P.M.							
12:30							
1:00							
1:30							
2:00							
2:30							
3:00							
3:30							
4:00							
4:30							
5:00							
5:30							
6:00							
6:30							
7:00							
7:30							
8:00							
8:30							
9:00							
9:30							
10:00							
10:30							
11:00							
11:30							
12:00 A.M.							

1

Whole Numbers

Real-World Application

Races in which runners climb the steps inside a building are called "run-up" races. There are 2058 steps in the International Towerthon, Kuala Lumpur, Malaysia. Write expanded notation for the number of steps.

This problem appears as Exercise 13 in Section 1.1.

1.1 STANDARD NOTATION

Objectives

a Give the meaning of digits in standard notation.

b Convert from standard notation to expanded notation.

c Convert between standard notation and word names.

What does the digit 2 mean in each number?

1. 526,555

2. 265,789

3. 42,789,654

4. 24,789,654

5. 8924

6. 5,643,201

To the student:

In the preface, at the front of the text, you will find a Student Organizer card. This pullout card will help you keep track of important dates and useful contact information. You can also use it to plan time for class, study, work, and relaxation. By managing your time wisely, you will provide yourself the best possible opportunity to be successful in this course.

Answers on page A-1

We study mathematics in order to be able to solve problems. In this section, we study how numbers are named. We begin with the concept of place value.

a Place Value

Consider the numbers in the following table.

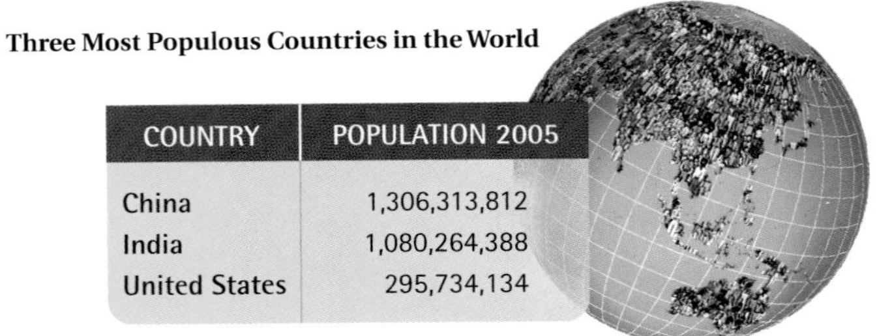

Three Most Populous Countries in the World

COUNTRY	POPULATION 2005
China	1,306,313,812
India	1,080,264,388
United States	295,734,134

Source: The World Factbook, July 2005 estimates

A **digit** is a number 0, 1, 2, 3, 4, 5, 6, 7, 8, or 9 that names a place-value location. For large numbers, digits are separated by commas into groups of three, called **periods.** Each period has a name: *ones, thousands, millions, billions, trillions,* and so on. To understand the population of China in the table above, we can use a **place-value chart,** as shown below.

PLACE-VALUE CHART

Periods →	Trillions			Billions			Millions			Thousands			Ones		
						1	3	0	6	3	1	3	8	1	2
	Hundreds	Tens	Ones	Hundreds	Tens	Ones	Hundreds	Tens	Ones	Hundreds	Tens	Ones	Hundreds	Tens	Ones
				1 billion			306 millions			313 thousands			812 ones		

EXAMPLES What does the digit 8 mean in each number?

1. 278,342	8 thousands
2. 872,342	8 hundred thousands
3. 28,343,399,223	8 billions
4. 1,023,850	8 hundreds
5. 98,413,099	8 millions
6. 6328	8 ones

Do Margin Exercises 1–6.

EXAMPLE 7 *American Red Cross.* In 2003, private donations to the American Red Cross totaled about \$587,492,000. What does each digit name?

Source: *The Chronicle of Philanthropy*

5 8 7, 4 9 2, 0 0 0

- 5: hundred millions
- 8: ten millions
- 7: millions
- 4: hundred thousands
- 9: ten thousands
- 2: thousands
- 0: hundreds
- 0: tens
- 0: ones

AMERICAN RED CROSS

7. **Presidential Library.** In the first year of operation, 280,219 people visited the Ronald Reagan Presidential Library in Simi Valley, California. What does each digit name?

   Source: *USA Today* research by Bruce Rosenstein; National Archives & Records Administration; Associated Press

Do Exercise 7.

b Converting from Standard Notation to Expanded Notation

To answer questions such as "How many?", "How much?", and "How tall?", we use whole numbers. The set, or collection, of **whole numbers** is

$$0, 1, 2, 3, 4, 5, 6, 7, 8, 9, 10, 11, 12, \ldots.$$

The set goes on indefinitely. There is no largest whole number, and the smallest whole number is 0. Each whole number can be named using various notations. The set $1, 2, 3, 4, 5, \ldots$, without 0, is called the set of **natural numbers.**

Let's look at the data from the bar graph shown here.

Fewer Computer Majors

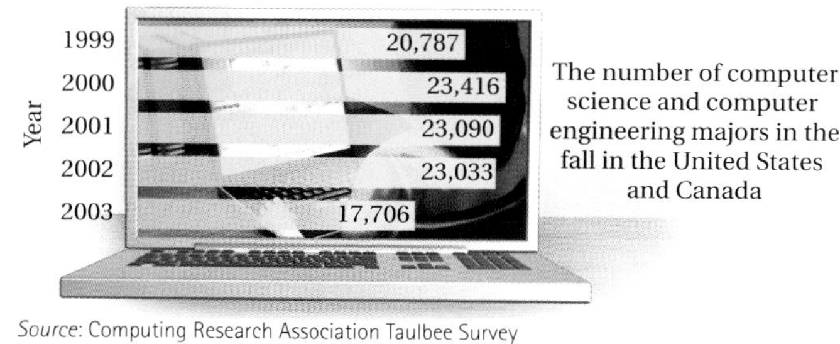

Source: Computing Research Association Taulbee Survey

The number of computer majors in 2003 was 17,706. **Standard notation** for the number of computer majors is 17,706. We write **expanded notation** for 17,706 as follows:

$$17{,}706 = 1 \text{ ten thousand} + 7 \text{ thousands} + 7 \text{ hundreds} + 0 \text{ tens} + 6 \text{ ones}.$$

Write expanded notation.

8. 1895

9. 23,416, the number of computer majors in 2000

10. 3031 mi (miles), the diameter of Mercury

Answers on page A-1

11. 4180 mi, the length of the Nile River, the longest river in the world

12. 154,616, the number of Labrador retrievers registered in 2002
Source: The American Kennel Club

Write a word name. (Refer to the figure below right.)

13. 49, the total number of medals won by Australia

14. 16, the number of silver medals won by Germany

15. 38, the number of bronze medals won by Russia

Answers on page A-1

EXAMPLE 8 Write expanded notation for 4218 mi, the diameter of Mars.

4218 = 4 thousands + 2 hundreds + 1 ten + 8 ones

EXAMPLE 9 Write expanded notation for 3400.

3400 = 3 thousands + 4 hundreds + 0 tens + 0 ones, or
3 thousands + 4 hundreds

EXAMPLE 10 Write expanded notation for 563,384, the population of Washington, D.C.

563,384 = 5 hundred thousands + 6 ten thousands
+ 3 thousands + 3 hundreds + 8 tens + 4 ones

Do Exercises 8–12 (8–10 are on the preceding page).

C Converting Between Standard Notation and Word Names

We often use **word names** for numbers. When we pronounce a number, we are speaking its word name. Russia won 92 medals in the 2004 Summer Olympics in Athens, Greece. A word name for 92 is "ninety-two." Word names for some two-digit numbers like 27, 39, and 92 use hyphens. Others like 17 use only one word, "seventeen."

TOP COUNTRIES IN SUMMER OLYMPICS 2004	MEDAL COUNT GOLD	SILVER	BRONZE	TOTAL
United States of America	35	39	29	103
Russia	27	27	38	92
People's Republic of China	32	17	14	63
Australia	17	16	16	49
Germany	14	16	18	48

Source: 2004 Olympics, Athens, Greece

EXAMPLES Write a word name.

11. 35, the total number of gold medals won by the United States

Thirty-five

12. 17, the number of silver medals won by the People's Republic of China

Seventeen

Do Exercises 13–15 on the preceding page.

For word names for larger numbers, we begin at the left with the largest period. The number named in the period is followed by the name of the period; then a comma is written and the next period is named.

EXAMPLE 13 Write a word name for 46,605,314,732.

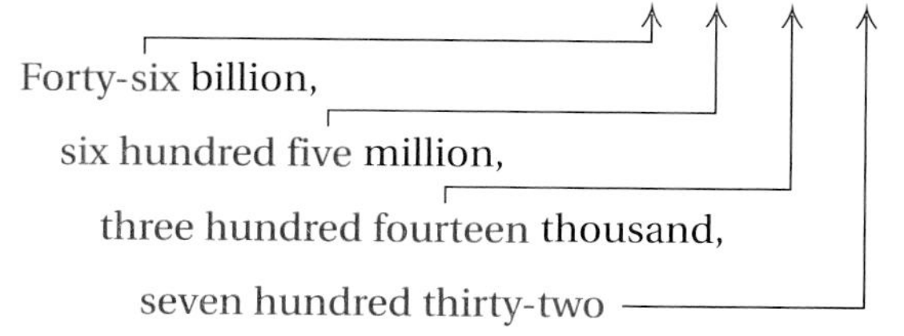

The word "and" *should not* appear in word names for whole numbers. Although we commonly hear such expressions as "two hundred *and* one," the use of "and" is not, strictly speaking, correct in word names for whole numbers. For decimal notation, it is appropriate to use "and" for the decimal point. For example, 317.4 is read as "three hundred seventeen *and* four tenths."

Do Exercises 16–19.

EXAMPLE 14 Write standard notation.

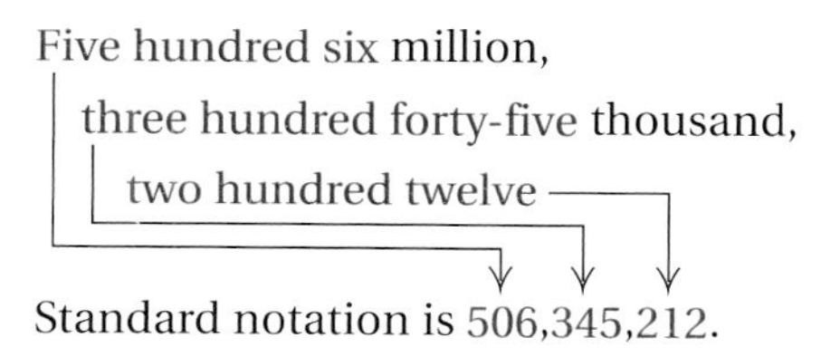

Five hundred six million,
three hundred forty-five thousand,
two hundred twelve

Standard notation is 506,345,212.

Do Exercise 20.

Write a word name.

16. 204

17. $44,155, the average salary in 2001 for those who have a bachelor's degree or more

Source: U.S. Bureau of the Census

18. 1,879,204

19. 6,449,000,000, the world population in 2005

Source: U.S. Bureau of the Census

20. Write standard notation.

Two hundred thirteen million,
one hundred five thousand,
three hundred twenty-nine

Answers on page A-1

1.1 EXERCISE SET

For Extra Help

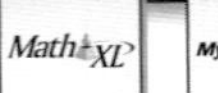
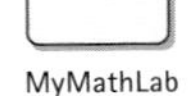

MathXL | MyMathLab | InterAct Math | Math Tutor Center | Digital Video Tutor CD 1 Videotape 1 | Student's Solutions Manual

a What does the digit 5 mean in each case?

1. 235,888

2. 253,777

3. 1,488,526

4. 500,736

Used Cars. 1,582,370 certified used cars were sold in 2004 in the United States. **Source:** *Motor Trend,* April 2005, p. 26

In the number 1,582,370, what digit names the number of:

5. thousands?

6. ones?

7. millions?

8. hundred thousands?

b Write expanded notation.

9. 5702

10. 3097

11. 93,986

12. 38,453

Step-Climbing Races. Races in which runners climb the steps inside a building are called "run-up" races. The graph below shows the number of steps in four buildings. In Exercises 13–16, write expanded notation for the number of steps in each race.

Step-Climbing Races

Source: New York Road Runners Club

13. 2058 steps in the International Towerthon, Kuala Lumpur, Malaysia

14. 1776 steps in the CN Tower Run-Up, Toronto, Ontario, Canada

15. 1268 steps in the World Financial Center, New York City, New York

16. 1081 steps in the Skytower Run-Up, Auckland, New Zealand

Overseas Travelers. The chart below shows the residence and number of overseas travelers to the United States in 2003. In Exercises 17–22, write expanded notation for the number of travelers from each country.

OVERSEAS TRAVELERS TO U.S., 2003

Australia	405,698
Brazil	348,945
Germany	1,180,212
India	272,161
Japan	3,169,682
Spain	284,031

Source: U.S. Department of Commerce, ITA, Office of Travel and Tourism Industries, 2004

17. 405,698 from Australia

18. 3,169,682 from Japan

19. 272,161 from India

20. 284,031 from Spain

21. 1,180,212 from Germany

22. 348,945 from Brazil

Write a word name.

23. 85

24. 48

25. 88,000

26. 45,987

27. 123,765

28. 111,013

29. 7,754,211,577

30. 43,550,651,808

Write standard notation.

31. Two million, two hundred thirty-three thousand, eight hundred twelve

32. Three hundred fifty-four thousand, seven hundred two

33. Eight billion

34. Seven hundred million

Write a word name for the number in each sentence.

35. *Great Pyramid.* The area of the base of the Great Pyramid in Egypt is 566,280 square feet.

36. *Population of the United States.* The population of the United States in July 2005 was estimated to be 295,734,134.

Source: *The World Factbook*

37. *Busiest Airport.* In 2003, the world's busiest airport, Hartsfield in Atlanta, had 76,086,792 passengers.

Source: Airports Council International World Headquarters, Geneva, Switzerland

38. *Prisoners.* There were 2,078,570 total prisoners, federal, state, and local, in the United States in 2003.

Source: *Prison and Jail Inmates at Midyear 2003,* U.S. Bureau of Justice Statistics

Write standard notation for the number in each sentence.

39. Light travels nine trillion, four hundred sixty billion kilometers in one year.

40. The distance from the sun to Pluto is three billion, six hundred sixty-four million miles.

41. *Pacific Ocean.* The area of the Pacific Ocean is sixty-four million, one hundred eighty-six thousand square miles.

42. *Internet Users.* In a recent year, there were fifty-four million, five hundred thousand Internet users in China.

Source: Computer Industry Almanac, Inc.

To the student and the instructor: The Discussion and Writing exercises are meant to be answered with one or more sentences. They can be discussed and answered collaboratively by the entire class or by small groups. Because of their open-ended nature, the answers to these exercises do not appear at the back of the book. They are denoted by the symbol **DW**.

43. **DW** Explain why we use commas when writing large numbers.

44. **DW** Write an English sentence in which the number 370,000,000 is used.

SYNTHESIS

To the student and the instructor: The Synthesis exercises found at the end of every exercise set challenge students to combine concepts or skills studied in that section or in preceding parts of the text. Exercises marked with a ▦ symbol are meant to be solved using a calculator.

45. How many whole numbers between 100 and 400 contain the digit 2 in their standard notation?

46. ▦ What is the largest number that you can name on your calculator? How many digits does that number have? How many periods?

1.2 ADDITION

Objectives

a Add whole numbers.

b Use addition in finding perimeter.

a Addition of Whole Numbers

Addition of whole numbers corresponds to combining or putting things together.

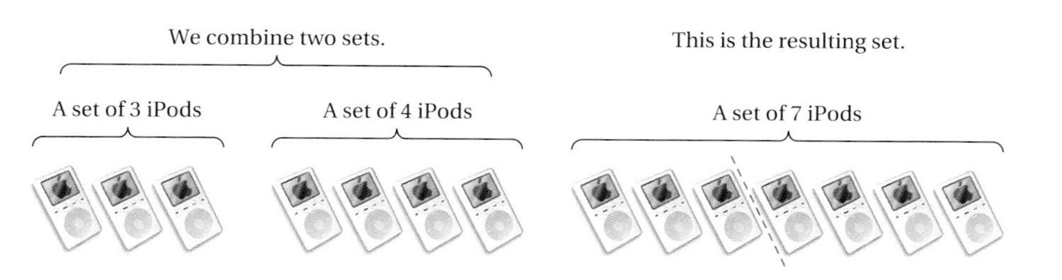

The addition that corresponds to the figure above is

$$\underset{\text{Addend}}{3} + \underset{\text{Addend}}{4} = \underset{\text{Sum}}{7}.$$

We say that the **sum** of 3 and 4 is 7. The numbers added are called **addends.**

Addition also corresponds to moving distances on a number line. The number line below is marked with tick marks at equal distances of 1 *unit.* The sum $3 + 4$ is shown. We first move 3 units from 0, and then 4 more units, and end up at 7. The addition that corresponds to the situation is $3 + 4 = 7$.

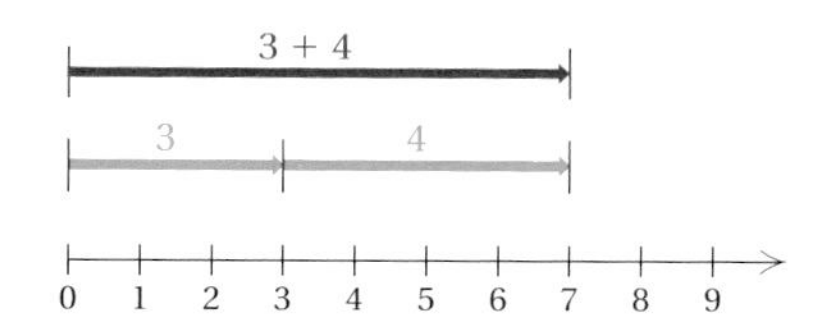

How do we do an addition of three numbers, like $2 + 3 + 6$? We do it by adding 3 and 6, and then 2. We can show this with parentheses:

$2 + (3 + 6) = 2 + 9 = 11.$ Parentheses tell what to do first.

We could also add 2 and 3, and then 6:

$(2 + 3) + 6 = 5 + 6 = 11.$

Either way we get 11. It does not matter how we group the numbers. This illustrates the **associative law of addition,** $a + (b + c) = (a + b) + c$. We can also add whole numbers in any order. That is, $2 + 3 = 3 + 2$. This illustrates the **commutative law of addition,** $a + b = b + a$. Together, the commutative and associative laws tell us that to add more than two numbers, we can use any order and grouping we wish. Adding 0 to a number does not change the number: $a + 0 = 0 + a = a$. That is, $6 + 0 = 0 + 6 = 6$, or $198 + 0 = 0 + 198 = 198$. We say that 0 is the **additive identity.**

To add whole numbers, we add the ones digits first, then the tens, then the hundreds, then the thousands, and so on.

EXAMPLE 1 Add: 6878 + 4995.

Place values are lined up in columns

$$\begin{array}{rcccc} & & & 1 & \\ & 6 & 8 & 7 & 8 \\ + & 4 & 9 & 9 & 5 \\ \hline & & & & 3 \end{array}$$

Add ones. We get 13 ones, or 1 ten + 3 ones. Write 3 in the ones column and 1 above the tens. This is called *carrying*, or *regrouping*.

$$\begin{array}{rcccc} & & 1 & 1 & \\ & 6 & 8 & 7 & 8 \\ + & 4 & 9 & 9 & 5 \\ \hline & & & 7 & 3 \end{array}$$

Add tens. We get 17 tens, or 1 hundred + 7 tens. Write 7 in the tens column and 1 above the hundreds.

$$\begin{array}{rcccc} & 1 & 1 & 1 & \\ & 6 & 8 & 7 & 8 \\ + & 4 & 9 & 9 & 5 \\ \hline & & 8 & 7 & 3 \end{array}$$

Add hundreds. We get 18 hundreds, or 1 thousand + 8 hundreds. Write 8 in the hundreds column and 1 above the thousands.

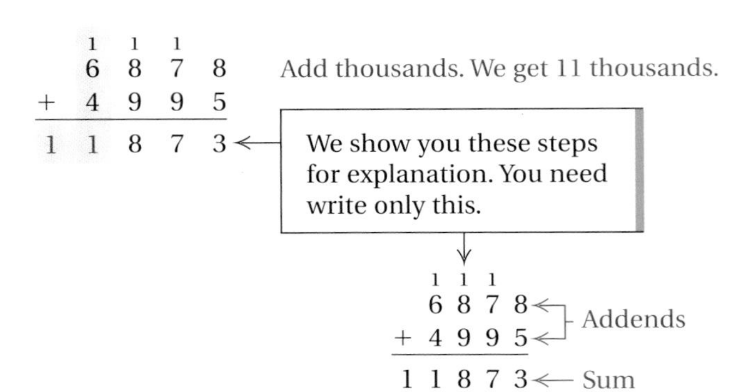

EXAMPLE 2 Add: 391 + 276 + 789 + 498.

$$\begin{array}{rccc} & & 2 & \\ & 3 & 9 & 1 \\ & 2 & 7 & 6 \\ & 7 & 8 & 9 \\ + & 4 & 9 & 8 \\ \hline & & & 4 \end{array}$$

Add ones. We get 24, so we have 2 tens + 4 ones. Write 4 in the ones column and 2 above the tens.

$$\begin{array}{rccc} & 3 & 2 & \\ & 3 & 9 & 1 \\ & 2 & 7 & 6 \\ & 7 & 8 & 9 \\ + & 4 & 9 & 8 \\ \hline & & 5 & 4 \end{array}$$

Add tens. We get 35 tens, so we have 30 tens + 5 tens. This is also 3 hundreds + 5 tens. Write 5 in the tens column and 3 above the hundreds.

$$\begin{array}{rccc} & 3 & 2 & \\ & 3 & 9 & 1 \\ & 2 & 7 & 6 \\ & 7 & 8 & 9 \\ + & 4 & 9 & 8 \\ \hline 1 & 9 & 5 & 4 \end{array}$$

Add hundreds. We get 19 hundreds.

Do Exercises 1–4.

Add.

1.
$$\begin{array}{r} 7968 \\ +\ 5497 \\ \hline \end{array}$$

2. 6203 + 3542

3.
$$\begin{array}{r} 9804 \\ +\ 6378 \\ \hline \end{array}$$

4.
$$\begin{array}{r} 1932 \\ 6723 \\ 9878 \\ +\ 8941 \\ \hline \end{array}$$

Answers on page A-1

b Finding Perimeter

Addition can be used when finding perimeter.

PERIMETER

The distance around an object is its **perimeter.**

EXAMPLE 3 Find the perimeter of the figure.

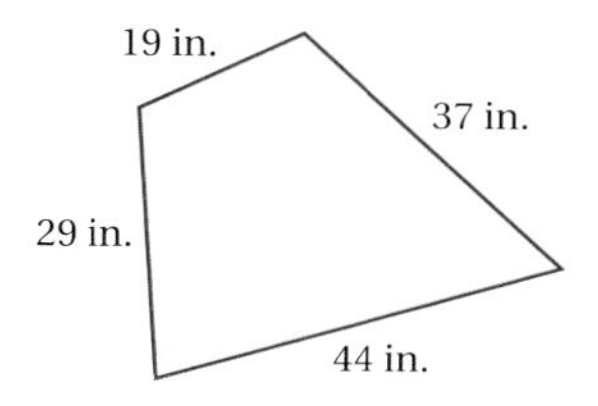

We add the lengths of the sides:

Perimeter = 29 in. + 19 in. + 37 in. + 44 in.

We carry out the addition as follows.

$$\begin{array}{r} \scriptstyle 2 \\ 29 \\ 19 \\ 37 \\ +\ 44 \\ \hline 129 \end{array}$$

The perimeter of the figure is 129 in.

Do Exercises 5 and 6.

EXAMPLE 4 Find the perimeter of the octagonal (eight-sided) resort swimming pool.

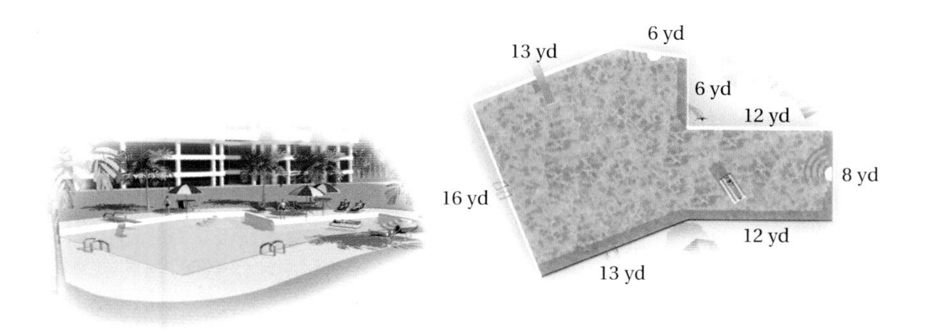

Perimeter = 13 yd + 6 yd + 6 yd + 12 yd + 8 yd
+ 12 yd + 13 yd + 16 yd

The perimeter of the pool is 86 yd.

Do Exercise 7.

Find the perimeter of each figure.

5.

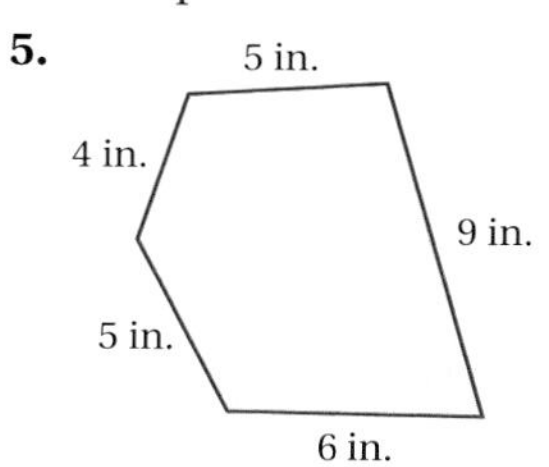

6.

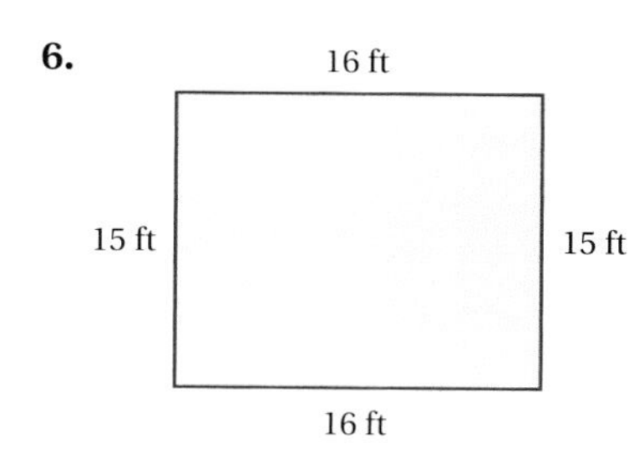

Solve.

7. **Index Cards.** Two standard sizes for index cards are 3 in. (inches) by 5 in. and 5 in. by 8 in. Find the perimeter of each card.

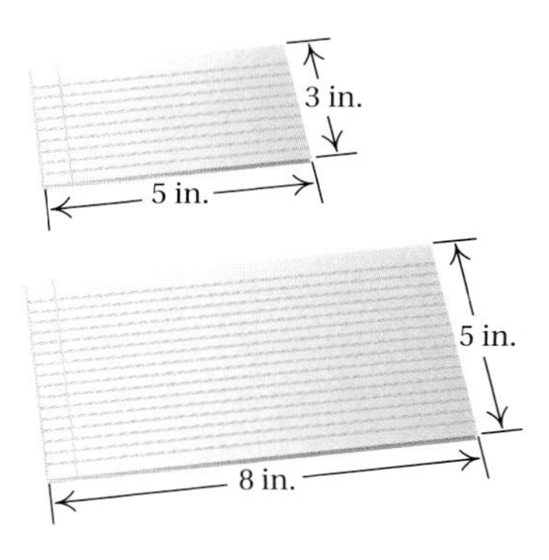

Answers on page A-1

CALCULATOR CORNER

Adding Whole Numbers *To the student and the instructor:* This is the first of a series of *optional* discussions on using a calculator. A calculator is *not* a requirement for this textbook. There are many kinds of calculators and different instructions for their usage. We have included instructions here for a minimum-cost calculator. Be sure to consult your user's manual as well. Also, check with your instructor about whether you are allowed to use a calculator in the course.

To add whole numbers on a calculator, we use the [+] and [=] keys. For example, to add 57 and 34, we press [5] [7] [+] [3] [4] [=]. The calculator displays [91], so $57 + 34 = 91$. To find $314 + 259 + 478$, we press [3] [1] [4] [+] [2] [5] [9] [+] [4] [7] [8] [=]. The display reads [1051], so $314 + 259 + 478 = 1051$.

Exercises: Use a calculator to find each sum.

1. $19 + 36$
2. $73 + 48$
3. $925 + 677$
4. $276 + 458$
5. $\begin{array}{r} 826 \\ 415 \\ +\ 691 \\ \hline \end{array}$
6. $\begin{array}{r} 253 \\ 490 \\ +\ 121 \\ \hline \end{array}$

Study Tips

USING THIS TEXTBOOK

Throughout this textbook, you will find a feature called "Study Tips." One of the most important ways in which to improve your math study skills is to learn the proper use of the textbook. Here we highlight a few points that we consider most helpful.

- **Be sure to note the symbols [a], [b], [c], and so on, that correspond to the objectives you are to master in each section.** The first time you see them is in the margin at the beginning of the section; the second time is in the subheadings of each section; and the third time is in the exercise set for the section. You will also find symbols like [1.1a] or [1.2c] next to the skill maintenance exercises in each exercise set and the review exercises at the end of the chapter, as well as in the answers to the chapter tests and the cumulative reviews. These objective symbols allow you to refer to the appropriate place in the text when you need to review a topic.
- **Read and study each step of each example.** The examples include important side comments that explain each step. These examples and annotations have been carefully chosen so that you will be fully prepared to do the exercises.
- **Stop and do the margin exercises as you study a section.** This gives you immediate reinforcement of each concept as it is introduced and is one of the most effective ways to master the mathematical skills in this text. Don't deprive yourself of this benefit!
- **Note the icons listed at the top of each exercise set.** These refer to the many distinctive multimedia study aids that accompany the book.

1.2 EXERCISE SET

For Extra Help

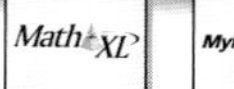

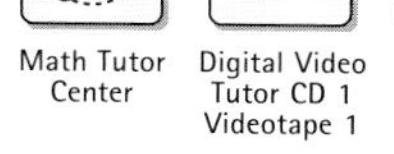

MathXL | MyMathLab | InterAct Math | Math Tutor Center | Digital Video Tutor CD 1 Videotape 1 | Student's Solutions Manual

a Add.

1. $\begin{array}{r} 364 \\ +\ 23 \\ \hline \end{array}$

2. $\begin{array}{r} 1521 \\ +\ 348 \\ \hline \end{array}$

3. $\begin{array}{r} 1716 \\ +\ 3482 \\ \hline \end{array}$

4. $\begin{array}{r} 7503 \\ +\ 2683 \\ \hline \end{array}$

5. $\begin{array}{r} 86 \\ +\ 78 \\ \hline \end{array}$

6. $\begin{array}{r} 73 \\ +\ 69 \\ \hline \end{array}$

7. $\begin{array}{r} 99 \\ +\ 1 \\ \hline \end{array}$

8. $\begin{array}{r} 999 \\ +\ 11 \\ \hline \end{array}$

9. $8113 + 390$

10. $271 + 3338$

11. $356 + 4910$

12. $280 + 34{,}702$

13. $3870 + 92 + 7 + 497$

14. $10{,}120 + 12{,}989 + 5738$

15. $\begin{array}{r} 4825 \\ +\ 1783 \\ \hline \end{array}$

16. $\begin{array}{r} 3654 \\ +\ 2700 \\ \hline \end{array}$

17. $\begin{array}{r} 23{,}443 \\ +\ 10{,}989 \\ \hline \end{array}$

18. $\begin{array}{r} 45{,}879 \\ +\ 21{,}786 \\ \hline \end{array}$

19. $\begin{array}{r} 77{,}543 \\ +\ 23{,}767 \\ \hline \end{array}$

20. $\begin{array}{r} 99{,}999 \\ +\ 112 \\ \hline \end{array}$

21. $\begin{array}{r} 45 \\ 25 \\ 36 \\ 44 \\ +\ 80 \\ \hline \end{array}$

22. $\begin{array}{r} 38 \\ 27 \\ 32 \\ 14 \\ +\ 76 \\ \hline \end{array}$

23. $\begin{array}{r} 12{,}070 \\ 2{,}954 \\ +\ 3{,}400 \\ \hline \end{array}$

24. $\begin{array}{r} 42{,}487 \\ 83{,}141 \\ +\ 36{,}712 \\ \hline \end{array}$

25. $\begin{array}{r} 4835 \\ 729 \\ 9204 \\ 8986 \\ +\ 7931 \\ \hline \end{array}$

26. $\begin{array}{r} 989 \\ 566 \\ 834 \\ 920 \\ +\ 703 \\ \hline \end{array}$

b Find the perimeter of each figure.

27.

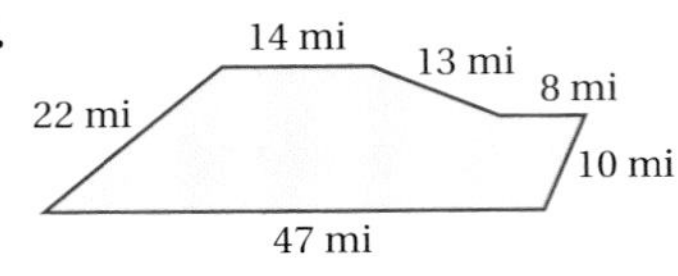

28.

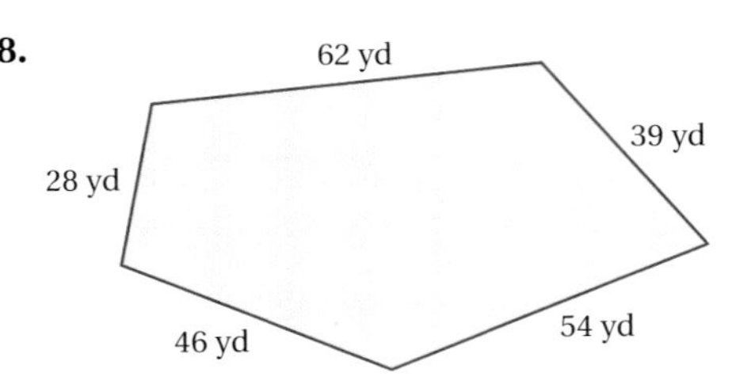

29. Find the perimeter of a standard hockey rink.

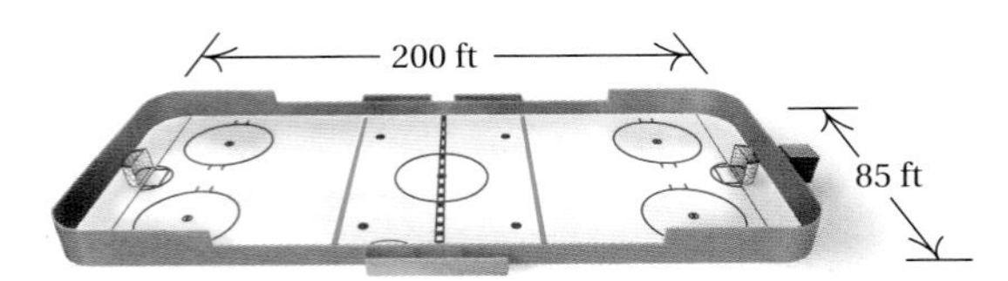

30. In Major League Baseball, how far does a batter travel in circling the bases when a home run has been hit?

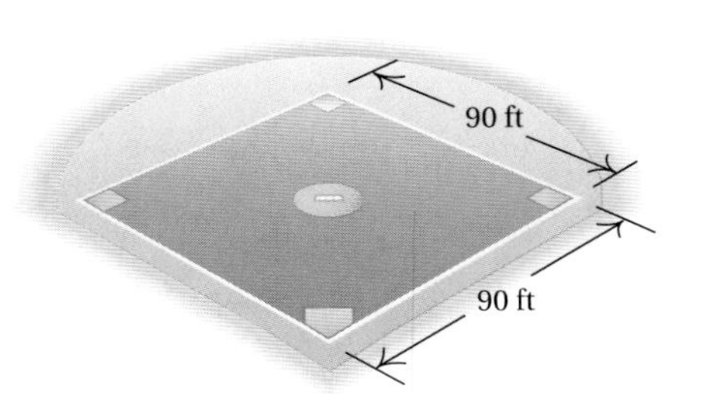

31. **DW** Explain in your own words what the associative law of addition means.

32. **DW** Describe a situation that corresponds to this mathematical expression:

80 mi + 245 mi + 336 mi.

SKILL MAINTENANCE

The exercises that follow begin an important feature called *Skill Maintenance exercises.* These exercises provide an ongoing review of topics previously covered in the book. You will see them in virtually every exercise set. It has been found that this kind of continuing review can significantly improve your performance on a final examination.

33. What does the digit 8 mean in 486,205? [1.1a]

34. Write a word name for the number in the following sentence: [1.1c]

In fiscal year 2004, Starbucks Corporation had total net revenues of $5,294,247,000.

Source: Starbucks Corporation

SYNTHESIS

35. A fast way to add all the numbers from 1 to 10 inclusive is to pair 1 with 9, 2 with 8, and so on. Use a similar approach to add all numbers from 1 to 100 inclusive.

1.3 SUBTRACTION

Objectives

a Convert between addition sentences and subtraction sentences.

b Subtract whole numbers.

a Subtraction and Related Sentences

TAKE AWAY

Subtraction of whole numbers applies to two kinds of situations. The first is called "take away." Consider the following example.

A bowler starts with 10 pins and knocks down 8 of them.

From 10 pins, the bowler "takes away" 8 pins. There are 2 pins left. The subtraction is $10 - 8 = 2$.

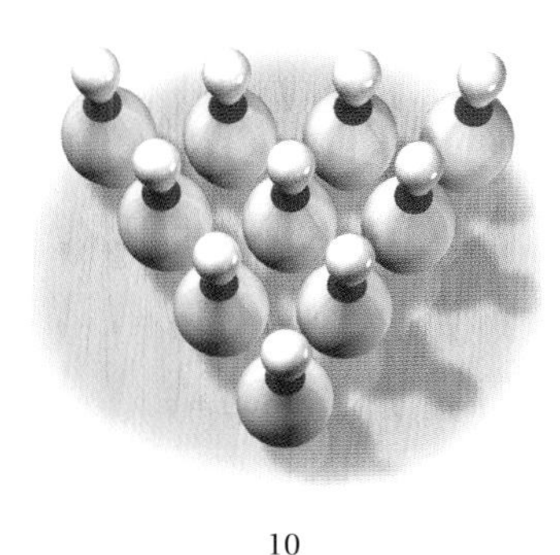

10

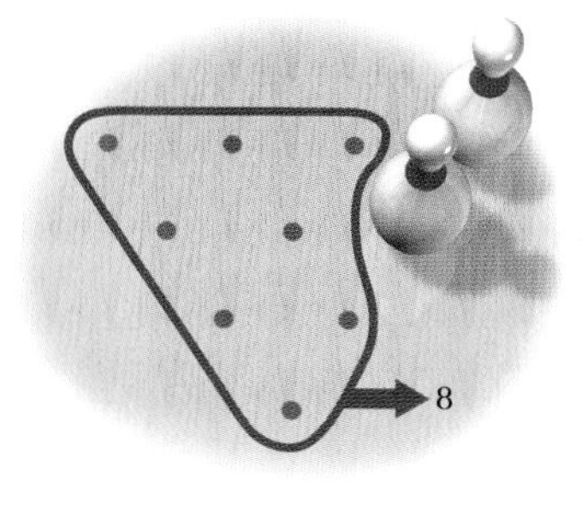

$10 - 8 = 2$

We use the following terminology with subtraction:

$$\underset{\text{Minuend}}{10} - \underset{\text{Subtrahend}}{8} = \underset{\text{Difference}}{2}.$$

The **minuend** is the number from which another number is being subtracted. The **subtrahend** is the number being subtracted. The **difference** is the result of subtracting the subtrahend from the minuend.

Subtraction also corresponds to moving distances on a number line. The number line below is marked with tick marks at equal distances of 1 unit. The difference $10 - 8$ is shown. We first move from 0 right 10 units, and then left 8 units, and end up at 2. The subtraction that corresponds to the situation is $10 - 8 = 2$.

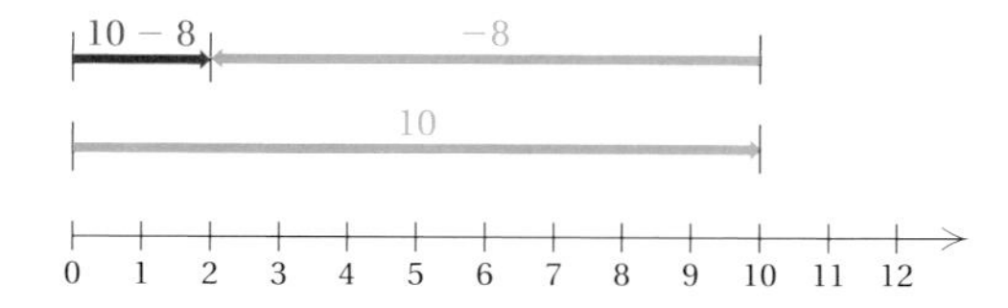

This leads us to the following definition of subtraction.

> **SUBTRACTION**
>
> The difference $a - b$ is that unique whole number c for which $a = c + b$.

For example, $13 - 4$ is the number 9 since $13 = 9 + 4$.

Study Tips

LEARNING RESOURCES

Are you aware of all the learning resources that exist for this textbook? Many details are given in the Preface.

- The *Student's Solutions Manual* contains worked-out solutions to the odd-numbered exercises in the exercise sets.
- An extensive set of *videotapes* supplements this text. These are available on CD-ROM by calling 1-800-282-0693.
- *Tutorial software* called Math XL Tutorials on CD is available with this text. If it is not available in the campus learning center, you can order it by calling 1-800-282-0693.
- The Addison-Wesley *Math Tutor Center* has experienced instructors available to help with the odd-numbered exercises. You can order this service by calling 1-800-824-7799.
- Extensive help is available online via MyMathLab and/or MathXL. Ask your instructor for information about these or visit MyMathLab.com and MathXL.com.

Write a related addition sentence.

1. $7 - 5 = 2$

2. $17 - 8 = 9$

Write two related subtraction sentences.

3. $5 + 8 = 13$

4. $11 + 3 = 14$

Answers on page A-1

RELATED SENTENCES

Subtraction is defined in terms of addition. For example, $5 - 2$ is that number which, when added to 2, gives 5. Thus for the subtraction sentence

$5 - 2 = 3$, Taking away 2 from 5 gives 3.

there is a *related addition sentence*

$5 = 3 + 2$. Putting back the 2 gives 5 again.

In fact, we know that answers we find to subtractions are correct only because of the related addition, which provides a handy way to *check* a subtraction.

EXAMPLE 1 Write a related addition sentence: $8 - 5 = 3$.

$8 - 5 = 3$

This number gets added.

$8 = 3 + 5$

By the commutative law of addition, there is also another addition sentence:

$8 = 5 + 3$.

The related addition sentence is $8 = 3 + 5$.

Do Exercises 1 and 2.

EXAMPLE 2 Write two related subtraction sentences: $4 + 3 = 7$.

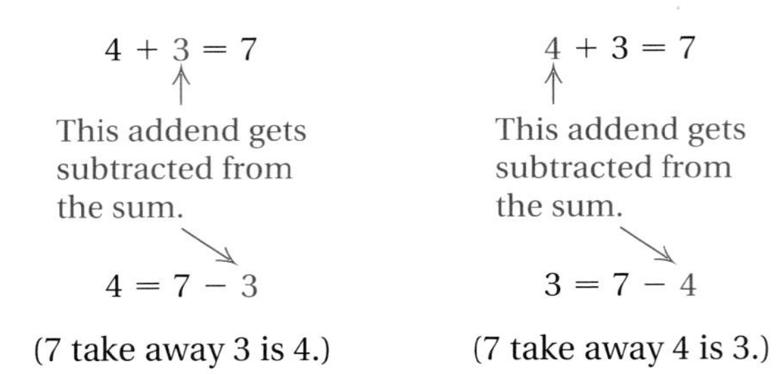

(7 take away 3 is 4.) (7 take away 4 is 3.)

The related subtraction sentences are $4 = 7 - 3$ and $3 = 7 - 4$.

Do Exercises 3 and 4.

HOW MANY MORE

The second kind of situation to which subtraction can apply is called "how many more." You have 2 notebooks, but you need 7. You can think of this as "how many do I need to add to 2 to get 7?" Finding the answer can be thought of as finding a missing addend, and can be found by subtracting 2 from 7.

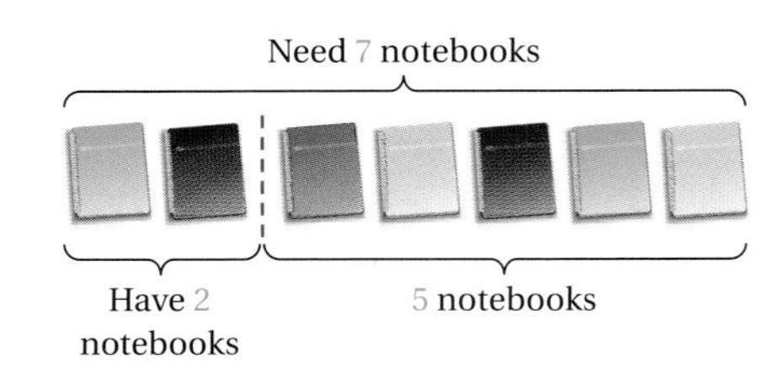

What must be added to 2 to get 7? The answer is 5.

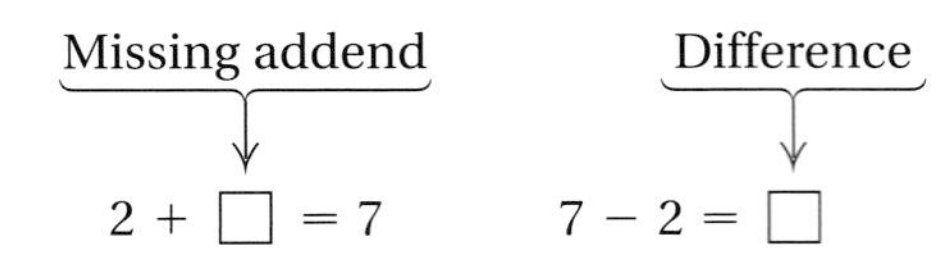

Let's look at the following example in which a missing addend occurs: Jillian wants to buy the roll-on luggage shown in this ad. She has \$30. She needs \$79. How much more does she need in order to buy the luggage?

Thinking of this situation in terms of a missing addend, we have:

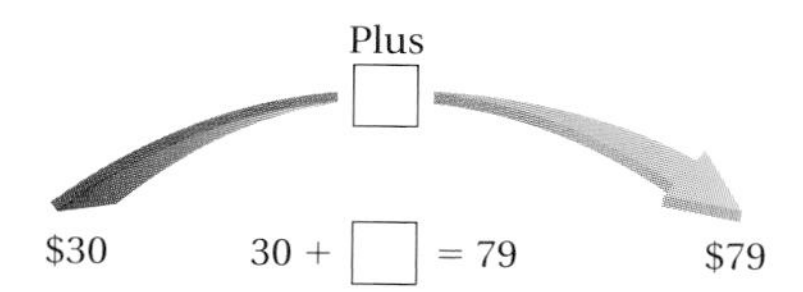

To find the answer, we think of the related subtraction sentence:

$$30 + \square = 79$$

$$\square = 79 - 30.$$

b Subtraction of Whole Numbers

To subtract numbers, we subtract the ones digits first, then the tens digits, then the hundreds, then the thousands, and so on.

EXAMPLE 3 Subtract: 9768 − 4320.

$$\begin{array}{r} 9\ 7\ 6\ 8 \\ -\ 4\ 3\ 2\ 0 \\ \hline 8 \end{array}$$ Subtract ones.

$$\begin{array}{r} 9\ 7\ 6\ 8 \\ -\ 4\ 3\ 2\ 0 \\ \hline 4\ 8 \end{array}$$ Subtract tens.

$$\begin{array}{r} 9\ 7\ 6\ 8 \\ -\ 4\ 3\ 2\ 0 \\ \hline 4\ 4\ 8 \end{array}$$ Subtract hundreds.

$$\begin{array}{r} 9\ 7\ 6\ 8 \\ -\ 4\ 3\ 2\ 0 \\ \hline 5\ 4\ 4\ 8 \end{array}$$ Subtract thousands.

This is for explanation.

$$\begin{array}{r} 9\ 7\ 6\ 8 \\ -\ 4\ 3\ 2\ 0 \\ \hline 5\ 4\ 4\ 8 \end{array}$$

You should write only this.

5. Subtract.

$$\begin{array}{r} 7\ 8\ 9\ 3 \\ -\ 4\ 0\ 9\ 2 \\ \hline \end{array}$$

Answer on page A-1

CALCULATOR CORNER

Subtracting Whole Numbers To subtract whole numbers on a calculator, we use the [−] and [=] keys. For example, to find 63 − 47, we press [6] [3] [−] [4] [7] [=]. The calculator displays [16], so 63 − 47 = 16. We can check this result by adding the subtrahend, 47, and the difference, 16. To do this, we press [1] [6] [+] [4] [7] [=]. The sum is the minuend, 63, so the subtraction is correct.

Exercises: Use a calculator to perform each subtraction. Check by adding.

1. 57 − 29
2. 81 − 34
3. 145 − 78
4. 612 − 493
5. $\begin{array}{r} 4976 \\ -\ 2848 \\ \hline \end{array}$
6. $\begin{array}{r} 12{,}406 \\ -\ 9813 \\ \hline \end{array}$

We have considered the subtraction 9768 − 4320 = □. That is, we have found the missing addend in the sentence 9768 = 4320 + □. If 5448 is indeed the missing addend, then if we add it to 4320, the answer should be 9768. The related addition sentence is the basis for adding as a *check*.

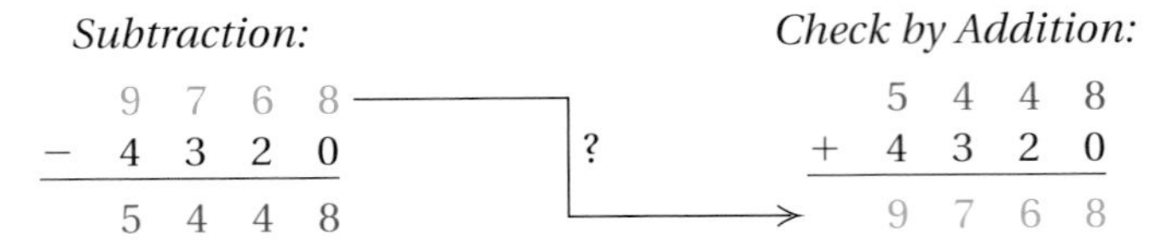

Do Exercise 5 on the preceding page.

EXAMPLE 4 Subtract: 348 − 165.

We have

$$\begin{array}{rcl} 3 \text{ hundreds} + 4 \text{ tens} + 8 \text{ ones} & = & 2 \text{ hundreds} + 14 \text{ tens} + 8 \text{ ones} \\ -\ 1 \text{ hundred} - 6 \text{ tens} - 5 \text{ ones} & = & -\ 1 \text{ hundred} - 6 \text{ tens} - 5 \text{ ones} \\ \hline & = & 1 \text{ hundred} + 8 \text{ tens} + 3 \text{ ones} \\ & = & 183. \end{array}$$

Note that in this case, although we can subtract the ones (8 − 5 = 3), we cannot do so with the tens, because 4 − 6 is *not* a whole number. To see why, consider

4 − 6 = □ and the related addition sentence 4 = □ + 6.

There is no whole number that when added to 6 gives 4. To complete the subtraction, we must *borrow* 1 hundred from 3 hundreds and regroup it with 4 tens. Then we can do the subtraction 14 tens − 6 tens = 8 tens. Below we consider a shortened form.

$\begin{array}{r} 3\,4\,8 \\ -\ 1\,6\,5 \\ \hline 3 \end{array}$ Subtract ones.

$\begin{array}{r} \scriptstyle 2\ \ 14 \\ \not{3}\ \not{4}\ 8 \\ -\ 1\ 6\ 5 \\ \hline 3 \end{array}$ Borrow one hundred. That is, 1 hundred = 10 tens, and 10 tens + 4 tens = 14 tens. Write 2 above the hundreds column and 14 above the tens.

$\begin{array}{r} \scriptstyle 2\ \ 14 \\ \not{3}\ \not{4}\ 8 \\ -\ 1\ 6\ 5 \\ \hline 1\ 8\ 3 \end{array}$ Subtract tens; subtract hundreds.

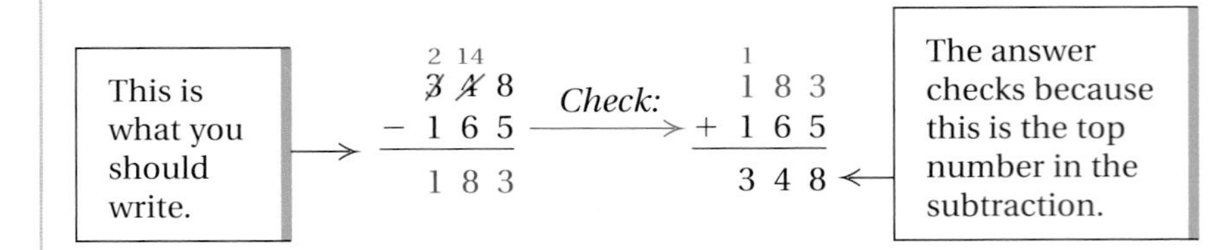

Answers on page A-1

EXAMPLE 5 Subtract: 6246 − 1879.

$$\begin{array}{r} \scriptstyle 3\ 16 \\ 6\ 2\ \not4\ \not6 \\ -\ 1\ 8\ 7\ 9 \\ \hline 7 \end{array}$$

We cannot subtract 9 ones from 6 ones, but we can subtract 9 ones from 16 ones. We borrow 1 ten to get 16 ones.

$$\begin{array}{r} \scriptstyle 13\quad \\ \scriptstyle 1\ \not3\ 16 \\ 6\ \not2\ \not4\ \not6 \\ -\ 1\ 8\ 7\ 9 \\ \hline 6\ 7 \end{array}$$

We cannot subtract 7 tens from 3 tens, but we can subtract 7 tens from 13 tens. We borrow 1 hundred to get 13 tens.

$$\begin{array}{r} \scriptstyle 11\ 13\quad \\ \scriptstyle 5\ \not1\ \not3\ 16 \\ \not6\ \not2\ \not4\ \not6 \\ -\ 1\ 8\ 7\ 9 \\ \hline 4\ 3\ 6\ 7 \end{array}$$

We cannot subtract 8 hundreds from 1 hundred, but we can subtract 8 hundreds from 11 hundreds. We borrow 1 thousand to get 11 hundreds.

This is what you should write. →

$$\begin{array}{r} \scriptstyle 11\ 13\quad \\ \scriptstyle 5\ \not1\ \not3\ 16 \\ \not6\ \not2\ \not4\ \not6 \\ -\ 1\ 8\ 7\ 9 \\ \hline 4\ 3\ 6\ 7 \end{array} \quad \textit{Check:} \longrightarrow \quad \begin{array}{r} \scriptstyle 1\ 1\ 1\ \ \\ 4\ 3\ 6\ 7 \\ +\ 1\ 8\ 7\ 9 \\ \hline 6\ 2\ 4\ 6 \end{array}$$

← The answer checks because this is the top number in the subtraction.

Do Exercises 6 and 7.

EXAMPLE 6 Subtract: 902 − 477.

$$\begin{array}{r} \scriptstyle 8\ 9\ 12 \\ \not9\,\not0\ \not2 \\ -\ 4\ 7\ 7 \\ \hline 4\ 2\ 5 \end{array}$$

We cannot subtract 7 ones from 2 ones. We have 9 hundreds, or 90 tens. We borrow 1 ten to get 12 ones. We then have 89 tens.

Do Exercises 8 and 9.

EXAMPLE 7 Subtract: 8003 − 3667.

$$\begin{array}{r} \scriptstyle 7\ 9\ 9\ 13 \\ \not8\,\not0\,\not0\ \not3 \\ -\ 3\ 6\ 6\ 7 \\ \hline 4\ 3\ 3\ 6 \end{array}$$

We have 8 thousands, or 800 tens. We borrow 1 ten to get 13 ones. We then have 799 tens.

EXAMPLES

8. Subtract: 6000 − 3762.

$$\begin{array}{r} \scriptstyle 5\ 9\ 9\ 10 \\ \not6\,\not0\,\not0\ \not0 \\ -\ 3\ 7\ 6\ 2 \\ \hline 2\ 2\ 3\ 8 \end{array}$$

9. Subtract: 6024 − 2968.

$$\begin{array}{r} \scriptstyle 11\quad \\ \scriptstyle 5\ 9\ \not1\ 14 \\ \not6\,\not0\ \not2\ \not4 \\ -\ 2\ 9\ 6\ 8 \\ \hline 3\ 0\ 5\ 6 \end{array}$$

Do Exercises 10–12.

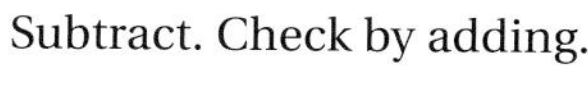

Subtract. Check by adding.

6. $\begin{array}{r} 8\ 6\ 8\ 6 \\ -\ 2\ 3\ 5\ 8 \\ \hline \end{array}$

7. $\begin{array}{r} 7\ 1\ 4\ 5 \\ -\ 2\ 3\ 9\ 8 \\ \hline \end{array}$

Subtract.

8. $\begin{array}{r} 7\ 0 \\ -\ 1\ 4 \\ \hline \end{array}$

9. $\begin{array}{r} 5\ 0\ 3 \\ -\ 2\ 9\ 8 \\ \hline \end{array}$

Subtract.

10. $\begin{array}{r} 7\ 0\ 0\ 7 \\ -\ 6\ 3\ 4\ 9 \\ \hline \end{array}$

11. $\begin{array}{r} 6\ 0\ 0\ 0 \\ -\ 3\ 1\ 4\ 9 \\ \hline \end{array}$

12. $\begin{array}{r} 9\ 0\ 3\ 5 \\ -\ 7\ 4\ 8\ 9 \\ \hline \end{array}$

Answers on page A-1

1.3 EXERCISE SET

For Extra Help

MathXL | MyMathLab | InterAct Math | Math Tutor Center | Digital Video Tutor CD 1 Videotape 1 | Student's Solutions Manual

a Write a related addition sentence.

1. $7 - 4 = 3$

2. $12 - 5 = 7$

3. $13 - 8 = 5$

4. $9 - 9 = 0$

5. $23 - 9 = 14$

6. $20 - 8 = 12$

7. $43 - 16 = 27$

8. $51 - 18 = 33$

Write two related subtraction sentences.

9. $6 + 9 = 15$

10. $7 + 9 = 16$

11. $8 + 7 = 15$

12. $8 + 0 = 8$

13. $17 + 6 = 23$

14. $11 + 8 = 19$

15. $23 + 9 = 32$

16. $42 + 10 = 52$

b Subtract.

17. $\begin{array}{r} 65 \\ -\ 21 \\ \hline \end{array}$

18. $\begin{array}{r} 87 \\ -\ 34 \\ \hline \end{array}$

19. $\begin{array}{r} 866 \\ -\ 333 \\ \hline \end{array}$

20. $\begin{array}{r} 526 \\ -\ 323 \\ \hline \end{array}$

21. $86 - 47$

22. $73 - 28$

23. $981 - 747$

24. $887 - 698$

25. $\begin{array}{r} 7769 \\ -\ 2387 \\ \hline \end{array}$

26. $\begin{array}{r} 6431 \\ -\ 2896 \\ \hline \end{array}$

27. $\begin{array}{r} 7640 \\ -\ 3809 \\ \hline \end{array}$

28. $\begin{array}{r} 8003 \\ -\ 599 \\ \hline \end{array}$

29. $\begin{array}{r} 12{,}647 \\ -\ 4{,}899 \\ \hline \end{array}$

30. $\begin{array}{r} 16{,}222 \\ -\ 5{,}888 \\ \hline \end{array}$

31. $90{,}237 - 47{,}209$

32. $84{,}703 - 298$

33. $\begin{array}{r} 80 \\ -\ 24 \\ \hline \end{array}$

34. $\begin{array}{r} 90 \\ -\ 78 \\ \hline \end{array}$

35. $\begin{array}{r} 690 \\ -\ 236 \\ \hline \end{array}$

36. $\begin{array}{r} 803 \\ -\ 418 \\ \hline \end{array}$

37. $\begin{array}{r} 6808 \\ -\ 3059 \\ \hline \end{array}$

38. $\begin{array}{r} 6408 \\ -\ \ \ 258 \\ \hline \end{array}$

39. $\begin{array}{r} 2300 \\ -\ \ \ 109 \\ \hline \end{array}$

40. $\begin{array}{r} 6007 \\ -\ 1589 \\ \hline \end{array}$

41. 101,734 − 5760

42. 15,017 − 7809

43. 10,008 − 19

44. 21,043 − 8909

45. $\begin{array}{r} 7000 \\ -\ 2794 \\ \hline \end{array}$

46. $\begin{array}{r} 8001 \\ -\ 6543 \\ \hline \end{array}$

47. $\begin{array}{r} 48,000 \\ -\ 37,695 \\ \hline \end{array}$

48. $\begin{array}{r} 17,043 \\ -\ 11,598 \\ \hline \end{array}$

49. **DW** Describe two situations that correspond to the subtraction \$20 − \$17, one "take away" and one "how many more."

50. **DW** Is subtraction commutative (is there a commutative law of subtraction)? Why or why not?

SKILL MAINTENANCE

Add. [1.2a]

51. $\begin{array}{r} 946 \\ +\ \ \ 78 \\ \hline \end{array}$

52. $\begin{array}{r} 9078 \\ +\ 3654 \\ \hline \end{array}$

53. $\begin{array}{r} 57,877 \\ +\ 32,406 \\ \hline \end{array}$

54. $\begin{array}{r} 8004 \\ 6789 \\ 7720 \\ +\ 6851 \\ \hline \end{array}$

55. 567 + 778

56. 901 + 23

57. 12,885 + 9807

58. 9909 + 1011

59. Write a word name for 6,375,602. [1.1c]

60. What does the digit 7 mean in 6,375,602? [1.1a]

SYNTHESIS

61. Fill in the missing digits to make the subtraction true:

9,□48,621 − 2,097,□81 = 7,251,140.

62. ▦ Subtract: 3,928,124 − 1,098,947.

1.4 ROUNDING AND ESTIMATING; ORDER

Objectives

a Round to the nearest ten, hundred, or thousand.

b Estimate sums and differences by rounding.

c Use < or > for □ to write a true sentence in a situation like 6 □ 10.

a Rounding

We round numbers in various situations when we do not need an exact answer. For example, we might round to see if we are being charged the correct amount in a store. We might also round to check if an answer to a problem is reasonable or to check a calculation done by hand or on a calculator.

To understand how to round, we first look at some examples using number lines, even though this is not the way we generally do rounding.

EXAMPLE 1 Round 47 to the nearest ten.

Here is a part of a number line; 47 is between 40 and 50. Since 47 is closer to 50, we round up to 50.

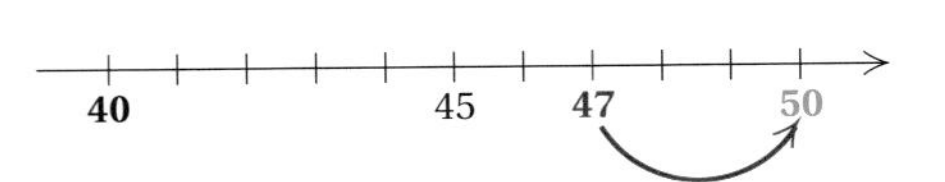

EXAMPLE 2 Round 42 to the nearest ten.

42 is between 40 and 50. Since 42 is closer to 40, we round down to 40.

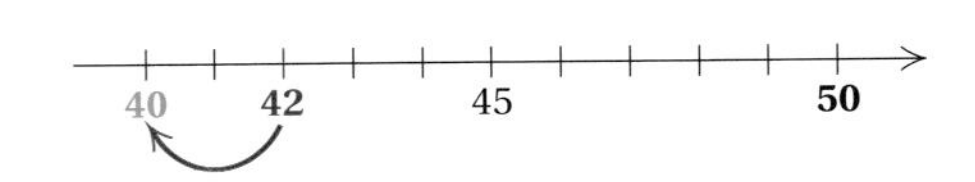

Do Exercises 1–4.

EXAMPLE 3 Round 45 to the nearest ten.

45 is halfway between 40 and 50. We could round 45 down to 40 or up to 50. We agree to round up to 50.

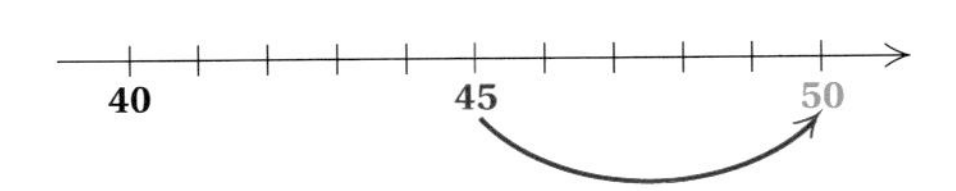

When a number is halfway between rounding numbers, round up.

Do Exercises 5–7.

Here is a rule for rounding.

ROUNDING WHOLE NUMBERS

To round to a certain place:

a) Locate the digit in that place.
b) Consider the next digit to the right.
c) If the digit to the right is 5 or higher, round up. If the digit to the right is 4 or lower, round down.
d) Change all digits to the right of the rounding location to zeros.

Round to the nearest ten.

1. 37

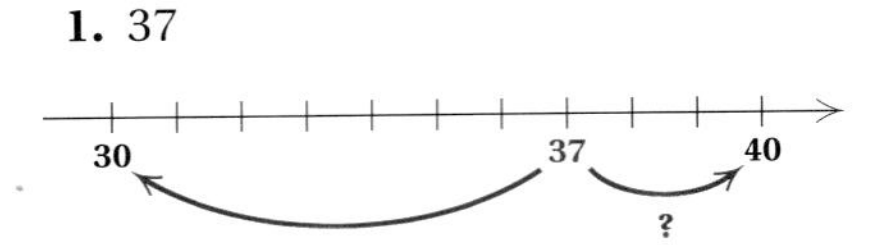

2. 52

3. 73

4. 98

Round to the nearest ten.

5. 35

6. 75

7. 85

Answers on page A-2

EXAMPLE 4 Round 6485 to the nearest ten.

a) Locate the digit in the tens place, 8.

6 4 8 5
↑

b) Consider the next digit to the right, 5.

6 4 8 5
↑

c) Since that digit, 5, is 5 or higher, round 8 tens up to 9 tens.

d) Change all digits to the right of the tens digit to zeros.

6 4 9 0 ← This is the answer.

EXAMPLE 5 Round 6485 to the nearest hundred.

a) Locate the digit in the hundreds place, 4.

6 4 8 5
↑

b) Consider the next digit to the right, 8.

6 4 8 5
↑

c) Since that digit, 8, is 5 or higher, round 4 hundreds up to 5 hundreds.

d) Change all digits to the right of hundreds to zeros.

6 5 0 0 ← This is the answer.

EXAMPLE 6 Round 6485 to the nearest thousand.

a) Locate the digit in the thousands place, 6.

6 4 8 5
↑

b) Consider the next digit to the right, 4.

6 4 8 5
↑

c) Since that digit, 4, is 4 or lower, round down, meaning that 6 thousands stays as 6 thousands.

d) Change all digits to the right of thousands to zeros.

6 0 0 0 ← This is the answer.

Do Exercises 8–19.

CAUTION!

7000 is not a correct answer to Example 6. It is incorrect to round from the ones digit over, as follows:

6485, → 6490, → 6500, → 7000.

Note that 6485 is closer to 6000 than it is to 7000.

Round to the nearest ten.

8. 137

9. 473

10. 235

11. 285

Round to the nearest hundred.

12. 641

13. 759

14. 750

15. 9325

Round to the nearest thousand.

16. 7896

17. 8459

18. 19,343

19. 68,500

Answers on page A-2

20. Round 48,968 to the nearest ten, hundred, and thousand.

21. Round 269,582 to the nearest ten, hundred, and thousand.

Refer to the chart on the next page to answer Margin Exercises 22 and 23.

22. By eliminating options, find a way that Ethan and Olivia can buy the ION·2 and stay within their $16,500 budget.

23. Tara and Alex are shopping for a new car. They are considering a Saturn ION·3 and have allowed a budget of $19,000.

a) Estimate by first rounding to the nearest hundred the cost of an ION·3 with all the options.

b) Can they afford this car with a budget of $19,000?

Answers on page A-2

Sometimes rounding involves changing more than one digit in a number.

EXAMPLE 7 Round 78,595 to the nearest ten.

a) Locate the digit in the tens place, 9.

7 8,5 9 5
↑

b) Consider the next digit to the right, 5.

7 8,5 9 5
↑

c) Since that digit, 5, is 5 or higher, round 9 tens to 10 tens. To carry this out, we think of 10 tens as 1 hundred + 0 tens, and increase the hundreds digit by 1, to get 6 hundreds + 0 tens. We then write 6 in the hundreds place and 0 in the tens place.

d) Change the digit to the right of the tens digit to zero.

7 8,6 0 0 ← This is the answer.

Note that if we round this number to the nearest hundred, we get the same answer.

Do Exercises 20 and 21.

b Estimating

Estimating can be done in many ways. In general, an estimate made by rounding to the nearest ten is more accurate than one rounded to the nearest hundred, and an estimate rounded to the nearest hundred is more accurate than one rounded to the nearest thousand, and so on.

In the following example, we see how estimation can be used in making a purchase.

EXAMPLE 8 *Estimating the Cost of an Automobile Purchase.* Ethan and Olivia Benson are shopping for a new car. They are considering a Saturn ION. There are three basic models of this car, and each has options beyond the basic price, as shown in the chart on the following page. Ethan and Olivia have allowed themselves a budget of $16,500. They look at the list of options and want to make a quick estimate of the cost of model ION·2 with all the options.

Estimate by rounding to the nearest hundred the cost of the ION·2 with all the options and decide whether it will fit into their budget.

MODEL ION·1 SEDAN (4 DOOR) 2.2-LITER ENGINE, 4-SPEED AUTOMATIC TRANSMISSION	MODEL ION·2 SEDAN (4 DOOR) 2.2-LITER ENGINE, 5-SPEED MANUAL TRANSMISSION	MODEL ION·3 SEDAN (4 DOOR) 2.2-LITER ENGINE, 5-SPEED MANUAL TRANSMISSION
Base Price: $12,975	Base Price: $14,945	Base Price: $16,470
Each of these vehicles comes with several options. Note that some of the options are standard on certain models. Others are not available for all models.		
Antilock Braking System with Traction Control:	$400	
Head Curtain Side Air Bags:	$395	
Power Sunroof (Not available for ION·1):	$725	
Rear Spoiler (Not available for ION·1):	$250	
Air Conditioning with Dust and Pollen Filtration (Standard on ION·2 and ION·3):	$960	
CD/MP3 Player with AM/FM Stereo and 4 Coaxial Speakers (Standard on ION·3):	ION·1—$510 ION·2—$220	
Power Package: Power Windows, Power Exterior Mirrors, Remote Keyless Entry, and Cruise Control (Not available for ION·1 and Standard for ION·3):	$825	

Source: Saturn

First, we list the base price of the ION·2 and then the cost of each of the options. We then round each number to the nearest hundred and add.

$$\begin{array}{rr} 14,945 & 14,900 \\ 400 & 400 \\ 395 & 400 \\ 725 & 700 \\ 250 & 300 \\ 220 & 200 \\ +\ 825 & +\ 800 \\ \hline & 17,700 \end{array} \leftarrow \text{Estimated answer}$$

Air conditioning is standard on the ION·2, so we do not include that cost. The estimated cost is $17,700. Since Ethan and Olivia have allowed themselves a budget of $16,500 for the car, they will need to forgo some options.

Do Exercises 22 and 23 on the preceding page.

EXAMPLE 9 Estimate this sum by first rounding to the nearest ten:

78 + 49 + 31 + 85.

We round each number to the nearest ten. Then we add.

$$\begin{array}{rr} 78 & 80 \\ 49 & 50 \\ 31 & 30 \\ +\ 85 & +\ 90 \\ \hline & 250 \end{array} \leftarrow \text{Estimated answer}$$

Do Exercises 24 and 25.

24. Estimate the sum by first rounding to the nearest ten. Show your work.

$$\begin{array}{r} 74 \\ 23 \\ 35 \\ +\ 66 \\ \hline \end{array}$$

25. Estimate the sum by first rounding to the nearest hundred. Show your work.

$$\begin{array}{r} 650 \\ 685 \\ 238 \\ +\ 168 \\ \hline \end{array}$$

Answers on page A-2

EXAMPLE 10 Estimate the difference by first rounding to the nearest thousand: 9324 − 2849.

We have

$$\begin{array}{r} 9\,3\,2\,4 \\ -\ 2\,8\,4\,9 \\ \hline \end{array} \qquad \begin{array}{r} 9\,0\,0\,0 \\ -\ 3\,0\,0\,0 \\ \hline 6\,0\,0\,0 \end{array}$$

Do Exercises 26 and 27.

The sentence $7 - 5 = 2$ says that $7 - 5$ is the same as 2. When we round, the result is rarely the same as the number we started with. Thus we use the symbol $\approx$ when rounding. This symbol means "**is approximately equal to.**" For example, when 687 is rounded to the nearest ten, we can write $687 \approx 690$.

C Order

We know that 2 is not the same as 5. We express this by the sentence $2 \neq 5$. We also know that 2 is less than 5. We symbolize this by the expression $2 < 5$. We can see this order on the number line: 2 is to the left of 5. The number 0 is the smallest whole number.

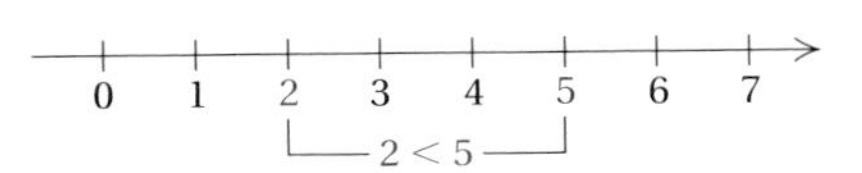

ORDER OF WHOLE NUMBERS

For any whole numbers a and b:

1. $a < b$ (read "a is less than b") is true when a is to the left of b on the number line.
2. $a > b$ (read "a is greater than b") is true when a is to the right of b on the number line.

We call $<$ and $>$ **inequality symbols.**

EXAMPLE 11 Use $<$ or $>$ for □ to write a true sentence: 7 □ 11.

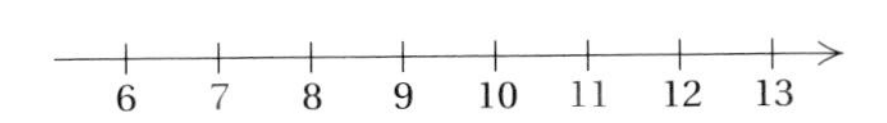

Since 7 is to the left of 11 on the number line, $7 < 11$.

EXAMPLE 12 Use $<$ or $>$ for □ to write a true sentence: 92 □ 87.

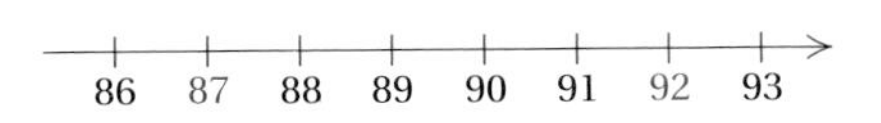

Since 92 is to the right of 87 on the number line, $92 > 87$.

A sentence like $8 + 5 = 13$ is called an **equation.** It is a *true* equation. The equation $4 + 8 = 11$ is a *false* equation. A sentence like $7 < 11$ is called an **inequality.** The sentence $7 < 11$ is a *true* inequality. The sentence $23 > 69$ is a *false* inequality.

Do Exercises 28–33.

26. Estimate the difference by first rounding to the nearest hundred. Show your work.

$$\begin{array}{r} 9\,2\,8\,5 \\ -\ 6\,7\,3\,9 \\ \hline \end{array}$$

27. Estimate the difference by first rounding to the nearest thousand. Show your work.

$$\begin{array}{r} 2\,3{,}2\,7\,8 \\ -\ 1\,1{,}6\,9\,8 \\ \hline \end{array}$$

Use $<$ or $>$ for □ to write a true sentence. Draw a number line if necessary.

28. 8 □ 12

29. 12 □ 8

30. 76 □ 64

31. 64 □ 76

32. 217 □ 345

33. 345 □ 217

Answers on page A-2

EXERCISE SET

For Extra Help

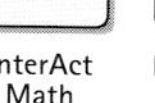

MathXL · MyMathLab · InterAct Math · Math Tutor Center · Digital Video Tutor CD 1 Videotape 1 · Student's Solutions Manual

a Round to the nearest ten.

1. 48 **2.** 532 **3.** 467 **4.** 8945

5. 731 **6.** 17 **7.** 895 **8.** 798

Round to the nearest hundred.

9. 146 **10.** 874 **11.** 957 **12.** 650

13. 9079 **14.** 4645 **15.** 32,850 **16.** 198,402

Round to the nearest thousand.

17. 5876 **18.** 4500 **19.** 7500 **20.** 2001

21. 45,340 **22.** 735,562 **23.** 373,405 **24.** 6,713,855

b Estimate the sum or difference by first rounding to the nearest ten. Show your work.

25. $\begin{array}{r} 78 \\ +\ 97 \\ \hline \end{array}$

26. $\begin{array}{r} 62 \\ 97 \\ 46 \\ +\ 88 \\ \hline \end{array}$

27. $\begin{array}{r} 8074 \\ -\ 2347 \\ \hline \end{array}$

28. $\begin{array}{r} 673 \\ -\ \ 28 \\ \hline \end{array}$

Estimate the sum by first rounding to the nearest ten. Do any of the given sums seem to be incorrect when compared to the estimate? Which ones?

29. $\begin{array}{r} 45 \\ 77 \\ 25 \\ +\ 56 \\ \hline 343 \end{array}$

30. $\begin{array}{r} 41 \\ 21 \\ 55 \\ +\ 60 \\ \hline 177 \end{array}$

31. $\begin{array}{r} 622 \\ 78 \\ 81 \\ +\ 111 \\ \hline 932 \end{array}$

32. $\begin{array}{r} 836 \\ 374 \\ 794 \\ +\ 938 \\ \hline 3947 \end{array}$

Estimate the sum or difference by first rounding to the nearest hundred. Show your work.

33. $\begin{array}{r} 7348 \\ +\ 9247 \\ \hline \end{array}$

34. $\begin{array}{r} 568 \\ 472 \\ 938 \\ +\ 402 \\ \hline \end{array}$

35. $\begin{array}{r} 6852 \\ -\ 1748 \\ \hline \end{array}$

36. $\begin{array}{r} 9438 \\ -\ 2787 \\ \hline \end{array}$

Planning a Kitchen. Perfect Kitchens offers custom kitchen packages with three choices for each of four items: cabinets, countertops, appliances, and flooring. The chart below lists the price for each choice.

Perfect Kitchens lets you customize your kitchen with one choice from each of the following four features:

CABINETS	TYPE	PRICE
(a)	Oak	$7450
(b)	Cherry	8820
(c)	Painted	9630
COUNTERTOPS	**TYPE**	**PRICE**
(d)	Laminate	$1595
(e)	Solid surface	2870
(f)	Granite	3528
APPLIANCES	**PRICE RANGE**	**PRICE**
(g)	Low	$1540
(h)	Medium	3575
(i)	High	6245
FLOORING	**TYPE**	**PRICE**
(j)	Vinyl	$ 625
(k)	Ceramic tile	985
(l)	Hardwood	1160

37. Estimate the cost of remodeling a kitchen with choices (a), (d), (g), and (j) by rounding to the nearest hundred dollars.

38. Estimate the cost of a kitchen with choices (c), (f), (i), and (l) by rounding to the nearest hundred dollars.

39. Sara and Ben are planning to remodel their kitchen and have a budget of $17,700. Estimate by rounding to the nearest hundred dollars the cost of their kitchen remodeling project if they choose options (b), (e), (i), and (k). Can they afford their choices?

40. The Davidsons must make a final decision on the kitchen choices for their new home. The allotted kitchen budget is $16,000. Estimate by rounding to the nearest hundred dollars the kitchen cost if they choose options (a), (f), (h), and (l). Does their budget allotment cover the cost?

41. Suppose you are planning a new kitchen and must stay within a budget of $14,500. Decide on the options you would like and estimate the cost by rounding to the nearest hundred dollars. Does your budget support your choices?

42. Suppose you are planning a new kitchen and must stay within a budget of $18,500. Decide on the options you would like and estimate the cost by rounding to the nearest hundred dollars. Does your budget support your choices?

Estimate the sum by first rounding to the nearest hundred. Do any of the given sums seem to be incorrect when compared to the estimate? Which ones?

43.
$$\begin{array}{r} 216 \\ 84 \\ 745 \\ +\ 595 \\ \hline 1640 \end{array}$$

44.
$$\begin{array}{r} 481 \\ 702 \\ 623 \\ +\ 1043 \\ \hline 1849 \end{array}$$

45.
$$\begin{array}{r} 750 \\ 428 \\ 63 \\ +\ 205 \\ \hline 1446 \end{array}$$

46.
$$\begin{array}{r} 326 \\ 275 \\ 758 \\ +\ 943 \\ \hline 2302 \end{array}$$

Estimate the sum or difference by first rounding to the nearest thousand. Show your work.

47.
$$\begin{array}{r} 9643 \\ 4821 \\ 8943 \\ +\ 7004 \\ \hline \end{array}$$

48.
$$\begin{array}{r} 7648 \\ 9348 \\ 7842 \\ +\ 2222 \\ \hline \end{array}$$

49.
$$\begin{array}{r} 92{,}149 \\ -\ 22{,}555 \\ \hline \end{array}$$

50.
$$\begin{array}{r} 84{,}890 \\ -\ 11{,}110 \\ \hline \end{array}$$

C Use $<$ or $>$ for $\square$ to write a true sentence. Draw a number line if necessary.

51. $0 \ \square \ 17$

52. $32 \ \square \ 0$

53. $34 \ \square \ 12$

54. $28 \ \square \ 18$

55. $1000 \ \square \ 1001$

56. $77 \ \square \ 117$

57. $133 \ \square \ 132$

58. $999 \ \square \ 997$

59. $460 \ \square \ 17$

60. $345 \ \square \ 456$

61. $37 \ \square \ 11$

62. $12 \ \square \ 32$

Daily Newspapers. The top three daily newspapers in the United States are *USA Today*, the *Wall Street Journal*, and the *New York Times*. The daily circulation of each as of September 30, 2002, is listed in the table below. Use this table when answering Exercises 63 and 64.

NEWSPAPER	DAILY CIRCULATION
USA Today	2,136,068
Wall Street Journal	1,800,607
New York Times	1,113,000

Source: Editor and Publisher International Year Book 2003

63. Use an inequality to compare the daily circulation of the *Wall Street Journal* and *USA Today*.

64. Use an inequality to compare the daily circulation of *USA Today* and the *New York Times*.

65. *Pedestrian Fatalities.* The annual number of pedestrians killed when hit by a motor vehicle has declined from 6482 in 1990 to 4827 in 2003. Use an inequality to compare these annual totals of pedestrian fatalities.

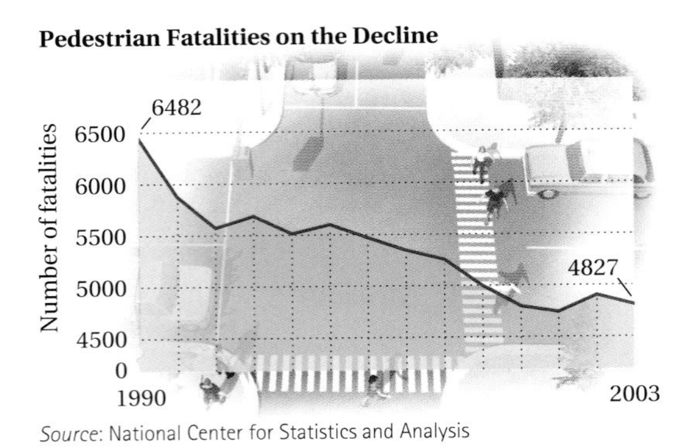

Source: National Center for Statistics and Analysis

66. *Life Expectancy.* The life expectancy of a female in 2010 is predicted to be about 82 yr and of a male about 76 yr. Use an inequality to compare these life expectancies.

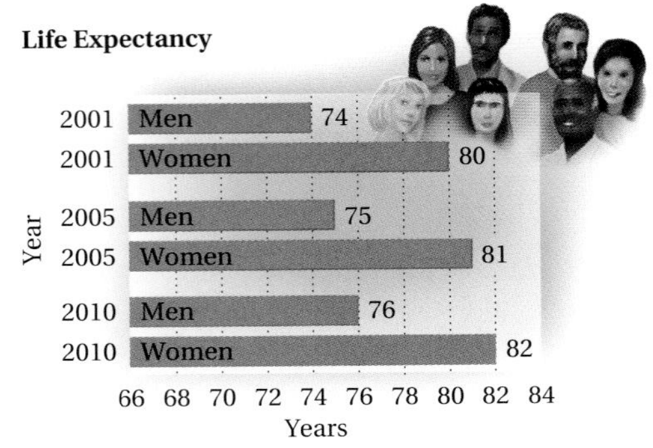

Source: U.S. Bureau of the Census

67. D~W~ Explain how estimating and rounding can be useful when shopping for groceries.

68. D~W~ When rounding 748 to the nearest hundred, a student rounds to 750 and then to 800. What mistake is he making?

SKILL MAINTENANCE

Write expanded notation.

69. 7992 [1.1b]

70. 23,000,000 [1.1b]

71. Write a word name for 246,605,004,032. [1.1c]

72. Write a word name for 1,005,100. [1.1c]

Add. [1.2a]

73. $\begin{array}{r} 67,789 \\ +\ 18,965 \\ \hline \end{array}$

74. $\begin{array}{r} 9002 \\ +\ 4587 \\ \hline \end{array}$

Subtract. [1.3b]

75. $\begin{array}{r} 67,789 \\ -\ 18,965 \\ \hline \end{array}$

76. $\begin{array}{r} 9002 \\ -\ 4587 \\ \hline \end{array}$

SYNTHESIS

77.–80. ▦ Use a calculator to find the sums and differences in each of Exercises 47–50. Then compare your answers with those found using estimation. Even when using a calculator it is possible to make an error if you press the wrong buttons, so it is a good idea to check by estimating.

1.5 MULTIPLICATION

Objectives

a Multiply whole numbers.

b Estimate products by rounding.

c Use multiplication in finding area.

a Multiplication of Whole Numbers

REPEATED ADDITION

The multiplication 3×5 corresponds to this repeated addition:

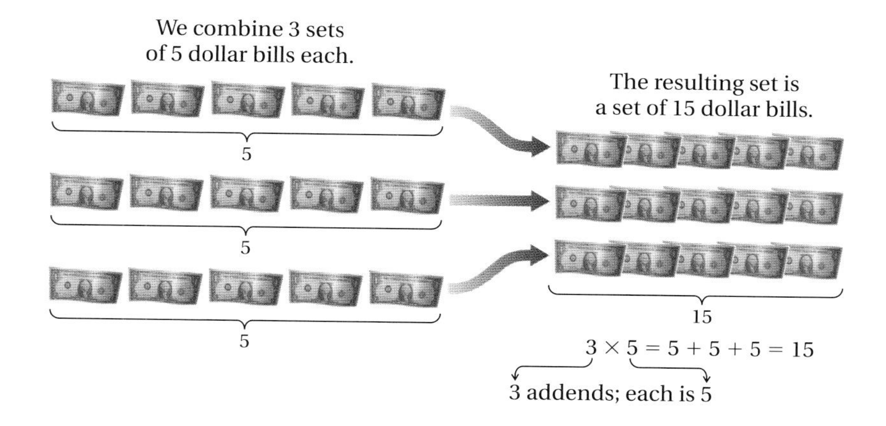

The numbers that we multiply are called **factors.** The result of the multiplication is called a **product.**

$$\underset{\text{Factor}}{3} \times \underset{\text{Factor}}{5} = \underset{\text{Product}}{15}$$

RECTANGULAR ARRAYS

Multiplications can also be thought of as rectangular arrays. Each of the following corresponds to the multiplication 3×5.

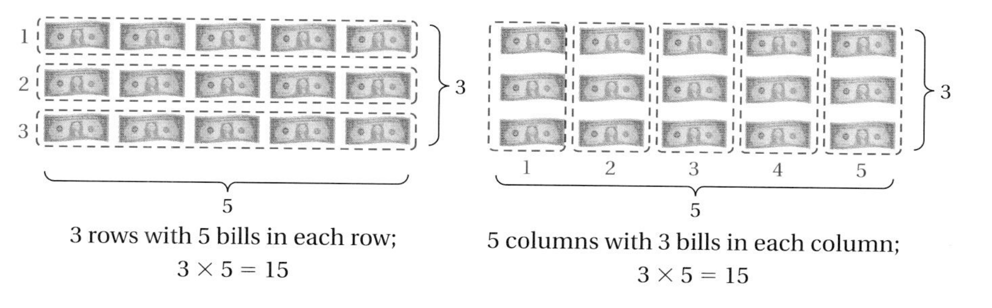

3 rows with 5 bills in each row; $3 \times 5 = 15$

5 columns with 3 bills in each column; $3 \times 5 = 15$

When you write a multiplication sentence corresponding to a real-world situation, you should think of either a rectangular array or repeated addition. In some cases, it may help to think both ways.

We have used an "×" to denote multiplication. A dot "·" is also commonly used. (Use of the dot is attributed to the German mathematician Gottfried Wilhelm von Leibniz in 1698.) Parentheses are also used to denote multiplication. For example,

$$3 \times 5 = 3 \cdot 5 = (3)(5) = 3(5) = 15.$$

Multiply.

1. $\begin{array}{r} 58 \\ \times \quad 2 \\ \hline \end{array}$

2. $\begin{array}{r} 37 \\ \times \quad 4 \\ \hline \end{array}$

3. $\begin{array}{r} 823 \\ \times \quad 6 \\ \hline \end{array}$

4. $\begin{array}{r} 1348 \\ \times \quad 5 \\ \hline \end{array}$

Answers on page A-2

The product of 0 and any whole number is 0: $0 \cdot a = a \cdot 0 = 0$. Multiplying a number by 1 does not change the number: $1 \cdot a = a \cdot 1 = a$. We say that 1 is the **multiplicative identity.** For example, $0 \cdot 3 = 3 \cdot 0 = 0$ and $1 \cdot 3 = 3 \cdot 1 = 3$.

EXAMPLE 1 Multiply: 5×734.

We have

$$\begin{array}{rl} 734 & \\ \times \quad 5 & \\ \hline 20 & \leftarrow \text{Multiply the 4 ones by 5: } 5 \times 4 = 20. \\ 150 & \leftarrow \text{Multiply the 3 tens by 5: } 5 \times 30 = 150. \\ 3500 & \leftarrow \text{Multiply the 7 hundreds by 5: } 5 \times 700 = 3500. \\ \hline 3670 & \leftarrow \text{Add.} \end{array}$$

Instead of writing each product on a separate line, we can use a shorter form.

$$\begin{array}{r} {}^{2} \\ 734 \\ \times \quad 5 \\ \hline 0 \end{array}$$

Multiply the ones by 5: $5 \cdot (4 \text{ ones}) = 20 \text{ ones} = 2 \text{ tens} + 0 \text{ ones}$. Write 0 in the ones column and 2 above the tens.

$$\begin{array}{r} {}^{1\,2} \\ 734 \\ \times \quad 5 \\ \hline 70 \end{array}$$

Multiply the 3 tens by 5 and add 2 tens: $5 \cdot (3 \text{ tens}) = 15 \text{ tens}$, $15 \text{ tens} + 2 \text{ tens} = 17 \text{ tens} = 1 \text{ hundred} + 7 \text{ tens}$. Write 7 in the tens column and 1 above the hundreds.

$$\begin{array}{r} {}^{1\,2} \\ 734 \\ \times \quad 5 \\ \hline 3670 \end{array}$$

Multiply the 7 hundreds by 5 and add 1 hundred: $5 \cdot (7 \text{ hundreds}) = 35 \text{ hundreds}$, $35 \text{ hundreds} + 1 \text{ hundred} = 36 \text{ hundreds}$.

$$\left.\begin{array}{r} {}^{1\,2} \\ 734 \\ \times \quad 5 \\ \hline 3670 \end{array}\right\}$$

You should write only this.

Do Exercises 1–4.

Let's find the product

$$\begin{array}{r} 54 \\ \times 32 \\ \hline \end{array}$$

To do this, we multiply 54 by 2, then 54 by 30, and then add.

$$\begin{array}{r} 54 \\ \times \quad 2 \\ \hline 108 \end{array} \qquad \begin{array}{r} {}^{1} \\ 54 \\ \times \quad 30 \\ \hline 1620 \end{array}$$

Since we are going to add the results, let's write the work this way.

$$\begin{array}{rl} 54 & \\ \times 32 & \\ \hline 108 & \text{Multiplying by 2} \\ 1620 & \text{Multiplying by 30} \\ \hline 1728 & \text{Adding to obtain the product} \end{array}$$

The fact that we can do this is based on a property called the **distributive law.** It says that to multiply a number by a sum, $a \cdot (b + c)$, we can multiply each addend by a and then add like this: $(a \cdot b) + (a \cdot c)$. Thus, $a \cdot (b + c) = (a \cdot b) + (a \cdot c)$. Applied to the example above, the distributive law gives us

$$54 \cdot 32 = 54 \cdot (30 + 2) = (54 \cdot 30) + (54 \cdot 2).$$

EXAMPLE 2 Multiply: 43×57.

```
    2
    5 7
  × 4 3
  -----
  1 7 1      Multiplying by 3
```

```
    2
    2
    5 7
  × 4 3
 ------
  1 7 1
2 2 8 0      Multiplying by 40. (We write a 0 and then multiply 57 by 4).
```

> You may have learned that such a 0 does not have to be written. You may omit it if you wish. If you do omit it, remember, when multiplying by tens, to start writing the answer in the tens place.

```
    2
    2
    5 7
  × 4 3
  1 7 1
2 2 8 0
-------
2 4 5 1      Adding to obtain the product
```

Do Exercises 5 and 6.

EXAMPLE 3 Multiply: 457×683.

```
      5 2
      6 8 3
    × 4 5 7
    -------
    4 7 8 1      Multiplying 683 by 7
```

```
      4 1
      5 2
      6 8 3
    × 4 5 7
  ---------
    4 7 8 1
  3 4 1 5 0      Multiplying 683 by 50
```

```
      3 1
      4 1
      5 2
      6 8 3
    × 4 5 7
    -------
    4 7 8 1
  3 4 1 5 0
2 7 3 2 0 0      Multiplying 683 by 400
-----------
3 1 2,1 3 1      Adding
```

Do Exercises 7 and 8.

Multiply.

5.
```
   4 5
 × 2 3
```

6. 48×63

Multiply.

7.
```
   7 4 6
 ×   6 2
```

8. 245×837

Multiply.

9.
```
   4 7 2
 × 3 0 6
```

10. 408×704

11.
```
   2 3 4 4
 × 6 0 0 5
```

Answers on page A-2

Multiply.

12. $\begin{array}{r} 472 \\ \times\ 830 \\ \hline \end{array}$

13. $\begin{array}{r} 2344 \\ \times\ 7400 \\ \hline \end{array}$

14. 100×562

15. 1000×562

16. a) Find $23 \cdot 47$.

b) Find $47 \cdot 23$.

c) Compare your answers to parts (a) and (b).

Multiply.

17. $5 \cdot 2 \cdot 4$

18. $5 \cdot 1 \cdot 3$

Answers on page A-2

EXAMPLE 4 Multiply: 306×274.

Note that $306 = 3$ hundreds $+ 6$ ones.

$$\begin{array}{rl} 274 & \\ \times\ 306 & \\ \hline 1644 & \text{Multiplying by 6} \\ 82200 & \text{Multiplying by 3 hundreds. (We write 00 and then multiply 274 by 3.)} \\ \hline 83,844 & \text{Adding} \end{array}$$

Do Exercises 9–11 on the preceding page.

EXAMPLE 5 Multiply: 360×274.

Note that $360 = 3$ hundreds $+ 6$ tens.

$$\begin{array}{rl} 274 & \\ \times\ 360 & \\ \hline 16440 & \leftarrow \text{Multiplying by 6 tens. (We write 0 and then multiply 274 by 6.)} \\ 82200 & \leftarrow \text{Multiplying by 3 hundreds. (We write 00 and then multiply 274 by 3.)} \\ \hline 98,640 & \text{Adding} \end{array}$$

Do Exercises 12–15.

Check on your own that $17 \cdot 37 = 629$ and that $37 \cdot 17 = 629$. This illustrates the **commutative law of multiplication.** It says that we can multiply two numbers in any order, $a \cdot b = b \cdot a$, and get the same answer.

Do Exercise 16.

To multiply three or more numbers, we generally group them so that we multiply two at a time. Consider $2 \cdot (3 \cdot 4)$ and $(2 \cdot 3) \cdot 4$. The parentheses tell what to do first:

$2 \cdot (3 \cdot 4) = 2 \cdot (12) = 24.$ We multiply 3 and 4, then 2.

We can also multiply 2 and 3, then 4:

$(2 \cdot 3) \cdot 4 = (6) \cdot 4 = 24.$

Either way we get 24. It does not matter how we group the numbers. This illustrates the **associative law of multiplication:** $a \cdot (b \cdot c) = (a \cdot b) \cdot c$.

Do Exercises 17 and 18.

b Estimating Products by Rounding

EXAMPLE 6 *Lawn Tractors.* Leisure Lawn Care is buying new lawn tractors that cost $1534 each. By rounding to the nearest ten, estimate the cost of purchasing 18 tractors.

Exact	*Nearest ten*
1 5 3 4	1 5 3 0
× 1 8	× 2 0
2 7,6 1 2	3 0,6 0 0

The lawn tractors will cost about $30,600.

Do Exercise 19.

EXAMPLE 7 Estimate the following product by first rounding to the nearest ten and to the nearest hundred: 683 × 457.

Nearest ten	*Nearest hundred*	*Exact*
6 8 0	7 0 0	6 8 3
× 4 6 0	× 5 0 0	× 4 5 7
4 0 8 0 0	3 5 0,0 0 0	4 7 8 1
2 7 2 0 0 0		3 4 1 5 0
3 1 2,8 0 0		2 7 3 2 0 0
		3 1 2,1 3 1

Do Exercise 20.

c Finding Area

The area of a rectangular region can be considered to be the number of square units needed to fill it. Here is a rectangle 4 cm (centimeters) long and 3 cm wide. It takes 12 square centimeters (sq cm) to fill it.

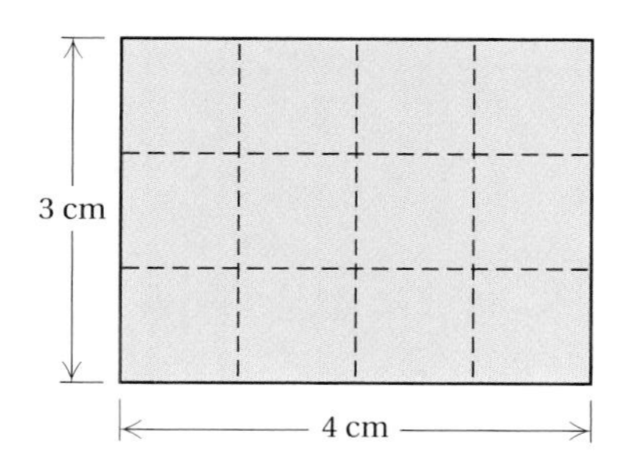

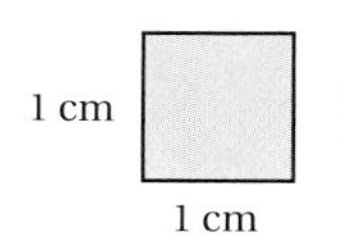

This is a square centimeter (a square unit).

In this case, we have a rectangular array of 3 rows, each of which contains 4 squares. The number of square units is given by 3 · 4, or 12. That is, $A = l \cdot w = 3 \text{ cm} \cdot 4 \text{ cm} = 12$ sq cm.

CALCULATOR CORNER

Multiplying Whole Numbers To multiply whole numbers on a calculator, we use the [×] and [=] keys. For example, to find 13 × 47, we press [1] [3] [×] [4] [7] [=]. The calculator displays 611, so 13 × 47 = 611.

Exercises: Use a calculator to find each product.

1. 56 × 8
2. 845 × 26
3. 5 · 1276
4. 126(314)
5. 3 7 6 0 × 4 8
6. 5 2 1 8 × 4 5 3

19. Lawn Tractors. By rounding to the nearest ten, estimate the cost to Leisure Lawn Care of 12 lawn tractors.

20. Estimate the product by first rounding to the nearest ten and to the nearest hundred. Show your work.

$$\begin{array}{r} 837 \\ \times\, 245 \\ \hline \end{array}$$

Answers on page A-2

21. **Table Tennis.** Find the area of a standard table tennis table that has dimensions of 9 ft by 5 ft.

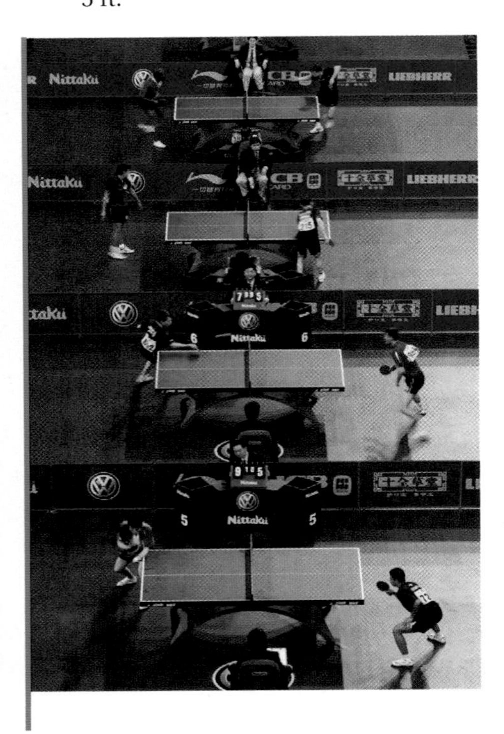

EXAMPLE 8 *Professional Pool Table.* The playing area of a standard pool table has dimensions of 50 in. by 100 in. (There are rails 6 in. wide on the outside that are not included in the playing area.) Find the playing area.

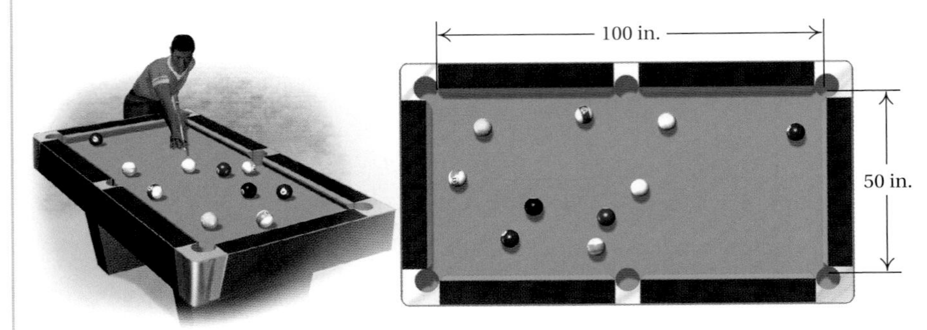

If we think of filling the rectangle with square inches, we have a rectangular array. The length $l = 100$ in. and the width $w = 50$ in. Thus the area A is given by the formula

$$A = l \cdot w = 100 \cdot 50 = 5000 \text{ sq in.}$$

Do Exercise 21.

Study Tips — TIME MANAGEMENT (PART 1)

Time is the most critical factor in your success in learning mathematics. Have reasonable expectations about the time you need to study math.

- **Juggling time.** Working 40 hours per week and taking 12 credit hours is equivalent to working two full-time jobs. Can you handle such a load? Your ratio of number of work hours to number of credit hours should be about 40/3, 30/6, 20/9, 10/12, or 5/14.
- **A rule of thumb on study time.** Budget about 2–3 hours for homework and study for every hour you spend in class each week.
- **Scheduling your time.** Use the Student Organizer card to make an hour-by-hour schedule of your typical week. Include work, school, home, sleep, study, and leisure times. Try to schedule time for study when you are most alert. Choose a setting that will enable you to focus and concentrate. Plan for success and it will happen!

"You cannot increase the quality or quantity of your achievement or performance except to the degree in which you increase your ability to use time effectively."

Brian Tracy, motivational/inspirational speaker

Answer on page A-2

1.5 EXERCISE SET

For Extra Help

MathXL | MyMathLab | InterAct Math | Math Tutor Center | Digital Video Tutor CD 1 Videotape 1 | Student's Solutions Manual

a Multiply.

1. $\begin{array}{r} 87 \\ \times 10 \\ \hline \end{array}$

2. $\begin{array}{r} 100 \\ \times 96 \\ \hline \end{array}$

3. $\begin{array}{r} 2340 \\ \times 1000 \\ \hline \end{array}$

4. $\begin{array}{r} 800 \\ \times 70 \\ \hline \end{array}$

5. $\begin{array}{r} 65 \\ \times 8 \\ \hline \end{array}$

6. $\begin{array}{r} 87 \\ \times 4 \\ \hline \end{array}$

7. $\begin{array}{r} 94 \\ \times 6 \\ \hline \end{array}$

8. $\begin{array}{r} 76 \\ \times 9 \\ \hline \end{array}$

9. $3 \cdot 509$

10. $7 \cdot 806$

11. $7(9229)$

12. $4(7867)$

13. $90(53)$

14. $60(78)$

15. $(47)(85)$

16. $(34)(87)$

17. $\begin{array}{r} 640 \\ \times 72 \\ \hline \end{array}$

18. $\begin{array}{r} 777 \\ \times 77 \\ \hline \end{array}$

19. $\begin{array}{r} 444 \\ \times 33 \\ \hline \end{array}$

20. $\begin{array}{r} 509 \\ \times 88 \\ \hline \end{array}$

21. $\begin{array}{r} 509 \\ \times 408 \\ \hline \end{array}$

22. $\begin{array}{r} 432 \\ \times 375 \\ \hline \end{array}$

23. $\begin{array}{r} 853 \\ \times 936 \\ \hline \end{array}$

24. $\begin{array}{r} 346 \\ \times 650 \\ \hline \end{array}$

25. $\begin{array}{r} 6428 \\ \times 3224 \\ \hline \end{array}$

26. $\begin{array}{r} 8928 \\ \times 3172 \\ \hline \end{array}$

27. $\begin{array}{r} 3482 \\ \times 104 \\ \hline \end{array}$

28. $\begin{array}{r} 6408 \\ \times 6064 \\ \hline \end{array}$

29. $\begin{array}{r} 5006 \\ \times 4008 \\ \hline \end{array}$

30. $\begin{array}{r} 6789 \\ \times 2330 \\ \hline \end{array}$

31. $\begin{array}{r} 5608 \\ \times 4500 \\ \hline \end{array}$

32. $\begin{array}{r} 4560 \\ \times 7890 \\ \hline \end{array}$

33. $\begin{array}{r} 876 \\ \times 345 \\ \hline \end{array}$

34. $\begin{array}{r} 355 \\ \times 299 \\ \hline \end{array}$

35. $\begin{array}{r} 7889 \\ \times 6224 \\ \hline \end{array}$

36. $\begin{array}{r} 6501 \\ \times 3449 \\ \hline \end{array}$

b Estimate the product by first rounding to the nearest ten. Show your work.

37. $\begin{array}{r} 45 \\ \times 67 \\ \hline \end{array}$

38. $\begin{array}{r} 51 \\ \times 78 \\ \hline \end{array}$

39. $\begin{array}{r} 34 \\ \times 29 \\ \hline \end{array}$

40. $\begin{array}{r} 63 \\ \times 54 \\ \hline \end{array}$

Estimate the product by first rounding to the nearest hundred. Show your work.

41. $\begin{array}{r} 876 \\ \times 345 \\ \hline \end{array}$

42. $\begin{array}{r} 355 \\ \times 299 \\ \hline \end{array}$

43. $\begin{array}{r} 432 \\ \times 199 \\ \hline \end{array}$

44. $\begin{array}{r} 789 \\ \times 434 \\ \hline \end{array}$

45. *Toyota Sienna.* A pharmaceutical company buys a Toyota Sienna for each of its 112 sales representatives. Each car costs $27,896 plus an additional $540 per car in destination charges.

a) Estimate the total cost of the purchase by rounding the cost of each car, the destination charge, and the number of sales representatives to the nearest hundred.

b) Estimate the total cost of the purchase by rounding the cost of each car to the nearest thousand and the destination charge and the number of reps to the nearest hundred.

Source: Toyota

46. A travel club of 176 people decides to fly from Los Angeles to Tokyo. The cost of a round-trip ticket is $643.

a) Estimate the total cost of the trip by rounding the cost of the airfare and the number of travelers to the nearest ten.

b) Estimate the total cost of the trip by rounding the cost of the airfare and the number of travelers to the nearest hundred.

C Find the area of the region.

47.

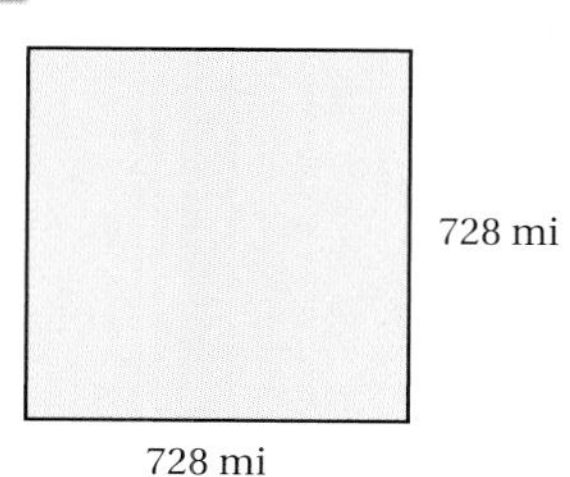

48.

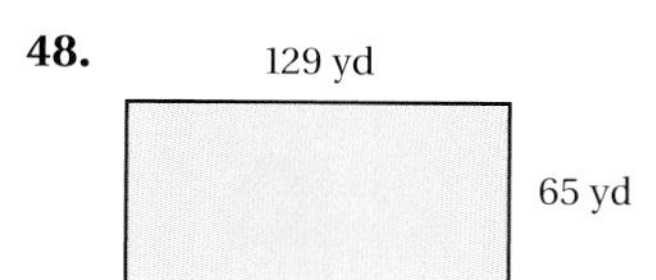

49. Find the area of the region formed by the base lines on a Major League Baseball diamond.

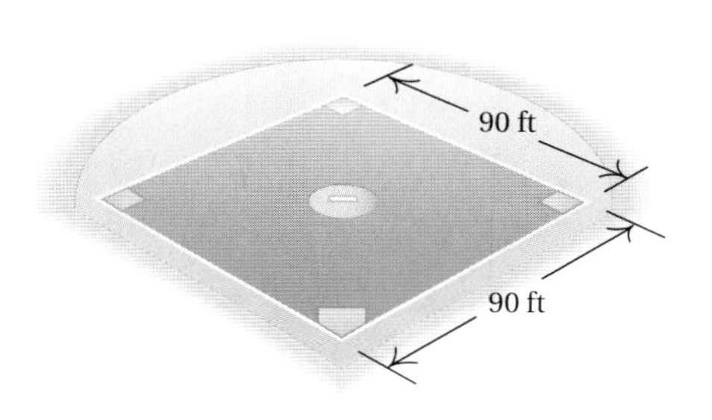

50. Find the area of a standard-sized hockey rink.

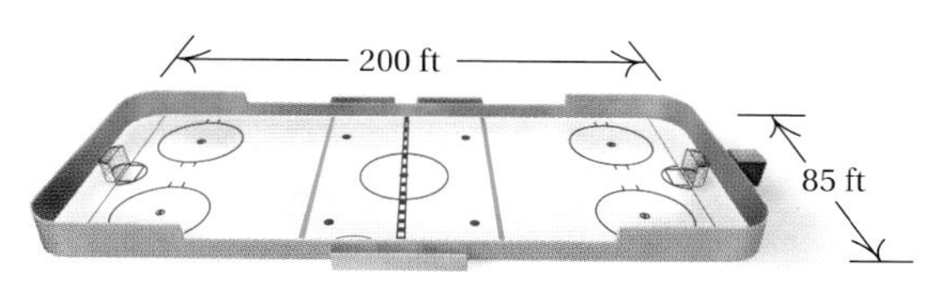

51. D_W Describe a situation that corresponds to each multiplication: $4 \cdot \$150$; $\$4 \cdot 150$.

52. D_W Explain the multiplication illustrated in the diagram below.

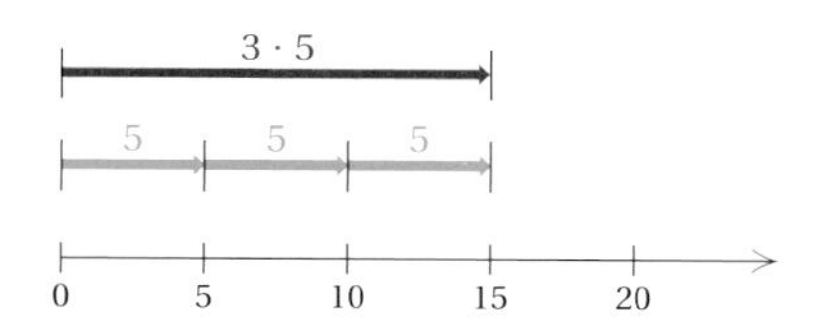

SKILL MAINTENANCE

Add. [1.2a]

53.
$$\begin{array}{r} 4908 \\ 5667 \\ +\ 2110 \\ \hline \end{array}$$

54.
$$\begin{array}{r} 9876 \\ 876 \\ 76 \\ +\quad 6 \\ \hline \end{array}$$

55.
$$\begin{array}{r} 340{,}798 \\ +\ 86{,}679 \\ \hline \end{array}$$

56.
$$\begin{array}{r} 88{,}777 \\ +\ 22{,}333 \\ \hline \end{array}$$

Subtract. [1.3b]

57.
$$\begin{array}{r} 4908 \\ -\ 3667 \\ \hline \end{array}$$

58.
$$\begin{array}{r} 9876 \\ -\quad 987 \\ \hline \end{array}$$

59.
$$\begin{array}{r} 340{,}798 \\ -\ 86{,}679 \\ \hline \end{array}$$

60.
$$\begin{array}{r} 88{,}777 \\ -\ 22{,}333 \\ \hline \end{array}$$

61. Round 6,375,602 to the nearest thousand. [1.4a]

62. Round 6,375,602 to the nearest ten. [1.4a]

SYNTHESIS

63. An 18-story office building is box-shaped. Each floor measures 172 ft by 84 ft with a 20-ft by 35-ft rectangular area lost to an elevator and a stairwell. How much area is available as office space?

1.6 DIVISION

Objectives

a Convert between division sentences and multiplication sentences.

b Divide whole numbers.

1. Consider $54 \div 6 = 9$. Express this division in two other ways.

a Division and Related Sentences

REPEATED SUBTRACTION

Division of whole numbers applies to two kinds of situations. The first is repeated subtraction. Suppose we have 20 notebooks in a pile, and we want to find out how many sets of 5 there are. One way to do this is to repeatedly subtract sets of 5 as follows.

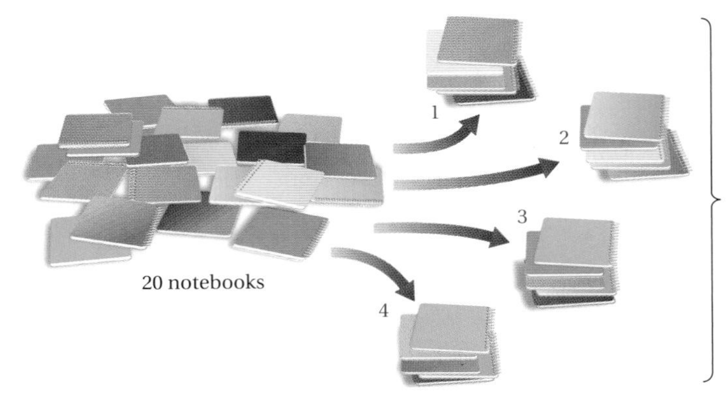

Since there are 4 sets of 5 notebooks each, we have

$$\underset{\text{Dividend}}{20} \div \underset{\text{Divisor}}{5} = \underset{\text{Quotient}}{4}.$$

The division $20 \div 5$, read "20 divided by 5," corresponds to the figure above. We say that the **dividend** is 20, the **divisor** is 5, and the **quotient** is 4. We divide the *dividend* by the *divisor* to get the *quotient.*

We can also express the division $20 \div 5 = 4$ as

$$\frac{20}{5} = 4 \quad \text{or} \quad 5\overline{)20}\ \text{(quotient 4)}$$

Do Exercise 1.

RECTANGULAR ARRAYS

We can also think of division in terms of rectangular arrays. Consider again the pile of 20 notebooks and division by 5. We can arrange the notebooks in a rectangular array with 5 rows and ask, "How many are in each row?"

We can also consider a rectangular array with 5 notebooks in each column and ask, "How many columns are there?" The answer is still 4.

In each case, we are asking, "What do we multiply 5 by in order to get 20?"

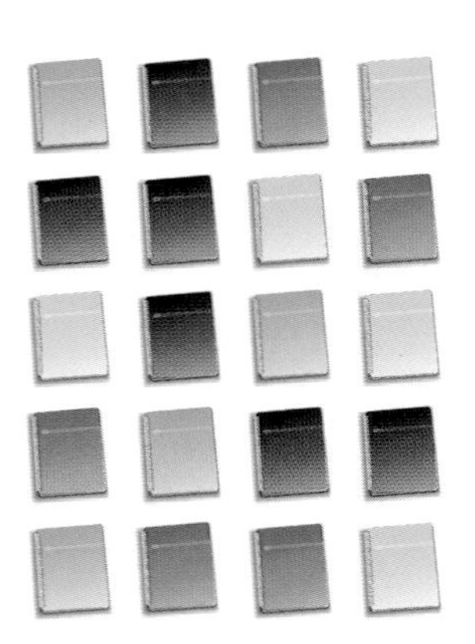

Missing factor: $5 \cdot \square = 20$ Quotient: $20 \div 5 = \square$

Answer on page A-2

This leads us to the following definition of division.

DIVISION

The quotient $a \div b$, where $b \neq 0$, is that unique number c for which $a = b \cdot c$.

RELATED SENTENCES

By looking at rectangular arrays, we can see how multiplication and division are related. The array of notebooks on the preceding page shows that $5 \cdot 4 = 20$. The array also shows the following:

$$20 \div 5 = 4 \quad \text{and} \quad 20 \div 4 = 5.$$

The division $20 \div 5$ is defined to be the number that when multiplied by 5 gives 20. Thus, for every division sentence, there is a related multiplication sentence.

$20 \div 5 = 4$ Division sentence

$20 = 4 \cdot 5$ Related multiplication sentence

To get the related multiplication sentence, we use
Dividend = Quotient · Divisor.

EXAMPLE 1 Write a related multiplication sentence: $12 \div 6 = 2$.

We have

$12 \div 6 = 2$ Division sentence

$12 = 2 \cdot 6.$ Related multiplication sentence

The related multiplication sentence is $12 = 2 \cdot 6$.

By the commutative law of multiplication, there is also another multiplication sentence: $12 = 6 \cdot 2$.

Do Exercises 2 and 3.

For every multiplication sentence, we can write related division sentences.

EXAMPLE 2 Write two related division sentences: $7 \cdot 8 = 56$.

We have

$7 \cdot 8 = 56$ — This factor (7) becomes a divisor. → $7 = 56 \div 8.$

$7 \cdot 8 = 56$ — This factor (8) becomes a divisor. → $8 = 56 \div 7.$

The related division sentences are $7 = 56 \div 8$ and $8 = 56 \div 7$.

Do Exercises 4 and 5.

Write a related multiplication sentence.

2. $15 \div 3 = 5$

3. $72 \div 8 = 9$

Write two related division sentences.

4. $6 \cdot 2 = 12$

5. $7 \cdot 6 = 42$

Answers on page A-2

b Division of Whole Numbers

Before we consider division with remainders, let's recall four basic facts about division.

DIVIDING BY 1

Any number divided by 1 is that same number:

$$a \div 1 = \frac{a}{1} = a.$$

DIVIDING A NUMBER BY ITSELF

Any nonzero number divided by itself is 1:

$$\frac{a}{a} = 1, \quad a \neq 0.$$

DIVIDENDS OF 0

Zero divided by any nonzero number is 0:

$$\frac{0}{a} = 0, \quad a \neq 0.$$

EXCLUDING DIVISION BY 0

Division by 0 is not defined. (We agree not to divide by 0.)

$\frac{a}{0}$ is **not defined.**

Why can't we divide by 0? Suppose the number 4 could be divided by 0. Then if $\square$ were the answer,

$$4 \div 0 = \square$$

and since 0 times any number is 0, we would have

$$4 = \square \cdot 0 = 0. \qquad \text{False!}$$

Thus, $a \div 0$ would be some number $\square$ such that $a = \square \cdot 0 = 0$. So the only possible number that could be divided by 0 would be 0 itself.

But such a division would give us any number we wish, for

$$\left.\begin{array}{lll} 0 \div 0 = 8 & \text{because} & 0 = 8 \cdot 0; \\ 0 \div 0 = 3 & \text{because} & 0 = 3 \cdot 0; \\ 0 \div 0 = 7 & \text{because} & 0 = 7 \cdot 0. \end{array}\right\} \text{All true!}$$

We avoid the preceding difficulties by agreeing to exclude division by 0.

Study Tips

HIGHLIGHTING

- **Highlight important points.** You are probably used to highlighting key points as you study. If that works for you, continue to do so. But you will notice many design features throughout this book that already highlight important points. Thus you may not need to highlight as much as you generally do.
- **Highlight points that you do not understand.** Use a unique mark to indicate trouble spots that can lead to questions to be asked during class, in a tutoring session, or when calling or contacting the AW Math Tutor Center.

Suppose we have 18 cans of soda and want to pack them in cartons of 6 cans each. How many cartons will we fill? We can determine this by repeated subtraction. We keep track of the number of times we subtract. We stop when the number of objects remaining, the **remainder,** is smaller than the divisor.

EXAMPLE 3 Divide by repeated subtraction: $18 \div 6$.

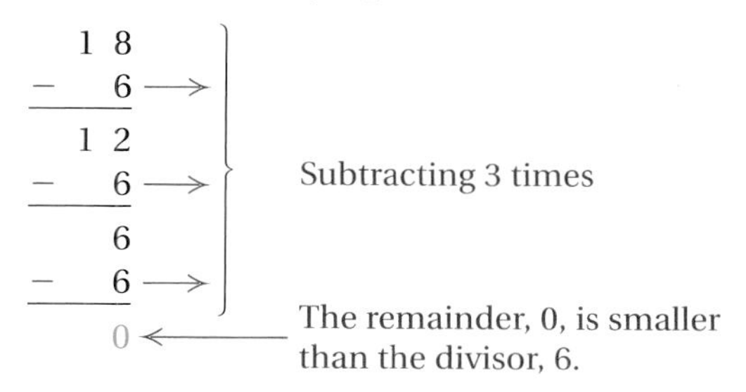

1 8
− 6
1 2
− 6
6
− 6
0
Subtracting 3 times
The remainder, 0, is smaller than the divisor, 6.

Thus, $18 \div 6 = 3$.

Suppose we have 22 cans of soda and want to pack them in cartons of 6 cans each. We end up with 3 cartons with 4 cans left over.

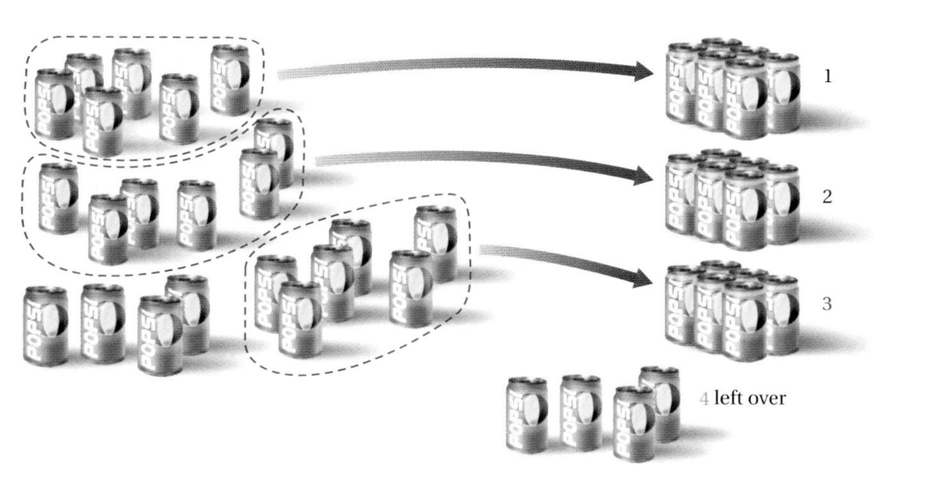

EXAMPLE 4 Divide by repeated subtraction: $22 \div 6$.

$$\begin{array}{r} 22 \\ -\ \ 6 \\ \hline 16 \\ -\ \ 6 \\ \hline 10 \\ -\ \ 6 \\ \hline 4 \end{array}$$

Subtracting 3 times

4 ← Remainder

Check: $3 \cdot 6 = 18$,
$18 + 4 = 22$.

We can write this division as follows.

$$\begin{array}{r} 3 \\ 6\overline{)22} \\ 18 \\ \hline 4 \end{array}$$

Divide by repeated subtraction. Then check.

6. $54 \div 9$

7. $61 \div 9$

8. $53 \div 12$

9. $157 \div 24$

Answers on page A-2

Divide and check.

10. $4\overline{)239}$

11. $6\overline{)8855}$

12. $5\overline{)5075}$

Note that

Quotient · Divisor + Remainder = Dividend.

We write answers to a division sentence as follows.

$$22 \div 6 = 3 \text{ R } 4$$

Dividend, Divisor, Quotient, Remainder

Do Exercises 6–9 on the preceding page.

EXAMPLE 5 Divide and check: $3642 \div 5$.

$5\overline{)3642}$, quotient digit: ?

1. Find the number of thousands in the quotient. Consider 3 thousands ÷ 5 and think 3 ÷ 5. Since 3 ÷ 5 is not a whole number, move to hundreds.

$$\begin{array}{r} 7 \\ 5\overline{)3642} \\ \underline{3500} \\ 142 \end{array}$$

2. Find the number of hundreds in the quotient. Consider 36 hundreds ÷ 5 and think 36 ÷ 5. The estimate is about 7 hundreds. Multiply 700 by 5 and subtract.

142 ← The remainder is larger than the divisor.

$$\begin{array}{r} 72 \\ 5\overline{)3642} \\ \underline{3500} \\ 142 \\ \underline{100} \\ 42 \end{array}$$

3. Find the number of tens in the quotient using 142, the first remainder. Consider 14 tens ÷ 5 and think 14 ÷ 5. The estimate is about 2 tens. Multiply 20 by 5 and subtract. (If our estimate had been 3 tens, we could not have subtracted 150 from 142.)

42 ← The remainder is larger than the divisor.

$$\begin{array}{r} 728 \\ 5\overline{)3642} \\ \underline{3500} \\ 142 \\ \underline{100} \\ 42 \\ \underline{40} \\ 2 \end{array}$$

4. Find the number of ones in the quotient using 42, the second remainder. Consider 42 ones ÷ 5 and think 42 ÷ 5. The estimate is about 8 ones. Multiply 8 by 5 and subtract. The remainder, 2, is less than the divisor, 5, so we are finished.

2 ← The remainder is less than the divisor.

You may have learned to divide like this, not writing the extra zeros. You may omit them if desired.

$$\begin{array}{r} 728 \\ 5\overline{)3642} \\ \underline{35} \\ 14 \\ \underline{10} \\ 42 \\ \underline{40} \\ 2 \end{array}$$

Check: $728 \cdot 5 = 3640$,
$3640 + 2 = 3642$.

The answer is 728 R 2.

Do Exercises 10–12.

Answers on page A-2

We can summarize our division procedure as follows.

> To do division of whole numbers:
>
> **a)** Estimate.
> **b)** Multiply.
> **c)** Subtract.

Sometimes rounding the divisor helps us find estimates.

EXAMPLE 6 Divide: 8904 ÷ 42.

We mentally round 42 to 40.

$$\begin{array}{r} 2 \\ 42\,\overline{)\,8904} \\ \underline{8400} \\ 504 \end{array}$$

← *Think*: 89 hundreds ÷ 40. Estimate 2 hundreds. Multiply 200 · 42 and subtract.

$$\begin{array}{r} 21 \\ 42\,\overline{)\,8904} \\ \underline{8400} \\ 504 \\ \underline{420} \\ 84 \end{array}$$

← *Think*: 50 tens ÷ 40. Estimate 1 ten. Multiply 10 · 42 and subtract.

$$\begin{array}{r} 212 \\ 42\,\overline{)\,8904} \\ \underline{8400} \\ 504 \\ \underline{420} \\ 84 \\ \underline{84} \\ 0 \end{array}$$

← *Think*: 84 ones ÷ 40. Estimate 2 ones. Multiply 2 · 42 and subtract.

Caution!

Be careful to keep the digits lined up correctly.

The answer is 212. *Remember*: If after estimating and multiplying you get a number that is larger than the number from which it is being subtracted, lower your estimate.

Do Exercises 13 and 14.

Divide.

13. $45\,\overline{)\,6030}$

14. $52\,\overline{)\,3288}$

Divide.

15. $6\,\overline{)\,4846}$

16. $7\,\overline{)\,7616}$

Answers on page A-2

Divide.

17. $27\overline{)9724}$

18. $56\overline{)44{,}847}$

Answers on page A-2

Study Tips

EXERCISES

- **Odd-numbered exercises.** Usually an instructor assigns some odd-numbered exercises as homework. When you complete these, you can check your answers at the back of the book. If you miss any, check your work in the *Student's Solutions Manual* or ask your instructor for guidance.
- **Even-numbered exercises.** Whether or not your instructor assigns the even-numbered exercises, always do some on your own. Remember, there are no answers given for the chapter tests, so you need to practice doing exercises without answers. Check your answers later with a friend or your instructor.

ZEROS IN QUOTIENTS

EXAMPLE 7 Divide: 6341 ÷ 7.

$$\begin{array}{r} 9 \\ 7\overline{)6341} \\ \underline{6300} \\ 41 \end{array}$$

← *Think*: 63 hundreds ÷ 7. Estimate 9 hundreds. Multiply 900 · 7 and subtract.

$$\begin{array}{r} 90 \\ 7\overline{)6341} \\ \underline{6300} \\ 41 \end{array}$$

← Think: 4 tens ÷ 7. The tens digit of the quotient is 0.

$$\begin{array}{r} 905 \\ 7\overline{)6341} \\ \underline{6300} \\ 41 \\ \underline{35} \\ 6 \end{array}$$

41 ← *Think*: 41 ones ÷ 7. Estimate 5 ones. Multiply 5 · 7 and subtract.

6 ← The remainder, 6, is less than the divisor, 7.

The answer is 905 R 6.

Do Exercises 15 and 16 on the preceding page.

EXAMPLE 8 Divide: 8889 ÷ 37.

We round 37 to 40.

$$\begin{array}{r} 2 \\ 37\overline{)8889} \\ \underline{7400} \\ 1489 \end{array}$$

← *Think*: 37 ≈ 40; 88 hundreds ÷ 40. Estimate 2 hundreds. Multiply 200 · 37 and subtract.

$$\begin{array}{r} 24 \\ 37\overline{)8889} \\ \underline{7400} \\ 1489 \\ \underline{1480} \\ 9 \end{array}$$

1489 ← *Think*: 148 tens ÷ 40. Estimate 4 tens. Multiply 40 · 37 and subtract.

$$\begin{array}{r} 240 \\ 37\overline{)8889} \\ \underline{7400} \\ 1489 \\ \underline{1480} \\ 9 \end{array}$$

9 ← The remainder, 9, is less than the divisor, 37.

The answer is 240 R 9.

Do Exercises 17 and 18.

CALCULATOR CORNER

Dividing Whole Numbers: Finding Remainders To divide whole numbers on a calculator, we use the [÷] and [=] keys. For example, to divide 711 by 9, we press [7] [1] [1] [÷] [9] [=]. The display reads [79], so 711 ÷ 9 = 79.

When we enter 453 ÷ 15, the display reads [30.2]. Note that the result is not a whole number. This tells us that there is a remainder. The number 30.2 is expressed in decimal notation. The symbol "." is called a decimal point. (Decimal notation will be studied in Chapter 4.) The number to the left of the decimal point, 30, is the quotient. We can use the remaining part of the result to find the remainder. To do this, first subtract 30 from 30.2. Then multiply the difference by the divisor, 15. We get 3. This is the remainder. Thus, 453 ÷ 15 = 30 R 3. The steps that we performed to find this result can be summarized as follows:

$$453 \div 15 = 30.2,$$
$$30.2 - 30 = .2,$$
$$0.2 \times 15 = 3.$$

To follow these steps on a calculator, we press [4] [5] [3] [÷] [1] [5] [=] and write the number that appears to the left of the decimal point. This is the quotient. Then we continue by pressing [−] [3] [0] [=] [×] [1] [5] [=]. The last number that appears is the remainder. In some cases, it will be necessary to round the remainder to the nearest one.

To check this result, we multiply the quotient by the divisor and then add the remainder.

$$30 \times 15 = 450,$$
$$450 + 3 = 453$$

Exercises: Use a calculator to perform each division. Check the results with a calculator also.

1. 92 ÷ 27
2. $19\overline{)532}$
3. $6\overline{)746}$
4. 3817 ÷ 29
5. $126\overline{)35,715}$
6. $308\overline{)259,831}$

1.6 EXERCISE SET

For Extra Help

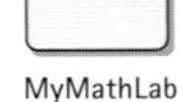

MathXL | MyMathLab | InterAct Math | Math Tutor Center | Digital Video Tutor CD 1 Videotape 1 | Student's Solutions Manual

a Write a related multiplication sentence.

1. $18 \div 3 = 6$

2. $72 \div 9 = 8$

3. $22 \div 22 = 1$

4. $32 \div 1 = 32$

5. $54 \div 6 = 9$

6. $90 \div 10 = 9$

7. $37 \div 1 = 37$

8. $28 \div 28 = 1$

Write two related division sentences.

9. $9 \times 5 = 45$

10. $2 \cdot 7 = 14$

11. $37 \cdot 1 = 37$

12. $4 \cdot 12 = 48$

13. $8 \times 8 = 64$

14. $9 \cdot 7 = 63$

15. $11 \cdot 6 = 66$

16. $1 \cdot 43 = 43$

b Divide, if possible. If not possible, write "not defined."

17. $72 \div 6$

18. $54 \div 9$

19. $\frac{23}{23}$

20. $\frac{37}{37}$

21. $22 \div 1$

22. $\frac{56}{1}$

23. $\frac{16}{0}$

24. $74 \div 0$

Divide.

25. $277 \div 5$

26. $699 \div 3$

27. $864 \div 8$

28. $869 \div 8$

29. $4\overline{)1228}$

30. $3\overline{)2124}$

31. $6\overline{)4521}$

32. $9\overline{)9110}$

33. $297 \div 4$

34. $389 \div 2$

35. $738 \div 8$

36. $881 \div 6$

37. $5\overline{)8515}$

38. $3\overline{)6027}$

39. $9\overline{)8888}$

40. $8\overline{)4139}$

41. $127{,}000 \div 10$

42. $127{,}000 \div 100$

43. $127{,}000 \div 1000$

44. $4260 \div 10$

45. $70\overline{)3692}$

46. $20\overline{)5798}$

47. $30\overline{)875}$

48. $40\overline{)987}$

49. $852 \div 21$

50. $942 \div 23$

51. $85\overline{)7672}$

52. $54\overline{)2729}$

53. $111\overline{)3219}$

54. $102\overline{)5612}$

55. $8\overline{)843}$

56. $7\overline{)749}$

57. $5\overline{)8047}$

58. $9\overline{)7273}$

59. $5\overline{)5036}$

60. $7\overline{)7074}$

61. $1058 \div 46$

62. $7242 \div 24$

63. $3425 \div 32$

64. $48\overline{)4899}$

65. $24\overline{)8880}$

66. $36\overline{)7563}$

67. $28\overline{)17{,}067}$

68. $36\overline{)28{,}929}$

69. $80\overline{)24{,}320}$

70. $90\overline{)88{,}560}$

71. $285\overline{)999{,}999}$

72. $306\overline{)888{,}888}$

73. $456\overline{)3{,}679{,}920}$

74. $803\overline{)5{,}622{,}606}$

75. DW Is division associative? Why or why not? Give an example.

76. DW Suppose a student asserts that "$0 \div 0 = 0$ because nothing divided by nothing is nothing." Devise an explanation to persuade the student that the assertion is false.

SKILL MAINTENANCE

VOCABULARY REINFORCEMENT

In each of Exercises 77–84, fill in the blank with the correct term from the given list. Some of the choices may not be used and some may be used more than once.

77. The distance around an object is its ______________. [1.2b]

78. A sentence like $10 - 3 = 7$ is called a(n) ______________; a sentence like $31 < 33$ is called a(n) ______________. [1.4c]

79. For large numbers, ______________ are separated by commas into groups of three, called ______________. [1.1a]

80. The number 0 is called the ______________ identity. [1.2a]

81. In the sentence $28 \div 7 = 4$, the ______________ is 28. [1.6a]

82. In the sentence $10 \times 1000 = 10{,}000$, 10 and 1000 are called ______________ and 10,000 is called the ______________. [1.5a]

83. The ______________ is the number from which another number is being subtracted. [1.3a]

84. The sentence $3 \times (6 \times 2) = (3 \times 6) \times 2$ illustrates the ______________ law of multiplication. [1.5a]

associative	minuend
commutative	subtrahend
addends	digits
factors	periods
perimeter	additive
dividend	multiplicative
quotient	equation
product	inequality

SYNTHESIS

85. Complete the following table.

a	b	$a \cdot b$	$a + b$
	68	3672	
84			117
		32	12

86. Find a pair of factors whose product is 36 and:
a) whose sum is 13.
b) whose difference is 0.
c) whose sum is 20.
d) whose difference is 9.

87. A group of 1231 college students is going to take buses for a field trip. Each bus can hold only 42 students. How many buses are needed?

88. Fill in the missing digits to make the equation true:
$34{,}584{,}132 \div 76\square = 4\square{,}386.$

1.7 SOLVING EQUATIONS

Objectives

a Solve simple equations by trial.

b Solve equations like $x + 28 = 54$, $28 \cdot x = 168$, and $98 \cdot 2 = y$.

Find a number that makes the sentence true.

1. $8 = 1 + \square$

2. $\square + 2 = 7$

3. Determine whether 7 is a solution of $\square + 5 = 9$.

4. Determine whether 4 is a solution of $\square + 5 = 9$.

Answers on page A-2

a Solutions by Trial

Let's find a number that we can put in the blank to make this sentence true:

$$9 = 3 + \square.$$

We are asking "9 is 3 plus what number?" The answer is 6.

$$9 = 3 + 6$$

Do Exercises 1 and 2.

A sentence with = is called an **equation.** A **solution** of an equation is a number that makes the sentence true. Thus, 6 is a solution of

$9 = 3 + \square$ because $9 = 3 + 6$ is true.

However, 7 is not a solution of

$9 = 3 + \square$ because $9 = 3 + 7$ is false.

Do Exercises 3 and 4.

We can use a letter instead of a blank. For example,

$$9 = 3 + n.$$

We call n a **variable** because it can represent any number. If a replacement for a variable makes an equation true, it is a **solution** of the equation.

SOLUTIONS OF AN EQUATION

A **solution** is a replacement for the variable that makes the equation true. When we find all the solutions, we say that we have **solved** the equation.

EXAMPLE 1 Solve $y + 12 = 27$ by trial.

We replace y with several numbers.

If we replace y with 13, we get a false equation: $13 + 12 = 27$.

If we replace y with 14, we get a false equation: $14 + 12 = 27$.

If we replace y with 15, we get a true equation: $15 + 12 = 27$.

No other replacement makes the equation true, so the solution is 15.

EXAMPLES Solve.

2. $7 + n = 22$
(7 plus what number is 22?)
The solution is 15.

3. $63 = 3 \cdot x$
(63 is 3 times what number?)
The solution is 21.

Do Exercises 5–8 on the following page.

b Solving Equations

We now begin to develop more efficient ways to solve certain equations. When an equation has a variable alone on one side, it is easy to see the solution or to compute it. When a calculation is on one side and the variable is alone on the other, we can find the solution by carrying out the calculation.

EXAMPLE 4 Solve: $x = 245 \times 34$.

To solve the equation, we carry out the calculation.

$$\begin{array}{r} 245 \\ \times\ 34 \\ \hline 980 \\ 7350 \\ \hline 8330 \end{array}$$

$x = 245 \times 34$

$x = 8330$

The solution is 8330.

Do Exercises 9–12.

Look at the equation

$x + 12 = 27.$

We can get x alone by writing a related subtraction sentence:

$x = 27 - 12$ 12 gets subtracted to find the related subtraction sentence.

$x = 15.$ Doing the subtraction

It is useful in our later study of algebra to think of this as "subtracting 12 *on both sides*." Thus

$x + 12 - 12 = 27 - 12$ Subtracting 12 on both sides

$x + 0 = 15$ Carrying out the subtraction

$x = 15.$

SOLVING $x + a = b$

To solve $x + a = b$, subtract a on both sides.

If we can get an equation in a form with the variable alone on one side, we can "see" the solution.

EXAMPLE 5 Solve: $t + 28 = 54$.

We have

$t + 28 = 54$

$t + 28 - 28 = 54 - 28$ Subtracting 28 on both sides

$t + 0 = 26$

$t = 26.$

Solve by trial.

5. $n + 3 = 8$

6. $x - 2 = 8$

7. $45 \div 9 = y$

8. $10 + t = 32$

Solve.

9. $346 \times 65 = y$

10. $x = 2347 + 6675$

11. $4560 \div 8 = t$

12. $x = 6007 - 2346$

Answers on page A-2

Solve. Be sure to check.

13. $x + 9 = 17$

14. $77 = m + 32$

15. Solve: $155 = t + 78$. Be sure to check.

Solve. Be sure to check.

16. $4566 + x = 7877$

17. $8172 = h + 2058$

Answers on page A-2

To check the answer, we substitute 26 for t in the original equation.

Check:
$$\begin{array}{c|c} t + 28 & 54 \\ \hline 26 + 28 \ ? & 54 \\ 54 & \end{array} \quad \text{TRUE}$$

The solution is 26.

Do Exercises 13 and 14.

EXAMPLE 6 Solve: $182 = 65 + n$.

We have

$$182 = 65 + n$$
$$182 - 65 = 65 + n - 65 \quad \text{Subtracting 65 on both sides}$$
$$117 = 0 + n \quad \text{65 plus } n \text{ minus 65 is } 0 + n.$$
$$117 = n.$$

Check:
$$\begin{array}{c|c} 182 & 65 + n \\ \hline 182 \ ? & 65 + 117 \\ & 182 \end{array} \quad \text{TRUE}$$

The solution is 117.

Do Exercise 15.

EXAMPLE 7 Solve: $7381 + x = 8067$.

We have

$$7381 + x = 8067$$
$$7381 + x - 7381 = 8067 - 7381 \quad \text{Subtracting 7381 on both sides}$$
$$x = 686.$$

The check is left to the student. The solution is 686.

Do Exercises 16 and 17.

We now learn to solve equations like $8 \cdot n = 96$. Look at

$$8 \cdot n = 96.$$

We can get n alone by writing a related division sentence:

$$n = 96 \div 8 = \frac{96}{8} \quad \text{96 is divided by 8.}$$
$$n = 12. \quad \text{Doing the division}$$

It is useful in our later study of algebra to think of the preceding as "dividing by 8 *on both sides*." Thus,

$$\frac{8 \cdot n}{8} = \frac{96}{8} \quad \text{Dividing by 8 on both sides}$$
$$n = 12. \quad \text{8 times } n \text{ divided by 8 is } n.$$

SOLVING $a \cdot x = b$

To solve $a \cdot x = b$, divide by a on both sides.

EXAMPLE 8 Solve: $10 \cdot x = 240$.

We have

$$10 \cdot x = 240$$

$$\frac{10 \cdot x}{10} = \frac{240}{10} \quad \text{Dividing by 10 on both sides}$$

$$x = 24.$$

Check:

$$\begin{array}{c|c} 10 \cdot x & = 240 \\ \hline 10 \cdot 24 & ?\ 240 \\ 240 & \end{array} \quad \text{TRUE}$$

The solution is 24.

EXAMPLE 9 Solve: $5202 = 9 \cdot t$.

We have

$$5202 = 9 \cdot t$$

$$\frac{5202}{9} = \frac{9 \cdot t}{9} \quad \text{Dividing by 9 on both sides}$$

$$578 = t.$$

The check is left to the student. The solution is 578.

Do Exercises 18–20.

EXAMPLE 10 Solve: $14 \cdot y = 1092$.

We have

$$14 \cdot y = 1092$$

$$\frac{14 \cdot y}{14} = \frac{1092}{14} \quad \text{Dividing by 14 on both sides}$$

$$y = 78.$$

The check is left to the student. The solution is 78.

EXAMPLE 11 Solve: $n \cdot 56 = 4648$.

We have

$$n \cdot 56 = 4648$$

$$\frac{n \cdot 56}{56} = \frac{4648}{56} \quad \text{Dividing by 56 on both sides}$$

$$n = 83.$$

The check is left to the student. The solution is 83.

Do Exercises 21 and 22.

Solve. Be sure to check.

18. $8 \cdot x = 64$

19. $144 = 9 \cdot n$

20. $5152 = 8 \cdot t$

Solve. Be sure to check.

21. $18 \cdot y = 1728$

22. $n \cdot 48 = 4512$

Answers on page A-2

1.7 EXERCISE SET

For Extra Help

MathXL | MyMathLab | InterAct Math | Math Tutor Center | Digital Video Tutor CD 1 Videotape 1 | Student's Solutions Manual

 Solve by trial.

1. $x + 0 = 14$

2. $x - 7 = 18$

3. $y \cdot 17 = 0$

4. $56 \div m = 7$

b Solve. Be sure to check.

5. $13 + x = 42$

6. $15 + t = 22$

7. $12 = 12 + m$

8. $16 = t + 16$

9. $3 \cdot x = 24$

10. $6 \cdot x = 42$

11. $112 = n \cdot 8$

12. $162 = 9 \cdot m$

13. $45 \cdot 23 = x$

14. $23 \cdot 78 = y$

15. $t = 125 \div 5$

16. $w = 256 \div 16$

17. $p = 908 - 458$

18. $9007 - 5667 = m$

19. $x = 12{,}345 + 78{,}555$

20. $5678 + 9034 = t$

21. $3 \cdot m = 96$

22. $4 \cdot y = 96$

23. $715 = 5 \cdot z$

24. $741 = 3 \cdot t$

25. $10 + x = 89$

26. $20 + x = 57$

27. $61 = 16 + y$

28. $53 = 17 + w$

29. $6 \cdot p = 1944$

30. $4 \cdot w = 3404$

31. $5 \cdot x = 3715$

32. $9 \cdot x = 1269$

33. $47 + n = 84$

34. $56 + p = 92$

35. $x + 78 = 144$

36. $z + 67 = 133$

37. $165 = 11 \cdot n$

38. $660 = 12 \cdot n$

39. $624 = t \cdot 13$

40. $784 = y \cdot 16$

41. $x + 214 = 389$

42. $x + 221 = 333$

43. $567 + x = 902$

44. $438 + x = 807$

45. $18 \cdot x = 1872$

46. $19 \cdot x = 6080$

47. $40 \cdot x = 1800$

48. $20 \cdot x = 1500$

49. $2344 + y = 6400$

50. $9281 = 8322 + t$

51. $8322 + 9281 = x$

52. $9281 - 8322 = y$

53. $234 \cdot 78 = y$

54. $10{,}534 \div 458 = q$

55. $58 \cdot m = 11{,}890$

56. $233 \cdot x = 22{,}135$

57. **D**$_W$ Describe a procedure that can be used to convert any equation of the form $a \cdot b = c$ to a related division equation.

58. **D**$_W$ Describe a procedure that can be used to convert any equation of the form $a + b = c$ to a related subtraction equation.

SKILL MAINTENANCE

59. Write two related subtraction sentences: $7 + 8 = 15$. [1.3a]

60. Write two related division sentences: $6 \cdot 8 = 48$. [1.6a]

Use $>$ or $<$ for $\square$ to write a true sentence. [1.4c]

61. $123 \square 789$

62. $342 \square 339$

63. $688 \square 0$

64. $0 \square 11$

Divide. [1.6b]

65. $1283 \div 9$

66. $1278 \div 9$

67. $17\overline{)5678}$

68. $17\overline{)5689}$

SYNTHESIS

Solve.

69. 🖩 $23{,}465 \cdot x = 8{,}142{,}355$

70. 🖩 $48{,}916 \cdot x = 14{,}332{,}388$

1.8 APPLICATIONS AND PROBLEM SOLVING

Objective

a Solve applied problems involving addition, subtraction, multiplication, or division of whole numbers.

a A Problem-Solving Strategy

Applications and problem solving are the most important uses of mathematics. To solve a problem using the operations on the whole numbers, we first look at the situation. We try to translate the problem to an equation. Then we solve the equation. We check to see if the solution of the equation is a solution of the original problem. We are using the following five-step strategy.

FIVE STEPS FOR PROBLEM SOLVING

1. *Familiarize* yourself with the situation.
   a) Carefully read and reread until you understand *what* you are being asked to find.
   b) Draw a diagram or see if there is a formula that applies to the situation.
   c) Assign a letter, or *variable,* to the unknown.
2. *Translate* the problem to an equation using the letter or variable.
3. *Solve* the equation.
4. *Check* the answer in the original wording of the problem.
5. *State* the answer to the problem clearly with appropriate units.

EXAMPLE 1 *International Adoptions.* The top ten countries of origin for United States international adoptions in 2003 and 2002 are listed in the table below. Find the total number of adoptions from China, South Korea, India, and Vietnam in 2003.

INTERNATIONAL ADOPTIONS

RANK	2003 COUNTRY OF ORIGIN	2003 NUMBER	2002 COUNTRY OF ORIGIN	2002 NUMBER
1	China (mainland)	6859	China (mainland)	5053
2	Russia	5209	Russia	4939
3	Guatemala	2328	Guatemala	2219
4	South Korea	1790	South Korea	1779
5	Kazakhstan	825	Ukraine	1106
6	Ukraine	702	Kazakhstan	819
7	India	472	Vietnam	766
8	Vietnam	382	India	466
9	Colombia	272	Colombia	334
10	Haiti	250	Bulgaria	260

Source: U.S. Department of State

1. Familiarize. We can make a drawing or at least visualize the situation.

6859	+	1790	+	472	+	382
from China		from South Korea		from India		from Vietnam

Since we are combining numbers of adoptions, addition can be used. First, we define the unknown. We let n = the total number of adoptions from China, South Korea, India, and Vietnam.

2. Translate. We translate to an equation:

$$6859 + 1790 + 472 + 382 = n.$$

3. Solve. We solve the equation by carrying out the addition.

$$\begin{array}{r} \overset{2}{6}\,\overset{3}{8}\,\overset{1}{5}\,9 \\ 1\,7\,9\,0 \\ 4\,7\,2 \\ +\quad 3\,8\,2 \\ \hline 9\,5\,0\,3 \end{array}$$

$$6859 + 1790 + 472 + 382 = n$$
$$9503 = n$$

4. Check. We check 9503 in the original problem. There are many ways in which this can be done. For example, we can repeat the calculation. (We leave this to the student.) Another way is to check whether the answer is reasonable. In this case, we would expect the total to be greater than the number of adoptions from any of the individual countries, which it is. We can also estimate by rounding. Here we round to the nearest hundred.

$$6859 + 1790 + 472 + 382 \approx 6900 + 1800 + 500 + 400$$
$$= 9600$$

Since $9600 \approx 9503$, we have a partial check. If we had an estimate like 4800 or 7500, we might be suspicious that our calculated answer is incorrect. Since our estimated answer is close to our calculation, we are further convinced that our answer checks.

5. State. The total number of adoptions from China, South Korea, India, and Vietnam in 2003 is 9503.

Do Exercises 1–3.

Refer to the table on the preceding page to answer Margin Exercises 1–3.

1. Find the total number of adoptions from China, Russia, Kazakhstan, and India in 2002.

2. Find the total number of adoptions from Russia, Guatemala, Vietnam, Colombia, and Haiti in 2003.

3. Find the total number of adoptions from the top ten countries in 2003.

Answers on page A-3

4. **Checking Account Balance.** The balance in Heidi's checking account is \$2003. She uses her debit card to buy the same Roto Zip Spiral Saw Combo, featured in Example 2, that Tyler did. Find the new balance in her checking account.

Answer on page A-3

EXAMPLE 2 *Checking Account Balance.* The balance in Tyler's checking account is \$528. He uses his debit card to buy the Roto Zip Spiral Saw Combo shown in this ad. Find the new balance in his checking account.

Source: Reproduced with permission from the copyright owner (©2004 by Robert Bosch Tool Corporation). Further reproductions strictly prohibited.

1. **Familiarize.** We first make a drawing or at least visualize the situation. We let M = the new balance in his account. This gives us the following:

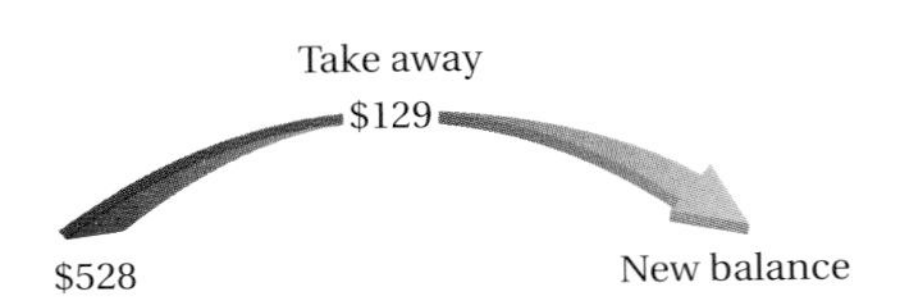

2. **Translate.** We can think of this as a "take-away" situation. We translate to an equation.

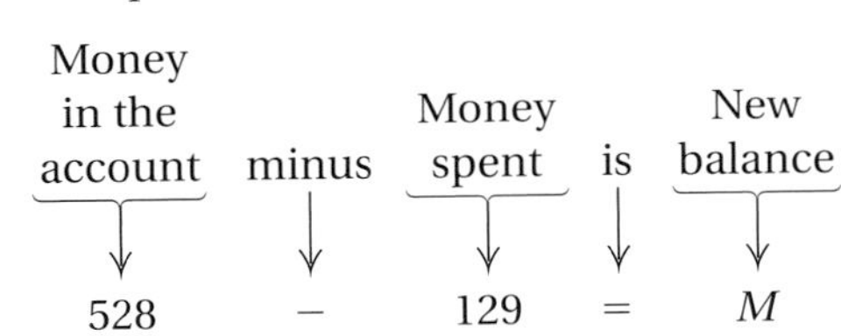

3. **Solve.** This sentence tells us what to do. We subtract.

$$\begin{array}{r} \overset{4}{\cancel{5}}\ \overset{\overset{11}{\cancel{1}}}{\cancel{2}}\ \overset{18}{\cancel{8}} \\ -\ 1\ 2\ 9 \\ \hline 3\ 9\ 9 \end{array} \qquad \begin{aligned} 528 - 129 &= M \\ 399 &= M \end{aligned}$$

4. **Check.** To check our answer of \$399, we can repeat the calculation. We note that the answer should be less than the original amount, \$528, which it is. We can add the difference, 399, to the subtrahend, 129: $129 + 399 = 528$. We can also estimate:

$$528 - 129 \approx 530 - 130 = 400 \approx 399.$$

5. **State.** Tyler has a new balance of \$399 in his checking account.

Do Exercise 4.

In the real world, problems may not be stated in written words. You must still become familiar with the situation before you can solve the problem.

EXAMPLE 3 *Travel Distance.* Abigail is driving from Indianapolis to Salt Lake City to interview for a news anchor position. The distance from Indianapolis to Salt Lake City is 1634 mi. She travels 1154 mi to Denver. How much farther must she travel?

1. **Familiarize.** We first make a drawing or at least visualize the situation. We let $x =$ the remaining distance to Salt Lake City.

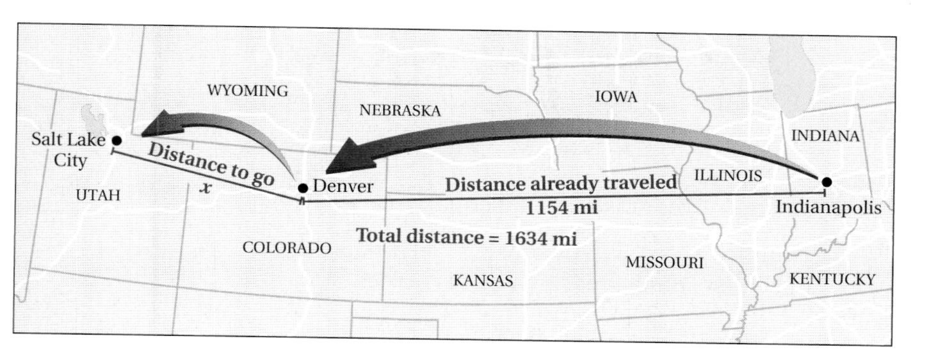

2. **Translate.** We see that this is a "how much more" situation. We translate to an equation.

Distance already traveled	plus	Distance to go	is	Total distance of trip
↓	↓	↓	↓	↓
1154	+	x	=	1634

3. **Solve.** To solve the equation, we subtract 1154 on both sides:

$$\begin{aligned} 1154 + x &= 1634 \\ 1154 + x - 1154 &= 1634 - 1154 \\ x &= 480. \end{aligned}$$

$$\begin{array}{r} {\scriptstyle 5\ 13} \\ 1\,\not{6}\,\not{3}\,4 \\ -\ 1\,1\,5\,4 \\ \hline 4\,8\,0 \end{array}$$

4. **Check.** We check our answer of 480 mi in the original problem. This number should be less than the total distance, 1634 mi, which it is. We can add the difference, 480, to the subtrahend, 1154: $1154 + 480 = 1634$. We can also estimate:

$$\begin{aligned} 1634 - 1154 &\approx 1600 - 1200 \\ &= 400 \approx 480. \end{aligned}$$

The answer, 480 mi, checks.

5. **State.** Abigail must travel 480 mi farther to Salt Lake City.

Do Exercise 5.

5. **Home Vacuum.** William has \$228. He wants to purchase the home vacuum shown in the ad below. How much more does he need?

Answer on page A-3

6. **Total Cost of Gas Grills.** What is the total cost of 14 gas grills, each with 520 sq in. of total cooking surface and porcelain cast-iron cooking grates, if each one costs $398?

EXAMPLE 4 *Total Cost of Chairs.* What is the total cost of 6 Logan side chairs from Restoration Hardware if each one costs $210?

1. **Familiarize.** We first make a drawing or at least visualize the situation. We let $T =$ the cost of 6 chairs. Repeated addition works well in this case.

2. **Translate.** We translate to an equation.

Number of chairs	times	Cost of each chair	is	Total cost
↓	↓	↓	↓	↓
6	$\times$	$210	$=$	T

3. **Solve.** This sentence tells us what to do. We multiply.

$$\begin{array}{r} 2\ 1\ 0 \\ \times \quad\ \ 6 \\ \hline 1\ 2\ 6\ 0 \end{array} \qquad \begin{aligned} 6 \times 210 &= T \\ 1260 &= T \end{aligned}$$

4. **Check.** We have an answer, 1260, that is much greater than the cost of any individual chair, which is reasonable. We can repeat our calculation. We can also check by estimating:

$$6 \times 210 \approx 6 \times 200 = 1200 \approx 1260.$$

The answer checks.

5. **State.** The total cost of 6 chairs is $1260.

Do Exercise 6.

EXAMPLE 5 *Truck Bed Cover.* The dimensions of a fiberglass truck bed cover for a pickup truck are 79 in. by 68 in. What is the area of the cover?

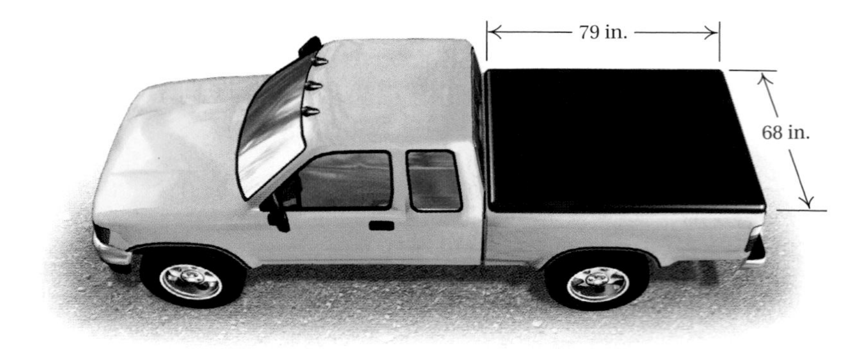

Answer on page A-3

1. **Familiarize.** The truck bed cover is a rectangle that measures 79 in. by 68 in. We let A = the area and use the area formula $A = l \cdot w$.
2. **Translate.** Using this formula, we have

$$A = \text{length} \cdot \text{width} = l \cdot w = 79 \cdot 68.$$

3. **Solve.** We carry out the multiplication.

$$\begin{array}{r} 79 \\ \times\ 68 \\ \hline 632 \\ 4740 \\ \hline 5372 \end{array} \qquad \begin{aligned} A &= 79 \cdot 68 \\ A &= 5372 \end{aligned}$$

4. **Check.** We repeat our calculation. We also note that the answer is greater than either the length or the width, which it should be. (This might not be the case if we were using fractions or decimals.) The answer checks.
5. **State.** The area of the truck bed cover is 5372 sq in.

Do Exercise 7.

EXAMPLE 6 *Cartons of Soda.* A bottling company produces 3304 cans of soda. How many 12-can cartons can be filled? How many cans will be left over?

1. **Familiarize.** We first make a drawing. We let n = the number of 12-can cartons that can be filled. The problem can be considered as repeated subtraction, taking successive sets of 12 cans and putting them into n cartons.

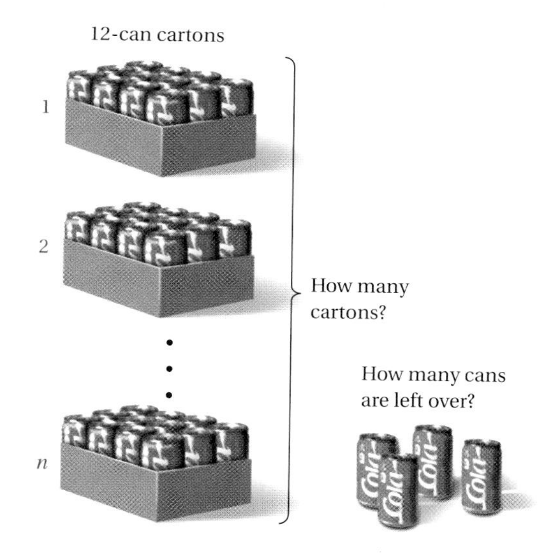

2. **Translate.** We translate to an equation.

Number of cans	divided by	Number in each carton	is	Number of cartons
↓	↓	↓	↓	↓
3304	÷	12	=	n

7. **Bed Sheets.** The dimensions of a flat sheet for a queen-size bed are 90 in. by 102 in. What is the area of the sheet?

Answer on page A-3

8. Cartons of Soda. The bottling company in Example 6 also uses 6-can cartons. How many 6-can cartons can be filled with 2269 cans of cola? How many will be left over?

Answer on page A-3

3. Solve. We solve the equation by carrying out the division.

$$\begin{array}{r} 275 \\ 12\overline{)3304} \\ \underline{2400} \\ 904 \\ \underline{840} \\ 64 \\ \underline{60} \\ 4 \end{array}$$

$$3304 \div 12 = n$$
$$275 \text{ R } 4 = n$$

4. Check. We can check by multiplying the number of cartons by 12 and adding the remainder, 4:

$12 \cdot 275 = 3300,$

$3300 + 4 = 3304.$

5. State. Thus, 275 twelve-can cartons can be filled. There will be 4 cans left over.

Do Exercise 8.

EXAMPLE 7 *Automobile Mileage.* The Pontiac G6 GT gets 21 miles to the gallon (mpg) in city driving. How many gallons will it use in 3843 mi of city driving?

Source: General Motors

1. Familiarize. We first make a drawing. It is often helpful to be descriptive about how we define a variable. In this case, we let g = the number of gallons ("g" comes from "gallons").

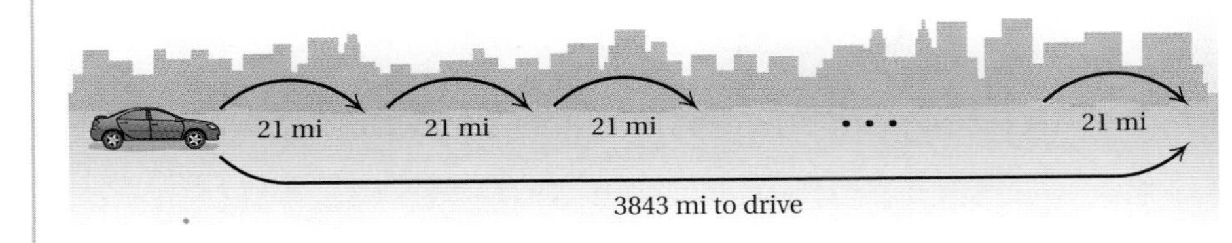

2. Translate. Repeated addition applies here. Thus the following multiplication applies to the situation.

Number of miles per gallon	times	Number of gallons needed	is	Number of miles to drive
↓	↓	↓	↓	↓
21	$\cdot$	g	$=$	3843

3. Solve. To solve the equation, we divide by 21 on both sides.

$$21 \cdot g = 3843$$
$$\frac{21 \cdot g}{21} = \frac{3843}{21}$$
$$g = 183$$

$$\begin{array}{r} 183 \\ 21\overline{)3843} \\ \underline{2100} \\ 1743 \\ \underline{1680} \\ 63 \\ \underline{63} \\ 0 \end{array}$$

4. **Check.** To check, we multiply 183 by 21: $21 \times 183 = 3843$.
5. **State.** The Pontiac G6 GT will use 183 gal.

Do Exercise 9.

9. **Automobile Mileage.** The Pontiac G6 GT gets 29 miles to the gallon (mpg) in highway driving. How many gallons will it take to drive 2291 mi of highway driving?
Source: General Motors

Multistep Problems

Sometimes we must use more than one operation to solve a problem, as in the following example.

EXAMPLE 8 *Aircraft Seating.* Boeing Corporation builds commercial aircraft. A Boeing 767 has a seating configuration with 4 rows of 6 seats across in first class and 35 rows of 7 seats across in economy class. Find the total seating capacity of the plane.
Sources: The Boeing Company; Delta Airlines

1. **Familiarize.** We first make a drawing.

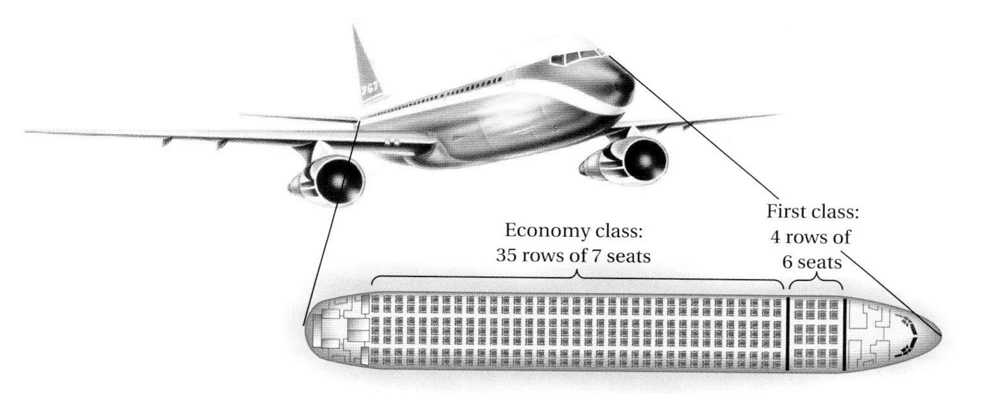

2. **Translate.** There are three parts to the problem. We first find the number of seats in each class. Then we add.

First-class: Repeated addition applies here. Thus the following multiplication corresponds to the situation. We let $F =$ the number of seats in first class.

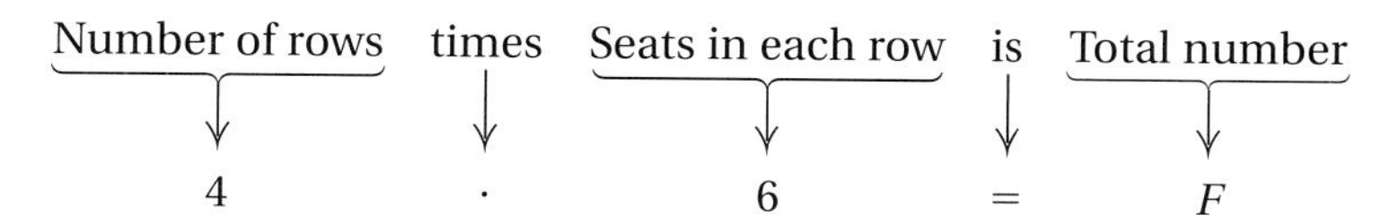

Number of rows times Seats in each row is Total number
4 · 6 = F

Economy class: Repeated addition applies here. Thus the following multiplication corresponds to the situation. We let $E =$ the number of seats in economy class.

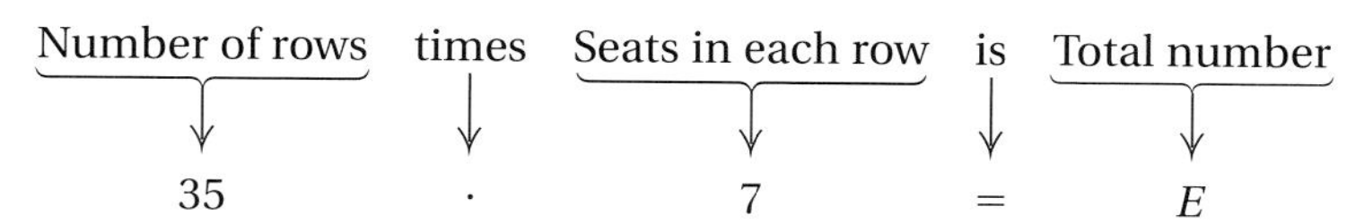

Number of rows times Seats in each row is Total number
35 · 7 = E

We let $T =$ the total number of seats in both classes.

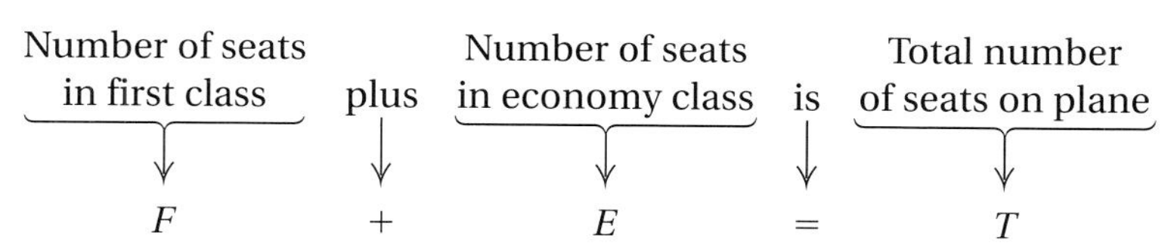

Answer on page A-3

10. Aircraft Seating. A Boeing 767 used for foreign travel has three classes of seats. First class has 3 rows of 5 seats across; business class has 6 rows with 6 seats across and 1 row with 2 seats on each of the outside aisles. Economy class has 18 rows with 7 seats across. Find the total seating capacity of the plane.

Sources: The Boeing Company; Delta Airlines

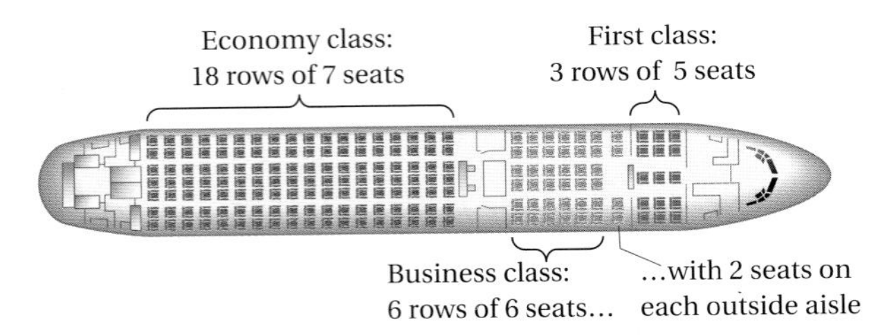

3. Solve. We solve each equation and add the solutions.

$$4 \cdot 6 = F \qquad 35 \cdot 7 = E \qquad F + E = T$$
$$24 = F \qquad 245 = E \qquad 24 + 245 = T$$
$$269 = T$$

4. Check. To check, we repeat our calculations. (We leave this to the student.) We could also check by rounding, multiplying, and adding.

5. State. There are 269 seats in a Boeing 767.

Do Exercise 10.

As you consider the following exercises, here are some words and phrases that may be helpful to look for when you are translating problems to equations.

KEY WORDS, PHRASES, AND CONCEPTS

ADDITION (+)	SUBTRACTION (−)
add	subtract
added to	subtracted from
sum	difference
total	minus
plus	less than
more than	decreased by
increased by	take away
	how much more
	missing addend
MULTIPLICATION (·)	**DIVISION (÷)**
multiply	divide
multiplied by	divided by
product	quotient
times	repeated subtraction
of	missing factor
repeated addition	finding equal quantities
rectangular arrays	

Answer on page A-3

Translating for Success

Brick-mason Expense. A commercial contractor is building 30 two-unit condominiums in a retirement community. The brick-mason expense for each building is \$10,860. What is the total cost of bricking the buildings?

Heights. Dean's sons are on the high school basketball team. Their heights are 73 in., 69 in., and 76 in. How much taller is the tallest son than the shortest son?

Account Balance. You have \$423 in your checking account. Then you deposit \$73 and use your debit card for purchases of \$76 and \$69. How much is left in your account?

Purchasing Camcorder. A camcorder is on sale for \$423. Jenny has only \$69. How much more does she need to buy the camcorder?

Purchasing Coffee Makers. Sara purchases 8 coffee makers for the newly remodeled bed-and-breakfast hotel that she manages. If she pays \$52 for each coffee maker, what is the total cost of her purchase?

The goal of these matching questions is to practice step (2), *Translate,* of the five-step problem-solving process. Translate each word problem to an equation and select a correct translation from equations A–O.

A. $8 \cdot 52 = n$

B. $69 \cdot n = 76$

C. $73 - 76 - 69 = n$

D. $423 + 73 - 76 - 69 = n$

E. $30 \cdot 10{,}860 = n$

F. $15 \cdot n = 195$

G. $69 + n = 423$

H. $n = 10{,}860 - 300$

I. $n = 423 \div 69$

J. $30 \cdot n = 10{,}860$

K. $15 \cdot 195 = n$

L. $n = 52 - 8$

M. $69 + n = 76$

N. $15 \div 195 = n$

O. $52 + n = 60$

Answers on page A-3

6. *Hourly Rate.* Miller Auto Repair charges \$52 an hour for labor. Jackson Auto Care charges \$60 per hour. How much more does Jackson charge than Miller?

7. *College Band.* A college band with 195 members marches in a 15-row formation in the homecoming halftime performance. How many members are in each row?

8. *Shoe Purchase.* A professional football team purchases 15 pairs of shoes at \$195 a pair. What is the total cost of this purchase?

9. *Loan Payment.* Kendra borrows \$10,860 for a new boat. The loan is to be paid off in 30 payments. How much is each payment?

10. *College Enrollment.* At the beginning of the fall term, the total enrollment in Lakeview Community College was 10,860. By the end of the first two weeks, 300 students withdrew. How many students were then enrolled?

a Solve.

Longest Broadway Run. The bar graph below lists the five broadway shows with the greatest number of performances as of May 3, 2004.

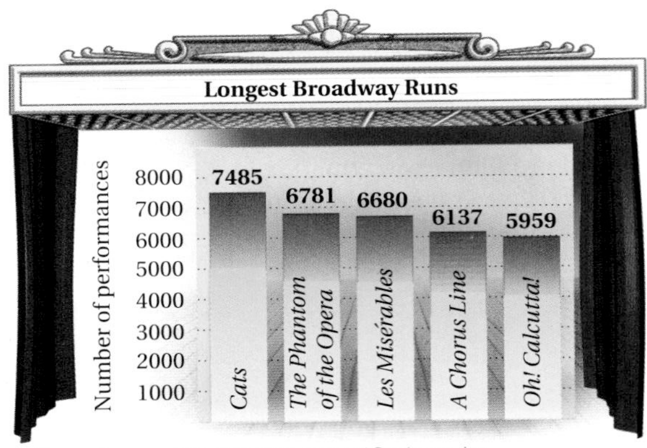

Source: League of American Theatres and Producers, Inc.

1. What was the total number of performances of all five shows?

2. What was the total number of performances of the top three shows?

3. How many more performances of *Cats* were there than performances of *The Phantom of the Opera*?

4. How many more performances of *Les Misérables* were there than performances of *Oh! Calcutta!*?

5. *Boundaries Between Countries.* The boundary between mainland United States and Canada including the Great Lakes is 3987 miles long. The length of the boundary between the United States and Mexico is 1933 miles. How much longer is the Canadian border?

Source: U.S. Geological Survey

6. *Caffeine.* Hershey's 6-oz milk chocolate almond bar contains 25 milligrams of caffeine. A 20-oz bottle of Coca-Cola has 32 more milligrams of caffeine than the Hershey bar. How many milligrams of caffeine does the 20-oz bottle of Coca-Cola have?

Source: *National Geographic,* "Caffeine," by T. R. Reid, January 2005

7. A carpenter drills 216 holes in a rectangular array in a pegboard. There are 12 holes in each row. How many rows are there?

8. Lou works as a CPA. He arranges 504 entries on a spreadsheet in a rectangular array that has 36 rows. How many entries are in each row?

Bachelor's Degree. The line graph below illustrates data about bachelor's degrees awarded to men and women from 1970 to 2002. Use this graph when answering Exercises 9–12.

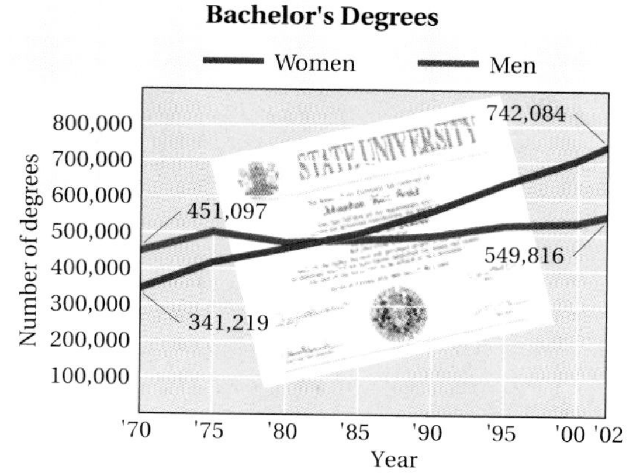

Source: U.S. Department of Education

9. Find the total number of bachelor's degrees awarded in 1970 and the total number awarded in 2002.

10. Determine how many more bachelor's degrees were awarded in 2002 than in 1970.

11. How many more bachelor's degrees were awarded to women than to men in 2002?

12. How many more bachelor's degrees were awarded to men than to women in 1970?

13. *Median Mortgage Debt.* The median mortgage debt in 2001 was $29,475 more than the median mortgage debt in 1989. The debt in 1989 was $39,802. What was the median mortgage debt in 2001?

Source: Federal Reserve Board Survey of Consumer Finances; Consumer Bank Association 2003 Home Equity Study; Freddie Mac

14. *Olympics in Athens.* In the first modern Olympics in Athens, Greece, in 1896, there were 43 events. In 2004, there were 258 more events in the Summer Olympics in Athens. How many events were there in 2004?

Source: *USA Today* research; The Olympic Games: Athens 1896–Athens 2004

15. *Longest Rivers.* The longest river in the world is the Nile in Egypt at 4100 mi. The longest river in the United States is the Missouri–Mississippi at 3860 mi. How much longer is the Nile?

16. *Speeds on Interstates.* Recently, speed limits on interstate highways in many Western states were raised from 65 mph to 75 mph. By how many miles per hour were they raised?

17. There are 24 hours (hr) in a day and 7 days in a week. How many hours are there in a week?

18. There are 60 min in an hour and 24 hr in a day. How many minutes are there in a day?

19. *Crossword.* The *USA Today* crossword puzzle is a rectangle containing 15 rows with 15 squares in each row. How many squares does the puzzle have altogether?

20. *Pixels.* A computer screen consists of small rectangular dots called *pixels.* How many pixels are there on a screen that has 600 rows with 800 pixels in each row?

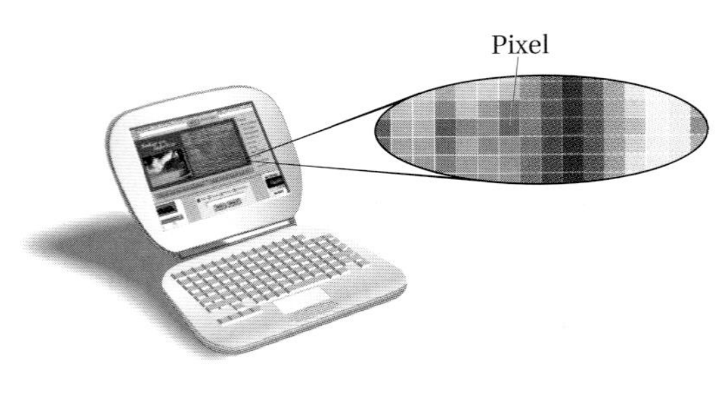

21. *Refrigerator Purchase.* Gourmet Deli has a chain of 24 restaurants. It buys a commercial refrigerator for each store at a cost of $1019 each. Find the total cost of the purchase.

22. *Microwave Purchase.* Bridgeway College is constructing new dorms, in which each room has a small kitchen. It buys 96 microwave ovens at $88 each. Find the total cost of the purchase.

23. *"Seinfeld" Episodes.* "Seinfeld" was a long-running television comedy with 177 episodes. A local station picks up the syndicated reruns. If the station runs 5 episodes per week, how many full weeks will pass before it must start over with past episodes? How many episodes will be left for the last week?

24. A lab technician separates a vial containing 70 cubic centimeters (cc) of blood into test tubes, each of which contains 3 cc of blood. How many test tubes can be filled? How much blood is left over?

25. *Automobile Mileage.* The 2005 Hyundai Tucson GLS gets 26 miles to the gallon (mpg) in highway driving. How many gallons will it use in 6136 mi of highway driving?

Source: Hyundai

26. *Automobile Mileage.* The 2005 Volkswagen Jetta (5 cylinder) gets 24 miles to the gallon (mpg) in city driving. How many gallons will it use in 3960 mi of city driving?

Source: Volkswagen of America, Inc.

Boeing Jets. Use the chart below in Exercises 27–32 to compare the Boeing 747 jet with its main competitor, the Boeing 777 jet.

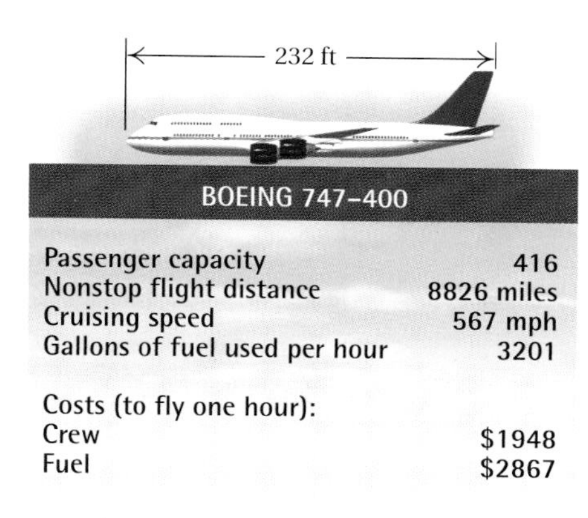

BOEING 747–400	
Passenger capacity	416
Nonstop flight distance	8826 miles
Cruising speed	567 mph
Gallons of fuel used per hour	3201
Costs (to fly one hour):	
Crew	$1948
Fuel	$2867

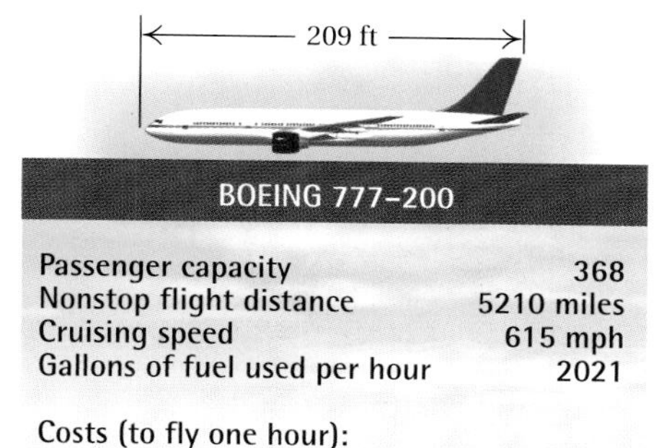

BOEING 777–200	
Passenger capacity	368
Nonstop flight distance	5210 miles
Cruising speed	615 mph
Gallons of fuel used per hour	2021
Costs (to fly one hour):	
Crew	$1131
Fuel	$1816

Source: Éclat Consulting; Boeing

27. The nonstop flight distance of the Boeing 747 jet is how much greater than the nonstop flight distance of the Boeing 777 jet?

28. How much larger is the passenger capacity of the Boeing 747 jet than the passenger capacity of the Boeing 777 jet?

29. How many gallons of fuel are needed for a 4-hour flight of the Boeing 747?

30. How much longer is the Boeing 747 than the Boeing 777?

31. What is the total cost for the crew and fuel for a 3-hour flight of the Boeing 747 jet?

32. What is the total cost for the crew and fuel for a 2-hour flight of the Boeing 777 jet?

33. Dana borrows $5928 for a used car. The loan is to be paid off in 24 equal monthly payments. How much is each payment (excluding interest)?

34. A family borrows $7824 to build a sunroom on the back of their home. The loan is to be paid off in equal monthly payments of $163 (excluding interest). How many months will it take to pay off the loan?

35. *High School Court.* The standard basketball court used by high school players has dimensions of 50 ft by 84 ft.

a) What is its area?

b) What is its perimeter?

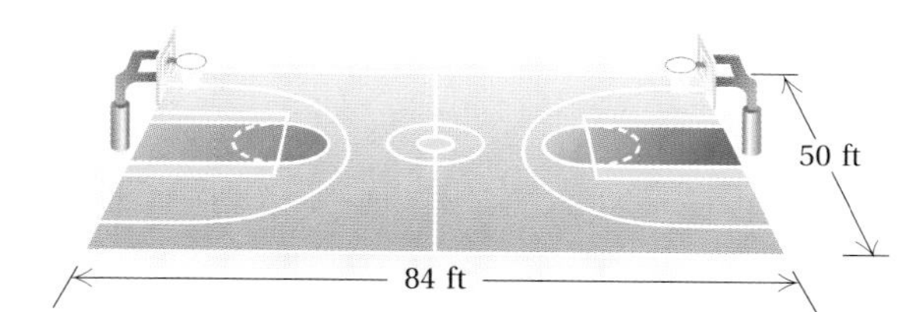

36. *NBA Court.* The standard basketball court used by college and NBA players has dimensions of 50 ft by 94 ft.

a) What is its area?

b) What is its perimeter?

c) How much greater is the area of an NBA court than a high school court? (See Exercise 35.)

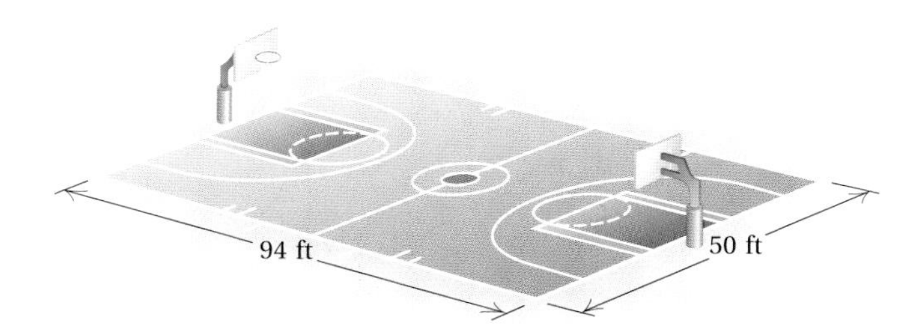

37. *Clothing Imports and Exports.* In the United States, the exports of clothing in 2003 totaled $2,596,000,000 while the imports totaled $31,701,000,000. How much more were the imports than the exports?

Source: U.S. Census Bureau, Foreign Trade Division

38. *Corn Imports and Exports.* In the United States, the exports of corn in 2003 totaled $2,264,000,000 while the imports totaled $130,000,000. How much more were the exports than the imports?

Source: U.S. Census Bureau, Foreign Trade Division

39. *Colonial Population.* Before the establishment of the U.S. Census in 1790, it was estimated that the Colonial population in 1780 was 2,780,400. This was an increase of 2,628,900 from the population in 1680. What was the Colonial population in 1680?

Source: *Time Almanac,* 2005

40. *Deaths by Firearms.* Deaths by firearms totaled 29,573 in 2001. This was a decrease of 10,022 from the deaths by firearms in 1993. How many deaths by firearms were there in 1993?

Source: Center for Disease Control and Prevention, *National Vital Statistics Reports,* vol 52, no. 3, September 18, 2003

41. *Hershey Bars.* Hershey Chocolate USA makes small, fun-size chocolate bars. How many 20-bar packages can be filled with 11,267 bars? How many bars will be left over?

42. *Reese's Peanut Butter Cups.* H. B. Reese Candy Co. makes small, fun-size peanut butter cups. The company manufactures 23,579 cups and fills 1025 packages. How many cups are in a package? How many cups will be left over?

43. *Map Drawing.* A map has a scale of 64 mi to the inch. How far apart *in reality* are two cities that are 6 in. apart on the map? How far apart *on the map* are two cities that, in reality, are 1728 mi apart?

0 64
Miles per inch
CANADA
MONTANA
NORTH DAKOTA
MINNESOTA
Bismarck
Fargo
SOUTH DAKOTA
5 83 29 94 83 281 94 212

44. *Map Drawing.* A map has a scale of 150 mi to the inch. How far apart *on the map* are two cities that, in reality, are 2400 mi apart? How far apart *in reality* are two cities that are 13 in. apart on the map?

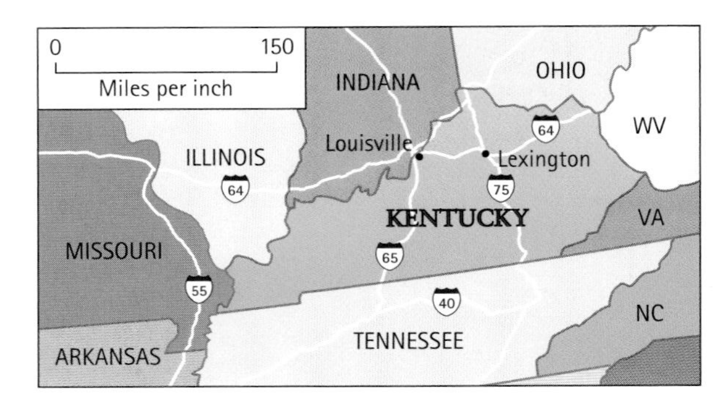

45. *Crossword.* The *Los Angeles Times* crossword puzzle is a rectangle containing 441 squares arranged in 21 rows. How many columns does the puzzle have?

46. *Sheet of Stamps.* A sheet of 100 stamps typically has 10 rows of stamps. How many stamps are in each row?

47. Copies of this book are generally shipped from the warehouse in cartons containing 24 books each. How many cartons are needed to ship 1355 books?

48. According to the H. J. Heinz Company, 16-oz bottles of catsup are generally shipped in cartons containing 12 bottles each. How many cartons are needed to ship 528 bottles of catsup?

49. Elena buys 5 video games at \$64 each and pays for them with \$10 bills. How many \$10 bills does it take?

50. Pedro buys 5 video games at \$64 each and pays for them with \$20 bills. How many \$20 bills does it take?

51. You have \$568 in your bank account. You use your debit card for \$46, \$87, and \$129. Then you deposit \$94 back in the account after the return of some books. How much is left in your account?

52. The balance in your bank account is \$749. You use your debit card for \$34 and \$65. Then you make a deposit of \$123 from your paycheck. What is your new balance?

Weight Loss. Many Americans exercise for weight control. It is known that one must burn off about 3500 calories in order to lose one pound. The chart shown here details how much of certain types of exercise is required to burn 100 calories. Use this chart for Exercises 53–56.

To burn off 100 calories, you must:

- Run for 8 minutes at a brisk pace, or
- Swim for 2 minutes at a brisk pace, or
- Bicycle for 15 minutes at 9 mph, or
- Do aerobic exercises for 15 minutes.

53. How long must you run at a brisk pace in order to lose one pound?

54. How long must you swim in order to lose one pound?

55. How long must you do aerobic exercises in order to lose one pound?

56. How long must you bicycle at 9 mph in order to lose one pound?

57. *Bones in the Hands and Feet.* There are 27 bones in each human hand and 26 bones in each human foot. How many bones are there in all in the hands and feet?

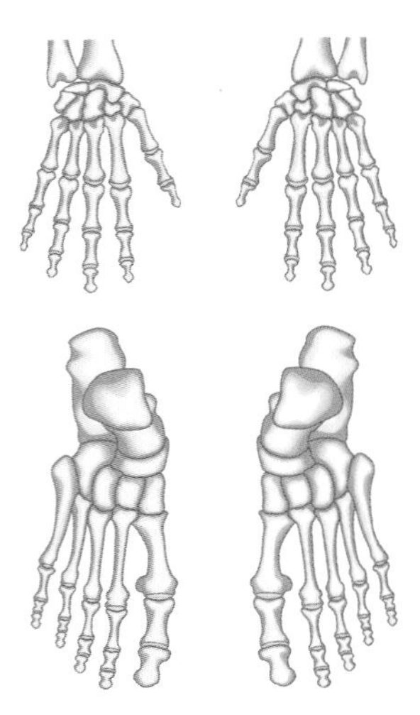

58. *Subway Travel.* The distance to Mars is about 303,000,000 miles. The number of miles that people in the United States traveled on subways in 2003 approximately equaled 22 round trips to Mars. What was the total distance traveled on subways?

Source: American Public Transportation Association, NASA

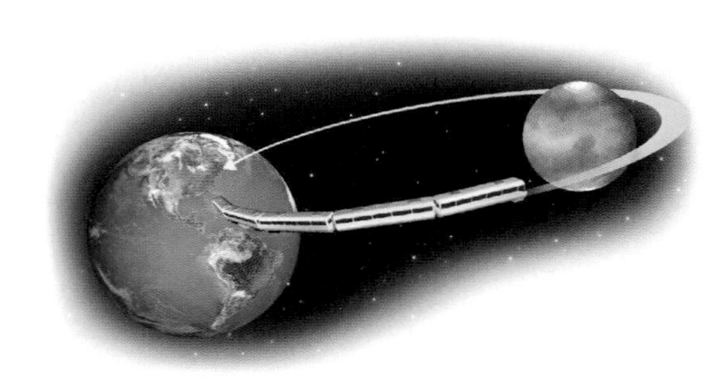

59. *Index Cards.* Index cards of dimension 3 in. by 5 in. are normally shipped in packages containing 100 cards each. How much writing area is available if one uses the front and back sides of a package of these cards?

60. An office for adjunct instructors at a community college has 6 bookshelves, each of which is 3 ft wide. The office is moved to a new location that has dimensions of 16 ft by 21 ft. Is it possible for the bookshelves to be put side by side on the 16-ft wall?

61. **DW** In the newspaper article "When Girls Play, Knees Fail," the author discusses the fact that female athletes have six times the number of knee injuries that male athletes have. What information would be needed if you were to write a math problem based on the article? What might the problem be?

Source: *The Arizona Republic,* 2/9/00, p. C1

62. **DW** Write a problem for a classmate to solve. Design the problem so that the solution is "The driver still has 329 mi to travel."

SKILL MAINTENANCE

Round 234,562 to the nearest: [1.4a]

63. Hundred.

64. Ten.

65. Thousand.

Estimate the computation by rounding to the nearest thousand. [1.4b]

66. $2783 + 4602 + 5797 + 8111$

67. $28{,}430 - 11{,}977$

68. $2100 + 5800$

69. $5800 - 2100$

Estimate the product by rounding to the nearest hundred. [1.5b]

70. $787 \cdot 363$

71. $887 \cdot 799$

72. $10{,}362 \cdot 4531$

SYNTHESIS

73. *Speed of Light.* Light travels about 186,000 miles per second (mi/sec) in a vacuum as in outer space. In ice it travels about 142,000 mi/sec, and in glass it travels about 109,000 mi/sec. In 18 sec, how many more miles will light travel in a vacuum than in ice? than in glass?

74. Carney Community College has 1200 students. Each professor teaches 4 classes and each student takes 5 classes. There are 30 students and 1 teacher in each classroom. How many professors are there at Carney Community College?

1.9 EXPONENTIAL NOTATION AND ORDER OF OPERATIONS

Objectives

a Write exponential notation for products such as $4 \cdot 4 \cdot 4$.

b Evaluate exponential notation.

c Simplify expressions using the rules for order of operations.

d Remove parentheses within parentheses.

a Writing Exponential Notation

Consider the product $3 \cdot 3 \cdot 3 \cdot 3$. Such products occur often enough that mathematicians have found it convenient to create a shorter notation, called **exponential notation,** for them. For example,

$\underbrace{3 \cdot 3 \cdot 3 \cdot 3}_{\text{4 factors}}$ is shortened to 3^4 ← exponent (3 is the base)

We read exponential notation as follows.

NOTATION	WORD DESCRIPTION
3^4	"three to the fourth power," or "the fourth power of three"
5^3	"five-cubed," or "the cube of five," or "five to the third power," or "the third power of five"
7^2	"seven squared," or "the square of seven," or "seven to the second power," or "the second power of seven"

The wording "seven squared" for 7^2 comes from the fact that a square with side s has area A given by $A = s^2$.

s

$A = s^2$ s

An expression like $3 \cdot 5^2$ is read "three times five squared," or "three times the square of five."

EXAMPLE 1 Write exponential notation for $10 \cdot 10 \cdot 10 \cdot 10 \cdot 10$.

Exponential notation is 10^5. 5 is the *exponent.* 10 is the *base.*

EXAMPLE 2 Write exponential notation for $2 \cdot 2 \cdot 2$.

Exponential notation is 2^3.

Do Exercises 1–4.

Write exponential notation.

1. $5 \cdot 5 \cdot 5 \cdot 5$

2. $5 \cdot 5 \cdot 5 \cdot 5 \cdot 5$

3. $10 \cdot 10$

4. $10 \cdot 10 \cdot 10 \cdot 10$

Evaluate.

5. 10^4

6. 10^2

7. 8^3

8. 2^5

Answers on page A-3

CALCULATOR CORNER

Exponential Notation
Many calculators have a $\boxed{y^x}$ or $\boxed{x^y}$ key for raising a base to a power. To find 16^3, for example, we press $\boxed{1}\ \boxed{6}\ \boxed{y^x}\ \boxed{3}\ \boxed{=}$ or $\boxed{1}\ \boxed{6}\ \boxed{x^y}\ \boxed{3}\ \boxed{=}$. The result is 4096.

Exercises: Use a calculator to find each of the following.

1. 3^5
2. 5^6
3. 12^4
4. 2^{11}

Simplify.

9. $93 - 14 \cdot 3$

10. $104 \div 4 + 4$

11. $25 \cdot 26 - (56 + 10)$

12. $75 \div 5 + (83 - 14)$

Answers on page A-3

b Evaluating Exponential Notation

We evaluate exponential notation by rewriting it as a product and computing the product.

EXAMPLE 3 Evaluate: 10^3.

$$10^3 = 10 \cdot 10 \cdot 10 = 1000$$

Caution!

10^3 does not mean $10 \cdot 3$.

EXAMPLE 4 Evaluate: 5^4.

$$5^4 = 5 \cdot 5 \cdot 5 \cdot 5 = 625$$

Do Exercises 5–8 on the preceding page.

c Simplifying Expressions

Suppose we have a calculation like the following:

$$3 + 4 \cdot 8.$$

How do we find the answer? Do we add 3 to 4 and then multiply by 8, or do we multiply 4 by 8 and then add 3? In the first case, the answer is 56. In the second, the answer is 35. We agree to compute as in the second case.

Consider the calculation

$$7 \cdot 14 - (12 + 18).$$

What do the parentheses mean? To deal with these questions, we must make some agreement regarding the order in which we perform operations. The rules are as follows.

RULES FOR ORDER OF OPERATIONS

1. Do all calculations within parentheses ( ), brackets [ ], or braces { } before operations outside.
2. Evaluate all exponential expressions.
3. Do all multiplications and divisions in order from left to right.
4. Do all additions and subtractions in order from left to right.

It is worth noting that these are the rules that computers and most scientific calculators use to do computations.

EXAMPLE 5 Simplify: $16 \div 8 \cdot 2$.

There are no parentheses or exponents, so we start with the third step.

$$\left.\begin{aligned} 16 \div 8 \cdot 2 &= 2 \cdot 2 \\ &= 4 \end{aligned}\right\}$$ Doing all multiplications and divisions in order from left to right

EXAMPLE 6 Simplify: $7 \cdot 14 - (12 + 18)$.

$$
\begin{aligned}
7 \cdot 14 - (12 + 18) &= 7 \cdot 14 - 30 && \text{Carrying out operations inside parentheses} \\
&= 98 - 30 && \text{Doing all multiplications and divisions} \\
&= 68 && \text{Doing all additions and subtractions}
\end{aligned}
$$

Do Exercises 9–12 on the preceding page.

EXAMPLE 7 Simplify and compare: $23 - (10 - 9)$ and $(23 - 10) - 9$.

We have

$$23 - (10 - 9) = 23 - 1 = 22;$$
$$(23 - 10) - 9 = 13 - 9 = 4.$$

We can see that $23 - (10 - 9)$ and $(23 - 10) - 9$ represent different numbers. Thus subtraction is not associative.

Do Exercises 13 and 14.

EXAMPLE 8 Simplify: $7 \cdot 2 - (12 + 0) \div 3 - (5 - 2)$.

$$
\begin{aligned}
7 \cdot 2 - (12 + 0) \div 3 - (5 - 2) &= 7 \cdot 2 - 12 \div 3 - 3 && \text{Carrying out operations inside parentheses} \\
&= 14 - 4 - 3 && \text{Doing all multiplications and divisions in order from left to right} \\
&= 7 && \text{Doing all additions and subtractions in order from left to right}
\end{aligned}
$$

Do Exercise 15.

EXAMPLE 9 Simplify: $15 \div 3 \cdot 2 \div (10 - 8)$.

$$
\begin{aligned}
15 \div 3 \cdot 2 \div (10 - 8) &= 15 \div 3 \cdot 2 \div 2 && \text{Carrying out operations inside parentheses} \\
&= 5 \cdot 2 \div 2 \\
&= 10 \div 2 \\
&= 5 && \text{Doing all multiplications and divisions in order from left to right}
\end{aligned}
$$

Do Exercises 16–18.

Simplify and compare.

13. $64 \div (32 \div 2)$ and $(64 \div 32) \div 2$

14. $(28 + 13) + 11$ and $28 + (13 + 11)$

15. Simplify:

$9 \times 4 - (20 + 4) \div 8 - (6 - 2)$.

Simplify.

16. $5 \cdot 5 \cdot 5 + 26 \cdot 71 - (16 + 25 \cdot 3)$

17. $30 \div 5 \cdot 2 + 10 \cdot 20 + 8 \cdot 8 - 23$

18. $95 - 2 \cdot 2 \cdot 2 \cdot 5 \div (24 - 4)$

Answers on page A-3

Simplify.

19. $5^3 + 26 \cdot 71 - (16 + 25 \cdot 3)$

EXAMPLE 10 Simplify: $4^2 \div (10 - 9 + 1)^3 \cdot 3 - 5$.

$4^2 \div (10 - 9 + 1)^3 \cdot 3 - 5$

$= 4^2 \div (1 + 1)^3 \cdot 3 - 5$ Subtracting inside parentheses

$= 4^2 \div 2^3 \cdot 3 - 5$ Adding inside parentheses

$= 16 \div 8 \cdot 3 - 5$ Evaluating exponential expressions

$= 2 \cdot 3 - 5$

$= 6 - 5$ Doing all multiplications and divisions in order from left to right

$= 1$ Subtracting

Do Exercises 19–21.

20. $(1 + 3)^3 + 10 \cdot 20 + 8^2 - 23$

EXAMPLE 11 Simplify: $2^9 \div 2^6 \cdot 2^3$.

$2^9 \div 2^6 \cdot 2^3 = 512 \div 64 \cdot 8$ There are no parentheses. Evaluating exponential expressions

$= 8 \cdot 8$

$= 64$ Doing all multiplications and divisions in order from left to right

Do Exercise 22.

21. $81 - 3^2 \cdot 2 \div (12 - 9)$

22. Simplify: $2^3 \cdot 2^8 \div 2^9$.

CALCULATOR CORNER

Order of Operations To determine whether a calculator is programmed to follow the rules for order of operations, we can enter a simple calculation that requires using those rules. For example, we enter [3] [+] [4] [×] [2] [=]. If the result is 11, we know that the rules for order of operations have been followed. That is, the multiplication $4 \times 2 = 8$ was performed first and then 3 was added to produce a result of 11. If the result is 14, we know that the calculator performs operations as they are entered rather than following the rules for order of operations. That means, in this case, that 3 and 4 were added first to get 7 and then that sum was multiplied by 2 to produce the result of 14. For such calculators, we would have to enter the operations in the order in which we want them performed. In this case, we would press [4] [×] [2] [+] [3] [=].

Many calculators have parenthesis keys that can be used to enter an expression containing parentheses. To enter $5(4 + 3)$, for example, we press [5] [(] [4] [+] [3] [)] [=]. The result is 35.

Exercises: Simplify.

1. $84 - 5 \cdot 7$
2. $80 + 50 \div 10$
3. $3^2 + 9^2 \div 3$
4. $4^4 \div 64 - 4$
5. $15 \cdot 7 - (23 + 9)$
6. $(4 + 3)^2$

Answers on page A-3

AVERAGES

In order to find the average of a set of numbers, we use addition and then division. For example, the average of 2, 3, 6, and 9 is found as follows.

$$\text{Average} = \frac{2 + 3 + 6 + 9}{4} = \frac{20}{4} = 5$$

The number of addends is 4.

Divide by 4.

The fraction bar acts as a pair of grouping symbols so

$$\frac{2 + 3 + 6 + 9}{4} \text{ is equivalent to } (2 + 3 + 6 + 9) \div 4.$$

Thus we are using order of operations when we compute an average.

AVERAGE

The **average** of a set of numbers is the sum of the numbers divided by the number of addends.

EXAMPLE 12 *Average Number of Career Hits.* The number of career hits of five Hall of Fame baseball players are given in the bar graph below. Find the average number of career hits of all five.

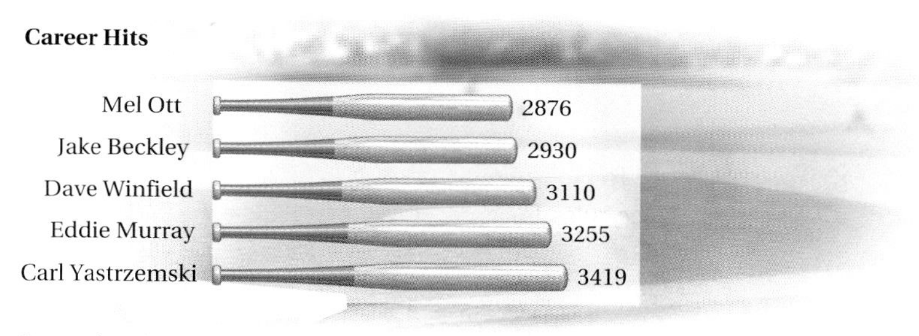

Sources: Associated Press; Major League Baseball

The average is given by

$$\frac{3419 + 3255 + 3110 + 2930 + 2876}{5} = \frac{15{,}590}{5} = 3118.$$

Thus the average number of career hits of these five Hall of Fame players is 3118.

Do Exercise 23.

d Removing Parentheses within Parentheses

When parentheses occur within parentheses, we can make them different shapes, such as [ ] (also called "brackets") and { } (also called "braces"). All of these have the same meaning. When parentheses occur within parentheses, computations in the innermost ones are to be done first.

23. World's Tallest Buildings. The heights, in feet, of the four tallest buildings in the world are given in the bar graph below. Find the average height of these buildings.

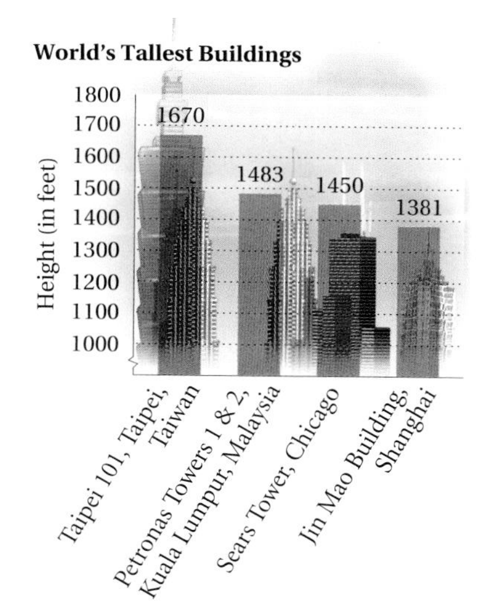

Source: Council on Tall Buildings and Urban Habitat, Lehigh University, 2004

Answer on page A-3

Simplify.

24. $9 \times 5 + \{6 \div [14 - (5 + 3)]\}$

25. $[18 - (2 + 7) \div 3] - (31 - 10 \times 2)$

Answers on page A-3

EXAMPLE 13 Simplify: $[25 - (4 + 3) \cdot 3] \div (11 - 7)$.

$[25 - (4 + 3) \cdot 3] \div (11 - 7)$

$= [25 - 7 \cdot 3] \div (11 - 7)$ Doing the calculations in the innermost parentheses first

$= [25 - 21] \div (11 - 7)$ Doing the multiplication in the brackets

$= 4 \div 4$ Subtracting

$= 1$ Dividing

EXAMPLE 14 Simplify: $16 \div 2 + \{40 - [13 - (4 + 2)]\}$.

$16 \div 2 + \{40 - [13 - (4 + 2)]\}$

$= 16 \div 2 + \{40 - [13 - 6]\}$ Doing the calculations in the innermost parentheses first

$= 16 \div 2 + \{40 - 7\}$ Again, doing the calculations in the innermost parentheses

$= 16 \div 2 + 33$ Subtracting inside the braces

$= 8 + 33$ Doing all multiplications and divisions in order from left to right

$= 41$ Adding

Do Exercises 24 and 25.

Study Tips

TEST PREPARATION

You are probably ready to begin preparing for your first test. Here are some test-taking study tips.

- **Make up your own test questions as you study.** After you have done your homework over a particular objective, write one or two questions on your own that you think might be on a test. You will be amazed at the insight this will provide.
- **Do an overall review of the chapter, focusing on the objectives and the examples.** This should be accompanied by a study of any class notes you may have taken.
- **Do the review exercises at the end of the chapter.** Check your answers at the back of the book. If you have trouble with an exercise, use the objective symbol as a guide to go back and do further study of that objective.
- **Call the AW Math Tutor Center at 1-888-777-0463 if you need extra help.**
- **Take the chapter test at the end of the chapter.** Check the answers and use the objective symbols at the back of the book as a reference for where to review.
- **Ask former students for old exams.** Working such exams can be very helpful and allows you to see what various professors think is important.
- **When taking a test, read each question carefully and try to do all the questions the first time through, but pace yourself.** Answer all the questions, and mark those to recheck if you have time at the end. Very often, your first hunch will be correct.
- **Try to write your test in a neat and orderly manner.** Very often, your instructor tries to give you partial credit when grading an exam. If your test paper is sloppy and disorderly, it is difficult to verify the partial credit. Doing your work neatly can ease such a task for the instructor.

1.9 EXERCISE SET

For Extra Help

MathXL | MyMathLab | InterAct Math | Math Tutor Center | Digital Video Tutor CD 1 Videotape 1 | Student's Solutions Manual

a Write exponential notation.

1. $3 \cdot 3 \cdot 3 \cdot 3$

2. $2 \cdot 2 \cdot 2 \cdot 2 \cdot 2$

3. $5 \cdot 5$

4. $13 \cdot 13 \cdot 13$

5. $7 \cdot 7 \cdot 7 \cdot 7 \cdot 7$

6. $10 \cdot 10$

7. $10 \cdot 10 \cdot 10$

8. $1 \cdot 1 \cdot 1 \cdot 1$

b Evaluate.

9. 7^2

10. 5^3

11. 9^3

12. 10^2

13. 12^4

14. 10^5

15. 11^2

16. 6^3

c Simplify.

17. $12 + (6 + 4)$

18. $(12 + 6) + 18$

19. $52 - (40 - 8)$

20. $(52 - 40) - 8$

21. $1000 \div (100 \div 10)$

22. $(1000 \div 100) \div 10$

23. $(256 \div 64) \div 4$

24. $256 \div (64 \div 4)$

25. $(2 + 5)^2$

26. $2^2 + 5^2$

27. $(11 - 8)^2 - (18 - 16)^2$

28. $(32 - 27)^3 + (19 + 1)^3$

29. $16 \cdot 24 + 50$

30. $23 + 18 \cdot 20$

31. $83 - 7 \cdot 6$

32. $10 \cdot 7 - 4$

33. $10 \cdot 10 - 3 \cdot 4$

34. $90 - 5 \cdot 5 \cdot 2$

35. $4^3 \div 8 - 4$

36. $8^2 - 8 \cdot 2$

37. $17 \cdot 20 - (17 + 20)$

38. $1000 \div 25 - (15 + 5)$

39. $6 \cdot 10 - 4 \cdot 10$

40. $3 \cdot 8 + 5 \cdot 8$

41. $300 \div 5 + 10$

42. $144 \div 4 - 2$

43. $3 \cdot (2 + 8)^2 - 5 \cdot (4 - 3)^2$

44. $7 \cdot (10 - 3)^2 - 2 \cdot (3 + 1)^2$

45. $4^2 + 8^2 \div 2^2$

46. $6^2 - 3^4 \div 3^3$

47. $10^3 - 10 \cdot 6 - (4 + 5 \cdot 6)$

48. $7^2 + 20 \cdot 4 - (28 + 9 \cdot 2)$

49. $6 \cdot 11 - (7 + 3) \div 5 - (6 - 4)$

50. $8 \times 9 - (12 - 8) \div 4 - (10 - 7)$

51. $120 - 3^3 \cdot 4 \div (5 \cdot 6 - 6 \cdot 4)$

52. $80 - 2^4 \cdot 15 \div (7 \cdot 5 - 45 \div 3)$

53. $2^3 \cdot 2^8 \div 2^6$

54. $2^7 \div 2^5 \cdot 2^4 \div 2^2$

55. Find the average of \$64, \$97, and \$121.

56. Find the average of four test grades of 86, 92, 80, and 78.

57. Find the average of 320, 128, 276, and 880.

58. Find the average of \$1025, \$775, \$2062, \$942, and \$3721.

d Simplify.

59. $8 \times 13 + \{42 \div [18 - (6 + 5)]\}$

60. $72 \div 6 - \{2 \times [9 - (4 \times 2)]\}$

61. $[14 - (3 + 5) \div 2] - [18 \div (8 - 2)]$

62. $[92 \times (6 - 4) \div 8] + [7 \times (8 - 3)]$

63. $(82 - 14) \times [(10 + 45 \div 5) - (6 \cdot 6 - 5 \cdot 5)]$

64. $(18 \div 2) \cdot \{[(9 \cdot 9 - 1) \div 2] - [5 \cdot 20 - (7 \cdot 9 - 2)]\}$

65. $4 \times \{(200 - 50 \div 5) - [(35 \div 7) \cdot (35 \div 7) - 4 \times 3]\}$

66. $15(23 - 4 \cdot 2)^3 \div (3 \cdot 25)$

67. $\{[18 - 2 \cdot 6] - [40 \div (17 - 9)]\} + \{48 - 13 \times 3 + [(50 - 7 \cdot 5) + 2]\}$

68. $(19 - 2^4)^5 - (141 \div 47)^2$

69. **DW** Consider the problem in Example 8 of Section 1.8. How can you translate the problem to a single equation involving what you have learned about order of operations? How does the single equation relate to how we solved the problem?

70. **DW** Consider the expressions $9 - (4 \cdot 2)$ and $(3 \cdot 4)^2$. Are the parentheses necessary in each case? Explain.

SKILL MAINTENANCE

Solve. [1.7b]

71. $x + 341 = 793$

72. $4197 + x = 5032$

73. $7 \cdot x = 91$

74. $1554 = 42 \cdot y$

75. $3240 = y + 898$

76. $6000 = 1102 + t$

77. $25 \cdot t = 625$

78. $10{,}000 = 100 \cdot t$

Solve. [1.8a]

79. *Colorado.* The state of Colorado is roughly the shape of a rectangle that is 273 mi by 382 mi. What is its area?

80. On a long four-day trip, a family bought the following amounts of gasoline for their motor home:

23 gallons, 24 gallons,
26 gallons, 25 gallons.

How much gasoline did they buy in all?

SYNTHESIS

Each of the answers in Exercises 81–83 is incorrect. First find the correct answer. Then place as many parentheses as needed in the expression in order to make the incorrect answer correct.

81. $1 + 5 \cdot 4 + 3 = 36$

82. $12 \div 4 + 2 \cdot 3 - 2 = 2$

83. $12 \div 4 + 2 \cdot 3 - 2 = 4$

84. Use one occurrence each of 1, 2, 3, 4, 5, 6, 7, 8, and 9 and any of the symbols $+$, $-$, $\times$, $\div$, and ( ) to represent 100.

1 Summary and Review

The review that follows is meant to prepare you for a chapter exam. It consists of two parts. The first part, Concept Reinforcement, is designed to increase understanding of the concepts through true/false exercises. The second part is the Review Exercises. These provide practice exercises for the exam, together with references to section objectives so you can go back and review. Before beginning, stop and look back over the skills you have obtained. What skills in mathematics do you have now that you did not have before studying this chapter?

CONCEPT REINFORCEMENT

Determine whether the statement is true or false. Answers are given at the back of the book.

________ **1.** The product of two natural numbers is always greater than either of the factors.

________ **2.** Zero divided by any nonzero number is 0.

________ **3.** Each member of the set of natural numbers is a member of the set of whole numbers

________ **4.** The sum of two natural numbers is always greater than either of the addends.

________ **5.** Any number divided by 1 is the number 1.

________ **6.** The number 0 is the smallest natural number.

Review Exercises

The review exercises that follow are for practice. Answers are given at the back of the book. If you miss an exercise, restudy the objective indicated in red next to the exercise or direction line that precedes it.

1. What does the digit 8 mean in 4,678,952? [1.1a]

2. In 13,768,940, what digit tells the number of millions? [1.1a]

Write expanded notation. [1.1b]

3. 2793

4. 56,078

5. 4,007,101

Write a word name. [1.1c]

6. 67,819

7. 2,781,427

8. 1,065,070,607, the population of India in 2004.
Source: U.S. Census Bureau, International Database

Write standard notation. [1.1c]

9. Four hundred seventy-six thousand, five hundred eighty-eight

10. *e-books.* The publishing industry predicts that sales of digital books will reach two billion, four hundred thousand by 2005.
Source: Andersen Consulting

Add. [1.2a]

11. $7304 + 6968$

12. $27,609 + 38,415$

13. $2703 + 4125 + 6004 + 8956$

14. $\begin{array}{r} 91,426 \\ +\ \ 7,495 \\ \hline \end{array}$

15. Write a related addition sentence: [1.3a]
$10 - 6 = 4.$

16. Write two related subtraction sentences: [1.3a]
$8 + 3 = 11.$

Subtract. [1.3b]

17. $8045 - 2897$

18. $9001 - 7312$

19. $6003 - 3729$

20. $\begin{array}{r} 37,405 \\ -\ 19,648 \\ \hline \end{array}$

Round 345,759 to the nearest: [1.4a]

21. Hundred.

22. Ten.

23. Thousand.

24. Hundred thousand.

Estimate the sum, difference, or product by first rounding to the nearest hundred. Show your work. [1.4b], [1.5b]

25. $41,348 + 19,749$

26. $38,652 - 24,549$

27. $396 \cdot 748$

Use $<$ or $>$ for $\square$ to write a true sentence. [1.4c]

28. $67 \square 56$

29. $1 \square 23$

Multiply. [1.5a]

30. $17,000 \cdot 300$

31. $7846 \cdot 800$

32. $726 \cdot 698$

33. $587 \cdot 47$

34. $\begin{array}{r} 8305 \\ \times\ \ 642 \\ \hline \end{array}$

35. Write a related multiplication sentence: [1.6a]
$56 \div 8 = 7.$

36. Write two related division sentences: [1.6a]
$13 \cdot 4 = 52.$

Divide. [1.6b]

37. $63 \div 5$

38. $80 \div 16$

39. $7\overline{)6394}$

40. $3073 \div 8$

41. $60\overline{)286}$

42. $4266 \div 79$

43. $38\overline{)17,176}$

44. $14\overline{)70,112}$

45. $52,668 \div 12$

Solve. [1.7b]

46. $46 \cdot n = 368$

47. $47 + x = 92$

48. $1 \cdot y = 58$

49. $24 = x + 24$

50. Write exponential notation: $4 \cdot 4 \cdot 4$. [1.9a]

Evaluate. [1.9b]

51. 10^4

52. 6^2

Simplify. [1.9c, d]

53. $8 \cdot 6 + 17$

54. $10 \cdot 24 - (18 + 2) \div 4 - (9 - 7)$

55. $7 + (4 + 3)^2$

56. $7 + 4^2 + 3^2$

57. $(80 \div 16) \times [(20 - 56 \div 8) + (8 \cdot 8 - 5 \cdot 5)]$

58. Find the average of 157, 170, and 168.

Solve. [1.8a]

59. *Computer Workstation.* Natasha has $196 and wants to buy a computer workstation for $698. How much more does she need?

Computer Workstation

Raised monitor platform, sliding keyboard shelf, and mobile CPU shelf. Locking 3-drawer file cabinet. Maple and honey finish with durable melamine work surface.

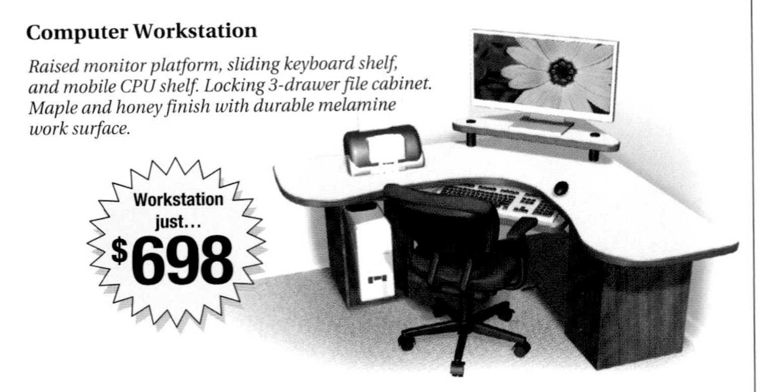

60. Toni has $406 in her checking account. She is paid $78 for a part-time job and deposits that in her checking account. How much is then in her account?

61. *Lincoln-Head Pennies.* In 1909, the first Lincoln-head pennies were minted. Seventy-three years later, these pennies were first minted with a decreased copper content. In what year was the copper content reduced?

62. A beverage company packed 228 cans of soda into 12-can cartons. How many cartons did they fill?

63. An apple farmer keeps bees in her orchard to help pollinate the apple blossoms so more apples will be produced. The bees from an average beehive can pollinate 30 surrounding trees during one growing season. A farmer has 420 trees. How many beehives does she need to pollinate them all?

Source: Jordan Orchards, Westminster, PA

64. An apartment builder bought 13 gas stoves at $425 each and 13 refrigerators at $620 each. What was the total cost?

65. A family budgets $7825 for food and clothing and $2860 for entertainment. The yearly income of the family was $38,283. How much of this income remained after these two allotments?

66. A chemist has 2753 mL of alcohol. How many 20-mL beakers can be filled? How much will be left over?

67. *Olympic Trampoline.* Shown below is an Olympic trampoline. Find the area and the perimeter of the trampoline. [1.2b], [1.5c]

Source: International Trampoline Industry Association, Inc.

68. D_W Write a problem for a classmate to solve. Design the problem so that the solution is "Each of the 144 bottles will contain 8 oz of hot sauce." [1.8a]

69. D_W Is subtraction associative? Why or why not? [1.3b]

SYNTHESIS

70. Determine the missing digit d. [1.5a]

$$\begin{array}{r} 9\ d \\ \times\ \ d\ 2 \\ \hline 8\ 0\ 3\ 6 \end{array}$$

71. Determine the missing digits a and b. [1.6b]

$$\begin{array}{r} 9\ a\ 1 \\ 2\ b\ 1\,\overline{)\,2\ 3\ 6{,}4\ 2\ 1} \end{array}$$

72. A mining company estimates that a crew must tunnel 2000 ft into a mountain to reach a deposit of copper ore. Each day the crew tunnels about 500 ft. Each night about 200 ft of loose rocks roll back into the tunnel. How many days will it take the mining company to reach the copper deposit? [1.8a]

1 Chapter Test

For Extra Help

Work It Out! Chapter Test Video on CD

1. In the number 546,789, which digit tells the number of hundred thousands?

2. Write expanded notation: 8843.

3. Write a word name: 38,403,277.

Add.

4. $\begin{array}{r} 6811 \\ +\ 3178 \\ \hline \end{array}$

5. $\begin{array}{r} 45{,}889 \\ +\ 17{,}902 \\ \hline \end{array}$

6. $\begin{array}{r} 12 \\ 8 \\ 3 \\ 7 \\ +\ 4 \\ \hline \end{array}$

7. $\begin{array}{r} 6203 \\ +\ 4312 \\ \hline \end{array}$

Subtract.

8. $\begin{array}{r} 7983 \\ -\ 4353 \\ \hline \end{array}$

9. $\begin{array}{r} 2974 \\ -\ 1935 \\ \hline \end{array}$

10. $\begin{array}{r} 8907 \\ -\ 2059 \\ \hline \end{array}$

11. $\begin{array}{r} 23{,}067 \\ -\ 17{,}892 \\ \hline \end{array}$

Multiply.

12. $\begin{array}{r} 4568 \\ \times\ 9 \\ \hline \end{array}$

13. $\begin{array}{r} 8876 \\ \times\ 600 \\ \hline \end{array}$

14. $\begin{array}{r} 65 \\ \times\ 37 \\ \hline \end{array}$

15. $\begin{array}{r} 678 \\ \times\ 788 \\ \hline \end{array}$

Divide.

16. $15 \div 4$

17. $420 \div 6$

18. $89 \overline{)8633}$

19. $44 \overline{)35{,}428}$

Solve.

20. *Hostess Ding Dongs®.* Hostess packages its Ding Dong® snack products in 12-packs. It manufactures 22,231 cakes. How many 12-packs can it fill? How many will be left over?

21. *Largest States.* The following table lists the five largest states in terms of their land area. Find the total land area of these states.

STATE	AREA (In Square Miles)
Alaska	571,951
Texas	261,797
California	155,959
Montana	145,552
New Mexico	121,356

Source: Department of Commerce, Bureau of the Census

22. *Pool Tables.* The Hartford™ pool table made by Brunswick Billiards comes in three sizes of playing area, 50 in. by 100 in., 44 in. by 88 in., and 38 in. by 76 in.

a) Find the perimeter and the area of the playing area of each table.

b) By how much area does the large table exceed the small table?

Source: Brunswick Billiards

23. *Voting Early.* In the 2004 Presidential Election, 345,689 Nevada voters voted early. This was 139,359 more than in the 2000 election. How many voted early in 2000?

Source: National Association of Secretaries of State

24. A sack of oranges weighs 27 lb. A sack of apples weighs 32 lb. Find the total weight of 16 bags of oranges and 43 bags of apples.

25. A box contains 5000 staples. How many staplers can be filled from the box if each stapler holds 250 staples?

Solve.

26. $28 + x = 74$

27. $169 \div 13 = n$

28. $38 \cdot y = 532$

29. $381 = 0 + a$

Round 34,578 to the nearest:

30. Thousand.

31. Ten.

32. Hundred.

Estimate the sum, difference, or product by first rounding to the nearest hundred. Show your work.

33. $\begin{array}{r} 23{,}649 \\ +\ 54{,}746 \\ \hline \end{array}$

34. $\begin{array}{r} 54{,}751 \\ -\ 23{,}649 \\ \hline \end{array}$

35. $\begin{array}{r} 824 \\ \times\ 489 \\ \hline \end{array}$

Use $<$ or $>$ for $\square$ to write a true sentence.

36. $34 \ \square\ 17$

37. $117 \ \square\ 157$

38. Write exponential notation: $12 \cdot 12 \cdot 12 \cdot 12$.

Evaluate.

39. 7^3

40. 10^5

41. 25^2

Simplify.

42. $35 - 1 \cdot 28 \div 4 + 3$

43. $10^2 - 2^2 \div 2$

44. $(25 - 15) \div 5$

45. $8 \times \{(20 - 11) \cdot [(12 + 48) \div 6 - (9 - 2)]\}$

46. $2^4 + 24 \div 12$

47. Find the average of 97, 98, 87, and 86.

SYNTHESIS

48. An open cardboard container is 8 in. wide, 12 in. long, and 6 in. high. How many square inches of cardboard are used?

49. Use trials to find the single-digit number a for which

$$359 - 46 + a \div 3 \times 25 - 7^2 = 339.$$

50. Cara spends \$229 a month to repay her student loan. If she has already paid \$9160 on the 10-yr loan, how many payments remain?

Fraction Notation: Multiplication and Division

2

Real-World Application

The length of a rectangular ice-skating rink in the atrium of a shopping mall is $\frac{7}{100}$ mi. The width is $\frac{3}{100}$ mi. What is the area of the skating rink?

This problem appears as Example 10 in Section 2.4

2.1 FACTORIZATIONS

Objectives

a Determine whether one number is a factor of another, and find the factors of a number.

b Find some multiples of a number, and determine whether a number is divisible by another.

c Given a number from 1 to 100, tell whether it is prime, composite, or neither.

d Find the prime factorization of a composite number.

> To the student:
>
> In the preface, at the front of the text, you will find a Student Organizer card. This pullout card will help you keep track of important dates and useful contact information. You can also use it to plan time for class, study, work, and relaxation. By managing your time wisely, you will provide yourself the best possible opportunity to be successful in this course.

In this chapter, we begin our work with fractions and fraction notation. Certain skills make such work easier. For example, in order to simplify $\frac{12}{32}$, it is important that we be able to *factor* 12 and 32, as follows:

$$\frac{12}{32} = \frac{4 \cdot 3}{4 \cdot 8}.$$

Then we "remove" a factor of 1:

$$\frac{4 \cdot 3}{4 \cdot 8} = \frac{4}{4} \cdot \frac{3}{8} = 1 \cdot \frac{3}{8} = \frac{3}{8}.$$

Thus factoring is an important skill in working with fractions.

a Factors and Factorization

In Sections 2.1 and 2.2, we consider only the **natural numbers** 1, 2, 3, and so on.

Let's look at the product $3 \cdot 4 = 12$. We say that 3 and 4 are **factors** of 12. When we divide 12 by 3, we get a remainder of 0. We say that the divisor 3 is a **factor** of the dividend 12.

FACTOR

- In the product $a \cdot b$, a and b are called **factors.**
- If we divide Q by d and get a remainder of 0, then the divisor d is a **factor** of the dividend Q.

EXAMPLE 1 Determine by long division whether 6 is a factor of 72.

$$\begin{array}{r} 12 \\ 6\overline{)72} \\ \underline{60} \\ 12 \\ \underline{12} \\ 0 \end{array}$$

The remainder is 0, so 6 is a factor of 72. We sometimes say that 6 divides 72 "evenly" because there is a remainder of 0.

EXAMPLE 2 Determine by long division whether 15 is a factor of 7894.

$$\begin{array}{r} 526 \\ 15\overline{)7894} \\ \underline{7500} \\ 394 \\ \underline{300} \\ 94 \\ \underline{90} \\ 4 \end{array} \leftarrow \text{Not } 0$$

The remainder is *not* 0, so 15 is not a factor of 7894.

Do Exercises 1 and 2 on the following page.

Consider $12 = 3 \cdot 4$. We say that $3 \cdot 4$ is a **factorization** of 12. Similarly, $6 \cdot 2$, $12 \cdot 1$, $2 \cdot 2 \cdot 3$, and $1 \cdot 3 \cdot 4$ are also factorizations of 12. Since $a = a \cdot 1$, every number has a factorization, and every number has factors. For some numbers, the factors consist of only the number itself and 1. For example, the only factorization of 17 is $17 \cdot 1$, so the only factors of 17 are 17 and 1.

EXAMPLE 3 Find all the factors of 70.

We find as many "two-factor" factorizations as we can. We check sequentially the numbers 1, 2, 3, and so on, to see if we can form any factorizations:

70

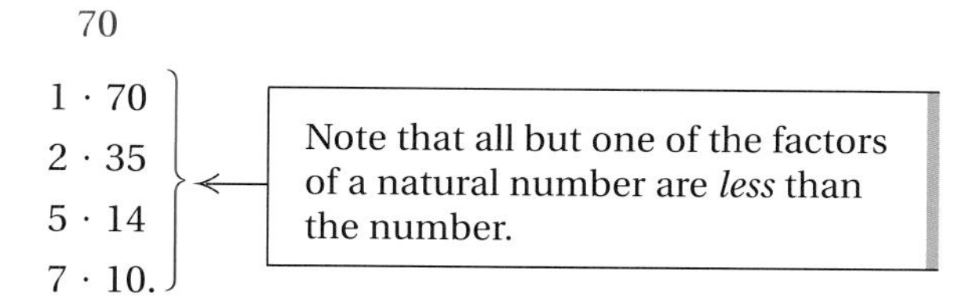

1 · 70
2 · 35
5 · 14
7 · 10.
Note that all but one of the factors of a natural number are *less* than the number.

Note that 3, 4, and 6 are not factors. If there are additional factors, they must be between 7 and 10. Since 8 and 9 are not factors, we are finished. The factors of 70 are 1, 2, 5, 7, 10, 14, 35, and 70.

Do Exercises 3–6.

b Multiples and Divisibility

A **multiple** of a natural number is a product of that number and some natural number. For example, some multiples of 2 are:

2 (because $2 = 1 \cdot 2$);
4 (because $4 = 2 \cdot 2$);
6 (because $6 = 3 \cdot 2$);
8 (because $8 = 4 \cdot 2$);
10 (because $10 = 5 \cdot 2$).

Note that all but one of the multiples of a number are *larger* than the number.

We find multiples of 2 by counting by twos: 2, 4, 6, 8, and so on. We can find multiples of 3 by counting by threes: 3, 6, 9, 12, and so on.

EXAMPLE 4 Show that each of the numbers 8, 12, 20, and 36 is a multiple of 4.

$8 = 2 \cdot 4 \qquad 12 = 3 \cdot 4 \qquad 20 = 5 \cdot 4 \qquad 36 = 9 \cdot 4$

Do Exercises 7 and 8.

EXAMPLE 5 Multiply by 1, 2, 3, and so on, to find ten multiples of 7.

$1 \cdot 7 = 7 \qquad 6 \cdot 7 = 42$
$2 \cdot 7 = 14 \qquad 7 \cdot 7 = 49$
$3 \cdot 7 = 21 \qquad 8 \cdot 7 = 56$
$4 \cdot 7 = 28 \qquad 9 \cdot 7 = 63$
$5 \cdot 7 = 35 \qquad 10 \cdot 7 = 70$

Do Exercise 9.

Determine whether the second number is a factor of the first.

1. 72; 8

2. 2384; 28

Find all the factors of the number.

3. 10

4. 45

5. 62

6. 24

7. Show that each of the numbers 5, 45, and 100 is a multiple of 5.

8. Show that each of the numbers 10, 60, and 110 is a multiple of 10.

9. Multiply by 1, 2, 3, and so on, to find ten multiples of 5.

Answers on page A-4

10. Determine whether 16 is divisible by 2.

11. Determine whether 125 is divisible by 5.

12. Determine whether 125 is divisible by 6.

Answers on page A-4

DIVISIBILITY

The number a is **divisible** by another number b if there exists a number c such that $a = b \cdot c$. The statements "a is **divisible** by b," "a is a **multiple** of b," and "b is a **factor** of a" all have the same meaning.

Thus 27 is *divisible* by 3 because $27 = 3 \cdot 9$. We can also say that 27 is a *multiple* of 3, and 3 is a *factor* of 27.

EXAMPLE 6 Determine whether 45 is divisible by 9.

We divide 45 by 9 to determine if 9 is a factor of 45. If so, then 45 is divisible by 9.

$$\begin{array}{r} 5 \\ 9\overline{)45} \\ \underline{45} \\ 0 \end{array}$$

Because the remainder is 0, 45 is divisible by 9.

EXAMPLE 7 Determine whether 98 is divisible by 4.

We divide 98 by 4:

$$\begin{array}{r} 24 \\ 4\overline{)98} \\ \underline{80} \\ 18 \\ \underline{16} \\ 2 \end{array} \leftarrow \text{Not } 0$$

Since the remainder is not 0, 98 is *not* divisible by 4.

Do Exercises 10–12.

CALCULATOR CORNER

Divisibility and Factors We can use a calculator to determine whether one number is divisible by another number or whether one number is a factor of another number. For example, to determine whether 387 is divisible by 18, we first press [3] [8] [7] [÷] [1] [8] [=]. The display reads [21.5]. Note that the result is not a natural number. (For a brief discussion of decimal notation, see the Calculator Corner on page 47. Decimal notation will be studied in detail in Chapter 4.) Thus we know that 387 is not a multiple of 18; that is, 387 is not divisible by 18 and 18 is not a factor of 387.

When we divide 387 by 9, the result is [43]. Since 43 is a natural number, we know that 387 is a multiple of 9; that is, $387 = 43 \cdot 9$. Thus, 387 is divisible by 9, and 9 is a factor of 387.

Exercises: For each pair of numbers, determine whether the first number is divisible by the second number.

1. 722; 19
2. 845; 7
3. 1047; 14
4. 5283; 9

For each pair of numbers, determine whether the second number is a factor of the first number.

5. 502; 8
6. 651; 21
7. 3875; 25
8. 8464; 12
9. 32,768; 256
10. 32,768; 864

c Prime and Composite Numbers

PRIME AND COMPOSITE NUMBERS

- A natural number that has exactly two *different* factors, only itself and 1, is called a **prime number.**
- The number 1 is *not* prime.
- A natural number, other than 1, that is not prime is **composite.**

EXAMPLE 8 Determine whether the numbers 1, 2, 3, 4, 5, 6, 7, 9, 10, 11, and 63 are prime, composite, or neither.

The number 1 is not prime. It does not have *two* different factors.

The number 2 is prime. It has only the factors 2 and 1.

The numbers 3, 5, 7, and 11 are prime. Each has only two factors, itself and 1.

The number 4 is not prime. It has the factors 1, 2, and 4 and is composite.

The numbers 6, 9, 10, and 63 are composite. Each has more than two factors.

Thus we have:

Prime: 2, 3, 5, 7, 11;
Composite: 4, 6, 9, 10, 63;
Neither: 1.

The number 2 is the *only* even prime number. It is also the smallest prime number. The number 0 is also neither prime nor composite, but 0 is *not* a natural number and thus is not considered here. We are considering only natural numbers.

Do Exercise 13.

The following is a table of the prime numbers from 2 to 157. There are more extensive tables, but these prime numbers will be the most helpful to you in this text.

A TABLE OF PRIMES FROM 2 TO 157

2, 3, 5, 7, 11, 13, 17, 19, 23, 29, 31, 37, 41, 43, 47, 53, 59, 61, 67, 71, 73, 79, 83, 89, 97, 101, 103, 107, 109, 113, 127, 131, 137, 139, 149, 151, 157

d Prime Factorizations

When we factor a composite number into a product of primes, we find the **prime factorization** of the number. To do this, we consider the primes

2, 3, 5, 7, 11, 13, 17, 19, 23, and so on,

and determine whether a given number is divisible by the primes.

13. Tell whether each number is prime, composite, or neither.

1, 2, 6, 12, 13, 19, 41, 65, 73, 99

Answer on page A-4

Study Tips

STUDY TIPS REVIEW

The following Study Tips were presented in Chapter 1. If you have not studied them, go back and do so now.

- Using This Textbook (Section 1.2)
- Learning Resources (Section 1.3)
- Time Management (Part 1) (Section 1.5)
- Highlighting (Section 1.6)
- Exercises (Section 1.6)
- Test Preparation (Section 1.9)

EXAMPLE 9 Find the prime factorization of 39.

a) We divide by the first prime, 2.

$$\begin{array}{r} 19 \quad R = 1 \\ 2\overline{)39} \\ \underline{38} \\ 1 \end{array}$$

Because the remainder is not 0, 2 is not a factor of 39, and 39 is not divisible by 2.

b) We divide by the next prime, 3.

$$\begin{array}{r} 13 \quad R = 0 \\ 3\overline{)39} \end{array}$$

The remainder is 0, so we know that $39 = 3 \cdot 13$. Because 13 is a prime, we are finished. The prime factorization is

$$39 = 3 \cdot 13.$$

EXAMPLE 10 Find the prime factorization of 220.

a) We divide by the first prime, 2.

$$\begin{array}{r} 110 \quad R = 0 \\ 2\overline{)220} \end{array} \qquad 220 = 2 \cdot 110$$

b) Because 110 is composite, we start with 2 again.

$$\begin{array}{r} 55 \quad R = 0 \\ 2\overline{)110} \end{array} \qquad 220 = 2 \cdot 2 \cdot 55$$

c) Since 55 is composite and is not divisible by 2 or 3, we divide by the next prime, 5.

$$\begin{array}{r} 11 \quad R = 0 \\ 5\overline{)55} \end{array} \qquad 220 = 2 \cdot 2 \cdot 5 \cdot 11$$

Because 11 is prime, we are finished. The prime factorization is

$$220 = 2 \cdot 2 \cdot 5 \cdot 11.$$

We abbreviate our procedure as follows.

$$\begin{array}{r} 11 \\ 5\overline{)55} \\ 2\overline{)110} \\ 2\overline{)220} \end{array}$$

$$220 = 2 \cdot 2 \cdot 5 \cdot 11$$

Multiplication is commutative so a factorization such as $2 \cdot 2 \cdot 5 \cdot 11$ could also be expressed as $5 \cdot 2 \cdot 2 \cdot 11$ or $2 \cdot 5 \cdot 11 \cdot 2$ (or, in exponential notation, as $2^2 \cdot 5 \cdot 11$ or $11 \cdot 2^2 \cdot 5$), but the prime factors are the same. For this reason, we agree that any of these is "the" prime factorization of 220.

Every number has just one (unique) prime factorization.

EXAMPLE 11 Find the prime factorization of 72.

We can do divisions "up" as follows:

3 ← Prime quotient
3)9
2)18
2)36
2)72 ← Begin here.

$72 = 2 \cdot 2 \cdot 2 \cdot 3 \cdot 3$

Or, we can do divisions "down":

2)72 ← Begin here.
2)36
2)18
3)9
3 ← Prime quotient

Another way to find the prime factorization of 72 uses a **factor tree** as follows. Begin by determining any factorization you can, and then continue factoring.

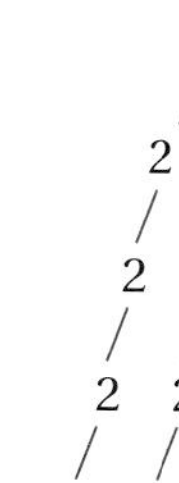

```
      72                72                 72
     /  \              /  \               /  \
    8    9            2    36            4    18
   / \   / \          |   /  \          / \   / \
  2   4 3   3         2  2    18       2   2 2   9
  |  / \ |   |        |  |   /  \      |   | |  / \
  2 2  2 3   3        2  2  2    9     2   2 2 3   3
                      |  |  |   / \
                      2  2  2  3   3
```

EXAMPLE 12 Find the prime factorization of 189.

We can use a string of successive divisions or a factor tree. Since 189 is not divisible by 2, we begin with 3.

7
3)21
3)63
3)189

$189 = 3 \cdot 3 \cdot 3 \cdot 7$

```
     189
    /   \
   3     63
   |    /  \
   3   7    9
   |   |   / \
   3   7  3   3
```

Caution!

Keep in mind the difference between finding all the factors of a number and finding the prime factorization. In Example 12, the prime factorization is $3 \cdot 3 \cdot 3 \cdot 7$. The factors of 189 are 1, 3, 7, 9, 21, 27, 63, and 189.

EXAMPLE 13 Find the prime factorization of 1638.

We can use a string of successive divisions.

13
7)91
3)273
3)819
2)1638

$1638 = 2 \cdot 3 \cdot 3 \cdot 7 \cdot 13$

Do Exercises 14–21.

Find the prime factorization of the number.

14. 6

15. 12

16. 45

17. 98

18. 126

19. 144

20. 1960

21. 1925

Answers on page A-4

a Determine whether the second number is a factor of the first.

1. 52; 14

2. 52; 13

3. 625; 25

4. 680; 16

Find all the factors of the number.

5. 18

6. 16

7. 54

8. 48

9. 4

10. 9

11. 1

12. 13

13. 98

14. 100

15. 255

16. 120

b Multiply by 1, 2, 3, and so on, to find ten multiples of the number.

17. 4

18. 11

19. 20

20. 50

21. 3

22. 5

23. 12

24. 13

25. 10

26. 6

27. 9

28. 14

29. Determine whether 26 is divisible by 6.

30. Determine whether 48 is divisible by 8.

31. Determine whether 1880 is divisible by 8.

32. Determine whether 4227 is divisible by 3.

33. Determine whether 256 is divisible by 16.

34. Determine whether 102 is divisible by 4.

35. Determine whether 4227 is divisible by 9.

36. Determine whether 200 is divisible by 25.

37. Determine whether 8650 is divisible by 16.

38. Determine whether 4143 is divisible by 7.

c Determine whether the number is prime, composite, or neither.

39. 1

40. 2

41. 9

42. 19

43. 11

44. 27

45. 29

46. 49

d Find the prime factorization of the number.

47. 8 **48.** 16 **49.** 14 **50.** 15 **51.** 42

52. 32 **53.** 25 **54.** 40 **55.** 50 **56.** 62

57. 169 **58.** 140 **59.** 100 **60.** 110 **61.** 35

62. 70 **63.** 72 **64.** 86 **65.** 77 **66.** 99

67. 2884 **68.** 484 **69.** 51 **70.** 91 **71.** 1200

72. 1800 **73.** 273 **74.** 675 **75.** 1122 **76.** 6435

77. D_W Is every natural number a multiple of 1? Explain.

78. D_W Explain a method for finding a composite number that contains exactly two factors other than itself and 1.

SKILL MAINTENANCE

Multiply. [1.5a]

79. $2 \cdot 13$ **80.** $8 \cdot 32$ **81.** $17 \cdot 25$ **82.** $25 \cdot 168$

Divide. [1.6b]

83. $0 \div 22$ **84.** $22 \div 1$ **85.** $22 \div 22$ **86.** $66 \div 22$

Solve. [1.8a]

87. Find the total cost of 7 shirts at $48 each and 4 pairs of pants at $69 each.

88. Sandy can type 62 words per minute. How long will it take her to type 12,462 words?

SYNTHESIS

89. *Factors and Sums.* The top number in each column of the table below can be factored as a product of two numbers whose sum is the bottom number in the column. For example, in the first column, 56 has been factored as $7 \cdot 8$, and $7 + 8 = 15$. Fill in the blank spaces in the table.

Product	56	63	36	72	140	96		168	110			
Factor	7									9	24	3
Factor	8						8	8		10	18	
Sum	15	16	20	38	24	20	14		21			24

2.2 DIVISIBILITY

Objective

a Determine whether a number is divisible by 2, 3, 4, 5, 6, 8, 9, or 10.

Suppose you are asked to find the simplest fraction notation for

$$\frac{117}{225}.$$

Since the numbers are quite large, you might feel that the task is difficult. However, both the numerator and the denominator are divisible by 9. If you knew this, you could factor and simplify quickly as follows:

$$\frac{117}{225} = \frac{9 \cdot 13}{9 \cdot 25} = \frac{9}{9} \cdot \frac{13}{25} = 1 \cdot \frac{13}{25} = \frac{13}{25}.$$

How did we know that both numbers have 9 as a factor? There is a simple test for determining this.

In this section, we learn quick ways to determine whether a number is divisible by 2, 3, 4, 5, 6, 8, 9, or 10. This will make simplifying fraction notation much easier.

a Rules for Divisibility

DIVISIBILITY BY 2

You may already know the test for divisibility by 2.

> **BY 2**
>
> A number is **divisible by 2** (is *even*) if it has a ones digit of 0, 2, 4, 6, or 8 (that is, it has an even ones digit).

Let's see why. Consider 354, which is

3 hundreds + 5 tens + 4.

Hundreds and tens are both multiples of 2. If the last digit is a multiple of 2, then the entire number is a multiple of 2.

EXAMPLES Determine whether the number is divisible by 2.

1. 355 is *not* a multiple of 2; 5 is *not* even.
2. 4786 is a multiple of 2; 6 is even.
3. 8990 is a multiple of 2; 0 is even.
4. 4261 is *not* a multiple of 2; 1 is *not* even.

Do Exercises 1–4.

Determine whether the number is divisible by 2.

1. 84

2. 59

3. 998

4. 2225

Answers on page A-4

DIVISIBILITY BY 3

BY 3

A number is **divisible by 3** if the sum of its digits is divisible by 3.

Let's illustrate why the test for divisibility by 3 works. Consider 852; since $852 = 3 \cdot 284$, 852 is divisible by 3.

$$
\begin{aligned}
852 &= 8 \cdot 100 + 5 \cdot 10 + 2 \cdot 1 \\
&= 8(99 + 1) + 5(9 + 1) + 2(1) \\
&= 8 \cdot 99 + 8 \cdot 1 + 5 \cdot 9 + 5 \cdot 1 + 2 \cdot 1
\end{aligned}
$$

Using the Distributive Law, $a(b + c) = a \cdot b + a \cdot c$

Since 99 and 9 are each a multiple of 3, we see that $8 \cdot 99$ and $5 \cdot 9$ are multiples of 3. This leaves $8 \cdot 1 + 5 \cdot 1 + 2 \cdot 1$, or $8 + 5 + 2$. If $8 + 5 + 2$, the sum of the digits, is divisible by 3, then 852 is divisible by 3.

EXAMPLES Determine whether the number is divisible by 3.

5. 18 $1 + 8 = 9$

6. 93 $9 + 3 = 12$

7. 201 $2 + 0 + 1 = 3$

Each is divisible by 3 because the sum of its digits is divisible by 3.

8. 256 $2 + 5 + 6 = 13$

The sum of the digits, 13, is *not* divisible by 3, so 256 is *not* divisible by 3.

Do Exercises 5–8.

DIVISIBILITY BY 6

A number divisible by 6 is a multiple of 6. But $6 = 2 \cdot 3$, so the number is also a multiple of 2 and 3. Thus we have the following.

BY 6

A number is **divisible by 6** if its ones digit is 0, 2, 4, 6, or 8 (is even) and the sum of its digits is divisible by 3.

EXAMPLES Determine whether the number is divisible by 6.

9. 720

Because 720 is even, it is divisible by 2. Also, $7 + 2 + 0 = 9$, so 720 is divisible by 3. Thus, 720 is divisible by 6.

720 ↑ Even $7 + 2 + 0 = 9$ ↑ Divisible by 3

10. 73

73 is *not* divisible by 6 because it is *not* even.

11. 256

Although 256 is even, it is *not* divisible by 6 because the sum of its digits, $2 + 5 + 6$, or 13, is *not* divisible by 3.

Do Exercises 9–12.

Determine whether the number is divisible by 3.

5. 111

6. 1111

7. 309

8. 17,216

Determine whether the number is divisible by 6.

9. 420

10. 106

11. 321

12. 444

Answers on page A-4

DIVISIBILITY BY 9

The test for divisibility by 9 is similar to the test for divisibility by 3.

BY 9

A number is **divisible by 9** if the sum of its digits is divisible by 9.

EXAMPLES Determine whether the number is divisible by 9.

12. 6984

Because $6 + 9 + 8 + 4 = 27$ and 27 is divisible by 9, 6984 is divisible by 9.

13. 322

Because $3 + 2 + 2 = 7$ and 7 is *not* divisible by 9, 322 is *not* divisible by 9.

Do Exercises 13–16.

Determine whether the number is divisible by 9.

13. 16

14. 117

15. 930

16. 29,223

DIVISIBILITY BY 10

BY 10

A number is **divisible by 10** if its ones digit is 0.

We know that this test works because the product of 10 and *any* number has a ones digit of 0.

EXAMPLES Determine whether the number is divisible by 10.

14. 3440 is divisible by 10 because the ones digit is 0.

15. 3447 is *not* divisible by 10 because the ones digit is not 0.

Do Exercises 17–20.

Determine whether the number is divisible by 10.

17. 305

18. 300

19. 847

20. 8760

DIVISIBILITY BY 5

BY 5

A number is **divisible by 5** if its ones digit is 0 or 5.

EXAMPLES Determine whether the number is divisible by 5.

16. 220 is divisible by 5 because the ones digit is 0.

17. 475 is divisible by 5 because the ones digit is 5.

18. 6514 is *not* divisible by 5 because the ones digit is neither 0 nor 5.

Do Exercises 21–24.

Let's see why the test for 5 works. Consider 7830:

$$7830 = 10 \cdot 783 = 5 \cdot 2 \cdot 783.$$

Since 7830 is divisible by 10 and 5 is a factor of 10, 7830 is divisible by 5.

Determine whether the number is divisible by 5.

21. 5780

22. 3427

23. 34,678

24. 7775

Answers on page A-4

Consider 6734:

6734 = 673 tens + 4.

Tens are multiples of 5, so the only number that must be checked is the ones digit. If the last digit is a multiple of 5, the entire number is. In this case, 4 is *not* a multiple of 5, so 6734 is *not* divisible by 5.

DIVISIBILITY BY 4

The test for divisibility by 4 is similar to the test for divisibility by 2.

BY 4

A number is **divisible by 4** if the number named by its last *two* digits is divisible by 4.

EXAMPLES Determine whether the number is divisible by 4.

19. 8212 is divisible by 4 because 12 is divisible by 4.

20. 5216 is divisible by 4 because 16 is divisible by 4.

21. 8211 is *not* divisible by 4 because 11 is *not* divisible by 4.

22. 7538 is *not* divisible by 4 because 38 is *not* divisible by 4.

Do Exercises 25–28.

To see why the test for divisibility by 4 works, consider 516:

516 = 5 hundreds + 16.

Hundreds are multiples of 4. If the number named by the last two digits is a multiple of 4, then the entire number is a multiple of 4.

DIVISIBILITY BY 8

The test for divisibility by 8 is an extension of the tests for divisibility by 2 and 4.

BY 8

A number is **divisible by 8** if the number named by its last *three* digits is divisible by 8.

EXAMPLES Determine whether the number is divisible by 8.

23. 5648 is divisible by 8 because 648 is divisible by 8.

24. 96,088 is divisible by 8 because 88 is divisible by 8.

25. 7324 is *not* divisible by 8 because 324 is *not* divisible by 8.

26. 13,420 is *not* divisible by 8 because 420 is *not* divisible by 8.

Do Exercises 29–32.

A NOTE ABOUT DIVISIBILITY BY 7

There are several tests for divisibility by 7, but all of them are more complicated than simply dividing by 7. So if you want to test for divisibility by 7, simply divide by 7, either by hand or using a calculator.

Determine whether the number is divisible by 4.

25. 216

26. 217

27. 5862

28. 23,524

Determine whether the number is divisible by 8.

29. 7564

30. 7864

31. 17,560

32. 25,716

Answers on page A-4

2.2 EXERCISE SET

For Extra Help

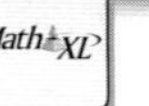

MathXL | MyMathLab | InterAct Math | Math Tutor Center | Digital Video Tutor CD 1 Videotape 2 | Student's Solutions Manual

a To answer Exercises 1–8, consider the following numbers.

46	300	85	256
224	36	711	8064
19	45,270	13,251	1867
555	4444	254,765	21,568

1. Which of the above are divisible by 2?

2. Which of the above are divisible by 3?

3. Which of the above are divisible by 4?

4. Which of the above are divisible by 5?

5. Which of the above are divisible by 6?

6. Which of the above are divisible by 8?

7. Which of the above are divisible by 9?

8. Which of the above are divisible by 10?

To answer Exercises 9–16, consider the following numbers.

56	200	75	35
324	42	812	402
784	501	2345	111,111
55,555	3009	2001	1005

9. Which of the above are divisible by 3?

10. Which of the above are divisible by 2?

11. Which of the above are divisible by 5?

12. Which of the above are divisible by 4?

13. Which of the above are divisible by 9?

14. Which of the above are divisible by 6?

15. Which of the above are divisible by 10?

16. Which of the above are divisible by 8?

To answer Exercises 17–24, consider the following numbers.

305	313,332	876	64,000
1101	7624	1110	9990
13,025	111,126	5128	126,111

17. Which of the above are divisible by 2?

18. Which of the above are divisible by 3?

19. Which of the above are divisible by 6?

20. Which of the above are divisible by 5?

21. Which of the above are divisible by 9?

22. Which of the above are divisible by 8?

23. Which of the above are divisible by 10?

24. Which of the above are divisible by 4?

25. D_W How can the divisibility tests be used to find prime factorizations?

26. D_W Which of the years from 2000 to 2020, if any, also happen to be prime numbers? Explain at least two ways in which you might go about solving this problem.

SKILL MAINTENANCE

Solve. [1.7b]

27. $56 + x = 194$

28. $y + 124 = 263$

29. $3008 = x + 2134$

30. $18 \cdot t = 1008$

31. $24 \cdot m = 624$

32. $338 = a \cdot 26$

Divide. [1.6b]

33. $2106 \div 9$

34. $45 \overline{)180{,}135}$

Solve. [1.8a]

35. An automobile with a 5-speed transmission gets 33 mpg in city driving. How many gallons of gas will it use to travel 1485 mi?

36. There are 60 min in 1 hr. How many minutes are there in 72 hr?

SYNTHESIS

Find the prime factorization of the number. Use divisibility tests where applicable.

37. 7800

38. 2520

39. 2772

40. 1998

41. Fill in the missing digits of the number

95, ☐☐ 8

so that it is divisible by 99.

42. A passenger in a taxicab asks for the driver's company number. The driver says abruptly, "Sure—you can have my number. Work it out: If you divide it by 2, 3, 4, 5, or 6, you will get a remainder of 1. If you divide it by 11, the remainder will be 0 and no driver has a company number that meets these requirements and is smaller than this one." Determine the number.

2.3 FRACTIONS AND FRACTION NOTATION

Objectives

a Identify the numerator and the denominator of a fraction and write fraction notation for part of an object or part of a set of objects and as a ratio.

b Simplify fraction notation like n/n to 1, $0/n$ to 0, and $n/1$ to n.

For each fraction, identify the numerator and the denominator.

1. $\dfrac{83}{100}$ of all scrap tires are reused or recycled.

Source: Rubber Manufacturers Association

2. $\dfrac{27}{50}$ of all kids prefer white bread.

Source: Bruce Horovitz, *USA Today*, 8/9/04

3. $\dfrac{11}{25}$ of all adults eat in restaurants on a given day.

Source: *AARP Bulletin*, November 2004

4. $\dfrac{21}{1000}$ of stay-at-home parents are men.

Source: U.S. Census Bureau

Answers on page A-4

The study of arithmetic begins with the set of whole numbers

0, 1, 2, 3, 4, 5, 6, 7, 8, 9, 10, 11, and so on.

But we also need to be able to use fractional parts of numbers such as halves, thirds, fourths, and so on. Here are some examples:

In 1950, about $\frac{3}{4}$ of the motor vehicles produced in the world were produced in the United States. In 2003, only $\frac{1}{5}$ were produced in the United States.

Source: American Automobile Manufacturers Association; Automotive News Data Center and Marketing Systems GmbH

More than $\frac{2}{5}$ of Americans are taking at least one prescription daily, and $\frac{1}{6}$ are taking three or more.

Source: Randolph Schmid, Associated Press, *Indianapolis Star*, 12/3/04

a Fractions and the Real World

Numbers like those above and the ones below are written in **fraction notation.** The top number is called the **numerator** and the bottom number is called the **denominator.**

$$\frac{1}{2},\quad \frac{3}{4},\quad \frac{8}{5},\quad \frac{11}{23}$$

EXAMPLE 1 Identify the numerator and the denominator.

$\dfrac{7}{8}$ 7 ← Numerator, 8 ← Denominator

Do Exercises 1–4.

Let's look at various situations that involve fractions.

FRACTIONS AS A PARTITION OF AN OBJECT DIVIDED INTO EQUAL PARTS

Consider a candy bar divided into 5 equal sections. If you eat 2 sections, you have eaten $\frac{2}{5}$ of the candy bar.

The denominator 5 tells us the unit, $\frac{1}{5}$. The numerator 2 tells us the number of equal parts we are considering, 2.

EXAMPLE 2 What part is shaded?

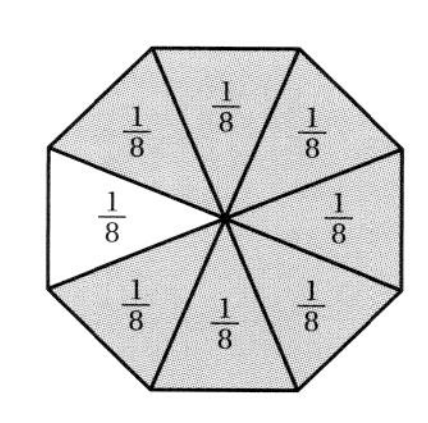

The equal parts are eighths. This tells us the unit, $\frac{1}{8}$. The *denominator* is 8. We have 7 of the units shaded. This tells us the *numerator*, 7. Thus,

$$\frac{7}{8} \begin{array}{l} \leftarrow \text{7 units are shaded.} \\ \leftarrow \text{The unit is } \frac{1}{8}. \end{array}$$

is shaded.

Do Exercises 5–9.

The markings on a ruler use fractions.

EXAMPLE 3 What part of an inch is highlighted?

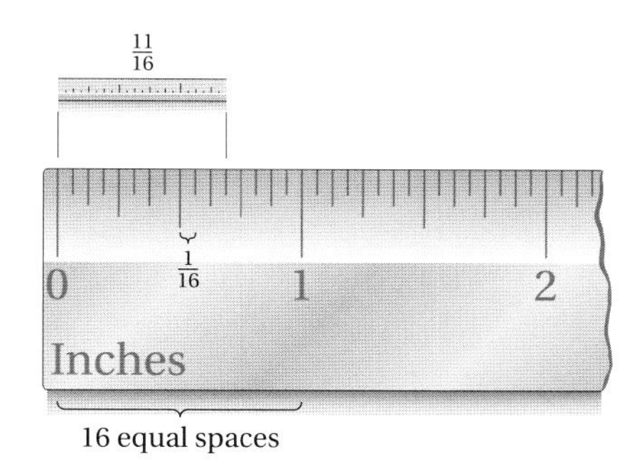

Each inch on the ruler shown above is divided into 16 equal parts. The highlighting extends to the 11th mark. Thus, $\frac{11}{16}$ of an inch is highlighted.

Do Exercise 10.

What part is shaded?

5.

6.

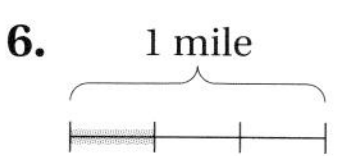

7.

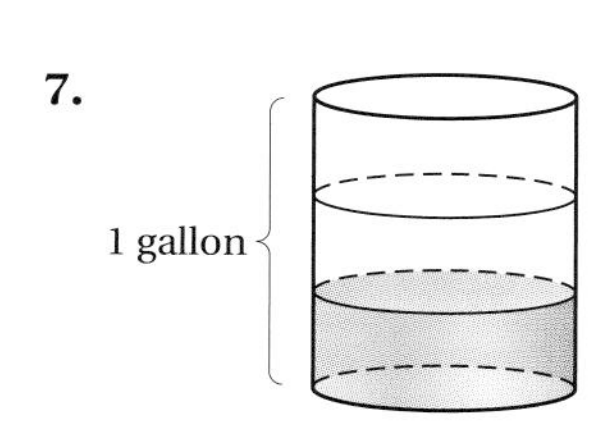

8.

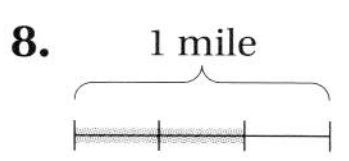

9.

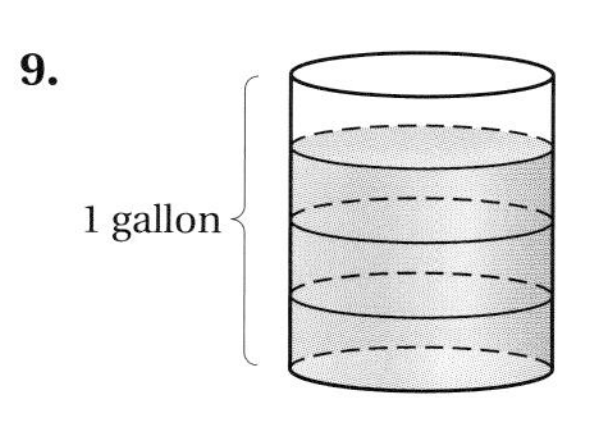

10. What part of an inch is highlighted?

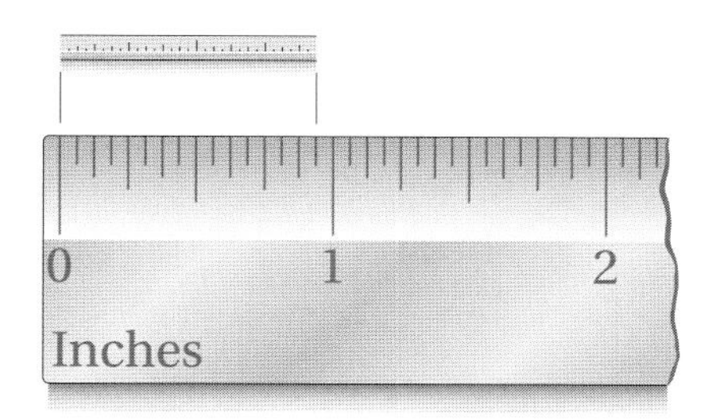

Answers on page A-4

11. What part is shaded?

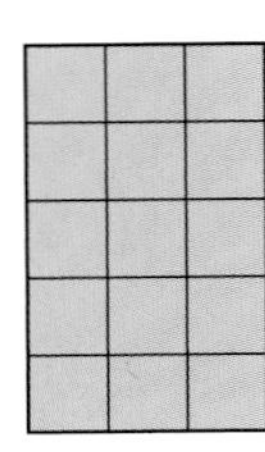

Fractions greater than or equal to 1, such as $\frac{24}{24}$, $\frac{10}{3}$, and $\frac{5}{4}$, correspond to situations like the following.

EXAMPLE 4 What part is shaded?

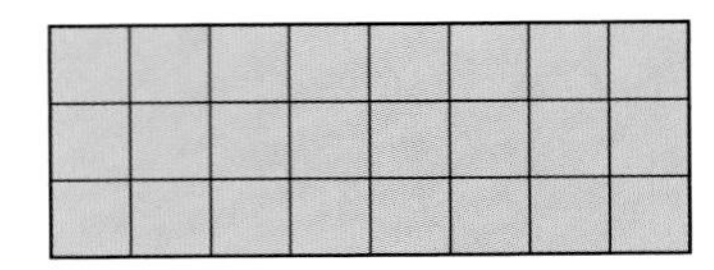

The rectangle is divided into 24 equal parts. The unit is $\frac{1}{24}$. The denominator is 24. All 24 equal parts are shaded. This tells us that the numerator is 24. Thus, $\frac{24}{24}$ are shaded.

Do Exercise 11.

What part is shaded?

12.

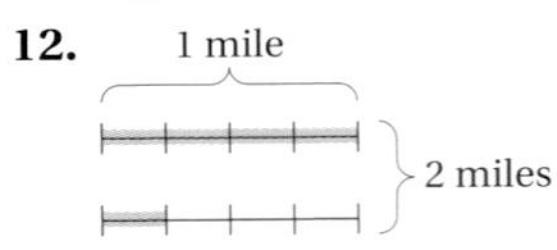

EXAMPLE 5 What part is shaded?

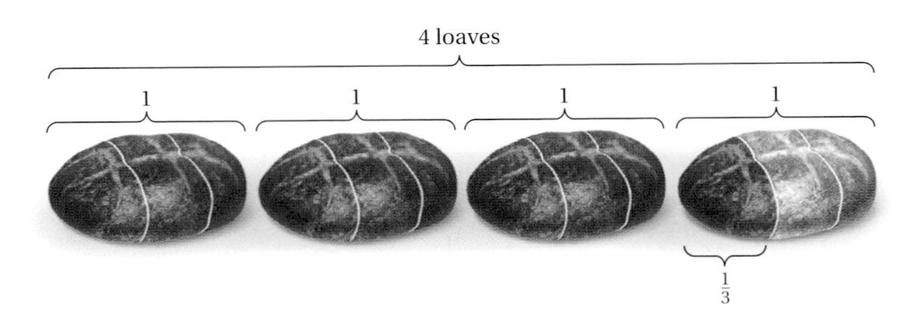

Each loaf of bread is divided into 3 equal parts. The unit is $\frac{1}{3}$. The *denominator* is 3. We have 10 of the units shaded. This tells us the *numerator* is 10. Thus, $\frac{10}{3}$ are shaded.

13.

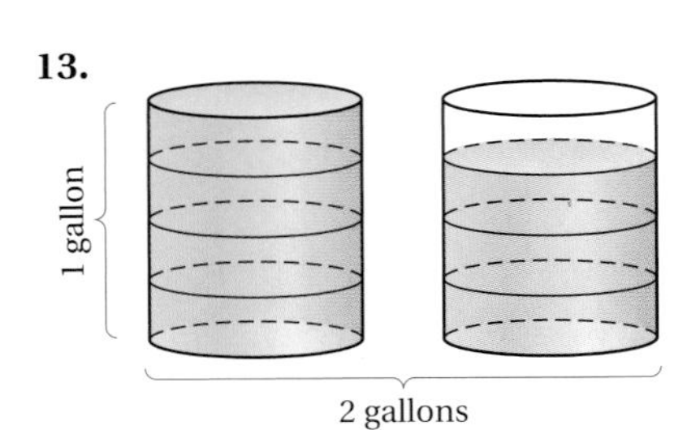

EXAMPLE 6 What part is shaded?

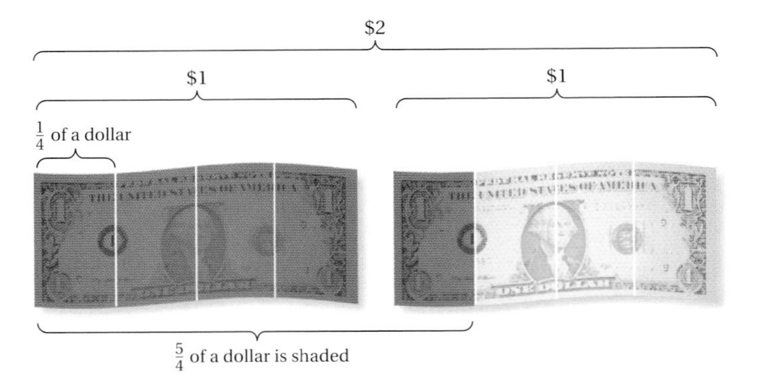

We can regard this as two objects of 4 parts each and take 5 of those parts. The unit is $\frac{1}{4}$. The denominator is 4, and the numerator is 5. Thus $\frac{5}{4}$ are shaded.

Do Exercises 12 and 13.

Fractions larger than or equal to 1, such as $\frac{13}{6}$ or $\frac{9}{9}$, are sometimes referred to as "improper" fractions. We will not use this terminology because notation such as $\frac{27}{8}$, $\frac{11}{3}$, and $\frac{4}{4}$ is quite "proper" and very common in algebra.

Answers on page A-4

FRACTIONS AS RATIOS

A **ratio** is a quotient of two quantities. We can express a ratio with fraction notation. (We will consider ratios in more detail in Chapter 5.)

EXAMPLE 7 What part of this set, or collection, of people are actresses who have won an Oscar? U.S. Senators?

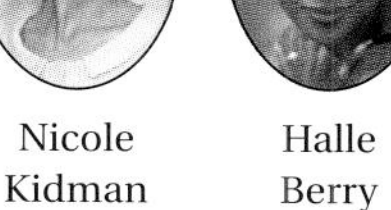

Nicole Kidman | Halle Berry | Elizabeth Dole | Hilary Swank | Gwyneth Paltrow | Julia Roberts | Diane Feinstein

There are 7 people in the set. We know that 5 of them, Nicole Kidman, Halle Berry, Julia Roberts, Hilary Swank, and Gwyneth Paltrow, are actresses who have won an Oscar. Thus, 5 of 7, or $\frac{5}{7}$, are Oscar-winning actresses. The 2 remaining are U.S. Senators. Thus, $\frac{2}{7}$ are U.S. Senators.

Do Exercise 14.

14. What part of this set, or collection, are clocks? thermometers?

EXAMPLE 8 *Baseball Standings.* The following are the final standings in the American League East for 2004, when the division was won by the New York Yankees. Find the ratio of Yankee wins to losses, wins to total games, and losses to total games.

EAST	W	L	Pct.	GB	STRK	LAST 10	vs. DIV.	HOME	AWAY	vs. NL
New York*	101	61	.623	—	W1	6-4	49-27	57-24	44-37	10-8
Boston	98	64	.605	3.0	L1	7-3	48-28	55-26	43-38	9-9
Baltimore	78	84	.481	23.0	W1	6-4	37-39	38-43	40-41	5-13
Tampa Bay	70	91	.435	30.5	W1	6-4	26-49	41-39	29-52	15-3
Toronto	67	94	.416	33.5	L1	4-6	29-46	40-41	27-53	8-10

*clinched division

Source: Major League Baseball

The Yankees won 101 games and lost 61 games. They played a total of 101 + 61, or 162 games. Thus we have the following.

The ratio of wins to losses is $\frac{101}{61}$.
The ratio of wins to total games is $\frac{101}{162}$.
The ratio of losses to total games is $\frac{61}{162}$.

Do Exercise 15.

15. Baseball Standings. Refer to the table in Example 8. The Boston Red Sox finished second in the American League East in 2004. Find the ratio of Red Sox wins to losses, wins to total games, and losses to total games.
Source: Major League Baseball

Answers on page A-4

Simplify.

16. $\frac{1}{1}$

17. $\frac{4}{4}$

18. $\frac{34}{34}$

19. $\frac{100}{100}$

20. $\frac{2347}{2347}$

21. $\frac{103}{103}$

Answers on page A-4

b Some Fraction Notation for Whole Numbers

FRACTION NOTATION FOR 1

The number 1 corresponds to situations like those shown here.

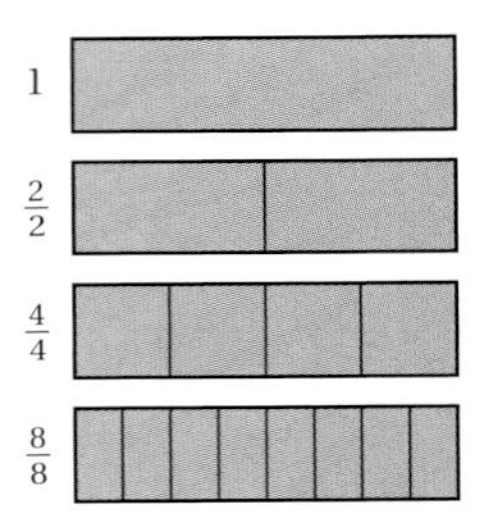

If we divide an object into n parts and take n of them, we get all of the object (1 whole object).

THE NUMBER 1 IN FRACTION NOTATION

$\frac{n}{n} = 1$, for any whole number n that is not 0.

EXAMPLES Simplify.

9. $\frac{5}{5} = 1$

10. $\frac{9}{9} = 1$

11. $\frac{23}{23} = 1$

Do Exercises 16–21.

FRACTION NOTATION FOR 0

Consider the fraction $\frac{0}{4}$. This corresponds to dividing an object into 4 parts and taking none of them. We get 0.

THE NUMBER 0 IN FRACTION NOTATION

$\frac{0}{n} = 0$, for any whole number n that is not 0.

EXAMPLES Simplify.

12. $\frac{0}{1} = 0$

13. $\frac{0}{9} = 0$

14. $\frac{0}{23} = 0$

Fraction notation with a denominator of 0, such as $n/0$, is meaningless because we cannot speak of an object being divided into *zero* parts. See also the discussion of excluding division by 0 in Section 1.6.

A DENOMINATOR OF 0

$\dfrac{n}{0}$ is not defined for any whole number n.

Do Exercises 22–27.

OTHER WHOLE NUMBERS

Consider the fraction $\frac{4}{1}$. This corresponds to taking 4 objects and dividing each into 1 part. (We do not divide them.) We have 4 objects.

$\frac{4}{1}$, or 4 objects

ANY WHOLE NUMBER IN FRACTION NOTATION

Any whole number divided by 1 is the whole number. That is,

$$\frac{n}{1} = n, \quad \text{for any whole number } n.$$

EXAMPLES Simplify.

15. $\dfrac{2}{1} = 2$

16. $\dfrac{9}{1} = 9$

17. $\dfrac{34}{1} = 34$

Do Exercises 28–31.

Simplify, if possible.

22. $\dfrac{0}{1}$

23. $\dfrac{0}{8}$

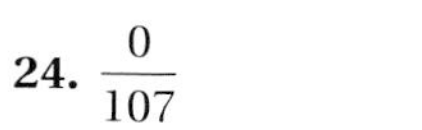

24. $\dfrac{0}{107}$

25. $\dfrac{4-4}{567}$

26. $\dfrac{15}{0}$

27. $\dfrac{0}{3-3}$

Simplify.

28. $\dfrac{8}{1}$

29. $\dfrac{10}{1}$

30. $\dfrac{346}{1}$

31. $\dfrac{24-1}{23-22}$

Answers on page A-4

2.3 EXERCISE SET

For Extra Help

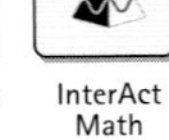

MathXL | MyMathLab | InterAct Math | Math Tutor Center | Digital Video Tutor CD 1 Videotape 2 | Student's Solutions Manual

a Identify the numerator and the denominator.

1. $\frac{3}{4}$

2. $\frac{9}{10}$

3. $\frac{11}{2}$

4. $\frac{18}{5}$

5. $\frac{0}{7}$

6. $\frac{1}{13}$

What part of the object or set of objects is shaded?

7.

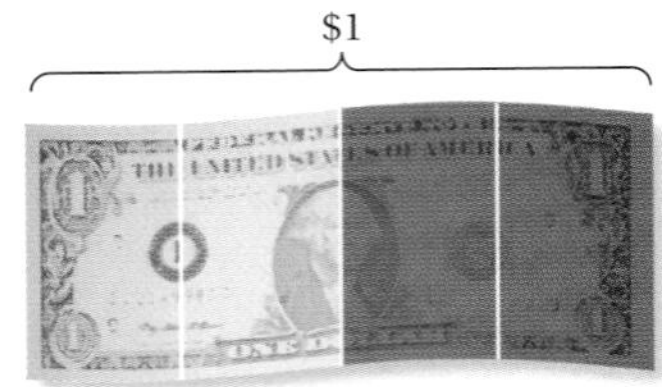

8.

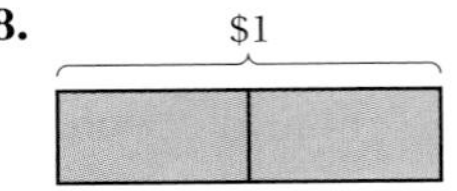

9.

1 yard

10.

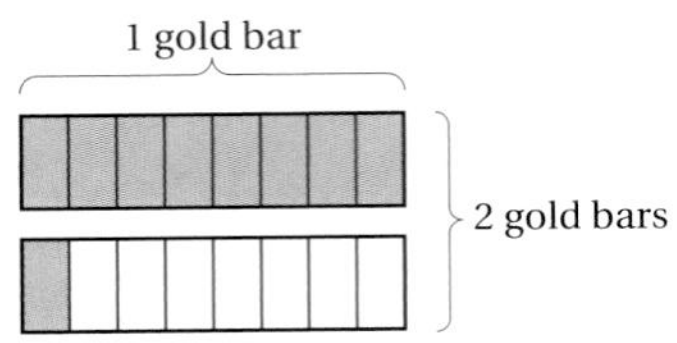

11.

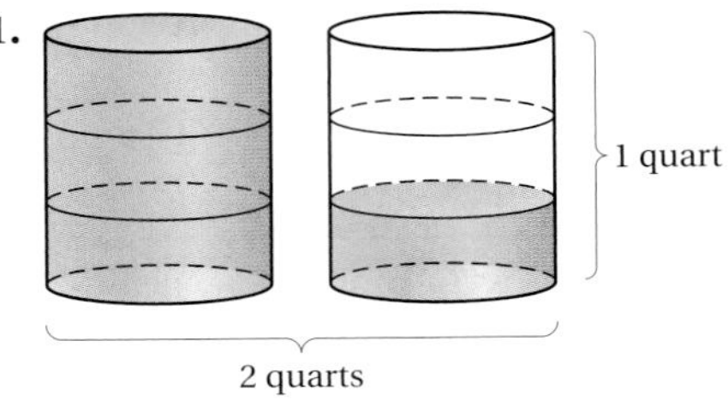

12.

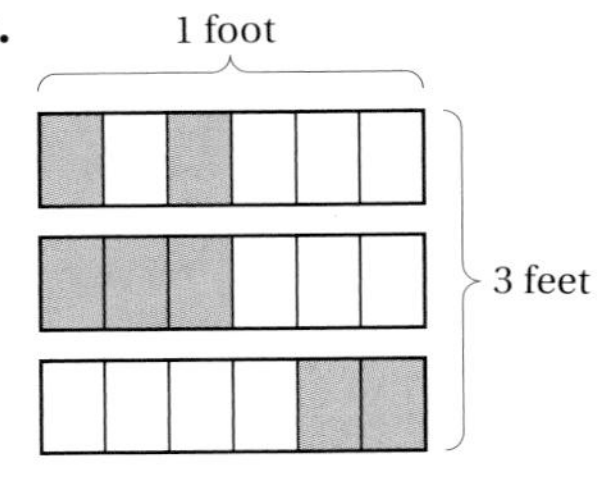

What part of an inch is highlighted?

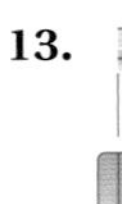

13.

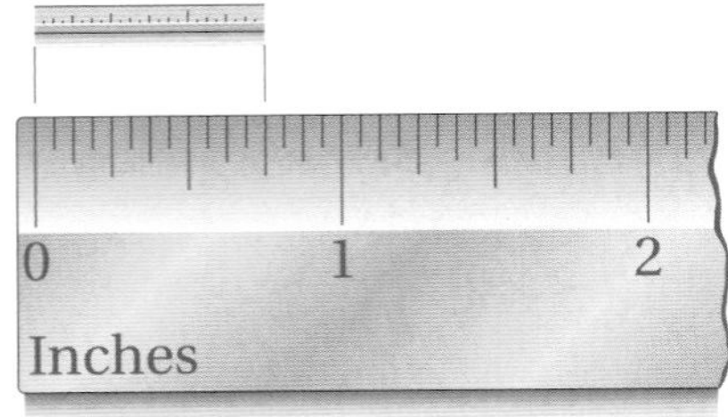

14.

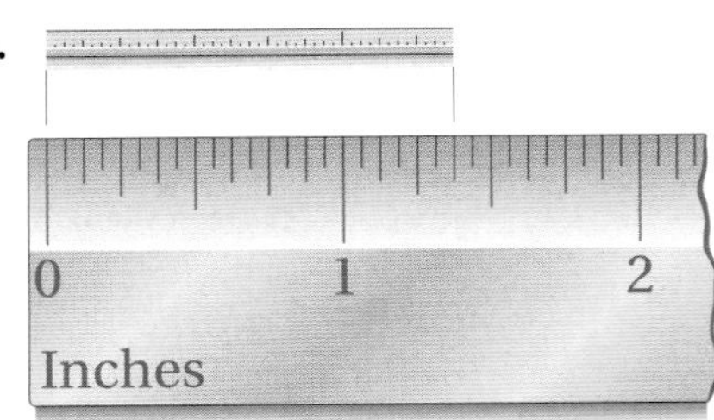

15.

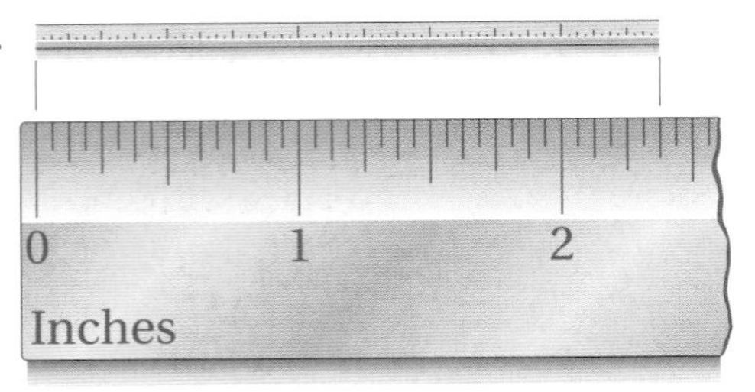

16.

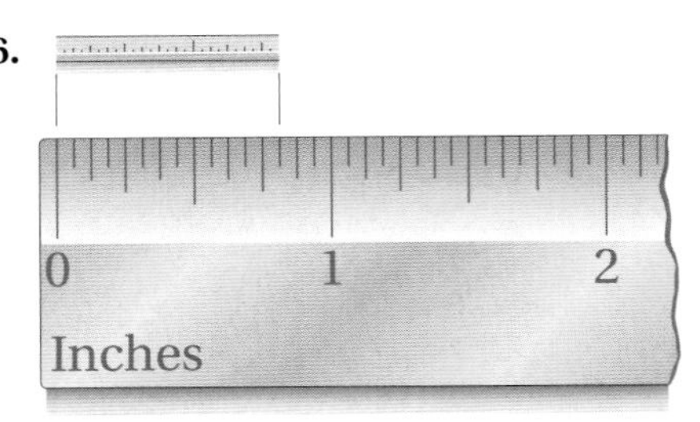

17. 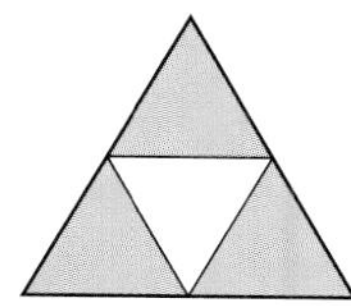

18. 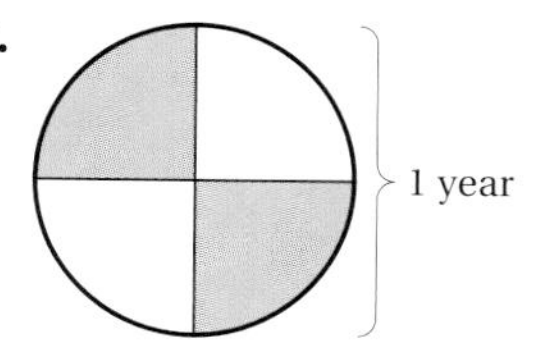

19. 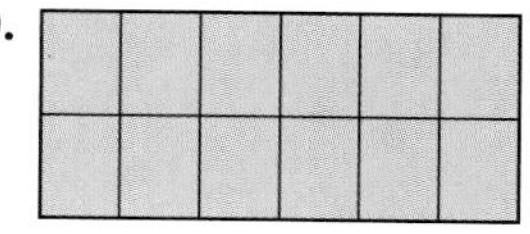

20. 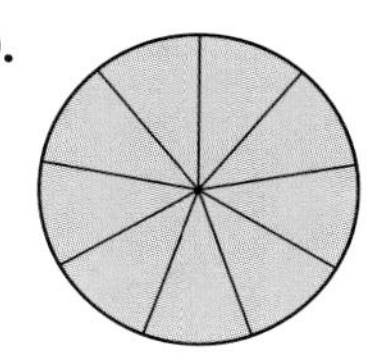

21. 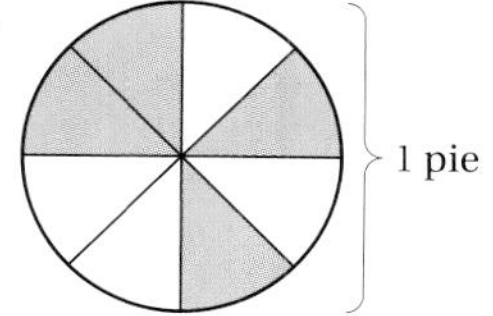

22. 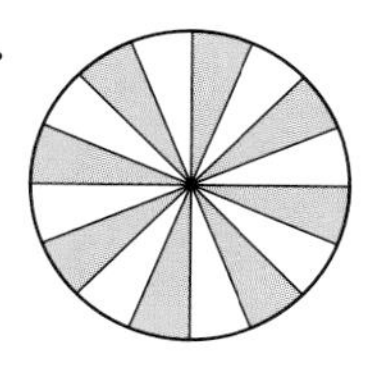

23. 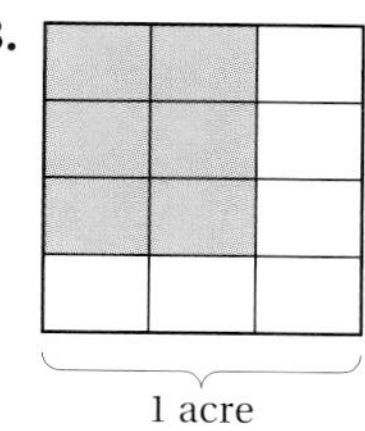

24. 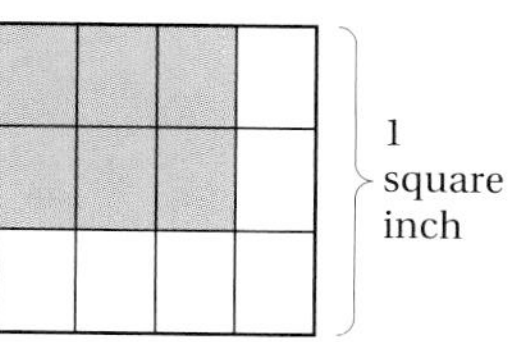

25. 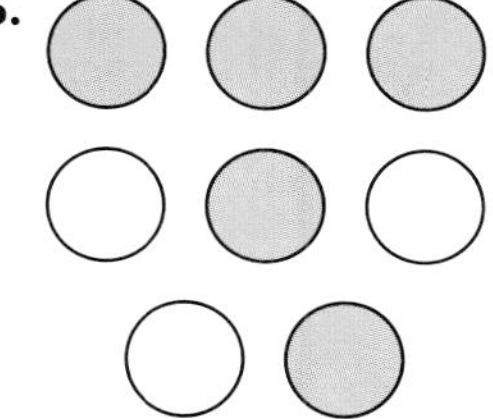

26. 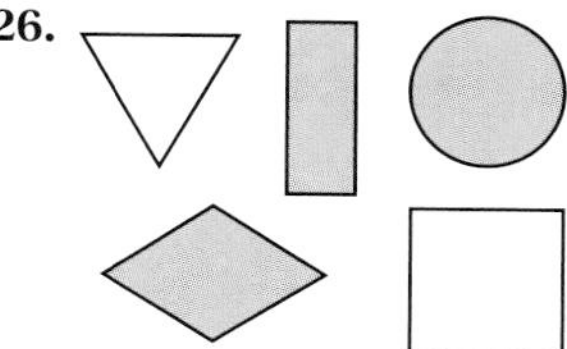

27. 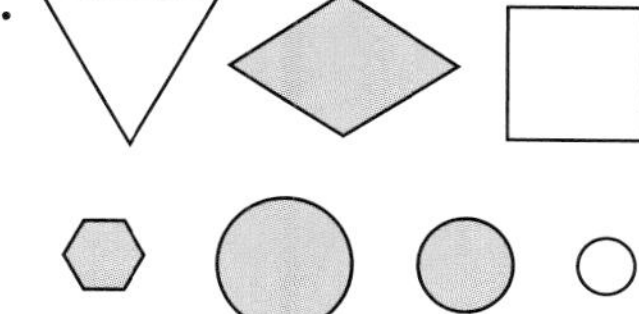

28. 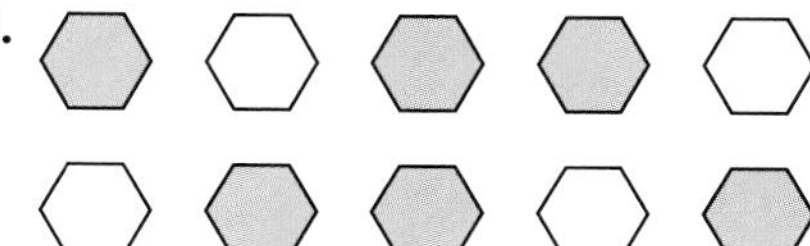

For each of Exercises 29–32, give fraction notation for the amount of gas (a) in the tank and (b) used from a full tank.

29. 

30. 

31. 

32. 

33. For the following set of people, what is the ratio of:

a) women to the total number of people?
b) women to men?
c) men to the total number of people?
d) men to women?

34. For the following set of nuts and bolts, what is the ratio of:

a) nuts to bolts?
b) bolts to nuts?
c) nuts to the total number of elements?
d) total number of elements to nuts?

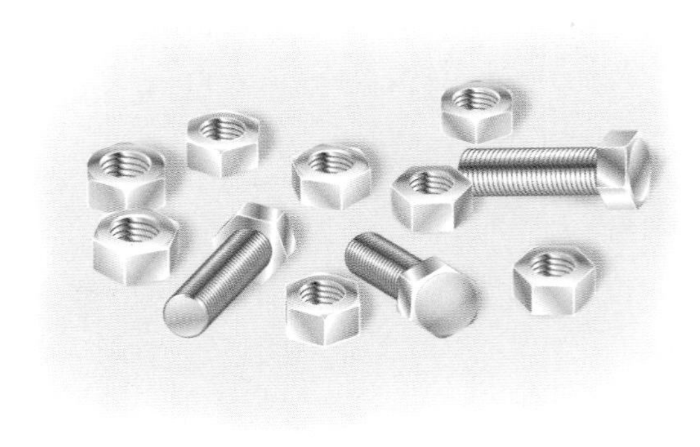

For Exercises 35 and 36, use the bar graph below that lists the police agency and the number of full-time police officers per 10,000 residents.

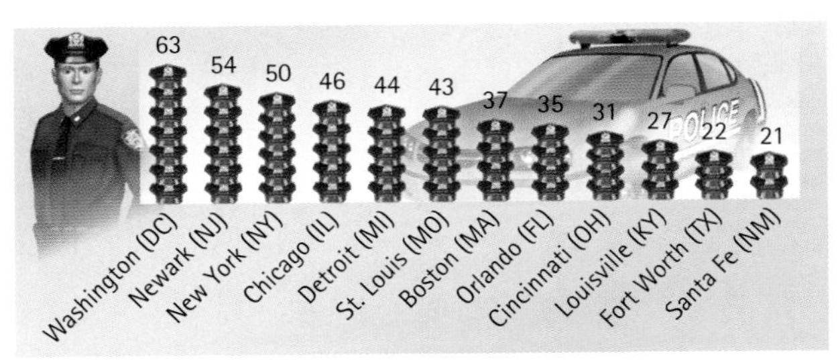

Source: Law Enforcement and Administrative Statistics, 2000

35. What is the ratio of police officers per 10,000 residents in the given city?

a) Orlando
b) New York
c) Detroit
d) Washington
e) St. Louis
f) Santa Fe

36. What is the ratio of police officers per 10,000 residents in the given city?

a) Chicago
b) Boston
c) Newark
d) Louisville
e) Cincinnati
f) Fort Worth

37. *Gas Mileage.* A 2005 Mini Cooper S will go 390 mi on 13 gal of gasoline. What is the ratio of:

a) miles driven to gasoline burned?
b) gasoline burned to miles driven?

38. Jake delivers car parts to auto service centers. On Thursday he had 15 deliveries. By noon he had delivered only 4 orders. What is the ratio of:

a) orders delivered to total number of orders?
b) orders delivered to orders not delivered?
c) orders not delivered to total number of orders?

39. *Veterinary Care.* Of every 1000 households that own dogs, 850 obtain veterinary care. What is the ratio of households who own dogs that seek veterinary care to all households that own dogs?

Source: *U.S. Pet Ownership and Demographics Sourcebook*, 2002

40. *Moviegoers.* Of every 1000 people who attend movies, 340 are in the 18–24 age group. What is the ratio of moviegoers in the 18–24 age group to all moviegoers?

Source: American Demographics

b Simplify.

41. $\frac{0}{8}$

42. $\frac{8}{8}$

43. $\frac{8-1}{9-8}$

44. $\frac{16}{1}$

45. $\frac{20}{20}$

46. $\frac{20}{1}$

47. $\frac{45}{45}$

48. $\frac{11-1}{10-9}$

49. $\frac{0}{238}$

50. $\frac{238}{1}$

51. $\frac{238}{238}$

52. $\frac{0}{16}$

53. $\frac{3}{3}$

54. $\frac{56}{56}$

55. $\frac{87}{87}$

56. $\frac{98}{98}$

57. $\frac{18}{18}$

58. $\frac{0}{18}$

59. $\frac{18}{1}$

60. $\frac{8-8}{1247}$

61. $\frac{729}{0}$

62. $\frac{1317}{0}$

63. $\frac{5}{6-6}$

64. $\frac{13}{10-10}$

65. DW Write a sentence to describe $\frac{5}{8}$ in each of the following situations.

a) As a part of a whole

b) As a part of a set

c) As a ratio

66. DW Explain in your own words why $n/n = 1$, for any natural number n.

SKILL MAINTENANCE

Round 34,562 to the nearest: [1.4a]

67. Ten.

68. Hundred.

69. Thousand.

70. Ten thousand.

Solve. [1.8a]

71. *Salaries and Education.* In 2001, the average annual salary for a person who completed only high school was $25,303. The average for someone with a bachelor's degree was $40,994. How much more did the person who had a bachelor's degree earn than the person who was only a high school graduate?

Source: U.S. Bureau of the Census

72. *Gas Mileage.* The Pontiac G6 GT gets 29 miles per gal (mpg) in highway driving. How many gallons will it use in 2784 miles of highway driving?

Subtract. [1.3b]

73. 9001 − 6798

74. 2037 − 1189

75. 67,113 − 29,874

76. 12,327 − 476

SYNTHESIS

What part of the object is shaded?

77.

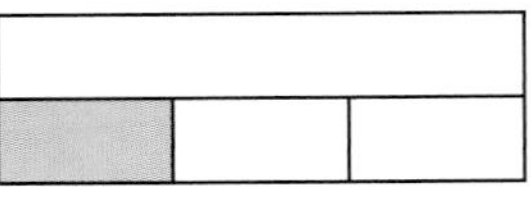

78.

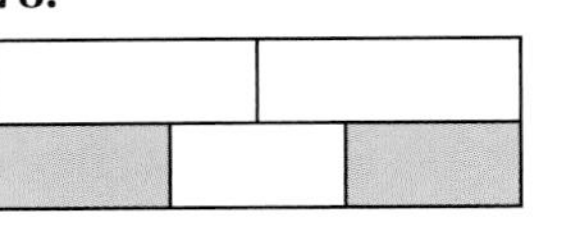

79.

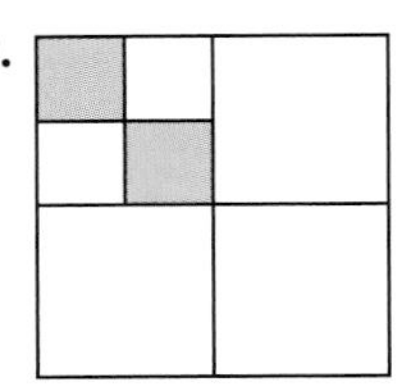

80.

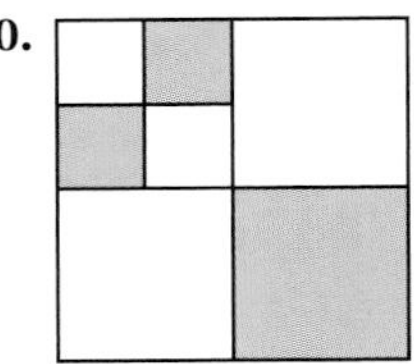

Shade or mark the figure to show $\frac{3}{5}$.

81.

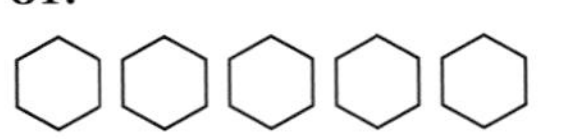

82.

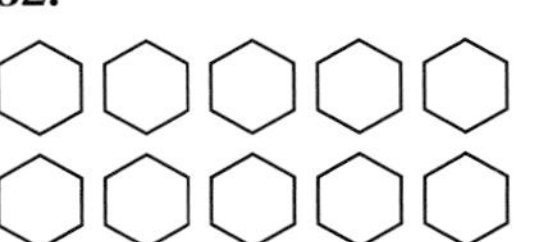

83.

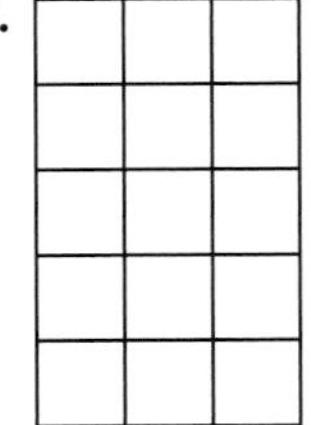

84.

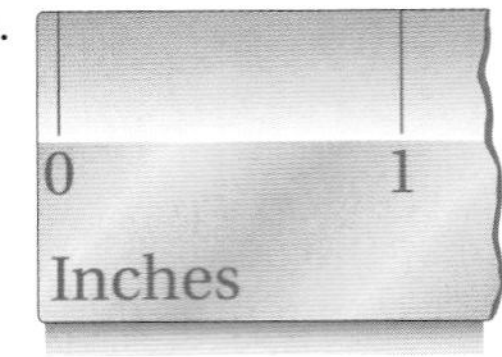

2.4 MULTIPLICATION AND APPLICATIONS

Objectives

a Multiply a whole number and a fraction.

b Multiply a fraction by a fraction.

c Solve applied problems involving multiplication of fractions.

a Multiplication by a Whole Number

We can find $3 \cdot \frac{1}{4}$ by thinking of repeated addition. We add three $\frac{1}{4}$'s. We see that $3 \cdot \frac{1}{4}$ is $\frac{3}{4}$.

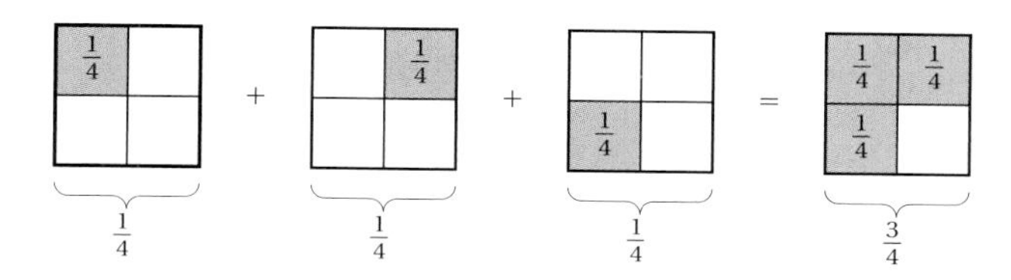

Do Exercises 1 and 2.

We find a product such as $6 \cdot \frac{4}{5}$ as follows.

To multiply a fraction by a whole number,

a) multiply the numerator by the whole number, and

$$6 \cdot \frac{4}{5} = \frac{6 \cdot 4}{5} = \frac{24}{5}$$

b) keep the same denominator.

EXAMPLES Multiply.

1. $5 \times \frac{3}{8} = \frac{5 \times 3}{8} = \frac{15}{8}$

2. $\frac{2}{5} \cdot 13 = \frac{2 \cdot 13}{5} = \frac{26}{5}$

3. $10 \cdot \frac{1}{3} = \frac{10}{3}$

Do Exercises 3–5.

b Multiplication Using Fraction Notation

When neither factor is a whole number, multiplication using fraction notation does not correspond to repeated addition. Let's visualize the product of two fractions. We consider the multiplication

$$\frac{3}{5} \cdot \frac{3}{4}.$$

This is equivalent to finding $\frac{3}{5}$ of $\frac{3}{4}$. We first consider an object and take $\frac{3}{4}$ of it. We divide it into 4 equal parts using vertical lines and take 3 of them. That is shown by the shading at right.

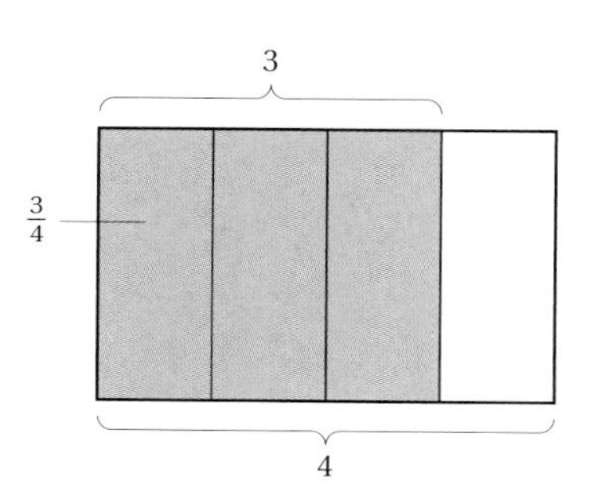

1. Find $2 \cdot \frac{1}{3}$.

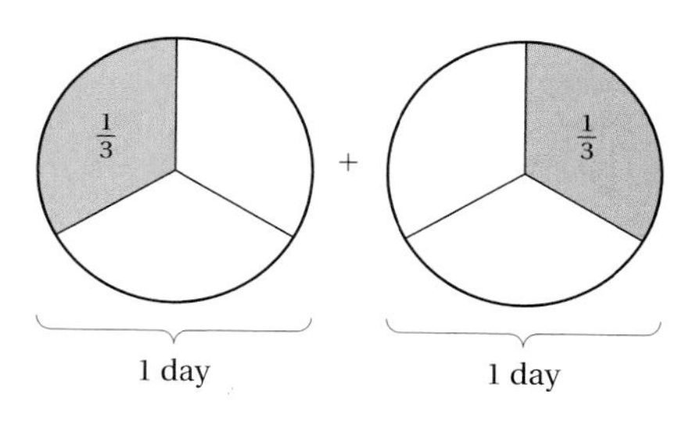

2. Find $5 \cdot \frac{1}{8}$.

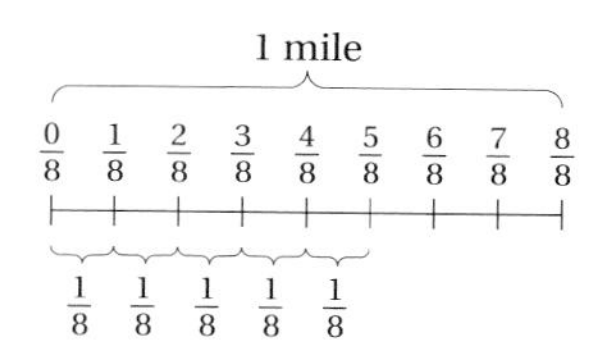

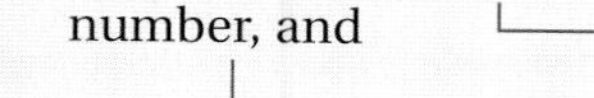

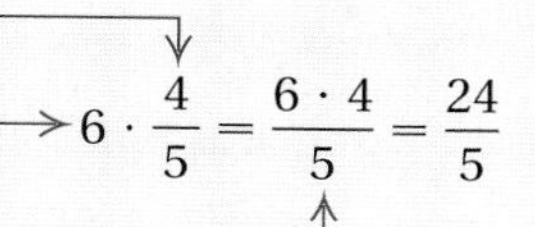

Multiply.

3. $5 \times \frac{2}{3}$

4. $11 \times \frac{3}{8}$

5. $23 \cdot \frac{2}{5}$

Answers on page A-5

6. Draw a diagram like the one at right to show the multiplication $\frac{2}{3} \cdot \frac{4}{5}$.

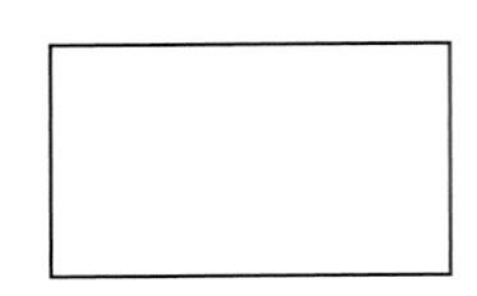

Multiply.

7. $\frac{3}{8} \cdot \frac{5}{7}$

8. $\frac{4}{3} \times \frac{8}{5}$

9. $\frac{3}{10} \cdot \frac{1}{10}$

10. $7 \cdot \frac{2}{3}$

Answers on page A-5

Next, we take $\frac{3}{5}$ of the shaded area. We divide it into 5 equal parts using horizontal lines and take 3 of them. That is shown by the darker shading below.

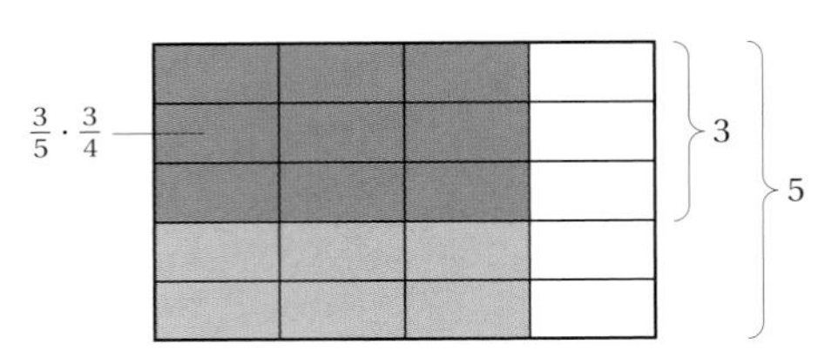

The entire object has been divided into 20 parts, and we have shaded 9 of them for a second time. Thus we see that $\frac{3}{5}$ of $\frac{3}{4}$ is $\frac{9}{20}$, or

$$\frac{3}{5} \cdot \frac{3}{4} = \frac{9}{20}.$$

The figure above shows a rectangular array inside a rectangular array. The number of pieces in the entire array is $5 \cdot 4$ (the product of the denominators). The number of pieces shaded a second time is $3 \cdot 3$ (the product of the numerators). And the product is represented by 9 pieces out of a set of 20, or $\frac{9}{20}$, or the product of the numerators over the product of the denominators. This leads us to a statement of the procedure for multiplying a fraction by a fraction.

Do Exercise 6.

We find a product such as $\frac{9}{7} \cdot \frac{3}{4}$ as follows.

To multiply a fraction by a fraction,

a) multiply the numerators to get the new numerator, and

b) multiply the denominators to get the new denominator.

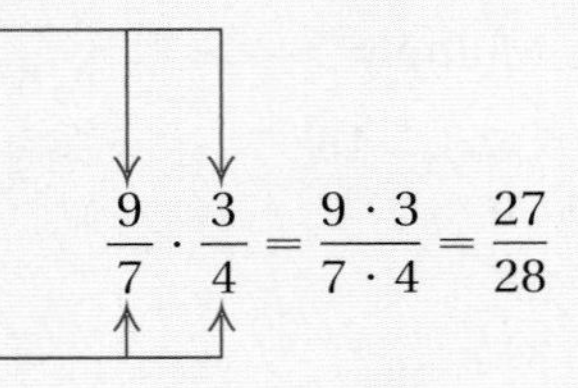

$$\frac{9}{7} \cdot \frac{3}{4} = \frac{9 \cdot 3}{7 \cdot 4} = \frac{27}{28}$$

EXAMPLES Multiply.

4. $\frac{5}{6} \times \frac{7}{4} = \frac{5 \times 7}{6 \times 4} = \frac{35}{24}$

Skip writing this step whenever you can.

5. $\frac{3}{5} \cdot \frac{7}{8} = \frac{3 \cdot 7}{5 \cdot 8} = \frac{21}{40}$

6. $\frac{3}{5} \cdot \frac{3}{4} = \frac{9}{20}$

7. $\frac{1}{4} \cdot \frac{1}{3} = \frac{1}{12}$

8. $6 \cdot \frac{4}{5} = \frac{6}{1} \cdot \frac{4}{5} = \frac{24}{5}$

Do Exercises 7–10.

C Applications and Problem Solving

Many problems that can be solved by multiplying fractions can be thought of in terms of rectangular arrays.

EXAMPLE 9 A real estate developer owns a plot of land that measures 1 square mile. He plans to use $\frac{4}{5}$ of the plot for a small strip mall and parking lot. Of this, $\frac{2}{3}$ will be needed for the parking lot. What part of the plot will be used for parking?

1. **Familiarize.** We first make a drawing to help familiarize ourselves with the problem. The land may not be rectangular. It could be in a shape like A or B below. But to think out the problem, we can think of it as a rectangle, as shown in shape C.

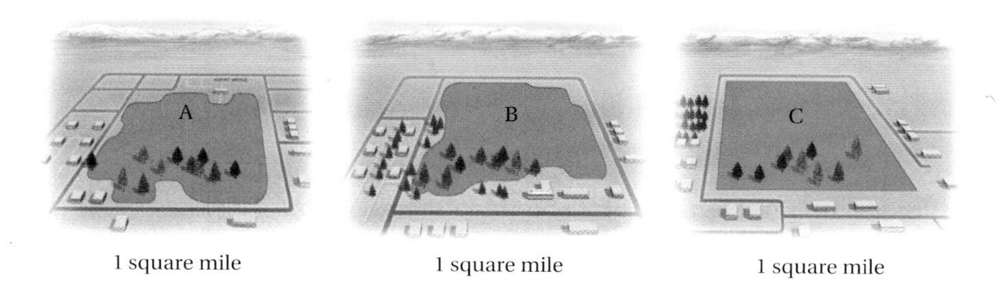

The strip mall including the parking lot uses $\frac{4}{5}$ of the plot. We shade $\frac{4}{5}$.

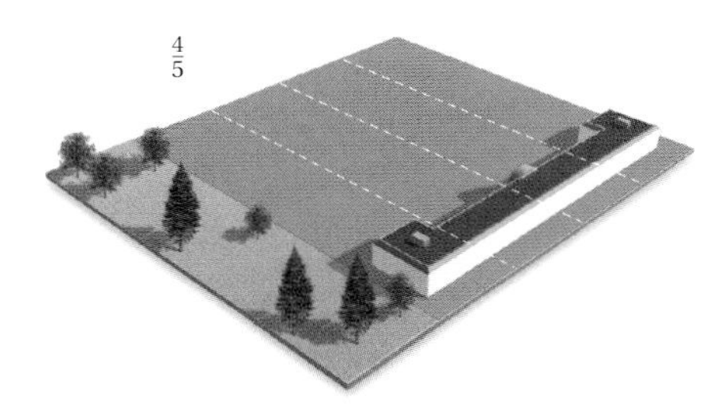

The parking lot alone takes $\frac{2}{3}$ of the preceding part. We shade that.

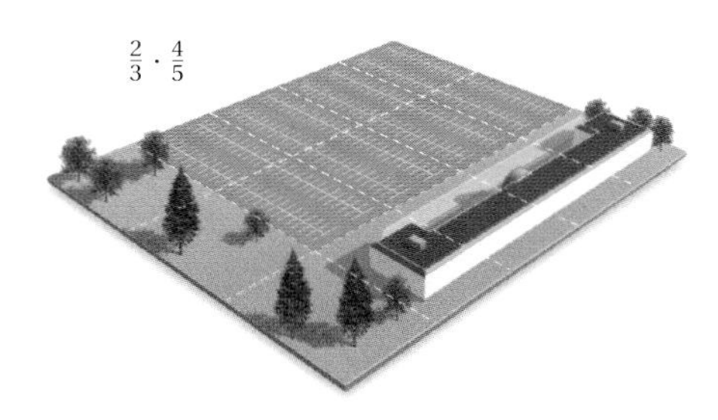

2. **Translate.** We let n = the part of the plot that is used for parking. We are taking "two-thirds of four-fifths." Recall from Section 1.8 that the word "of" corresponds to multiplication. Thus the following multiplication sentence corresponds to the situation:

$$\frac{2}{3} \cdot \frac{4}{5} = n.$$

11. A resort hotel uses $\frac{3}{4}$ of its extra land for recreational purposes. Of that, $\frac{1}{2}$ is used for swimming pools. What part of the land is used for swimming pools?

3. **Solve.** The number sentence tells us what to do. We multiply:

$$\frac{2}{3} \cdot \frac{4}{5} = \frac{2 \cdot 4}{3 \cdot 5} = \frac{8}{15}.$$

4. **Check.** We can check partially by noting that the answer is smaller than the original area, 1, which we expect since the developer is using only part of the original plot of land. Thus, $\frac{8}{15}$ is a reasonable answer. We can also check this in the figure above, where we see that 8 of 15 parts have been shaded a second time.

5. **State.** The parking lot takes $\frac{8}{15}$ of the square mile of land.

Do Exercise 11.

Example 9 and the preceding discussion indicate that the area of a rectangular region can be found by multiplying length by width. That is true whether length and width are whole numbers or not. Remember, the area of a rectangular region is given by the formula

$$A = l \cdot w.$$

12. **Area of a Ceramic Tile.** The length of a rectangular ceramic tile on an inlaid ceramic counter is $\frac{4}{9}$ ft. The width is $\frac{2}{9}$ ft. What is the area of one tile?

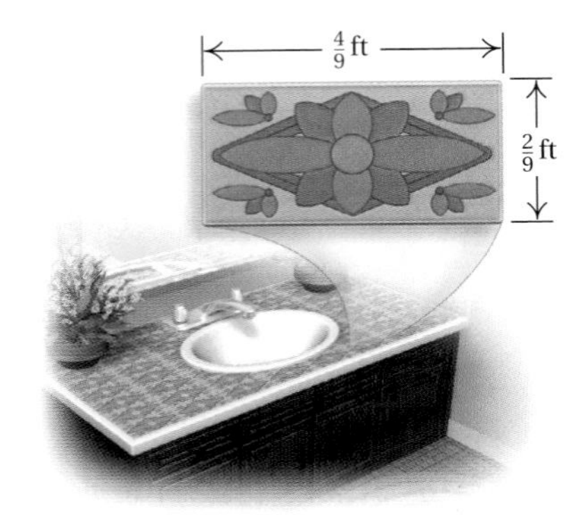

EXAMPLE 10 *Area of an Ice-Skating Rink.* The length of a rectangular ice-skating rink in the atrium of a shopping mall is $\frac{7}{100}$ mi. The width is $\frac{3}{100}$ mi. What is the area of the rink?

1. **Familiarize.** Recall that area is length times width. We make a drawing and let A = the area of the ice-skating rink.

2. **Translate.** Then we translate.

Area	is	Length	times	Width
↓	↓	↓	↓	↓
A	$=$	$\frac{7}{100}$	$\times$	$\frac{3}{100}$

3. **Solve.** The sentence tells us what to do. We multiply:

$$\frac{7}{100} \cdot \frac{3}{100} = \frac{7 \cdot 3}{100 \cdot 100} = \frac{21}{10{,}000}.$$

4. **Check.** We check by repeating the calculation. This is left to the student.

5. **State.** The area is $\frac{21}{10{,}000}$ mi^2.

Do Exercise 12.

Answers on page A-5

EXAMPLE 11 A recipe for Banana Oat Pancakes calls for $\frac{3}{4}$ cup of old fashioned oats. A chef is making $\frac{1}{2}$ of the recipe. How much oats should the chef use?

Source: Reprinted with permission from *Taste of Home* magazine, www.tasteofhome.com

1. **Familiarize.** We first make a drawing or at least visualize the situation. We let $n =$ the amount of oats the chef should use.

2. **Translate.** We are finding $\frac{1}{2}$ of $\frac{3}{4}$, so the multiplication sentence $\frac{1}{2} \cdot \frac{3}{4} = n$ corresponds to the situation.
3. **Solve.** We carry out the multiplication:

$$\frac{1}{2} \cdot \frac{3}{4} = \frac{1 \cdot 3}{2 \cdot 4} = \frac{3}{8}.$$

4. **Check.** We check by repeating the calculation. This is left to the student.
5. **State.** The chef should use $\frac{3}{8}$ cup of oats.

Do Exercise 13.

13. Of the students at Overton Junior College, $\frac{1}{8}$ participate in sports and $\frac{3}{5}$ of these play football. What fractional part of the students play football?

Answer on page A-5

Study Tips

BETTER TEST TAKING

How often do you make the following statement after taking a test: "I was able to do the homework, but I froze during the test"? Here are two tips to help you with this difficulty. Both are intended to make test taking less stressful by getting you to practice good test-taking habits on a daily basis.

- **Treat every homework exercise as if it were a test question.** If you had to work a problem at your job with no answer provided, what would you do? You would probably work it very deliberately, checking and rechecking every step. You might work it more than one time, or you might try to work it another way to check the result. Try to use this approach when doing your homework. Treat every exercise as though it were a test question with no answer at the back of the book.
- **Be sure that you do questions without answers as part of every homework assignment whether or not the instructor has assigned them!** One reason a test may seem such a different task is that questions on a test lack answers. That is the reason for taking a test: to see if you can do the questions without assistance. As part of your test preparation, be sure you do some exercises for which you do not have the answers. Thus when you take a test, you are doing a more familiar task.

The purpose of doing your homework using these approaches is to give you more test-taking practice beforehand. Let's make a sports analogy here. At a basketball game, the players take lots of practice shots before the game. They play the first half, go to the locker room, and come out for the second half. What do they do before the second half, even though they have just played 20 minutes of basketball? They shoot baskets again! We suggest the same approach here. Create more and more situations in which you practice taking test questions by treating each homework exercise like a test question and by doing exercises for which you have no answers. Good luck!

2.4 EXERCISE SET

For Extra Help

MathXL | MyMathLab | InterAct Math | Math Tutor Center | Digital Video Tutor CD 1 Videotape 2 | Student's Solutions Manual

a Multiply.

1. $3 \cdot \frac{1}{5}$

2. $2 \cdot \frac{1}{3}$

3. $5 \times \frac{1}{8}$

4. $4 \times \frac{1}{5}$

5. $\frac{2}{11} \cdot 4$

6. $\frac{2}{5} \cdot 3$

7. $10 \cdot \frac{7}{9}$

8. $9 \cdot \frac{5}{8}$

9. $\frac{2}{5} \cdot 1$

10. $\frac{3}{8} \cdot 1$

11. $\frac{2}{5} \cdot 3$

12. $\frac{3}{5} \cdot 4$

13. $7 \cdot \frac{3}{4}$

14. $7 \cdot \frac{2}{5}$

15. $17 \times \frac{5}{6}$

16. $\frac{3}{7} \cdot 40$

b Multiply.

17. $\frac{1}{2} \cdot \frac{1}{3}$

18. $\frac{1}{6} \cdot \frac{1}{4}$

19. $\frac{1}{4} \times \frac{1}{10}$

20. $\frac{1}{3} \times \frac{1}{10}$

21. $\frac{2}{3} \times \frac{1}{5}$

22. $\frac{3}{5} \times \frac{1}{5}$

23. $\frac{2}{5} \cdot \frac{2}{3}$

24. $\frac{3}{4} \cdot \frac{3}{5}$

25. $\frac{3}{4} \cdot \frac{3}{4}$

26. $\frac{3}{7} \cdot \frac{4}{5}$

27. $\frac{2}{3} \cdot \frac{7}{13}$

28. $\frac{3}{11} \cdot \frac{4}{5}$

29. $\frac{1}{10} \cdot \frac{7}{10}$

30. $\frac{3}{10} \cdot \frac{3}{10}$

31. $\frac{7}{8} \cdot \frac{7}{8}$

32. $\frac{4}{5} \cdot \frac{4}{5}$

33. $\frac{1}{10} \cdot \frac{1}{100}$

34. $\frac{3}{10} \cdot \frac{7}{100}$

35. $\frac{14}{15} \cdot \frac{13}{19}$

36. $\frac{12}{13} \cdot \frac{12}{13}$

c Solve.

37. *Tossed Salad.* The recipe for Cherry Brie Tossed Salad calls for $\frac{3}{4}$ cup of sliced almonds. How much is needed to make $\frac{1}{2}$ of the recipe?

Source: Reprinted with permission from *Taste of Home* magazine, www.tasteofhome.com

38. It takes $\frac{2}{3}$ yd of ribbon to make a bow. How much ribbon is needed to make 5 bows?

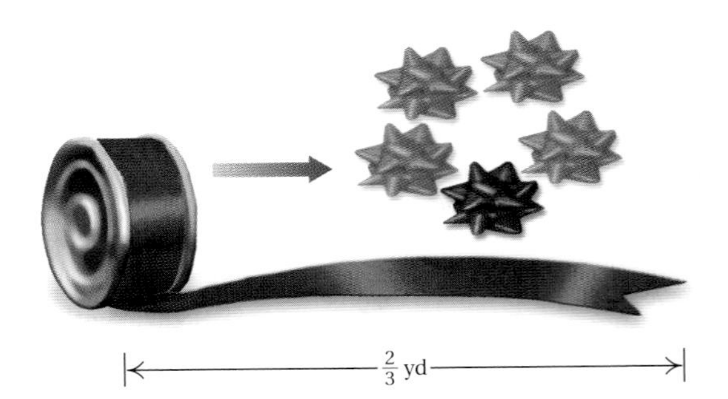

39. *Basement Carpet.* A basement floor is being covered with carpet. An area $\frac{7}{8}$ of the length and $\frac{3}{4}$ of the width is covered by lunch time. What fraction of the floor has been completed?

40. *Basketball: High School to Pro.* One of 35 high school basketball players plays college basketball. One of 75 college players plays professional basketball. What fractional part of high school players play professional basketball?

Source: Natic

41. A rectangular table top measures $\frac{4}{5}$ m long by $\frac{3}{5}$ m wide. What is its area?

42. If each piece of pie is $\frac{1}{6}$ of a pie, how much of the pie is $\frac{1}{2}$ of a piece?

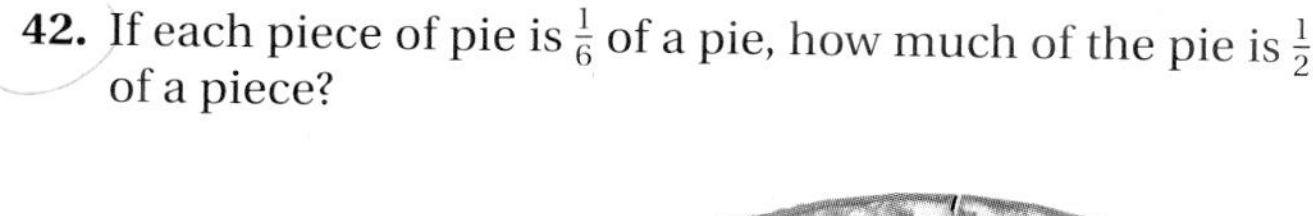

43. A gasoline can holds $\frac{7}{8}$ liter (L). How much will the can hold when it is $\frac{1}{2}$ full?

44. *Floor Tiling.* The floor of a room is being covered with tile. An area $\frac{3}{5}$ of the length and $\frac{3}{4}$ of the width is covered. What fraction of the floor has been tiled?

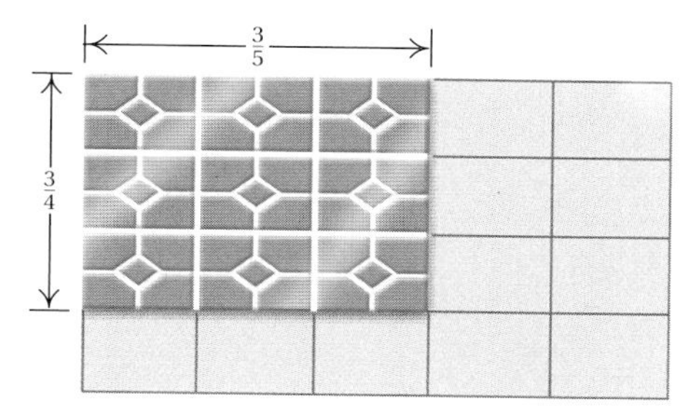

45. $\mathbf{D_W}$ Write a problem for a classmate to solve. Design the problem so that the solution is "About $\frac{1}{30}$ of the students are left-handed women."

46. $\mathbf{D_W}$ On pp. 115–116, we explained, using words and pictures, why $\frac{3}{5} \cdot \frac{3}{4}$ equals $\frac{9}{20}$. Present a similar explanation of why $\frac{2}{3} \cdot \frac{4}{7}$ equals $\frac{8}{21}$.

SKILL MAINTENANCE

Divide. [1.6b]

47. $7140 \div 35$

48. $32{,}200 \div 46$

49. $9\overline{)27{,}009}$

50. $35\overline{)7148}$

What does the digit 8 mean in each number? [1.1a]

51. 4,678,952

52. 8,473,901

53. 7148

54. 23,803

Simplify. [1.9c]

55. $12 - 3^2$

56. $(12 - 3)^2$

57. $8 \cdot 12 - (63 \div 9 + 13 \cdot 3)$

58. $(10 - 3)^4 + 10^3 \cdot 4 - 10 \div 5$

SYNTHESIS

Multiply. Write the answer using fraction notation.

59. 🖩 $\frac{341}{517} \cdot \frac{209}{349}$

60. 🖩 $\left(\frac{57}{61}\right)^3$

61. 🖩 $\left(\frac{2}{5}\right)^3\left(\frac{7}{9}\right)$

62. 🖩 $\left(\frac{1}{2}\right)^5\left(\frac{3}{5}\right)$

2.5 SIMPLIFYING

Objectives

a Use multiplying by 1 to find different fraction notation for a number.

b Simplify fraction notation.

c Use the test for equality to determine whether two fractions name the same number.

Multiply.

1. $\frac{1}{2} \cdot \frac{8}{8}$

2. $\frac{3}{5} \cdot \frac{10}{10}$

3. $\frac{13}{25} \cdot \frac{4}{4}$

4. $\frac{8}{3} \cdot \frac{25}{25}$

Answers on page A-5

a Multiplying by 1

Recall the following:

$$1 = \frac{1}{1} = \frac{2}{2} = \frac{3}{3} = \frac{4}{4} = \frac{10}{10} = \frac{45}{45} = \frac{100}{100} = \frac{n}{n}.$$

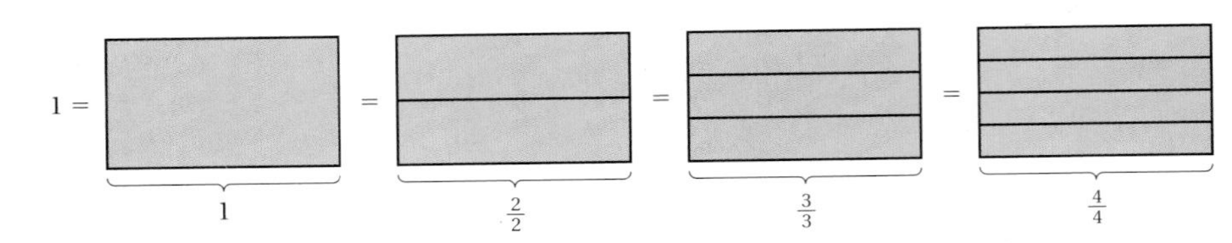

Any nonzero number divided by itself is 1. (See Section 1.6.)

Now recall the multiplicative identity from Section 1.5. For any whole number a, $1 \cdot a = a \cdot 1 = a$. This holds for numbers of arithmetic as well.

MULTIPLICATIVE IDENTITY FOR FRACTIONS

When we multiply a number by 1, we get the same number:

$$\frac{3}{5} = \frac{3}{5} \cdot 1 = \frac{3}{5} \cdot \frac{4}{4} = \frac{12}{20}.$$

Since $\frac{3}{5} = \frac{12}{20}$, we know that $\frac{3}{5}$ and $\frac{12}{20}$ are two names for the same number. We also say that $\frac{3}{5}$ and $\frac{12}{20}$ are **equivalent.**

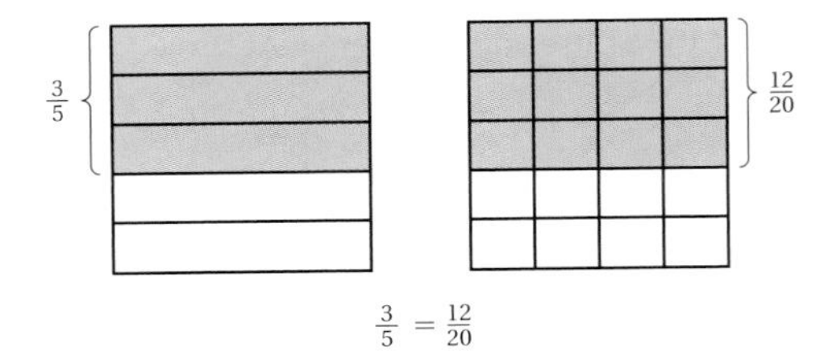

$\frac{3}{5} = \frac{12}{20}$

Do Exercises 1–4.

Suppose we want to find a name for $\frac{2}{3}$, but one that has a denominator of 9. We can multiply by 1 to find equivalent fractions:

$$\frac{2}{3} = \frac{2}{3} \cdot \frac{3}{3} = \frac{2 \cdot 3}{3 \cdot 3} = \frac{6}{9}.$$

Since $9 = 3 \cdot 3$, we chose $\frac{3}{3}$ for 1 in order to get a denominator of 9.

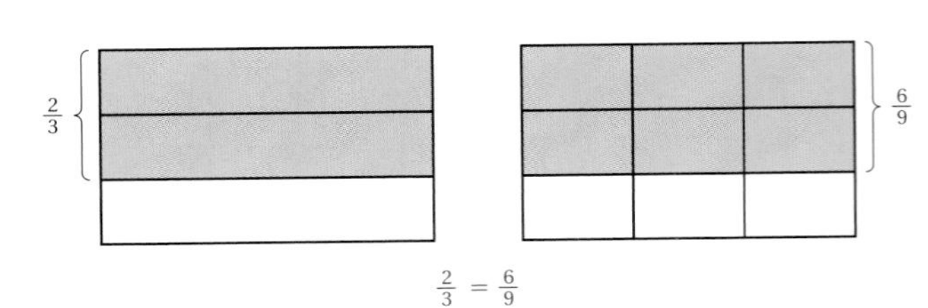

$\frac{2}{3} = \frac{6}{9}$

EXAMPLE 1 Find a name for $\frac{1}{4}$ with a denominator of 24.

Since $4 \cdot 6 = 24$, we multiply by $\frac{6}{6}$:

$$\frac{1}{4} = \frac{1}{4} \cdot \frac{6}{6} = \frac{1 \cdot 6}{4 \cdot 6} = \frac{6}{24}.$$

EXAMPLE 2 Find a name for $\frac{2}{5}$ with a denominator of 35.

Since $5 \cdot 7 = 35$, we multiply by $\frac{7}{7}$:

$$\frac{2}{5} = \frac{2}{5} \cdot \frac{7}{7} = \frac{2 \cdot 7}{5 \cdot 7} = \frac{14}{35}.$$

Do Exercises 5–9.

Find another name for the number, but with the denominator indicated. Use multiplying by 1.

5. $\frac{4}{3} = \frac{?}{9}$

6. $\frac{3}{4} = \frac{?}{24}$

7. $\frac{9}{10} = \frac{?}{100}$

8. $\frac{3}{15} = \frac{?}{45}$

9. $\frac{8}{7} = \frac{?}{49}$

b Simplifying Fraction Notation

All of the following are names for three-fourths:

$$\frac{3}{4}, \frac{6}{8}, \frac{9}{12}, \frac{12}{16}, \frac{15}{20}.$$

We say that $\frac{3}{4}$ is **simplest** because it has the smallest numerator and the smallest denominator. That is, the numerator and the denominator have no common factor other than 1.

To simplify, we reverse the process of multiplying by 1:

$$\frac{12}{18} = \frac{2 \cdot 6}{3 \cdot 6}$$ ← Factoring the numerator ← Factoring the denominator

$$= \frac{2}{3} \cdot \frac{6}{6}$$ Factoring the fraction

$$= \frac{2}{3} \cdot 1$$ $\frac{6}{6} = 1$

$$= \frac{2}{3}.$$ Removing a factor of 1: $\frac{2}{3} \cdot 1 = \frac{2}{3}$

EXAMPLES Simplify.

3. $\frac{8}{20} = \frac{2 \cdot 4}{5 \cdot 4} = \frac{2}{5} \cdot \frac{4}{4} = \frac{2}{5}$

4. $\frac{2}{6} = \frac{1 \cdot 2}{3 \cdot 2} = \frac{1}{3} \cdot \frac{2}{2} = \frac{1}{3}$

The number 1 allows for pairing of factors in the numerator and the denominator.

5. $\frac{30}{6} = \frac{5 \cdot 6}{1 \cdot 6} = \frac{5}{1} \cdot \frac{6}{6} = \frac{5}{1} = 5$

We could also simplify $\frac{30}{6}$ by doing the division $30 \div 6$. That is, $\frac{30}{6} = 30 \div 6 = 5$.

Do Exercises 10–13.

Simplify.

10. $\frac{2}{8}$

11. $\frac{10}{12}$

12. $\frac{40}{8}$

13. $\frac{24}{18}$

Answers on page A-5

Simplify.

14. $\frac{35}{40}$

15. $\frac{801}{702}$

16. $\frac{24}{21}$

17. $\frac{75}{300}$

18. $\frac{280}{960}$

19. $\frac{1332}{2880}$

20. Simplify each fraction in this circle graph.

Days Spent Shopping for the Holidays

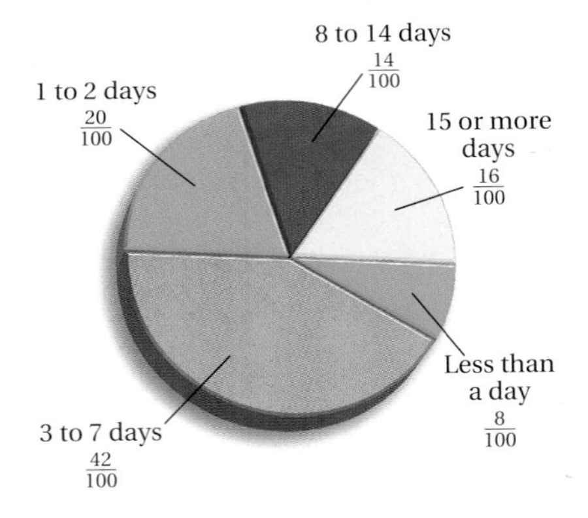

Answers on page A-5

The use of prime factorizations can be helpful for simplifying when numerators and/or denominators are larger numbers.

EXAMPLE 6 Simplify: $\frac{90}{84}$.

$\frac{90}{84} = \frac{2 \cdot 3 \cdot 3 \cdot 5}{2 \cdot 2 \cdot 3 \cdot 7}$ — Factoring the numerator and the denominator into primes

$= \frac{2 \cdot 3 \cdot 3 \cdot 5}{2 \cdot 3 \cdot 2 \cdot 7}$ — Changing the order so that like primes are above and below each other

$= \frac{2}{2} \cdot \frac{3}{3} \cdot \frac{3 \cdot 5}{2 \cdot 7}$ — Factoring the fraction

$= 1 \cdot 1 \cdot \frac{3 \cdot 5}{2 \cdot 7}$

$= \frac{3 \cdot 5}{2 \cdot 7}$ — Removing factors of 1

$= \frac{15}{14}$

The tests for divisibility (Section 2.2) are very helpful in simplifying fraction notation. We could have shortened the preceding example had we noted that 6 is a factor of both the numerator and the denominator. Then we have

$$\frac{90}{84} = \frac{6 \cdot 15}{6 \cdot 14} = \frac{6}{6} \cdot \frac{15}{14} = \frac{15}{14}.$$

EXAMPLE 7 Simplify: $\frac{603}{207}$.

At first glance this looks difficult. But note, using the test for divisibility by 9 (sum of digits divisible by 9), that both the numerator and the denominator are divisible by 9. Thus we can factor 9 from both numbers:

$$\frac{603}{207} = \frac{9 \cdot 67}{9 \cdot 23} = \frac{9}{9} \cdot \frac{67}{23} = \frac{67}{23}.$$

EXAMPLE 8 Simplify: $\frac{660}{1140}$.

Using the tests for divisibility, we have

$\frac{660}{1140} = \frac{10 \cdot 66}{10 \cdot 114} = \frac{10}{10} \cdot \frac{66}{114} = \frac{66}{114}$ — Both 660 and 1140 are divisible by 10.

$= \frac{6 \cdot 11}{6 \cdot 19} = \frac{6}{6} \cdot \frac{11}{19} = \frac{11}{19}.$ — Both 66 and 114 are divisible by 6.

Do Exercises 14–20.

CANCELING

Canceling is a shortcut that you may have used for removing a factor of 1 when working with fraction notation. With *great* concern, we mention it as a possibility for speeding up your work. Canceling may be done only when removing common factors in numerators and denominators. Each such pair allows us to remove a factor of 1 in a product.

Our concern is that canceling be done with care and understanding. In effect, slashes are used to indicate factors of 1 that have been removed. For instance, Example 6 might have been done faster as follows:

$$\frac{90}{84} = \frac{2 \cdot 3 \cdot 3 \cdot 5}{2 \cdot 2 \cdot 3 \cdot 7}$$ Factoring the numerator and the denominator

$$= \frac{\not{2} \cdot \not{3} \cdot 3 \cdot 5}{2 \cdot \not{2} \cdot \not{3} \cdot 7}$$ When a factor of 1 is noted, it is "canceled" as shown: $\frac{2 \cdot 3}{2 \cdot 3} = 1$.

$$= \frac{3 \cdot 5}{2 \cdot 7} = \frac{15}{14}.$$

Caution!

The difficulty with canceling is that it is often applied incorrectly in situations like the following:

$$\frac{\not{2} + 3}{\not{2}} = 3; \qquad \frac{\not{4} + 1}{\not{4} + 2} = \frac{1}{2}; \qquad \frac{1\not{5}}{\not{5}4} = \frac{1}{4}.$$

Wrong! Wrong! Wrong!

The correct answers are

$$\frac{2 + 3}{2} = \frac{5}{2}; \qquad \frac{4 + 1}{4 + 2} = \frac{5}{6}; \qquad \frac{15}{54} = \frac{3 \cdot 5}{3 \cdot 18} = \frac{3}{3} \cdot \frac{5}{18} = \frac{5}{18}.$$

In each situation, the number canceled was not a factor of 1. Factors are parts of products. For example, in $2 \cdot 3$, 2 and 3 are factors, but in $2 + 3$, 2 and 3 are *not* factors. Canceling may not be done when sums or differences are in numerators or denominators, as shown here. **If you cannot factor, you cannot cancel! If in doubt, do not cancel!**

C A Test for Equality

When denominators are the same, we say that fractions have a **common denominator.** When fractions have a common denominator, we can compare them by comparing numerators. Suppose we want to compare $\frac{3}{6}$ and $\frac{2}{4}$. First we find a common denominator. To do this, we multiply each fraction by 1, using the denominator of the other fraction to form the symbol for 1.

The "unit" is $\frac{1}{6}$.

$$\frac{3}{6} = \frac{3}{6} \cdot \frac{4}{4} = \frac{3 \cdot 4}{6 \cdot 4} = \frac{12}{24}$$

$$\frac{2}{4} = \frac{2}{4} \cdot \frac{6}{6} = \frac{2 \cdot 6}{4 \cdot 6} = \frac{12}{24}$$

The "unit" is $\frac{1}{4}$.

Both "units" are $\frac{1}{24}$.

We multiply $\frac{3}{6}$ by $\frac{4}{4}$ and $\frac{2}{4}$ by $\frac{6}{6}$.

Once we have a common denominator, 24, we compare the numerators. And since these numerators are both 12, the fractions are equal:

$$\frac{3}{6} = \frac{2}{4}.$$

CALCULATOR CORNER

Simplifying Fraction Notation Fraction calculators are equipped with a key, often labeled $\boxed{a^{b}/_{c}}$, that allows for simplification with fraction notation. To simplify

$$\frac{208}{256}$$

with such a fraction calculator, the following keystrokes can be used.

$\boxed{2}\boxed{0}\boxed{8}\boxed{a^{b}/_{c}}$
$\boxed{2}\boxed{5}\boxed{6}\boxed{=}$.

The display that appears

$\boxed{13 \lrcorner 16.}$

represents simplified fraction notation $\frac{13}{16}$.

Exercises: Use a fraction calculator to simplify each of the following.

1. $\frac{84}{90}$
2. $\frac{35}{40}$
3. $\frac{690}{835}$
4. $\frac{42}{150}$

Use = or ≠ for □ to write a true sentence.

21. $\frac{2}{6} \square \frac{3}{9}$

22. $\frac{2}{3} \square \frac{14}{20}$

Note in the preceding that if

$$\frac{3}{6} = \frac{2}{4}, \quad \text{then} \quad 3 \cdot 4 = 6 \cdot 2.$$

This tells us that we need to check only the products $3 \cdot 4$ and $6 \cdot 2$ to compare the fractions.

A TEST FOR EQUALITY

We multiply these two numbers: $3 \cdot 4$.

We multiply these two numbers: $6 \cdot 2$.

$$\frac{3}{6} \square \frac{2}{4}$$

We call $3 \cdot 4$ and $6 \cdot 2$ **cross products.** Since the cross products are the same, that is, $3 \cdot 4 = 6 \cdot 2$, we know that

$$\frac{3}{6} = \frac{2}{4}.$$

If a sentence $a = b$ is true, it means that a and b name the same number. If a sentence $a \neq b$ (read "a is not equal to b") is true, it means that a and b do *not* name the same number.

EXAMPLE 9 Use = or ≠ for □ to write a true sentence:

$$\frac{6}{7} \square \frac{7}{8}.$$

We multiply these two numbers: $6 \cdot 8 = 48$.

We multiply these two numbers: $7 \cdot 7 = 49$.

$$\frac{6}{7} \square \frac{7}{8}$$

Because $48 \neq 49$, $\frac{6}{7}$ and $\frac{7}{8}$ do not name the same number. Thus,

$$\frac{6}{7} \neq \frac{7}{8}.$$

EXAMPLE 10 Use = or ≠ for □ to write a true sentence:

$$\frac{6}{10} \square \frac{3}{5}.$$

We multiply these two numbers: $6 \cdot 5 = 30$.

We multiply these two numbers: $10 \cdot 3 = 30$

$$\frac{6}{10} \square \frac{3}{5}.$$

Because the cross products are the same, we have

$$\frac{6}{10} = \frac{3}{5}.$$

Do Exercises 21 and 22.

Answers on page A-5

2.5 EXERCISE SET

For Extra Help

MathXL | MyMathLab | InterAct Math | Math Tutor Center | Digital Video Tutor CD 1 Videotape 2 | Student's Solutions Manual

a Find another name for the given number, but with the denominator indicated. Use multiplying by 1.

1. $\frac{1}{2} = \frac{?}{10}$

2. $\frac{1}{6} = \frac{?}{18}$

3. $\frac{5}{8} = \frac{?}{32}$

4. $\frac{2}{9} = \frac{?}{18}$

5. $\frac{9}{10} = \frac{?}{30}$

6. $\frac{5}{6} = \frac{?}{48}$

7. $\frac{7}{8} = \frac{?}{32}$

8. $\frac{2}{5} = \frac{?}{25}$

9. $\frac{5}{12} = \frac{?}{48}$

10. $\frac{3}{8} = \frac{?}{56}$

11. $\frac{17}{18} = \frac{?}{54}$

12. $\frac{11}{16} = \frac{?}{256}$

13. $\frac{5}{3} = \frac{?}{45}$

14. $\frac{11}{5} = \frac{?}{30}$

15. $\frac{7}{22} = \frac{?}{132}$

16. $\frac{10}{21} = \frac{?}{126}$

b Simplify.

17. $\frac{2}{4}$

18. $\frac{4}{8}$

19. $\frac{6}{8}$

20. $\frac{8}{12}$

21. $\frac{3}{15}$

22. $\frac{8}{10}$

23. $\frac{24}{8}$

24. $\frac{36}{9}$

25. $\frac{18}{24}$

26. $\frac{42}{48}$

27. $\frac{14}{16}$

28. $\frac{15}{25}$

29. $\frac{12}{10}$

30. $\frac{16}{14}$

31. $\frac{16}{48}$

32. $\frac{100}{20}$

33. $\frac{150}{25}$

34. $\frac{19}{76}$

35. $\frac{17}{51}$

36. $\frac{425}{525}$

37. $\frac{220}{4125}$

38. $\frac{540}{810}$

39. $\frac{1575}{3920}$

40. $\frac{1000}{1080}$

c Use = or ≠ for □ to write a true sentence.

41. $\frac{3}{4}$ □ $\frac{9}{12}$

42. $\frac{4}{8}$ □ $\frac{3}{6}$

43. $\frac{1}{5}$ □ $\frac{2}{9}$

44. $\frac{1}{4}$ □ $\frac{2}{9}$

45. $\frac{3}{8}$ □ $\frac{6}{16}$

46. $\frac{2}{6}$ □ $\frac{6}{18}$

47. $\frac{2}{5}$ □ $\frac{3}{7}$

48. $\frac{1}{3}$ □ $\frac{1}{4}$

49. $\frac{12}{9} \square \frac{8}{6}$

50. $\frac{16}{14} \square \frac{8}{7}$

51. $\frac{5}{2} \square \frac{17}{7}$

52. $\frac{3}{10} \square \frac{7}{24}$

53. $\frac{3}{10} \square \frac{30}{100}$

54. $\frac{700}{1000} \square \frac{70}{100}$

55. $\frac{5}{10} \square \frac{520}{1000}$

56. $\frac{49}{100} \square \frac{50}{1000}$

57. D_W Explain in your own words when it *is* possible to "cancel" and when it *is not* possible to "cancel."

58. D_W Can fraction notation be simplified if its numerator and its denominator are two different prime numbers? Why or why not?

SKILL MAINTENANCE

Solve. [1.8a]

59. A playing field is 78 ft long and 64 ft wide. What is its area? its perimeter?

60. A landscaper buys 13 small maple trees and 17 small oak trees for a project. A maple costs $23 and an oak costs $37. How much is spent altogether for the trees?

Subtract. [1.3b]

61. $34 - 23$

62. $50 - 18$

63. $803 - 617$

64. $8344 - 5607$

Solve. [1.7b]

65. $30 \cdot x = 150$

66. $10{,}947 = 123 \cdot y$

67. $5280 = 1760 + t$

68. $x + 2368 = 11{,}369$

SYNTHESIS

Simplify. Use the list of prime numbers on p. 93.

69. $\frac{2603}{2831}$

70. $\frac{3197}{3473}$

71. *Shy People.* Sociologists have found that 4 of 10 people are shy. Write fraction notation for the part of the population that is shy; the part that is not shy. Simplify.

72. *Left-Handed People.* Sociologists estimate that 3 of 20 people are left-handed. In a crowd of 460 people, how many would you expect to be left-handed?

73. *Batting Averages.* For the 2004 season, Ichiro Suzuki, of the Seattle Mariners, won the American League batting title with 262 hits in 704 times at bat. Barry Bonds, of the San Francisco Giants, won the National League title with 135 hits in 373 times at bat. Did they have the same fraction of hits in times at bat (batting average)? Why or why not?

Source: Major League Baseball

74. On a test of 82 questions, a student got 63 correct. On another test of 100 questions, she got 77 correct. Did she get the same portion of each test correct? Why or why not?

2.6 MULTIPLYING, SIMPLIFYING, AND APPLICATIONS

Objectives

a Multiply and simplify using fraction notation.

b Solve applied problems involving multiplication of fractions.

a Multiplying and Simplifying Using Fraction Notation

It is often possible to simplify after we multiply. To make such simplifying easier, it is generally best not to carry out the products in the numerator and the denominator, but to factor and simplify first. Consider the product

$$\frac{3}{8} \cdot \frac{4}{9}.$$

We proceed as follows:

$\frac{3}{8} \cdot \frac{4}{9} = \frac{3 \cdot 4}{8 \cdot 9}$ — We write the products in the numerator and the denominator, but we do not carry them out.

$= \frac{3 \cdot 2 \cdot 2}{2 \cdot 2 \cdot 2 \cdot 3 \cdot 3}$ — Factoring the numerator and the denominator

$= \frac{3 \cdot 2 \cdot 2 \cdot 1}{2 \cdot 2 \cdot 2 \cdot 3 \cdot 3}$ — Using the identity property of 1 to insert the number 1 as a factor

$= \frac{3 \cdot 2 \cdot 2}{3 \cdot 2 \cdot 2} \cdot \frac{1}{2 \cdot 3}$ — Factoring the fraction

$= 1 \cdot \frac{1}{2 \cdot 3}$

$= \frac{1}{2 \cdot 3}$ — Removing a factor of 1

$= \frac{1}{6}.$

The procedure could have been shortened had we noticed that 4 is a factor of the 8 in the denominator:

$$\frac{3}{8} \cdot \frac{4}{9} = \frac{3 \cdot 4}{8 \cdot 9} = \frac{3 \cdot 4}{4 \cdot 2 \cdot 3 \cdot 3} = \frac{3 \cdot 4}{3 \cdot 4} \cdot \frac{1}{2 \cdot 3} = 1 \cdot \frac{1}{2 \cdot 3} = \frac{1}{2 \cdot 3} = \frac{1}{6}.$$

To multiply and simplify:

a) Write the products in the numerator and the denominator, but do not carry out the products.
b) Factor the numerator and the denominator.
c) Factor the fraction to remove a factor of 1.
d) Carry out the remaining products.

EXAMPLES Multiply and simplify.

1. $\frac{2}{3} \cdot \frac{9}{4} = \frac{2 \cdot 9}{3 \cdot 4} = \frac{2 \cdot 3 \cdot 3}{3 \cdot 2 \cdot 2} = \frac{2 \cdot 3}{2 \cdot 3} \cdot \frac{3}{2} = 1 \cdot \frac{3}{2} = \frac{3}{2}$

2. $\frac{6}{7} \cdot \frac{5}{3} = \frac{6 \cdot 5}{7 \cdot 3} = \frac{3 \cdot 2 \cdot 5}{7 \cdot 3} = \frac{3}{3} \cdot \frac{2 \cdot 5}{7} = 1 \cdot \frac{2 \cdot 5}{7} = \frac{2 \cdot 5}{7} = \frac{10}{7}$

3. $40 \cdot \frac{7}{8} = \frac{40 \cdot 7}{8} = \frac{8 \cdot 5 \cdot 7}{8 \cdot 1} = \frac{8}{8} \cdot \frac{5 \cdot 7}{1} = 1 \cdot \frac{5 \cdot 7}{1} = \frac{5 \cdot 7}{1} = 35$

Multiply and simplify.

1. $\frac{2}{3} \cdot \frac{7}{8}$

2. $\frac{4}{5} \cdot \frac{5}{12}$

3. $16 \cdot \frac{3}{8}$

4. $\frac{5}{8} \cdot 4$

5. A landscaper uses $\frac{2}{3}$ lb of peat moss for a rosebush. How much will be needed for 21 rosebushes?

Answers on page A-5

Caution!

Canceling can be used as follows for these examples.

1. $\frac{2}{3} \cdot \frac{9}{4} = \frac{2 \cdot 9}{3 \cdot 4} = \frac{\not{2} \cdot \not{3} \cdot 3}{\not{3} \cdot \not{2} \cdot 2} = \frac{3}{2}$ Removing a factor of 1: $\frac{2 \cdot 3}{2 \cdot 3} = 1$

2. $\frac{6}{7} \cdot \frac{5}{3} = \frac{6 \cdot 5}{7 \cdot 3} = \frac{\not{3} \cdot 2 \cdot 5}{7 \cdot \not{3}} = \frac{2 \cdot 5}{7} = \frac{10}{7}$ Removing a factor of 1: $\frac{3}{3} = 1$

3. $40 \cdot \frac{7}{8} = \frac{40 \cdot 7}{8} = \frac{\not{8} \cdot 5 \cdot 7}{\not{8} \cdot 1} = \frac{5 \cdot 7}{1} = 35$ Removing a factor of 1: $\frac{8}{8} = 1$

Remember, if you can't factor, you can't cancel!

Do Exercises 1–4.

b Applications and Problem Solving

EXAMPLE 4 Elite Elegance is preparing souvenir favors for a charity fund-raising dinner. How many pounds of caramels will be needed to fill 235 boxes if each box contains $\frac{2}{5}$ lb?

1. Familiarize. We first make a drawing or at least visualize the situation. Repeated addition will work here.

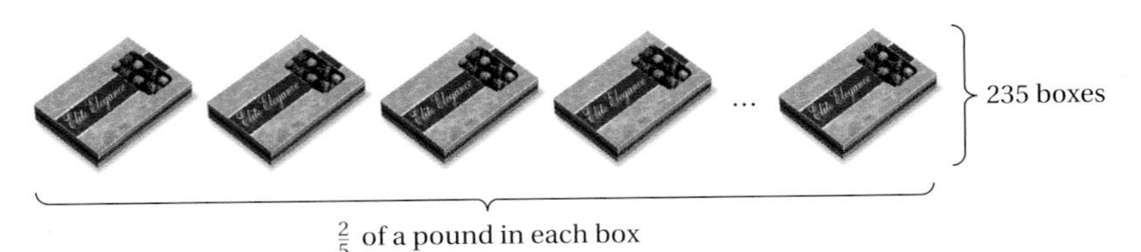

We let n = the number of pounds of caramels.

2. Translate. The problem translates to the following equation:

$$n = 235 \cdot \frac{2}{5}.$$

3. Solve. To solve the equation, we carry out the multiplication:

$$n = 235 \cdot \frac{2}{5} = \frac{235 \cdot 2}{5} \quad \text{Multiplying}$$
$$= \frac{5 \cdot 47 \cdot 2}{5 \cdot 1}$$
$$= \frac{5}{5} \cdot \frac{47 \cdot 2}{1}$$
$$= 94. \quad \text{Simplifying}$$

4. Check. We check by repeating the calculation. (The check is left to the student.) We can also think about the reasonableness of the answer. We are putting less than a pound of caramels in each box, so the answer should be less than 235. Since 94 is less than 235, we have a partial check of the reasonableness of the answer. The number 94 checks.

5. State. Thus, 94 lb of caramels will be needed.

Do Exercise 5.

2.6 EXERCISE SET

For Extra Help

MathXL | MyMathLab | InterAct Math | Math Tutor Center | Digital Video Tutor CD 1 Videotape 2 | Student's Solutions Manual

a Multiply and simplify.

Don't forget to simplify!

1. $\frac{2}{3} \cdot \frac{1}{2}$

2. $\frac{3}{8} \cdot \frac{1}{3}$

3. $\frac{7}{8} \cdot \frac{1}{7}$

4. $\frac{4}{9} \cdot \frac{1}{4}$

5. $\frac{1}{8} \cdot \frac{4}{5}$

6. $\frac{2}{5} \cdot \frac{1}{6}$

7. $\frac{1}{4} \cdot \frac{2}{3}$

8. $\frac{4}{6} \cdot \frac{1}{6}$

9. $\frac{12}{5} \cdot \frac{9}{8}$

10. $\frac{16}{15} \cdot \frac{5}{4}$

11. $\frac{10}{9} \cdot \frac{7}{5}$

12. $\frac{25}{12} \cdot \frac{4}{3}$

13. $9 \cdot \frac{1}{9}$

14. $4 \cdot \frac{1}{4}$

15. $\frac{1}{3} \cdot 3$

16. $\frac{1}{6} \cdot 6$

17. $\frac{7}{10} \cdot \frac{10}{7}$

18. $\frac{8}{9} \cdot \frac{9}{8}$

19. $\frac{7}{5} \cdot \frac{5}{7}$

20. $\frac{2}{11} \cdot \frac{11}{2}$

21. $\frac{1}{4} \cdot 8$

22. $\frac{1}{3} \cdot 18$

23. $24 \cdot \frac{1}{6}$

24. $16 \cdot \frac{1}{2}$

25. $12 \cdot \frac{3}{4}$

26. $18 \cdot \frac{5}{6}$

27. $\frac{3}{8} \cdot 24$

28. $\frac{2}{9} \cdot 36$

29. $13 \cdot \frac{2}{5}$

30. $15 \cdot \frac{1}{6}$

31. $\frac{7}{10} \cdot 28$

32. $\frac{5}{8} \cdot 34$

33. $\frac{1}{6} \cdot 360$

34. $\frac{1}{3} \cdot 120$

35. $240 \cdot \frac{1}{8}$

36. $150 \cdot \frac{1}{5}$

37. $\frac{4}{10} \cdot \frac{5}{10}$

38. $\frac{7}{10} \cdot \frac{34}{150}$

39. $\frac{8}{10} \cdot \frac{45}{100}$

40. $\frac{3}{10} \cdot \frac{8}{10}$

41. $\frac{11}{24} \cdot \frac{3}{5}$

42. $\frac{15}{22} \cdot \frac{4}{7}$

43. $\frac{10}{21} \cdot \frac{3}{4}$

44. $\frac{17}{18} \cdot \frac{3}{5}$

b Solve.

The *pitch* of a screw is the distance between its threads. With each complete rotation, the screw goes in or out a distance equal to its pitch. Use this information to answer Exercises 45 and 46.

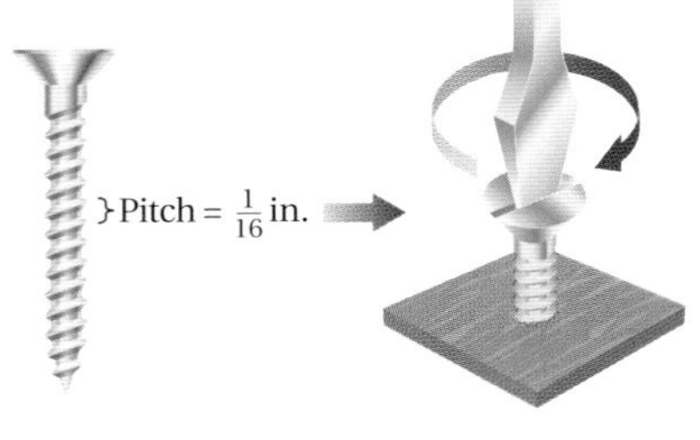

Each rotation moves the screw in or out $\frac{1}{16}$ in.

45. The pitch of a screw is $\frac{1}{16}$ in. How far will it go into a piece of oak when it is turned 10 complete rotations clockwise?

46. The pitch of a screw is $\frac{3}{32}$ in. How far will it go out of a piece of plywood when it is turned 10 complete rotations counterclockwise?

47. *Running Speeds.* The maximum running speed of an elk over approximately quarter-mile distances is about 45 mph. The maximum speed of a grizzly bear is about $\frac{2}{3}$ that of an elk. Find the running speed of a grizzly bear.

Source: *Natural History Magazine*, ©American Museum of Natural History

48. After Jack completes 60 hr of teacher training at college, he can earn $75 for working a full day as a substitute teacher. How much will he receive for working $\frac{1}{5}$ of a day?

49. *Mailing-List Addresses.* Business people have determined that $\frac{1}{4}$ of the addresses on a mailing list will change in one year. A business has a mailing list of 2500 people. After one year, how many addresses on that list will be incorrect?

50. *Shy People.* Sociologists have determined that $\frac{2}{5}$ of the people in the world are shy. A sales manager is interviewing 650 people for an aggressive sales position. How many of these people might be shy?

51. A recipe for piecrust calls for $\frac{2}{3}$ cup of flour. A chef is making $\frac{1}{2}$ of the recipe. How much flour should the chef use?

52. Of the students in the freshman class, $\frac{2}{5}$ have digital cameras; $\frac{1}{4}$ of these students also join the college photography club. What fraction of the students in the freshman class join the photography club?

53. A house worth $154,000 is assessed for $\frac{3}{4}$ of its value. What is the assessed value of the house?

54. Roxanne's tuition was $4600. A loan was obtained for $\frac{3}{4}$ of the tuition. How much was the loan?

55. *Map Scaling.* On a map, 1 in. represents 240 mi. How much does $\frac{2}{3}$ in. represent?

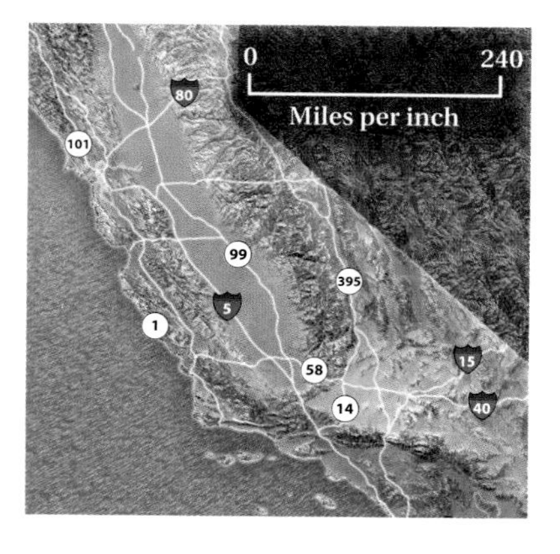

56. *Map Scaling.* On a map, 1 in. represents 120 mi. How much does $\frac{3}{4}$ in. represent?

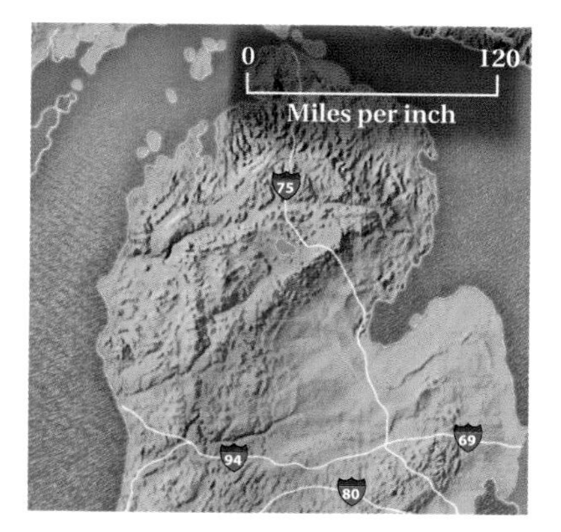

57. *Household Budgets.* A family has an annual income of \$36,000. Of this, $\frac{1}{4}$ is spent for food, $\frac{1}{5}$ for housing, $\frac{1}{10}$ for clothing, $\frac{1}{9}$ for savings, $\frac{1}{4}$ for taxes, and the rest for other expenses. How much is spent for each?

58. *Household Budgets.* A family has an annual income of \$29,700. Of this, $\frac{1}{4}$ is spent for food, $\frac{1}{5}$ for housing, $\frac{1}{10}$ for clothing, $\frac{1}{9}$ for savings, $\frac{1}{4}$ for taxes, and the rest for other expenses. How much is spent for each?

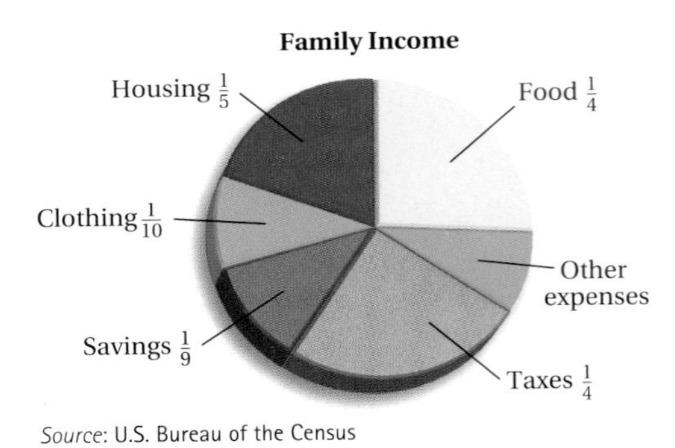

Source: U.S. Bureau of the Census

59. **D**W When multiplying using fraction notation, we form products in the numerator and the denominator, but do not immediately calculate the products. Why?

60. **D**W If a fraction's numerator and denominator have no factors (other than 1) in common, can the fraction be simplified? Why or why not?

SKILL MAINTENANCE

Solve. [1.7b]

61. $48 \cdot t = 1680$

62. $74 \cdot x = 6290$

63. $3125 = 25 \cdot t$

64. $2880 = 24 \cdot y$

65. $t + 28 = 5017$

66. $456 + x = 9002$

67. $8797 = y + 2299$

68. $10{,}000 = 3593 + m$

Subtract. [1.3b]

69. $\begin{array}{r} 9060 \\ -\ 4387 \\ \hline \end{array}$

70. $\begin{array}{r} 7800 \\ -\ 2462 \\ \hline \end{array}$

SYNTHESIS

Multiply and simplify. Use the list of prime numbers on p. 93 or a fraction calculator.

71. $\frac{201}{535} \cdot \frac{4601}{6499}$

72. $\frac{5767}{3763} \cdot \frac{159}{395}$

73. *College Profile.* Of students entering a college, $\frac{7}{8}$ have completed high school and $\frac{2}{3}$ are older than 20. If $\frac{1}{7}$ of all students are left-handed, what fraction of students entering the college are left-handed high school graduates over the age of 20?

74. *College Profile.* Refer to the information in Exercise 73. If 480 students are entering the college, how many of them are left-handed high school graduates 20 yr old or younger?

75. *College Profile.* Refer to Exercise 73. What fraction of students entering the college did not graduate high school, are 20 yr old or younger, and are left-handed?

2.7 DIVISION AND APPLICATIONS

Objectives

a Find the reciprocal of a number.

b Divide and simplify using fraction notation.

c Solve equations of the type $a \cdot x = b$ and $x \cdot a = b$, where a and b may be fractions.

d Solve applied problems involving division of fractions.

a Reciprocals

Look at these products:

$$8 \cdot \frac{1}{8} = \frac{8 \cdot 1}{8} = \frac{8}{8} = 1; \qquad \frac{2}{3} \cdot \frac{3}{2} = \frac{2 \cdot 3}{3 \cdot 2} = \frac{6}{6} = 1.$$

RECIPROCALS

If the product of two numbers is 1, we say that they are **reciprocals** of each other. To find a reciprocal of a fraction, interchange the numerator and the denominator.

$$\text{Number} \longrightarrow \frac{3}{4} \rightarrow \frac{4}{3} \longleftarrow \text{Reciprocal}$$

EXAMPLES Find the reciprocal.

1. The reciprocal of $\frac{4}{5}$ is $\frac{5}{4}$. $\quad \frac{4}{5} \cdot \frac{5}{4} = \frac{20}{20} = 1$

2. The reciprocal of $\frac{8}{7}$ is $\frac{7}{8}$. $\quad \frac{8}{7} \cdot \frac{7}{8} = \frac{56}{56} = 1$

3. The reciprocal of 24 is $\frac{1}{24}$. $\quad$ Think of 24 as $\frac{24}{1}$: $\frac{24}{1} \cdot \frac{1}{24} = \frac{24}{24} = 1.$

4. The reciprocal of $\frac{1}{3}$ is 3. $\quad \frac{1}{3} \cdot 3 = \frac{3}{3} = 1$

Do Exercises 1–4.

Find the reciprocal.

1. $\frac{2}{5}$

2. $\frac{10}{7}$

3. 9

4. $\frac{1}{5}$

Answers on page A-5

Does 0 have a reciprocal? If it did, it would have to be a number x such that

$$0 \cdot x = 1.$$

But 0 times any number is 0. Thus we have the following.

0 HAS NO RECIPROCAL

The number 0, or $\frac{0}{n}$, has no reciprocal. $\left(\text{Recall that } \frac{n}{0} \text{ is not defined.}\right)$

b Division

Consider the division $\frac{3}{4} \div \frac{1}{8}$. We are asking how many $\frac{1}{8}$'s are in $\frac{3}{4}$. We can answer this by looking at the figure below.

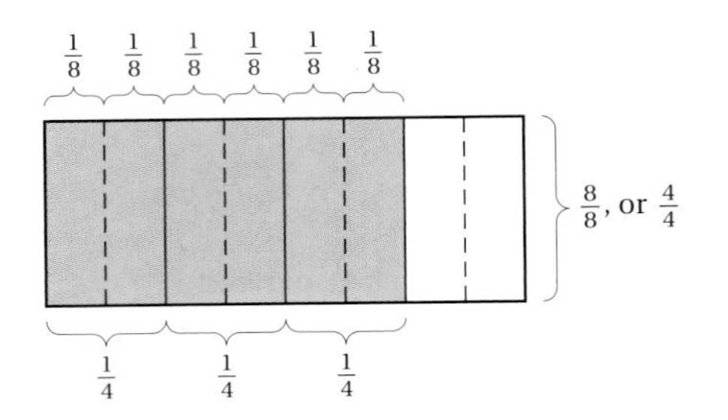

We see that there are six $\frac{1}{8}$'s in $\frac{3}{4}$. Thus,

$$\frac{3}{4} \div \frac{1}{8} = 6.$$

We can check this by multiplying:

$$6 \cdot \frac{1}{8} = \frac{6}{8} = \frac{3}{4}.$$

Here is a faster way to do this division:

$$\frac{3}{4} \div \frac{1}{8} = \frac{3}{4} \cdot \frac{8}{1} = \frac{24}{4} = 6.$$ Multiplying by the reciprocal of the divisor

To divide fractions, multiply the dividend by the reciprocal of the divisor:

$$\frac{2}{5} \div \frac{3}{4} = \frac{2}{5} \cdot \frac{4}{3} = \frac{2 \cdot 4}{5 \cdot 3} = \frac{8}{15}.$$

Multiply by the reciprocal of the divisor.

EXAMPLES Divide and simplify.

5. $\frac{5}{6} \div \frac{2}{3} = \frac{5}{6} \cdot \frac{3}{2} = \frac{5 \cdot 3}{6 \cdot 2} = \frac{5 \cdot 3}{3 \cdot 2 \cdot 2} = \frac{3}{3} \cdot \frac{5}{2 \cdot 2} = \frac{5}{2 \cdot 2} = \frac{5}{4}$

6. $\frac{7}{8} \div \frac{1}{16} = \frac{7}{8} \cdot 16 = \frac{7 \cdot 16}{8} = \frac{7 \cdot 2 \cdot 8}{8 \cdot 1} = \frac{8}{8} \cdot \frac{7 \cdot 2}{1} = \frac{7 \cdot 2}{1} = 14$

7. $\frac{2}{5} \div 6 = \frac{2}{5} \cdot \frac{1}{6} = \frac{2 \cdot 1}{5 \cdot 6} = \frac{2 \cdot 1}{5 \cdot 2 \cdot 3} = \frac{2}{2} \cdot \frac{1}{5 \cdot 3} = \frac{1}{5 \cdot 3} = \frac{1}{15}$

8. $\frac{3}{5} \div \frac{1}{2} = \frac{3}{5} \cdot 2 = \frac{3 \cdot 2}{5} = \frac{6}{5}$

Caution!

Canceling can be used as follows for Examples 5–7.

5. $\dfrac{5}{6} \div \dfrac{2}{3} = \dfrac{5}{6} \cdot \dfrac{3}{2} = \dfrac{5 \cdot 3}{6 \cdot 2} = \dfrac{5 \cdot \cancel{3}}{\cancel{3} \cdot 2 \cdot 2} = \dfrac{5}{2 \cdot 2} = \dfrac{5}{4}$ Removing a factor of 1: $\frac{3}{3} = 1$

6. $\dfrac{7}{8} \div \dfrac{1}{16} = \dfrac{7}{8} \cdot 16 = \dfrac{7 \cdot 16}{8} = \dfrac{7 \cdot \cancel{8} \cdot 2}{\cancel{8} \cdot 1} = \dfrac{7 \cdot 2}{1} = 14$ Removing a factor of 1: $\frac{8}{8} = 1$

7. $\dfrac{2}{5} \div 6 = \dfrac{2}{5} \cdot \dfrac{1}{6} = \dfrac{2 \cdot 1}{5 \cdot 6} = \dfrac{\cancel{2} \cdot 1}{5 \cdot \cancel{2} \cdot 3} = \dfrac{1}{5 \cdot 3} = \dfrac{1}{15}$ Removing a factor of 1: $\frac{2}{2} = 1$

Remember, if you can't factor, you can't cancel!

Do Exercises 5–9.

What is the explanation for multiplying by a reciprocal when dividing? Let's consider $\frac{2}{3} \div \frac{7}{5}$. We multiply by 1. The name for 1 that we will use is (5/7)/(5/7); it comes from the reciprocal of $\frac{7}{5}$.

$$\frac{2}{3} \div \frac{7}{5} = \frac{\frac{2}{3}}{\frac{7}{5}}$$ Writing fraction notation for the division

$$= \frac{\frac{2}{3}}{\frac{7}{5}} \cdot 1$$ Multiplying by 1

$$= \frac{\frac{2}{3}}{\frac{7}{5}} \cdot \frac{\frac{5}{7}}{\frac{5}{7}}$$ Multiplying by 1; $\frac{5}{7}$ is the reciprocal of $\frac{7}{5}$ and $\frac{\frac{5}{7}}{\frac{5}{7}} = 1$

$$= \frac{\frac{2}{3} \cdot \frac{5}{7}}{\frac{7}{5} \cdot \frac{5}{7}}$$ Multiplying the numerators and the denominators

$$= \frac{\frac{2}{3} \cdot \frac{5}{7}}{1}$$ After we multiplied, we got 1 for the denominator. The numerator shows the multiplication by the reciprocal.

$$= \frac{2}{3} \cdot \frac{5}{7} = \frac{10}{21}$$

Thus,

$$\frac{2}{3} \div \frac{7}{5} = \frac{2}{3} \cdot \frac{5}{7} = \frac{10}{21}.$$

Do Exercise 10.

Divide and simplify.

5. $\dfrac{6}{7} \div \dfrac{3}{4}$

6. $\dfrac{2}{3} \div \dfrac{1}{4}$

7. $\dfrac{4}{5} \div 8$

8. $60 \div \dfrac{3}{5}$

9. $\dfrac{3}{5} \div \dfrac{3}{5}$

10. Divide by multiplying by 1:

$$\frac{\frac{4}{5}}{\frac{6}{7}}.$$

Answers on page A-5

Solve.

11. $\frac{5}{6} \cdot y = \frac{2}{3}$

12. $\frac{3}{4} \cdot n = 24$

c Solving Equations

Now let's solve equations $a \cdot x = b$ and $x \cdot a = b$, where a and b may be fractions. We proceed as we did with equations involving whole numbers. We divide by a on both sides.

EXAMPLE 9 Solve: $\frac{4}{3} \cdot x = \frac{6}{7}$.

We have

$$\frac{4}{3} \cdot x = \frac{6}{7}$$

$$\frac{\frac{4}{3} \cdot x}{\frac{4}{3}} = \frac{\frac{6}{7}}{\frac{4}{3}} \quad \text{Dividing by } \tfrac{4}{3} \text{ on both sides}$$

$$x = \frac{6}{7} \cdot \frac{3}{4} \quad \text{Multiplying by the reciprocal}$$

$$= \frac{6 \cdot 3}{7 \cdot 4} = \frac{2 \cdot 3 \cdot 3}{7 \cdot 2 \cdot 2} = \frac{2}{2} \cdot \frac{3 \cdot 3}{7 \cdot 2} = \frac{3 \cdot 3}{7 \cdot 2} = \frac{9}{14}.$$

The solution is $\frac{9}{14}$.

EXAMPLE 10 Solve: $t \cdot \frac{4}{5} = 80$.

Dividing by $\frac{4}{5}$ on both sides, we get

$$t = 80 \div \frac{4}{5} = 80 \cdot \frac{5}{4} = \frac{80 \cdot 5}{4} = \frac{4 \cdot 20 \cdot 5}{4 \cdot 1} = \frac{4}{4} \cdot \frac{20 \cdot 5}{1} = \frac{20 \cdot 5}{1} = 100.$$

The solution is 100.

Do Exercises 11 and 12.

d Applications and Problem Solving

EXAMPLE 11 *Syringes.* How many syringes, each containing $\frac{3}{5}$ mL, can a clinical pharmacist fill from a 30 mL vial?

1. **Familiarize.** We are asking the question, "How many $\frac{3}{5}$'s are in 30?" Repeated addition will apply here. We make a drawing. We let $n =$ the number of syringes in all.

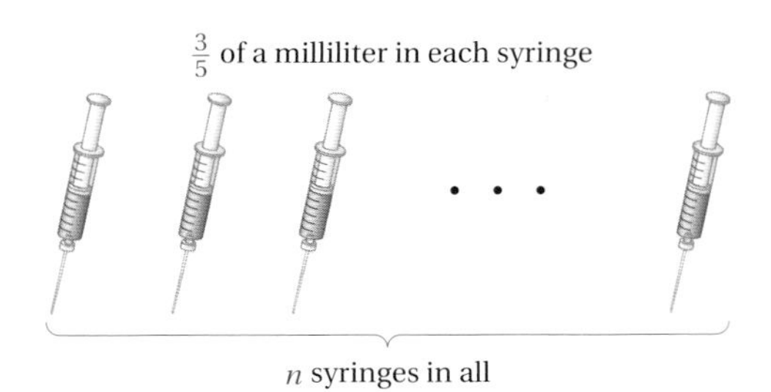

2. **Translate.** The equation that corresponds to the situation is

$$n = 30 \div \frac{3}{5}.$$

Answers on page A-5

3. Solve. We solve the equation by carrying out the division:

$$n = 30 \div \frac{3}{5} = 30 \cdot \frac{5}{3} = \frac{30 \cdot 5}{3} = \frac{3 \cdot 10 \cdot 5}{3 \cdot 1}$$
$$= \frac{3}{3} \cdot \frac{10 \cdot 5}{1} = 50.$$

4. Check. We check by repeating the calculation.

5. State. The pharmacist can fill 50 syringes.

Do Exercise 13.

EXAMPLE 12 *Hand-Knit Scarves.* Nayah knits winter scarves for the Quick-Knit Boutique. After she knits 63 inches of a scarf, $\frac{7}{8}$ of a scarf is complete. What is the total length of this scarf?

1. Familiarize. We ask: "63 in. is $\frac{7}{8}$ of what length?" We make a drawing or at least visualize the problem. We let s = the length of the scarf.

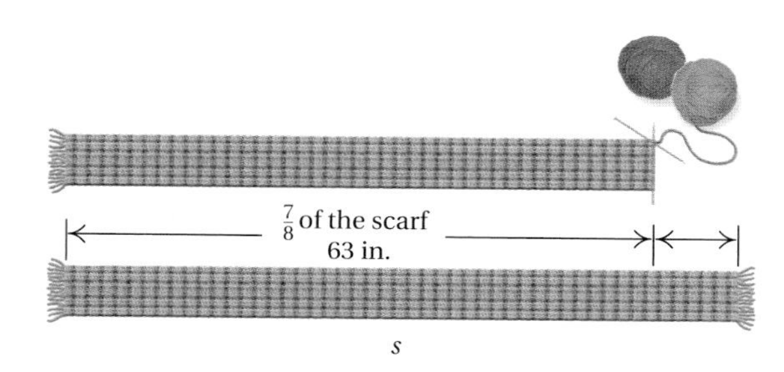

2. Translate. We translate to an equation.

Fraction completed	of	Total length of scarf	is	Amount already knitted
$\frac{7}{8}$	$\cdot$	s	$=$	63

3. Solve. The equation that corresponds to the situation is $\frac{7}{8} \cdot s = 63$. We divide by $\frac{7}{8}$ on both sides and carry out the division:

$$s = 63 \div \frac{7}{8} = 63 \cdot \frac{8}{7} = \frac{63 \cdot 8}{7} = \frac{7 \cdot 9 \cdot 8}{7} = \frac{7}{7} \cdot \frac{9 \cdot 8}{1} = 72.$$

4. Check. Since $\frac{7}{8} \cdot 72 = \frac{7 \cdot 8 \cdot 9}{8} = \frac{8}{8} \cdot \frac{7 \cdot 9}{1} = 63$, the answer, 72 checks.

5. State. The completed scarf is 72 in. long.

Do Exercise 14.

13. Each loop in a spring uses $\frac{3}{8}$ in. of wire. How many loops can be made from 120 in. of wire?

14. Sales Trip. Miles Lanosga sells soybean seeds to seed companies. After he had driven 210 mi, $\frac{5}{6}$ of his sales trip was completed. How long was the total trip?

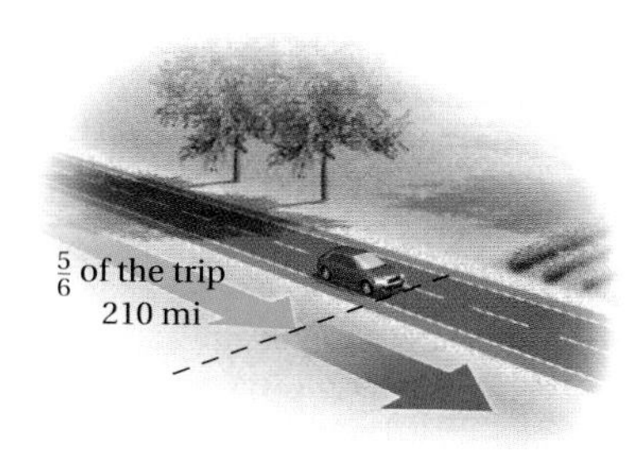

Answers on page A-5

Study Tips

STUDY RESOURCES

The new mathematical skills and concepts presented in class will be most easily learned if you begin the homework assignment as soon as possible after the lecture while they are fresh in your mind. Then if you have difficulty with any of the exercises, you have time to access supplementary resources such as:

- *Student's Solutions Manual*
- Video Lectures
- MathXL Tutorials on CD
- AW Math Tutor Center
- MathXL
- MyMathLab

Translating for Success

1. *Valentine Boxes.* Jane's Fudge Shop is preparing Valentine boxes. How many pounds of fudge will be needed to fill 80 boxes if each box contains $\frac{5}{16}$ lb?

2. *Gallons of Gasoline.* On the third day of a business trip, a sales representative used $\frac{4}{5}$ of a full tank of gasoline. If the tank is a 20-gallon tank, how many gallons were used on the third day?

3. *Purchasing a Shirt.* Tom received \$36 for his birthday. If he spends $\frac{3}{4}$ of the gift on a new shirt, what is the cost of the shirt?

4. *Checkbook Balance.* The balance in Sam's checking account is \$1456. He writes a check for \$28 and makes a deposit of \$52. What is the new balance?

5. *Valentine Boxes.* Jane's Fudge Shop prepared 80 lb of fudge for Valentine boxes. If each box contains $\frac{5}{16}$ lb, how many boxes can be filled?

The goal of these matching questions is to practice step (2), *Translate,* of the five-step problem-solving process. Translate each word problem to an equation and select a correct translation from equations A–O.

A. $x = \frac{3}{4} \cdot 36$

B. $28 \cdot x = 52$

C. $x = 80 \cdot \frac{5}{16}$

D. $x = 1456 \div 28$

E. $x = \frac{5}{4} \cdot 20$

F. $20 = \frac{4}{5} \cdot x$

G. $x = 12 \cdot 28$

H. $x = \frac{4}{5} \cdot 20$

I. $\frac{3}{4} \cdot x = 36$

J. $x = 1456 - 52 - 28$

K. $x \div 28 = 1456$

L. $x = 52 - 28$

M. $x = 52 \cdot 28$

N. $x = 1456 - 28 + 52$

O. $\frac{5}{16} \cdot x = 80$

Answers on page A-5

6. *Gasoline Tank.* A gasoline t contains 20 gal when it is $\frac{4}{5}$ fu How many gallons can it hol when full?

7. *Knitting a Scarf.* It takes Ra 36 hr to knit a scarf. She can only $\frac{3}{4}$ hr per day because sh taking 16 hr of college classe How many days will it take h to knit the scarf?

8. *Bicycle Trip.* On a recent 52-mile bicycle trip, David stopped to make a cellphone call after completing 28 mile How many more miles does bicycle after the call?

9. *Crème de Menthe Thins.* Ar Candies L.P. makes Crème d Menthe Thins. How many 28-piece packages can be fill with 1456 pieces?

10. *Cereal Donations.* The Williams family donates 28 boxes of cereal weekly to local Family in Crisis Center. How many boxes does this family donate in one year?

2.7 EXERCISE SET

For Extra Help

MathXL | MyMathLab | InterAct Math | Math Tutor Center | Digital Video Tutor CD 1 Videotape 2 | Student's Solutions Manual

a Find the reciprocal.

1. $\frac{5}{6}$

2. $\frac{7}{8}$

3. 6

4. 4

5. $\frac{1}{6}$

6. $\frac{1}{4}$

7. $\frac{10}{3}$

8. $\frac{17}{4}$

b Divide and simplify.

Don't forget to simplify!

9. $\frac{3}{5} \div \frac{3}{4}$

10. $\frac{2}{3} \div \frac{3}{4}$

11. $\frac{3}{5} \div \frac{9}{4}$

12. $\frac{6}{7} \div \frac{3}{5}$

13. $\frac{4}{3} \div \frac{1}{3}$

14. $\frac{10}{9} \div \frac{1}{3}$

15. $\frac{1}{3} \div \frac{1}{6}$

16. $\frac{1}{4} \div \frac{1}{5}$

17. $\frac{3}{8} \div 3$

18. $\frac{5}{6} \div 5$

19. $\frac{12}{7} \div 4$

20. $\frac{18}{5} \div 2$

21. $12 \div \frac{3}{2}$

22. $24 \div \frac{3}{8}$

23. $28 \div \frac{4}{5}$

24. $40 \div \frac{2}{3}$

25. $\frac{5}{8} \div \frac{5}{8}$

26. $\frac{2}{5} \div \frac{2}{5}$

27. $\frac{8}{15} \div \frac{4}{5}$

28. $\frac{6}{13} \div \frac{3}{26}$

29. $\frac{9}{5} \div \frac{4}{5}$

30. $\frac{5}{12} \div \frac{25}{36}$

31. $120 \div \frac{5}{6}$

32. $360 \div \frac{8}{7}$

c Solve.

33. $\frac{4}{5} \cdot x = 60$

34. $\frac{3}{2} \cdot t = 90$

35. $\frac{5}{3} \cdot y = \frac{10}{3}$

36. $\frac{4}{9} \cdot m = \frac{8}{3}$

37. $x \cdot \frac{25}{36} = \frac{5}{12}$

38. $p \cdot \frac{4}{5} = \frac{8}{15}$

39. $n \cdot \frac{8}{7} = 360$

40. $y \cdot \frac{5}{6} = 120$

d Solve.

41. Benny uses $\frac{2}{5}$ gram (g) of toothpaste each time he brushes his teeth. If Benny buys a 30-gram tube, how many times will he be able to brush his teeth?

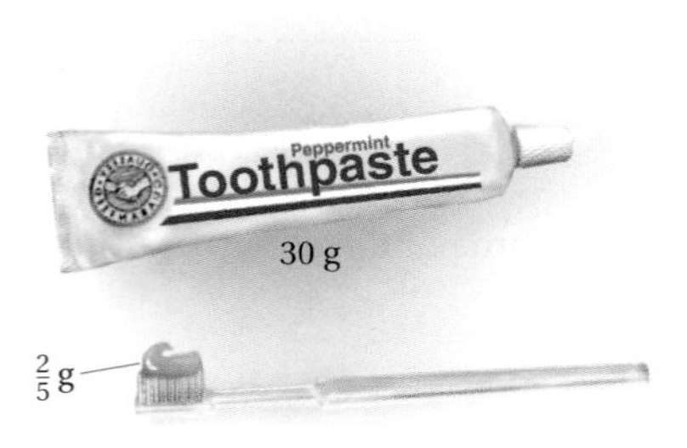

42. *Gasoline Tanker.* A tanker that delivers gasoline to gas stations had 1400 gal of gasoline when it was $\frac{7}{9}$ full. How much could the tanker hold when it is full?

43. A pair of basketball shorts requires $\frac{3}{4}$ yd of nylon. How many pairs of shorts can be made from 24 yd of nylon?

44. A child's baseball shirt requires $\frac{5}{6}$ yd of fabric. How many shirts can be made from 25 yd of the fabric?

45. How many $\frac{2}{3}$-cup sugar bowls can be filled from 16 cups of sugar?

46. How many $\frac{2}{3}$-cup cereal bowls can be filled from 10 cups of cornflakes?

47. A bucket had 12 L of water in it when it was $\frac{3}{4}$ full. How much could it hold altogether?

48. A tank had 20 L of gasoline in it when it was $\frac{4}{5}$ full. How much could it hold altogether?

49. Yoshi Teramoto sells hardware tools. After driving 180 kilometers (km), he completes $\frac{5}{8}$ of a sales trip. How long is the total trip? How many kilometers are left to drive?

50. A piece of coaxial cable $\frac{4}{5}$ meter (m) long is to be cut into 8 pieces of the same length. What is the length of each piece?

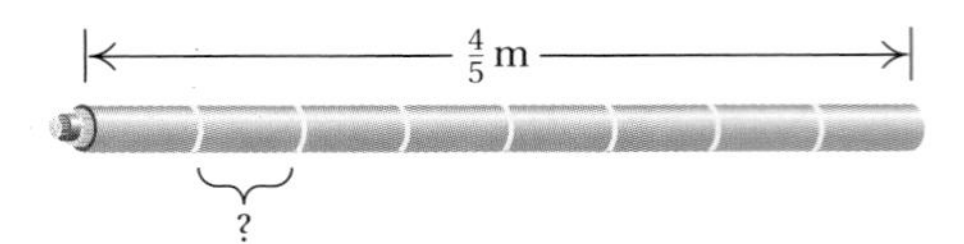

Pitch of a screw. Refer to Exercises 45 and 46 in Exercise Set 2.6.

51. After a screw has been turned 8 complete rotations, it is extended $\frac{1}{2}$ in. into a piece of wallboard. What is the pitch of the screw?

52. The pitch of a screw is $\frac{3}{32}$ in. How many complete rotations are necessary to drive the screw $\frac{3}{4}$ in. into a piece of pine wood?

53. $\mathbf{D_W}$ Without performing the division, explain why $5 \div \frac{1}{7}$ is a greater number than $5 \div \frac{2}{3}$.

54. $\mathbf{D_W}$ A student incorrectly insists that $\frac{2}{5} \div \frac{3}{4}$ is $\frac{15}{8}$. What mistake is he probably making?

SKILL MAINTENANCE

VOCABULARY REINFORCEMENT

In each of Exercises 55–62, fill in the blank with the correct term from the given list. Some of the choices may not be used.

55. The equation $14 + (2 + 30) = (14 + 2) + 30$ illustrates the ________________ law of addition. [1.2a]

56. In the product $10 \cdot \frac{3}{4}$, 10 and $\frac{3}{4}$ are called ________________. [2.1a]

57. A natural number that has exactly two different factors, only itself and 1, is called a ________________ number. [2.1c]

58. In the fraction $\frac{4}{17}$, we call 17 the ________________. [2.3a]

59. For any number a, $a + 0 = a$. The number 0 is the ________________ identity. [1.2a]

60. The product of 6 and $\frac{1}{6}$ is 1; we say that 6 and $\frac{1}{6}$ are ________________ of each other. [2.7a]

61. The set of ________________ numbers is 0, 1, 2, 3, 4, ... [1.1b]

62. A sentence with = is called a(n) ________________. [1.7a]

associative
commutative
additive
multiplicative
reciprocals
factors
prime
composite
numerator
denominator
equation
variables
whole
natural

SYNTHESIS

Simplify. Use the list of prime numbers on p. 93.

63. $\frac{711}{1957} \div \frac{10{,}033}{13{,}081}$

64. $\frac{8633}{7387} \div \frac{485}{581}$

65. $\left(\frac{9}{10} \div \frac{2}{5} \div \frac{3}{8}\right)^2$

66. $\dfrac{\left(\frac{3}{7}\right)^2 \div \frac{12}{5}}{\left(\frac{2}{9}\right)\left(\frac{9}{2}\right)}$

67. If $\frac{1}{3}$ of a number is $\frac{1}{4}$, what is $\frac{1}{2}$ of the number?

2 Summary and Review

The review that follows is meant to prepare you for a chapter exam. It consists of two parts. The first part, Concept Reinforcement, is designed to increase understanding of the concepts through true/false exercises. The second part is the Review Exercises. These provide practice exercises for the exam, together with references to section objectives so you can go back and review. Before beginning, stop and look back over the skills you have obtained. What skills in mathematics do you have now that you did not have before studying this chapter?

CONCEPT REINFORCEMENT

Determine whether the statement is true or false. Answers are given at the back of the book.

_______ **1.** A number a is divisible by another number b if b is a factor of a.

_______ **2.** If a number is not divisible by 6, then it is not divisible by 3.

_______ **3.** The number 1 is not prime.

_______ **4.** For any natural number n, $\frac{n}{n} > \frac{0}{n}$.

_______ **5.** A number is divisible by 10 if its ones digit is 0 or 5.

_______ **6.** If a number is divisible by 9, then it is also divisible by 3.

_______ **7.** The fraction $\frac{13}{6}$ is larger than the fraction $\frac{11}{6}$.

_______ **8.** The fraction $\frac{13}{7}$ is larger than the fraction $\frac{13}{6}$.

Review Exercises

Find all the factors of the number. [2.1a]

1. 60 **2.** 176

3. Multiply by 1, 2, 3, and so on, to find ten multiples of 8. [2.1b]

4. Determine whether 924 is divisible by 11. [2.1b]

5. Determine whether 1800 is divisible by 16. [2.1b]

Determine whether the number is prime, composite, or neither. [2.1c]

6. 37 **7.** 1 **8.** 91

Find the prime factorization of the number. [2.1d]

9. 70 **10.** 30

11. 45 **12.** 150

13. 648 **14.** 5250

To answer Exercises 15–22, consider the following numbers: [2.2a]

140	716	93	2802
95	2432	330	711
182	4344	255,555	
475	600	780	

Which of the above are divisible by the given number?

15. 3 **16.** 2

17. 4 **18.** 8

19. 5 **20.** 6

21. 9 **22.** 10

23. Identify the numerator and the denominator of $\frac{2}{7}$. [2.3a]

What part of the object or set of objects is shaded? [2.3a]

24.

25.

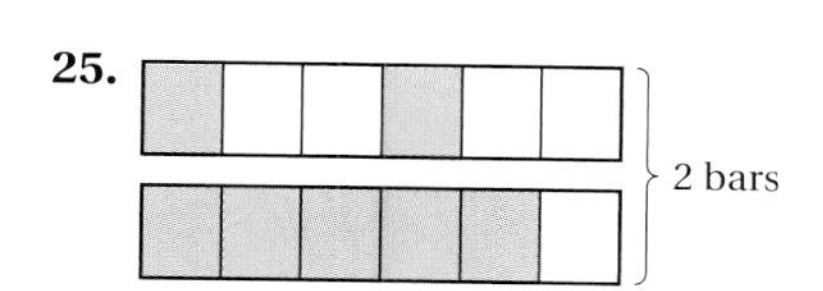

26.

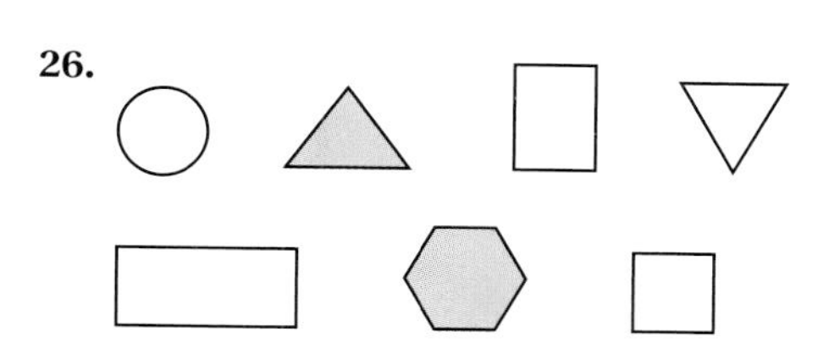

27. For a committee in the United States Senate that consists of 3 Democrats and 5 Republicans, what is the ratio of: [2.3a]

a) Democrats to Republicans?
b) Republicans to Democrats?
c) Democrats to the total number of members of the committee?

Simplify. [2.3b], [2.5b]

28. $\frac{0}{4}$ **29.** $\frac{23}{23}$ **30.** $\frac{48}{1}$

31. $\frac{48}{8}$ **32.** $\frac{10}{15}$ **33.** $\frac{7}{28}$

34. $\frac{21}{21}$ **35.** $\frac{0}{25}$ **36.** $\frac{12}{30}$

37. $\frac{18}{1}$ **38.** $\frac{32}{8}$ **39.** $\frac{9}{27}$

40. $\frac{18}{0}$ **41.** $\frac{5}{8-8}$ **42.** $\frac{88}{184}$

43. $\frac{140}{490}$ **44.** $\frac{1170}{1200}$ **45.** $\frac{288}{2025}$

46. Simplify, if possible, the fractions on this circle graph. [2.5b]

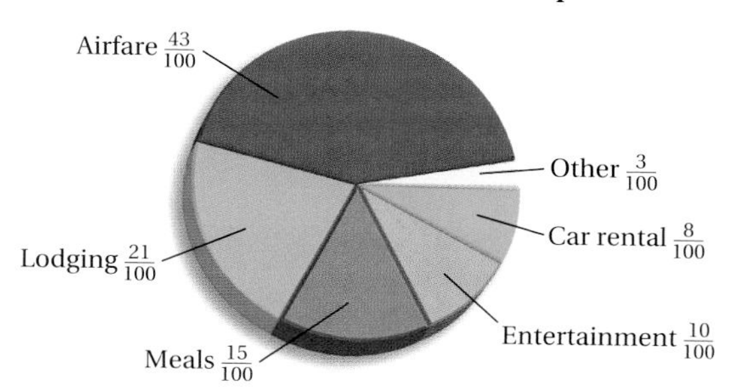

Use = or ≠ for □ to write a true sentence. [2.5c]

47. $\frac{3}{5}$ □ $\frac{4}{6}$ **48.** $\frac{4}{7}$ □ $\frac{8}{14}$

49. $\frac{4}{5}$ □ $\frac{5}{6}$ **50.** $\frac{4}{3}$ □ $\frac{28}{21}$

Multiply and simplify. [2.6a]

51. $4 \cdot \frac{3}{8}$ **52.** $\frac{7}{3} \cdot 24$

53. $9 \cdot \frac{5}{18}$ **54.** $\frac{6}{5} \cdot 20$

55. $\frac{3}{4} \cdot \frac{8}{9}$ **56.** $\frac{5}{7} \cdot \frac{1}{10}$

57. $\frac{3}{7} \cdot \frac{14}{9}$ **58.** $\frac{1}{4} \cdot \frac{2}{11}$

59. $\frac{4}{25} \cdot \frac{15}{16}$ **60.** $\frac{11}{3} \cdot \frac{30}{77}$

Find the reciprocal. [2.7a]

61. $\frac{4}{5}$

62. 3

63. $\frac{1}{9}$

64. $\frac{47}{36}$

Divide and simplify. [2.7b]

65. $6 \div \frac{4}{3}$

66. $\frac{5}{9} \div \frac{5}{18}$

67. $\frac{1}{6} \div \frac{1}{11}$

68. $\frac{3}{14} \div \frac{6}{7}$

69. $\frac{1}{4} \div \frac{1}{9}$

70. $180 \div \frac{3}{5}$

71. $\frac{23}{25} \div \frac{23}{25}$

72. $\frac{2}{3} \div \frac{3}{2}$

Solve. [2.7c]

73. $\frac{5}{4} \cdot t = \frac{3}{8}$

74. $x \cdot \frac{2}{3} = 160$

Solve. [2.6b], [2.7d]

75. A road crew repaves $\frac{1}{12}$ mi of road each day. How long will it take the crew to repave a $\frac{3}{4}$-mile stretch of road?

76. After driving 600 km, the Buxton family has completed $\frac{3}{5}$ of their vacation. How long is the total trip?

77. Molly is making a pepper steak recipe that calls for $\frac{2}{3}$ cup of green bell peppers. How much would be needed to make $\frac{1}{2}$ recipe? 3 recipes?

78. Bernardo usually earns \$105 for working a full day. How much does he receive for working $\frac{1}{7}$ of a day?

79. A book bag requires $\frac{4}{5}$ yd of fabric. How many bags can be made from 48 yd?

80. *Corn Production.* In 2003, the United States produced approximately $\frac{2}{5}$ of the world production of corn. The total world corn production was about 640,000,000 metric tons. How much corn did the United States produce?

Source: UN Food and Agriculture Organization

81. D_W Write, in your own words, a series of steps that can be used when simplifying fraction notation. [2.5b]

82. D_W A student claims that "taking $\frac{1}{2}$ of a number is the same as dividing by $\frac{1}{2}$." Explain the error in this reasoning. [2.7b]

83. D_W Use the number 9432 to explain why the test for divisibility by 9 works. [2.2a]

SYNTHESIS

84. In the division below, find a and b. [2.7b]

$$\frac{19}{24} \div \frac{a}{b} = \frac{187{,}853}{268{,}224}$$

85. A prime number that becomes a prime number when its digits are reversed is called a **palindrome prime.** For example, 17 is a palindrome prime because both 17 and 71 are primes. Which of the following numbers are palindrome primes? [2.1c]

13, 91, 16, 11, 15, 24, 29, 101, 201, 37

2 Chapter Test

For Extra Help

1. Find all the factors of 300.

Determine whether the number is prime, composite, or neither.

2. 41

3. 14

Find the prime factorization of the number.

4. 18

5. 60

6. Determine whether 1784 is divisible by 8.

7. Determine whether 784 is divisible by 9.

8. Determine whether 5552 is divisible by 5.

9. Determine whether 2322 is divisible by 6.

10. Identify the numerator and the denominator of $\frac{4}{5}$.

11. What part is shaded?

12. What part of the set is shaded?

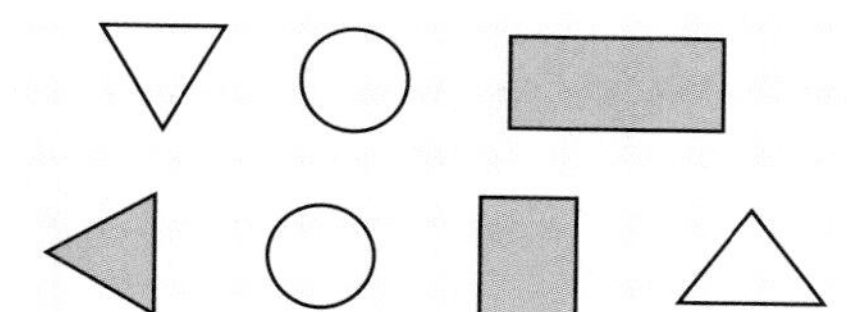

13. *Pass Completion Ratio.* In the 2004 regular season, Peyton Manning, of the Indianapolis Colts, completed 336 of 497 passes to become the No. 1 rated quarterback for the second year in a row as Associated Press NFL most valuable player.

a) What was the ratio of pass completions to attempts?
b) What was the ratio of incomplete passes to attempts?

Simplify.

14. $\frac{26}{1}$

15. $\frac{12}{12}$

16. $\frac{0}{16}$

17. $\frac{12}{24}$

18. $\frac{42}{7}$

19. $\frac{2}{28}$

20. $\frac{9}{0}$

21. $\frac{7}{2-2}$

22. $\frac{35}{140}$

23. $\frac{72}{108}$

Use $=$ or $\neq$ for $\square$ to write a true sentence.

24. $\frac{3}{4} \square \frac{6}{8}$

25. $\frac{5}{4} \square \frac{9}{7}$

Multiply and simplify.

26. $\frac{4}{3} \cdot 24$

27. $5 \cdot \frac{3}{10}$

28. $\frac{2}{3} \cdot \frac{15}{4}$

29. $\frac{3}{5} \cdot \frac{1}{6}$

30. $\frac{22}{15} \cdot \frac{5}{33}$

Find the reciprocal.

31. $\frac{5}{8}$

32. $\frac{1}{4}$

33. 18

Divide and simplify.

34. $\frac{3}{8} \div \frac{5}{4}$

35. $\frac{1}{5} \div \frac{1}{8}$

36. $12 \div \frac{2}{3}$

37. $\frac{24}{5} \div \frac{28}{15}$

Solve.

38. $\frac{7}{8} \cdot x = 56$

39. $t \cdot \frac{2}{5} = \frac{7}{10}$

40. There are 7000 students at La Poloma College, and $\frac{5}{8}$ of them live in dorms. How many live in dorms?

41. A strip of taffy $\frac{9}{10}$ m long is cut into 12 equal pieces. What is the length of each piece?

42. A thermos of iced tea had 3 qt of tea when it was $\frac{3}{5}$ full. How much tea could it hold when full?

43. The pitch of a screw is $\frac{1}{8}$ in. How far will it go into a piece of walnut when it is turned 6 complete rotations?

SYNTHESIS

44. A recipe for a batch of buttermilk pancakes calls for $\frac{3}{4}$ teaspoon (tsp) of salt. Jacqueline plans to cut the amount of salt in half for each of 5 batches of pancakes. How much salt will she need?

45. Grandma Hammons left $\frac{2}{3}$ of her $\frac{7}{8}$-acre apple farm to Karl. Karl gave $\frac{1}{4}$ of his share to his oldest daughter, Eileen. How much land did Eileen receive?

46. Simplify: $\left(\frac{3}{8}\right)^2 \div \frac{6}{7} \cdot \frac{2}{9} \div 5.$

47. Solve: $\frac{33}{38} \cdot \frac{34}{55} = \frac{17}{35} \cdot \frac{15}{19}x.$

Fraction Notation and Mixed Numerals

3

Real-World Application

Arie Luyendyk won the Indianapolis 500 in 1990 with the highest average speed of about 186 mph. This record high through 2004 is about $2\frac{12}{25}$ times the average speed of the first winner, Ray Harroun, in 1911. What was the average speed in the first Indianapolis 500?

Source: Indianapolis Motor Speedway

This problem appears as Example 9 in Section 3.6.

3.1 LEAST COMMON MULTIPLES

Objective

a Find the least common multiple, LCM, of two or more numbers.

1. By examining lists of multiples, find the LCM of 9 and 15.

In this chapter, we study addition and subtraction using fraction notation. Suppose we want to add $\frac{2}{3}$ and $\frac{1}{2}$. To do so, we rewrite the fractions with a common denominator. The number we choose for the common denominator is the least common multiple of the denominators, 6; $\frac{2}{3} + \frac{1}{2} = \frac{4}{6} + \frac{3}{6}$. Then we add the numerators and keep the common denominator. In order to do this, we must be able to find the **least common denominator (LCD),** or **least common multiple (LCM)** of the denominators. (A review of Section 2.1b might be helpful.)

a Finding Least Common Multiples

LEAST COMMON MULTIPLE, LCM

The **least common multiple,** or LCM, of two natural numbers is the smallest number that is a multiple of both numbers.

EXAMPLE 1 Find the LCM of 20 and 30.

First list some multiples of 20 by multiplying 20 by 1, 2, 3, and so on:

20, 40, 60, 80, 100, 120, 140, 160, 180, 200, 220, 240, ...

Then list some multiples of 30 by multiplying 30 by 1, 2, 3, and so on:

30, 60, 90, 120, 150, 180, 210, 240, ...

Now we determine the smallest number common to both lists. The LCM of 20 and 30 is 60.

Do Exercise 1.

Next we develop three methods that are more efficient for finding LCMs. You may choose to learn only one method (consult with your instructor), but if you are going to study algebra, you should definitely learn method 2.

METHOD 1: FINDING LCMS USING ONE LIST OF MULTIPLES

One method for finding LCMs uses *one* list of multiples. Let's consider finding the LCM of 9 and 12. The largest number, 12, is not a multiple of 9; so we check multiples of 12 until we find a number that is also a multiple of 9:

$1 \cdot 12 = 12$, not a multiple of 9;

$2 \cdot 12 = 24$, not a multiple of 9;

$3 \cdot 12 = 36$, a multiple of 9: $4 \cdot 9 = 36$.

The LCM of 9 and 12 is 36.

Answer on page A-6

Method 1. To find the LCM of a set of numbers using a list of multiples:

a) Determine whether the largest number is a multiple of the others. If it is, it is the LCM. That is, if the largest number has the others as factors, the LCM is that number.

b) If not, check multiples of the largest number until you get one that is a multiple of each of the others.

EXAMPLE 2 Find the LCM of 12 and 15.

a) 15 is not a multiple of 12.

b) Check multiples of 15:

$1 \cdot 15 = 15,$ Not a multiple of 12. When we divide 15 by 12, we get a nonzero remainder.

$2 \cdot 15 = 30,$ Not a multiple of 12

$3 \cdot 15 = 45,$ Not a multiple of 12

$4 \cdot 15 = 60.$ A multiple of 12: $5 \cdot 12 = 60$

The LCM = 60.

Do Exercise 2.

EXAMPLE 3 Find the LCM of 4 and 14.

a) 14 is not a multiple of 4.

b) Check multiples of 14:

$1 \cdot 14 = 14,$

$2 \cdot 14 = 28.$ A multiple of 4: $7 \cdot 4 = 28$

The LCM = 28.

EXAMPLE 4 Find the LCM of 8 and 32.

a) 32 is a multiple of 8 ($4 \cdot 8 = 32$), so the LCM = 32.

EXAMPLE 5 Find the LCM of 10, 100, and 1000.

a) 1000 is a multiple of 10 ($100 \cdot 10 = 1000$) and of 100 ($10 \cdot 100 = 1000$), so the LCM = 1000.

Do Exercises 3–6.

METHOD 2: FINDING LCMS USING PRIME FACTORIZATIONS

A second method for finding LCMs uses prime factorizations. Consider again 20 and 30. Their prime factorizations are $20 = 2 \cdot 2 \cdot 5$ and $30 = 2 \cdot 3 \cdot 5$. Let's look at these prime factorizations in order to find the LCM. Any multiple of 20 will have to have *two* 2's as factors and *one* 5 as a factor. Any multiple of 30 will have to have *one* 2, *one* 3, and *one* 5 as factors. The smallest number satisfying these conditions is

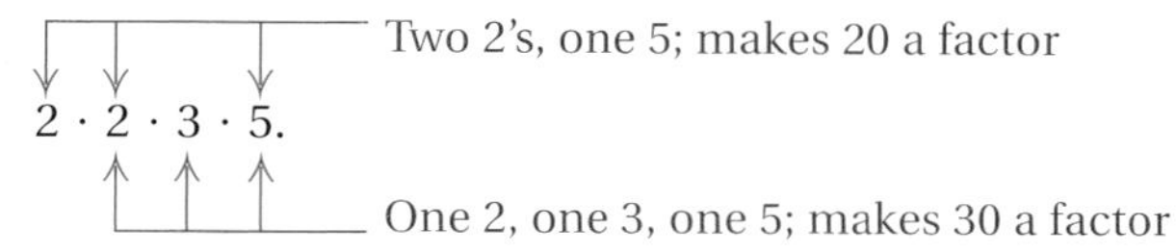

Two 2's, one 5; makes 20 a factor
2 · 2 · 3 · 5.
One 2, one 3, one 5; makes 30 a factor

2. By examining lists of multiples, find the LCM of 8 and 10.

Find the LCM.

3. 10, 15

4. 6, 8

5. 5, 10

6. 20, 40, 80

Answers on page A-6

Use prime factorizations to find the LCM.

7. 8, 10

8. 18, 40

9. 32, 54

Answers on page A-6

Thus the LCM of 20 and 30 is $2 \cdot 2 \cdot 3 \cdot 5$, or 60. It has all the factors of 20 and all the factors of 30, but the factors are not repeated when they are common to both numbers.

Observe that the greatest number of times 2 occurs as a factor of either 20 or 30 is two, and the LCM has 2 as a factor twice. The greatest number of times 3 occurs as a factor of either 20 or 30 is one, and the LCM has 3 as a factor once. The greatest number of times 5 occurs as a factor of either 20 or 30 is one, and the LCM has 5 as a factor once.

Method 2. To find the LCM of a set of numbers using prime factorizations:

a) Find the prime factorization of each number.
b) Create a product of factors, using each factor the greatest number of times that it occurs in any one factorization.

EXAMPLE 6 Find the LCM of 6 and 8.

a) Find the prime factorization of each number.

$$6 = 2 \cdot 3, \qquad 8 = 2 \cdot 2 \cdot 2$$

b) Create a product by writing factors that appear in the factorizations of 6 and 8, using each the greatest number of times that it occurs in any one factorization.

Consider the factor 2. The greatest number of times that 2 occurs in any one factorization is three. We write 2 as a factor three times.

$$2 \cdot 2 \cdot 2 \cdot ?$$

Consider the factor 3. The greatest number of times that 3 occurs in any one factorization is one. We write 3 as a factor one time.

$$2 \cdot 2 \cdot 2 \cdot 3 \cdot ?$$

Since there are no other prime factors in either factorization, the

LCM is $2 \cdot 2 \cdot 2 \cdot 3$, or 24.

EXAMPLE 7 Find the LCM of 24 and 36.

a) Find the prime factorization of each number.

$$24 = 2 \cdot 2 \cdot 2 \cdot 3, \qquad 36 = 2 \cdot 2 \cdot 3 \cdot 3$$

b) Create a product by writing factors, using each the greatest number of times that it occurs in any one factorization.

Consider the factor 2. The greatest number of times that 2 occurs in any one factorization is three. We write 2 as a factor three times:

$$2 \cdot 2 \cdot 2 \cdot ?$$

Consider the factor 3. The greatest number of times that 3 occurs in any one factorization is two. We write 3 as a factor two times:

$$2 \cdot 2 \cdot 2 \cdot 3 \cdot 3 \cdot ?$$

Since there are no other prime factors in either factorization, the

LCM is $2 \cdot 2 \cdot 2 \cdot 3 \cdot 3$, or 72.

Do Exercises 7–9.

EXAMPLE 8 Find the LCM of 27, 90, and 84.

a) Find the prime factorization of each number.

$27 = 3 \cdot 3 \cdot 3, \qquad 90 = 2 \cdot 3 \cdot 3 \cdot 5, \qquad 84 = 2 \cdot 2 \cdot 3 \cdot 7$

b) Create a product by writing factors, using each the greatest number of times that it occurs in any one factorization.

Consider the factor 2. The greatest number of times that 2 occurs in any one factorization is two. We write 2 as a factor two times:

$2 \cdot 2 \cdot ?$

Consider the factor 3. The greatest number of times that 3 occurs in any one factorization is three. We write 3 as a factor three times:

$2 \cdot 2 \cdot 3 \cdot 3 \cdot 3 \cdot ?$

Consider the factor 5. The greatest number of times that 5 occurs in any one factorization is one. We write 5 as a factor one time:

$2 \cdot 2 \cdot 3 \cdot 3 \cdot 3 \cdot 5 \cdot ?$

Consider the factor 7. The greatest number of times that 7 occurs in any one factorization is one. We write 7 as a factor one time:

$2 \cdot 2 \cdot 3 \cdot 3 \cdot 3 \cdot 5 \cdot 7 \cdot ?$

Since there are no other prime factors in any of the factorizations, the

LCM is $2 \cdot 2 \cdot 3 \cdot 3 \cdot 3 \cdot 5 \cdot 7$, or 3780.

Do Exercise 10.

The use of exponents might be helpful to you as an extension of the factorization method. Let's reconsider Example 8. We want to find the LCM of 27, 90, and 84. We factor and then convert to exponential notation:

$27 = 3 \cdot 3 \cdot 3 = 3^3,$

$90 = 2 \cdot 3 \cdot 3 \cdot 5 = 2^1 \cdot 3^2 \cdot 5^1,$ and

$84 = 2 \cdot 2 \cdot 3 \cdot 7 = 2^2 \cdot 3^1 \cdot 7^1.$

Note that in 84, the 2 in 2^2 is the largest exponent of 2 in any of the factorizations. It is also the exponent of 2 in the LCM. It indicates the greatest number of times that 2 occurs as a factor of any of the numbers. Similarly, in 27, the 3 in 3^3 is the largest exponent of 3 in any of the factorizations. It is also the exponent of 3 in the LCM. Likewise, the 1's in 5^1 and 7^1 tell us the exponents of 5 and 7 in the LCM. They indicate the greatest number of times that 5 and 7 occur as factors. Thus the

LCM is $2^2 \cdot 3^3 \cdot 5^1 \cdot 7^1$, or 3780.

Similarly, the LCMs in Examples 6 and 7 can be expressed with exponents as $2^3 \cdot 3$ and $2^3 \cdot 3^2$, respectively.

Do Exercises 11 and 12.

10. Find the LCM of 24, 35, and 45.

11. Use exponents to find the LCM of 24, 35, and 45.

12. Redo Margin Exercises 7–9 using exponents.

Answers on page A-6

Find the LCM.

13. 3, 18

14. 12, 24

EXAMPLE 9 Find the LCM of 7 and 21.

We find the prime factorization of each number. Because 7 is prime, it has no prime factorization.

$$7 = 7, \qquad 21 = 3 \cdot 7$$

Note that 7 is a factor of 21. We stated earlier that if one number is a factor of another, the LCM is the larger of the numbers. Thus the LCM is $7 \cdot 3$, or 21.

Do Exercises 13 and 14.

EXAMPLE 10 Find the LCM of 8 and 9.

We find the prime factorization of each number.

$$8 = 2 \cdot 2 \cdot 2, \qquad 9 = 3 \cdot 3$$

Note that the two numbers, 8 and 9, have no common prime factor. When this is the case, the LCM is just the product of the two numbers. Thus the LCM is $2 \cdot 2 \cdot 2 \cdot 3 \cdot 3$, or $8 \cdot 9$, or 72.

Do Exercises 15 and 16.

Find the LCM.

15. 4, 9

16. 5, 6, 7

Let's compare the two methods considered so far for finding LCMs: the multiples method and the factorization method.

Method 1, the **multiples method,** can be longer than the factorization method when the LCM is large or when there are more than two numbers. But this method can be faster and easier to use mentally for two numbers.

Method 2, the **factorization method,** works well for several numbers. It is just like a method used in algebra. If you are going to study algebra, you should definitely learn the factorization method.

METHOD 3: FINDING LCMS USING DIVISION BY PRIMES

Here is another method for finding LCMs that is especially useful for three or more numbers. For example, to find the LCM of 48, 72, and 80, we first look for any prime that divides any two of the numbers with no remainder. Then we divide as follows.

$$\begin{array}{r|rrr} 2 & 48 & 72 & 80 \\ \hline & 24 & 36 & 40 \end{array}$$

We repeat the process, bringing down any numbers not divisible by the prime, until we can divide no more, that is, until there are no two numbers divisible by the same prime:

$$\begin{array}{r|rrrl} 2 & 48 & 72 & 80 & \\ \hline 3 & 24 & 36 & 40 & \\ \hline 2 & 8 & 12 & 40 & \text{40 is not divisible by 3.} \\ \hline 2 & 4 & 6 & 20 & \\ \hline 2 & 2 & 3 & 10 & \\ \hline & 1 & 3 & 5 & \text{3 is not divisible by 2.} \end{array}$$

No two of 1, 3, and 5 are divisible by the same prime. We stop here. The LCM is

$$2 \cdot 3 \cdot 2 \cdot 2 \cdot 2 \cdot 1 \cdot 3 \cdot 5, \text{ or } 720.$$

Answers on page A-6

Method 3: To find the LCM using division by primes:

a) First look for any prime that divides at least two of the numbers with no remainder. Then divide, bringing down any numbers not divisible by the prime.

b) Repeat the process until you can divide no more, that is, until there are no two numbers divisible by the same prime.

EXAMPLE 11 Find the LCM of 24, 35, and 45.

```
5 | 24  35  45
3 | 24   7   9     24 is not divisible by 5.
     8   7   3     7 is not divisible by 3.
```

The LCM is $5 \cdot 3 \cdot 8 \cdot 7 \cdot 3$, or 2520.

EXAMPLE 12 Find the LCM of 12, 18, 20, and 21.

```
3 | 12  18  20  21
2 |  4   6  20   7
2 |  2   3  10   7
     1   3   5   7
```

The LCM is $3 \cdot 2 \cdot 2 \cdot 1 \cdot 3 \cdot 5 \cdot 7$, or 1260.

Do Exercises 17–19.

Find the LCM using division by primes.

17. 12, 75, 120

18. 27, 90, 84

19. 12, 24, 75, 120

Answers on page A-6

Study Tips

TIPS FROM A FORMER STUDENT

A former student of Professor Bittinger, Mike Rosenborg, earned a master's degree in mathematics and now teaches mathematics. Here are some of his study tips.

- Because working problems is the best way to learn math, instructors generally assign lots of problems. Never let yourself get behind in your math homework.
- If you are struggling with a math concept, do not give up. Ask for help from your friends and your instructor. Since each concept is built on previous concepts, any gaps in your understanding will follow you through the entire course, so make sure you understand each concept as you go along.
- Read your textbook! It will often contain the help and tips you need to solve any problem with which you are struggling. It may also bring out points that you missed in class or that your instructor may not have covered.
- Learn to use scratch paper to jot down your thoughts and to draw pictures. Don't try to figure everything out "in your head." You will think more clearly and accurately this way.
- When preparing for a test, it is often helpful to work at least two problems per section as practice: one easy and one difficult. Write out all the new rules and procedures your test will cover, and then read through them twice. Doing so will enable you to both learn and retain them better.
- Most schools have classrooms set up where you can get free help from math tutors. Take advantage of this, but be sure you do the work first. Don't let your tutor do all the work for you—otherwise you'll never learn the material.
- In math, as in many other areas of life, patience and persistence are virtues—cultivate them. "Cramming" for an exam will not help you learn and retain the material.

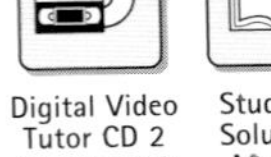

3.1 EXERCISE SET

For Extra Help

MathXL | MyMathLab | InterAct Math | Math Tutor Center | Digital Video Tutor CD 2 Videotape 3 | Student's Solutions Manual

a Find the LCM of the set of numbers.

1. 2, 4
2. 3, 15
3. 10, 25
4. 10, 15
5. 20, 40

6. 8, 12
7. 18, 27
8. 9, 11
9. 30, 50
10. 24, 36

11. 30, 40
12. 21, 27
13. 18, 24
14. 12, 18
15. 60, 70

16. 35, 45
17. 16, 36
18. 18, 20
19. 32, 36
20. 36, 48

21. 2, 3, 5
22. 5, 18, 3
23. 3, 5, 7
24. 6, 12, 18
25. 24, 36, 12

26. 8, 16, 22
27. 5, 12, 15
28. 12, 18, 40
29. 9, 12, 6
30. 8, 16, 12

31. 180, 100, 450, 60
32. 18, 30, 50, 48
33. 8, 48
34. 16, 32
35. 5, 50

36. 12, 72
37. 11, 13
38. 13, 14
39. 12, 35
40. 23, 25

41. 54, 63
42. 56, 72
43. 81, 90
44. 75, 100
45. 36, 54, 80

46. 22, 42, 51
47. 39, 91, 108, 26
48. 625, 75, 500, 25

Applications of LCMs: Planet Orbits. The earth, Jupiter, Saturn, and Uranus all revolve around the sun. The earth takes 1 yr, Jupiter 12 yr, Saturn 30 yr, and Uranus 84 yr to make a complete revolution. On a certain night, you look at those three distant planets and wonder how many years it will take before they have the same position again. (*Hint*: To find out, you find the LCM of 12, 30, and 84. It will be that number of years.)

Source: *The Handy Science Answer Book*

49. How often will Jupiter and Saturn appear in the same direction in the night sky as seen from the earth?

50. How often will Jupiter and Uranus appear in the same direction in the night sky as seen from the earth?

51. How often will Saturn and Uranus appear in the same direction in the night sky as seen from the earth?

52. How often will Jupiter, Saturn, and Uranus appear in the same direction in the night sky as seen from the earth?

53. $\mathbf{D_W}$ Use both methods 1 and 2 to find the LCM of each of the following sets of numbers.

a) 6, 8 **b)** 6, 7 **c)** 6, 21 **d)** 24, 36

Which method do you consider more efficient? Explain why.

54. $\mathbf{D_W}$ Is the LCM of two numbers always larger than either number? Why or why not?

SKILL MAINTENANCE

Solve.

55. Joy uses $\frac{1}{2}$ yd of dental floss each day. How long will a 45-yd container of dental floss last for Joy? [2.7d]

56. *Broadway Runs.* As of May 3, 2004, *Cats* had the longest Broadway run with 7485 performances from October 1982 to September 2000. From June 1988 to the present, *The Phantom of the Opera* has had 6781 performances. How many more performances of *The Phantom of the Opera* are needed to equal the record set by the *Cats* show? [1.8a]

Source: League of American Theatres and Producers, Inc.

57. Add: 23,456 + 5677 + 4002. [1.2a]

58. Subtract: 10,007 − 3068. [1.3b]

59. Multiply and simplify: $\frac{4}{5} \cdot \frac{10}{12}$. [2.6a]

60. Divide and simplify: $\frac{4}{5} \div \frac{7}{10}$. [2.7b]

SYNTHESIS

61. Find the LCM of 27, 90, 84, 210, 108, and 50.

62. Find the LCM of 18, 21, 24, 36, 63, 56, and 20.

63. A pencil company uses two sizes of boxes, 5 in. by 6 in. and 5 in. by 8 in. These boxes are packed in bigger cartons for shipping. Find the width and the length of the smallest carton that will accommodate boxes of either size without any room left over. (Each carton can contain only one type of box and all boxes must point in the same direction.)

64. Consider 8 and 12. Determine whether each of the following is the LCM of 8 and 12. Tell why or why not.

a) $2 \cdot 2 \cdot 3 \cdot 3$
b) $2 \cdot 2 \cdot 3$
c) $2 \cdot 3 \cdot 3$
d) $2 \cdot 2 \cdot 2 \cdot 3$

3.2 ADDITION AND APPLICATIONS

Objectives

a Add using fraction notation.

b Solve applied problems involving addition with fraction notation.

1. Find $\frac{1}{5} + \frac{3}{5}$.

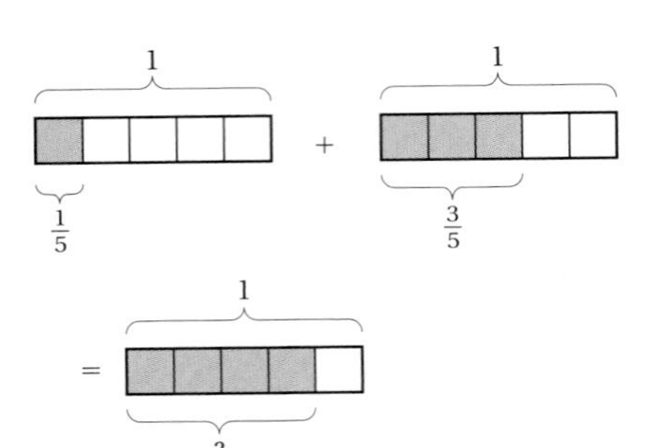

1
1
+
1/5
3/5
1
=
?

Add and simplify.

2. $\frac{1}{3} + \frac{2}{3}$

3. $\frac{5}{12} + \frac{1}{12}$

4. $\frac{9}{16} + \frac{3}{16}$

Answers on page A-6

a Addition Using Fraction Notation

LIKE DENOMINATORS

Addition using fraction notation corresponds to combining or putting like things together, just as addition with whole numbers does. For example,

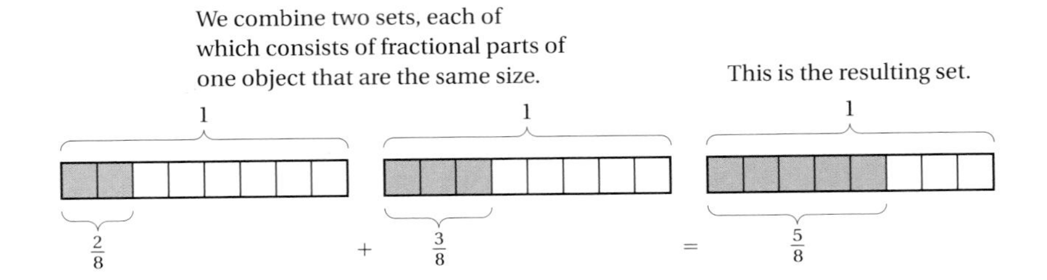

We combine two sets, each of which consists of fractional parts of one object that are the same size.
This is the resulting set.
1
1
1
2/8
+
3/8
=
5/8

2 eighths + 3 eighths = 5 eighths,

or $2 \cdot \frac{1}{8} + 3 \cdot \frac{1}{8} = 5 \cdot \frac{1}{8}$, or $\frac{2}{8} + \frac{3}{8} = \frac{5}{8}$.

We see that to add when denominators are the same, we add the numerators and keep the denominator.

Do Exercise 1.

To add when denominators are the same,

a) add the numerators,

b) keep the denominator, and

c) simplify, if possible.

$$\frac{2}{6} + \frac{5}{6} = \frac{2+5}{6} = \frac{7}{6}$$

EXAMPLES Add and simplify.

1. $\frac{2}{4} + \frac{1}{4} = \frac{2+1}{4} = \frac{3}{4}$ No simplifying is possible.

2. $\frac{11}{6} + \frac{3}{6} = \frac{11+3}{6} = \frac{14}{6} = \frac{2 \cdot 7}{2 \cdot 3} = \frac{2}{2} \cdot \frac{7}{3} = 1 \cdot \frac{7}{3} = \frac{7}{3}$ Here we simplified.

3. $\frac{3}{12} + \frac{5}{12} = \frac{3+5}{12} = \frac{8}{12} = \frac{4 \cdot 2}{4 \cdot 3} = \frac{4}{4} \cdot \frac{2}{3} = 1 \cdot \frac{2}{3} = \frac{2}{3}$

Do Exercises 2–4.

DIFFERENT DENOMINATORS

What do we do when denominators are different? We can find a common denominator by multiplying by 1. Consider adding $\frac{1}{6}$ and $\frac{3}{4}$. There are many common denominators that can be obtained. Let's look at two possibilities.

A. $\frac{1}{6}+\frac{3}{4}=\frac{1}{6}\cdot 1+\frac{3}{4}\cdot 1$

$=\frac{1}{6}\cdot\frac{4}{4}+\frac{3}{4}\cdot\frac{6}{6}$

$=\frac{4}{24}+\frac{18}{24}$

$=\frac{22}{24}$

$=\frac{11}{12}$

B. $\frac{1}{6}+\frac{3}{4}=\frac{1}{6}\cdot 1+\frac{3}{4}\cdot 1$

$=\frac{1}{6}\cdot\frac{2}{2}+\frac{3}{4}\cdot\frac{3}{3}$

$=\frac{2}{12}+\frac{9}{12}$

$=\frac{11}{12}$

We had to simplify in (A). We didn't have to simplify in (B). In (B), we used the least common multiple of the denominators, 12. That number is called the **least common denominator,** or **LCD.** As we will see in Example 6, we may still need to simplify when using the LCD, but it is usually easier.

To add when denominators are different:

a) Find the least common multiple of the denominators. That number is the least common denominator, LCD.
b) Multiply by 1, using an appropriate notation, n/n, to express each number in terms of the LCD.
c) Add the numerators, keeping the same denominator.
d) Simplify, if possible.

EXAMPLE 4 Add: $\frac{3}{4}+\frac{1}{8}$.

The LCD is 8. 4 is a factor of 8, so the LCM of 4 and 8 is 8.

$\frac{3}{4}+\frac{1}{8}=\frac{3}{4}\cdot 1+\frac{1}{8}$ ← This fraction already has the LCD as its denominator.

$=\frac{3}{4}\cdot\frac{2}{2}+\frac{1}{8}$ *Think*: $4\times\square=8$. The answer is 2, so we multiply by 1, using $\frac{2}{2}$.

$=\frac{6}{8}+\frac{1}{8}=\frac{7}{8}$

Do Exercise 5.

EXAMPLE 5 Add: $\frac{1}{9}+\frac{5}{6}$.

The LCD is 18. $9=3\cdot 3$ and $6=2\cdot 3$, so the LCM of 9 and 6 is $2\cdot 3\cdot 3$, or 18.

$\frac{1}{9}+\frac{5}{6}=\frac{1}{9}\cdot 1+\frac{5}{6}\cdot 1=\frac{1}{9}\cdot\frac{2}{2}+\frac{5}{6}\cdot\frac{3}{3}$

Think: $6\times\square=18$. The answer is 3, so we multiply by 1 using $\frac{3}{3}$.

Think: $9\times\square=18$. The answer is 2, so we multiply by 1 using $\frac{2}{2}$.

$=\frac{2}{18}+\frac{15}{18}=\frac{17}{18}$

Do Exercise 6.

5. Add. (Find the least common denominator.)

$\frac{2}{3}+\frac{1}{6}$

6. Add: $\frac{3}{8}+\frac{5}{6}$.

Answers on page A-6

7. Add: $\frac{1}{6} + \frac{7}{18}$.

8. Add: $\frac{4}{10} + \frac{1}{100} + \frac{3}{1000}$.

Add.

9. $\frac{7}{10} + \frac{2}{21} + \frac{1}{7}$

10. $\frac{7}{18} + \frac{5}{24} + \frac{11}{36}$

Answers on page A-6

EXAMPLE 6 Add: $\frac{5}{9} + \frac{11}{18}$.

The LCD is 18.

$$\frac{5}{9} + \frac{11}{18} = \frac{5}{9} \cdot \frac{2}{2} + \frac{11}{18} = \frac{10}{18} + \frac{11}{18}$$

$$= \frac{21}{18} = \frac{3 \cdot 7}{3 \cdot 6} = \frac{3}{3} \cdot \frac{7}{6} = \frac{7}{6} \quad \text{Simplifying}$$

Do Exercise 7.

EXAMPLE 7 Add: $\frac{1}{10} + \frac{3}{100} + \frac{7}{1000}$.

Since 10 and 100 are factors of 1000, the LCD is 1000. Then

$$\frac{1}{10} + \frac{3}{100} + \frac{7}{1000} = \frac{1}{10} \cdot \frac{100}{100} + \frac{3}{100} \cdot \frac{10}{10} + \frac{7}{1000}$$

$$= \frac{100}{1000} + \frac{30}{1000} + \frac{7}{1000} = \frac{137}{1000}.$$

Do Exercise 8.

When denominators are large, we most often use the prime factorization of each denominator to find the LCD. This is shown in Example 8. Using the prime factorization in this manner is similar to what is done in algebra.

EXAMPLE 8 Add: $\frac{13}{70} + \frac{11}{21} + \frac{6}{15}$.

We have

$$\frac{13}{70} + \frac{11}{21} + \frac{6}{15} = \frac{13}{2 \cdot 5 \cdot 7} + \frac{11}{3 \cdot 7} + \frac{6}{3 \cdot 5}. \quad \text{Factoring denominators}$$

The LCD is $2 \cdot 3 \cdot 5 \cdot 7$, or 210. Then

$$\frac{13}{70} + \frac{11}{21} + \frac{6}{15} = \frac{13}{2 \cdot 5 \cdot 7} \cdot \frac{3}{3} + \frac{11}{3 \cdot 7} \cdot \frac{2 \cdot 5}{2 \cdot 5} + \frac{6}{3 \cdot 5} \cdot \frac{7 \cdot 2}{7 \cdot 2}$$

> The LCM of 70, 21, and 15 is $2 \cdot 3 \cdot 5 \cdot 7$. In each case, think of which factors are needed to get the LCD. Then multiply by 1 to obtain the LCD in each denominator.

$$= \frac{13 \cdot 3}{2 \cdot 5 \cdot 7 \cdot 3} + \frac{11 \cdot 2 \cdot 5}{3 \cdot 7 \cdot 2 \cdot 5} + \frac{6 \cdot 7 \cdot 2}{3 \cdot 5 \cdot 7 \cdot 2}$$

$$= \frac{39}{3 \cdot 5 \cdot 7 \cdot 2} + \frac{110}{3 \cdot 5 \cdot 7 \cdot 2} + \frac{84}{3 \cdot 5 \cdot 7 \cdot 2}$$

$$= \frac{233}{3 \cdot 5 \cdot 7 \cdot 2}$$

$$= \frac{233}{210}. \quad \text{We left 210 factored until we knew we could not simplify.}$$

Do Exercises 9 and 10.

b Applications and Problem Solving

EXAMPLE 9 *Subflooring.* Matt Beecher Builders and Developers requires their subcontractors to use two layers of subflooring under a ceramic tile floor. First the contractors install a $\frac{3}{4}$-in. ($\frac{3}{4}$") layer of oriented strand board (OSB). Then a $\frac{1}{2}$-in. ($\frac{1}{2}$") sheet of cement board is mortared to the OSB. The mortar is $\frac{1}{8}$-in. thick. What is the total thickness of the two installed subfloors?

1. Familiarize. We first make a drawing. We let $T =$ the total thickness of the subfloors.

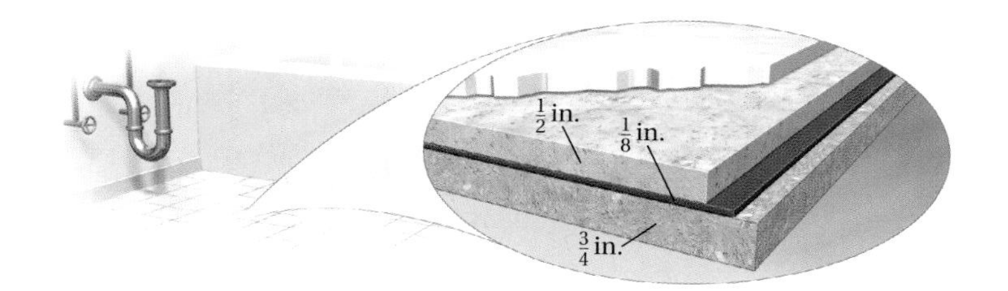

2. Translate. The problem can be translated to an equation as follows.

OSB	plus	Mortar	plus	Cement board	is	Total thickness
↓	↓	↓	↓	↓	↓	↓
$\frac{3}{4}$	$+$	$\frac{1}{8}$	$+$	$\frac{1}{2}$	$=$	T

3. Solve. To solve the equation, we carry out the addition. The LCM of the denominators is 8 because 2 and 4 are factors of 8. We multiply by 1 in order to obtain the LCD:

$$\frac{3}{4} + \frac{1}{8} + \frac{1}{2} = T$$

$$\frac{3}{4} \cdot \frac{2}{2} + \frac{1}{8} + \frac{1}{2} \cdot \frac{4}{4} = T$$

$$\frac{6}{8} + \frac{1}{8} + \frac{4}{8} = T$$

$$\frac{11}{8} = T.$$

4. Check. We check by repeating the calculation. We also note that the sum should be larger than any of the individual measurements, which it is. This tells us that the answer is reasonable.

5. State. The total thickness of the installed subfloors is $\frac{11}{8}$ in.

Do Exercise 11.

11. Sally jogs for $\frac{4}{5}$ mi, rests, and then jogs for another $\frac{1}{10}$ mi. How far does she jog in all?

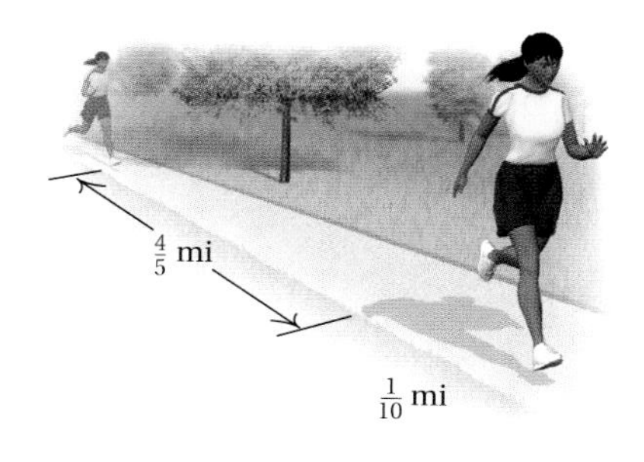

Answer on page A-6

3.2 EXERCISE SET

For Extra Help

MathXL | MyMathLab | InterAct Math | Math Tutor Center | Digital Video Tutor CD 2 Videotape 3 | Student's Solutions Manual

a Add and simplify.

1. $\frac{7}{8} + \frac{1}{8}$

2. $\frac{2}{5} + \frac{3}{5}$

3. $\frac{1}{8} + \frac{5}{8}$

4. $\frac{3}{10} + \frac{3}{10}$

5. $\frac{2}{3} + \frac{5}{6}$

6. $\frac{5}{6} + \frac{1}{9}$

7. $\frac{1}{8} + \frac{1}{6}$

8. $\frac{1}{6} + \frac{3}{4}$

9. $\frac{4}{5} + \frac{7}{10}$

10. $\frac{3}{4} + \frac{1}{12}$

11. $\frac{5}{12} + \frac{3}{8}$

12. $\frac{7}{8} + \frac{1}{16}$

13. $\frac{3}{20} + \frac{3}{4}$

14. $\frac{2}{15} + \frac{2}{5}$

15. $\frac{5}{6} + \frac{7}{9}$

16. $\frac{5}{8} + \frac{5}{6}$

17. $\frac{3}{10} + \frac{1}{100}$

18. $\frac{9}{10} + \frac{3}{100}$

19. $\frac{5}{12} + \frac{4}{15}$

20. $\frac{3}{16} + \frac{1}{12}$

21. $\frac{9}{10} + \frac{99}{100}$

22. $\frac{3}{10} + \frac{27}{100}$

23. $\frac{7}{8} + \frac{0}{1}$

24. $\frac{0}{1} + \frac{5}{6}$

25. $\frac{3}{8} + \frac{1}{6}$

26. $\frac{5}{8} + \frac{1}{6}$

27. $\frac{5}{12} + \frac{7}{24}$

28. $\frac{1}{18} + \frac{7}{12}$

29. $\frac{3}{16} + \frac{5}{16} + \frac{4}{16}$

30. $\frac{3}{8} + \frac{1}{8} + \frac{2}{8}$

31. $\frac{8}{10} + \frac{7}{100} + \frac{4}{1000}$

32. $\frac{1}{10} + \frac{2}{100} + \frac{3}{1000}$

33. $\frac{3}{8} + \frac{5}{12} + \frac{8}{15}$

34. $\frac{1}{2} + \frac{3}{8} + \frac{1}{4}$

35. $\frac{15}{24} + \frac{7}{36} + \frac{91}{48}$

36. $\frac{5}{7} + \frac{25}{52} + \frac{7}{4}$

b Solve.

37. Rene bought $\frac{1}{3}$ lb of orange pekoe tea and $\frac{1}{2}$ lb of English cinnamon tea. How many pounds of tea did he buy?

38. Stan bought $\frac{1}{4}$ lb of gumdrops and $\frac{1}{2}$ lb of caramels. How many pounds of candy did he buy?

39. Russ walked $\frac{7}{6}$ mi to a friend's dormitory, and then $\frac{3}{4}$ mi to class. How far did he walk?

40. Elaine walked $\frac{7}{8}$ mi to the student union, and then $\frac{2}{5}$ mi to class. How far did she walk?

41. *Concrete Mix.* A cubic meter of concrete mix contains 420 kilograms (kg) of cement, 150 kg of stone, and 120 kg of sand. What is the total weight of a cubic meter of the mix? What part is cement? stone? sand? Add these fractional amounts. What is the result?

42. *Punch Recipe.* A recipe for strawberry punch calls for $\frac{1}{5}$ quart (qt) of ginger ale and $\frac{3}{5}$ qt of strawberry soda. How much liquid is needed? If the recipe is doubled, how much liquid is needed? If the recipe is halved, how much liquid is needed?

43. *Carpentry.* To cut expenses, a carpenter sometimes glues two kinds of plywood together. He glues a $\frac{1}{4}$" piece of walnut plywood to a $\frac{3}{8}$" piece of less expensive plywood. What is the total thickness of these pieces?

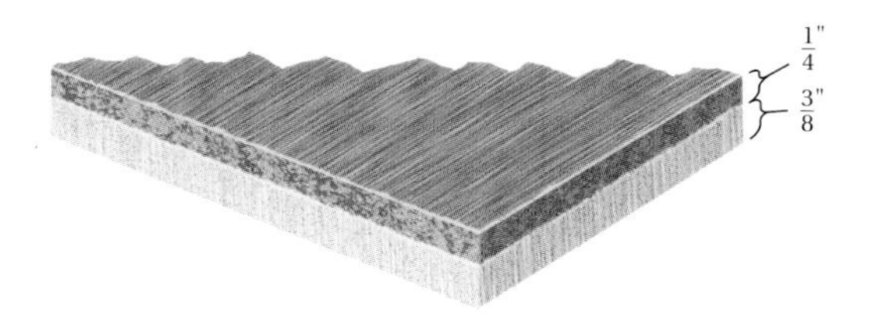

44. *Iced Cookies.* A chef prepared cookies for the freshmen orientation reception. He frosted the $\frac{5}{16}$" cookies with a $\frac{5}{32}$" layer of icing. What is the thickness of the iced cookie?

45. A baker used $\frac{1}{2}$ lb of flour for rolls, $\frac{1}{4}$ lb for donuts, and $\frac{1}{3}$ lb for cookies. How much flour was used?

46. A tile $\frac{5}{8}$ in. thick is glued to a board $\frac{7}{8}$ in. thick. The glue is $\frac{3}{32}$ in. thick. How thick is the result?

47. $\mathbf{D_W}$ Explain the role of multiplication when adding using fraction notation with different denominators.

48. $\mathbf{D_W}$ To add numbers with different denominators, a student consistently uses the product of the denominators as a common denominator. Is this correct? Why or why not?

SKILL MAINTENANCE

Multiply. [1.5a]

49. 408 · 516

50. 1125 · 3728

51. 423 · 8009

52. 2025 · 174

Presidential Elections. The results of the five closest presidential elections are listed in the following table. Use these data for Exercises 53–58. [1.3b]

DATE	PRESIDENT	ELECTORAL VOTES	POPULAR VOTES
1916	Woodrow Wilson (D)	277	9,129,606
	Charles E. Hughes (R)	254	8,538,221
1960	John F. Kennedy (D)	303	34,226,731
	Richard M. Nixon (R)	219	34,108,157
1968	Richard M. Nixon (R)	301	31,785,480
	Hubert H. Humphrey (D)	191	31,275,166
	George C. Wallace (American Independent)	46	9,906,473
1976	Jimmy Carter (D)	297	40,830,763
	Gerald R. Ford (R)	240	39,147,973
2000	George W. Bush (R)	271	50,455,156
	Albert A. Gore (D)	266	50,992,335

Source: *Time Almanac 2005*, p. 55

53. How many more popular votes did Albert A. Gore have than George W. Bush in the 2000 presidential election?

54. How many more electoral votes did George W. Bush have than Albert A. Gore in the 2000 presidential election?

55. In the 1960 presidential election, how many more electoral votes did John F. Kennedy have than Richard M. Nixon?

56. In the 1968 presidential election, how many more popular votes did Richard M. Nixon have than Hubert H. Humphrey?

57. How much greater was the total of the popular vote counts in the 2000 presidential election than in the 1976 election?

58. How much greater was the total of the popular vote counts in the 2000 presidential election than in the 1960 election?

59. Elsa has \$9 to spend on ride tickets at the school carnival. If the tickets cost 75¢, or $\$\frac{3}{4}$, each, how many tickets can she purchase? [2.7d]

60. The Bingham community garden is to be split into 16 equally sized plots. If the garden occupies $\frac{3}{4}$ acre of land, how large will each plot be? [2.7d]

SYNTHESIS

61. A guitarist's band is booked for Friday and Saturday nights at a local club. The guitarist is part of a trio on Friday and part of a quintet on Saturday. Thus the guitarist is paid one-third of one-half the weekend's pay for Friday and one-fifth of one-half the weekend's pay for Saturday. What fractional part of the band's pay did the guitarist receive for the weekend's work? If the band was paid \$1200, how much did the guitarist receive?

3.3 SUBTRACTION, ORDER, AND APPLICATIONS

Objectives

a Subtract using fraction notation.

b Use $<$ or $>$ with fraction notation to write a true sentence.

c Solve equations of the type $x + a = b$ and $a + x = b$, where a and b may be fractions.

d Solve applied problems involving subtraction with fraction notation.

a Subtraction Using Fraction Notation

LIKE DENOMINATORS

We can consider the difference $\frac{4}{8} - \frac{3}{8}$ as we did before, as either "take away" or "how many more." Let's consider "take away."

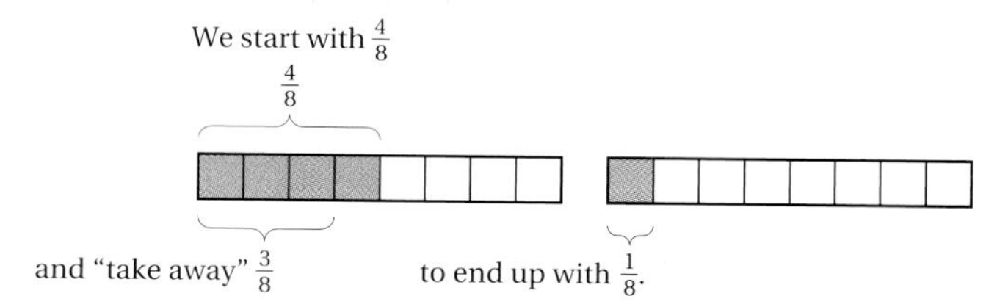

We start with $\frac{4}{8}$
$\frac{4}{8}$
and "take away" $\frac{3}{8}$
to end up with $\frac{1}{8}$.

We start with 4 eighths and take away 3 eighths:

$$4 \text{ eighths} - 3 \text{ eighths} = 1 \text{ eighth},$$

or $\quad 4 \cdot \frac{1}{8} - 3 \cdot \frac{1}{8} = \frac{1}{8}, \quad$ or $\quad \frac{4}{8} - \frac{3}{8} = \frac{1}{8}.$

To subtract when denominators are the same,

a) subtract the numerators,

b) keep the denominator, and

c) simplify, if possible.

$$\frac{7}{10} - \frac{4}{10} = \frac{7-4}{10} = \frac{3}{10}$$

EXAMPLES Subtract and simplify.

1. $\frac{7}{10} - \frac{3}{10} = \frac{7-3}{10} = \frac{4}{10} = \frac{2 \cdot 2}{5 \cdot 2} = \frac{2}{5} \cdot \frac{2}{2} = \frac{2}{5} \cdot 1 = \frac{2}{5}$

2. $\frac{8}{9} - \frac{2}{9} = \frac{8-2}{9} = \frac{6}{9} = \frac{2 \cdot 3}{3 \cdot 3} = \frac{2}{3} \cdot \frac{3}{3} = \frac{2}{3} \cdot 1 = \frac{2}{3}$

3. $\frac{32}{12} - \frac{25}{12} = \frac{32-25}{12} = \frac{7}{12}$

Do Exercises 1–3.

Subtract and simplify.

1. $\frac{7}{8} - \frac{3}{8}$

2. $\frac{10}{16} - \frac{4}{16}$

3. $\frac{8}{10} - \frac{3}{10}$

Answers on page A-6

DIFFERENT DENOMINATORS

The procedure for subtraction with different denominators is similar to the procedure for addition.

To subtract when denominators are different:

a) Find the least common multiple of the denominators. That number is the least common denominator, LCD.

b) Multiply by 1, using an appropriate notation, n/n, to express each number in terms of the LCD.

c) Subtract the numerators, keeping the same denominator.

d) Simplify, if possible.

4. Subtract: $\frac{3}{4} - \frac{2}{3}$.

Subtract.

5. $\frac{5}{6} - \frac{1}{9}$

6. $\frac{4}{5} - \frac{3}{10}$

7. Subtract: $\frac{11}{28} - \frac{5}{16}$.

8. Use < or > for $\square$ to write a true sentence:

$\frac{3}{8} \square \frac{5}{8}$.

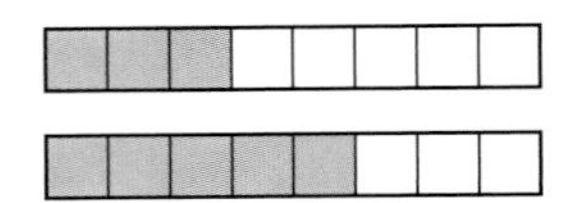

Answers on page A-6

EXAMPLE 4 Subtract: $\frac{2}{5} - \frac{3}{8}$.

The LCM of 5 and 8 is 40, so the LCD is 40.

$$\frac{2}{5} - \frac{3}{8} = \frac{2}{5} \cdot \frac{8}{8} - \frac{3}{8} \cdot \frac{5}{5}$$

Think: $8 \times \square = 40$. The answer is 5, so we multiply by 1, using $\frac{5}{5}$.

Think: $5 \times \square = 40$. The answer is 8, so we multiply by 1, using $\frac{8}{8}$.

$$= \frac{16}{40} - \frac{15}{40} = \frac{16 - 15}{40} = \frac{1}{40}$$

Do Exercise 4.

EXAMPLE 5 Subtract: $\frac{5}{6} - \frac{7}{12}$.

Since 12 is a multiple of 6, the LCM of 6 and 12 is 12. The LCD is 12.

$$\frac{5}{6} - \frac{7}{12} = \frac{5}{6} \cdot \frac{2}{2} - \frac{7}{12}$$

$$= \frac{10}{12} - \frac{7}{12} = \frac{10 - 7}{12} = \frac{3}{12}$$

$$= \frac{3 \cdot 1}{3 \cdot 4} = \frac{3}{3} \cdot \frac{1}{4} = \frac{1}{4}$$

Do Exercises 5 and 6.

EXAMPLE 6 Subtract: $\frac{17}{24} - \frac{4}{15}$.

We have

$$\frac{17}{24} - \frac{4}{15} = \frac{17}{3 \cdot 2 \cdot 2 \cdot 2} - \frac{4}{5 \cdot 3}.$$

The LCD is $3 \cdot 2 \cdot 2 \cdot 2 \cdot 5$, or 120. Then

$$\frac{17}{24} - \frac{4}{15} = \frac{17}{3 \cdot 2 \cdot 2 \cdot 2} \cdot \frac{5}{5} - \frac{4}{5 \cdot 3} \cdot \frac{2 \cdot 2 \cdot 2}{2 \cdot 2 \cdot 2}$$

The LCM of 24 and 15 is $2 \cdot 2 \cdot 2 \cdot 3 \cdot 5$. In each case, we multiply by 1 to obtain the LCD.

$$= \frac{17 \cdot 5}{3 \cdot 2 \cdot 2 \cdot 2 \cdot 5} - \frac{4 \cdot 2 \cdot 2 \cdot 2}{5 \cdot 3 \cdot 2 \cdot 2 \cdot 2}$$

$$= \frac{85}{120} - \frac{32}{120} = \frac{53}{120}.$$

Do Exercise 7.

b Order

We see from this figure that $\frac{4}{5}$ is greater than $\frac{3}{5}$, and $\frac{3}{5}$ is less than $\frac{4}{5}$. That is, $\frac{4}{5} > \frac{3}{5}$, and $\frac{3}{5} < \frac{4}{5}$.

$\frac{4}{5}$

$\frac{3}{5}$

To determine which of two numbers is greater when there is a common denominator, compare the numerators:

$$\frac{4}{5}, \frac{3}{5}; \quad 4 > 3; \quad \frac{4}{5} > \frac{3}{5}.$$

Do Exercises 8 and 9. (Exercise 8 is on the preceding page.)

When denominators are different, we cannot compare numerators. We multiply by 1 to make the denominators the same.

EXAMPLE 7 Use $<$ or $>$ for $\square$ to write a true sentence:

$$\frac{2}{5} \square \frac{3}{4}.$$

The LCD is 20. We have

$$\frac{2}{5} \cdot \frac{4}{4} = \frac{8}{20}; \quad \text{We multiply by 1 using } \frac{4}{4} \text{ to get the LCD.}$$

$$\frac{3}{4} \cdot \frac{5}{5} = \frac{15}{20}. \quad \text{We multiply by 1 using } \frac{5}{5} \text{ to get the LCD.}$$

Now that the denominators are the same, 20, we can compare the numerators. Since $8 < 15$, it follows that $\frac{8}{20} < \frac{15}{20}$, so

$$\frac{2}{5} < \frac{3}{4}.$$

EXAMPLE 8 Use $<$ or $>$ for $\square$ to write a true sentence:

$$\frac{9}{10} \square \frac{89}{100}.$$

The LCD is 100. We write $\frac{9}{10}$ with a denominator of 100 to make the denominators the same.

$$\frac{9}{10} \cdot \frac{10}{10} = \frac{90}{100} \quad \text{We multiply by } \frac{10}{10} \text{ to get the LCD.}$$

Since $90 > 89$, it follows that $\frac{90}{100} > \frac{89}{100}$, so

$$\frac{9}{10} > \frac{89}{100}.$$

Do Exercises 10–12.

C Solving Equations

Now let's solve equations of the form $x + a = b$ or $a + x = b$, where a and b may be fractions. Proceeding as we have before, we subtract a on both sides of the equation.

9. Use $<$ or $>$ for $\square$ to write a true sentence:

$$\frac{7}{10} \square \frac{6}{10}.$$

Use $<$ or $>$ for $\square$ to write a true sentence.

10. $\frac{2}{3} \square \frac{5}{8}$

11. $\frac{3}{4} \square \frac{8}{12}$

12. $\frac{5}{6} \square \frac{7}{8}$

Solve.

13. $x + \frac{2}{3} = \frac{5}{6}$

14. $\frac{3}{5} + t = \frac{7}{8}$

Answers on page A-6

15. Natasha has run for $\frac{2}{3}$ mi and will stop when she has run for $\frac{7}{8}$ mi. How much farther does she have to go?

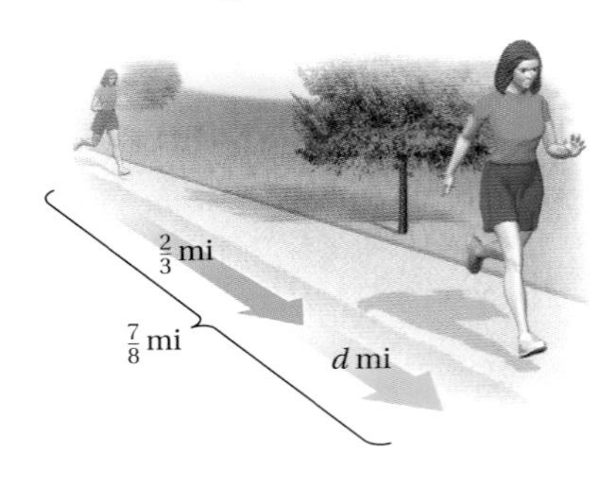

EXAMPLE 9 Solve: $x + \frac{1}{4} = \frac{3}{5}$.

$$x + \frac{1}{4} - \frac{1}{4} = \frac{3}{5} - \frac{1}{4}$$ Subtracting $\frac{1}{4}$ on both sides

$$x + 0 = \frac{3}{5} \cdot \frac{4}{4} - \frac{1}{4} \cdot \frac{5}{5}$$ The LCD is 20. We multiply by 1 to get the LCD.

$$x = \frac{12}{20} - \frac{5}{20}$$

$$x = \frac{7}{20}$$

Do Exercises 13 and 14 on the preceding page.

d Applications and Problem Solving

EXAMPLE 10 *Pendant Necklace.* At one time Coldwater Creek offered the pendant necklace illustrated at left. The sterling silver capping at the top measures $\frac{11}{32}$ in. and the total length of the pendant is $\frac{7}{8}$ in. Find the length, or diameter, w of the pearl ball on the pendant.

Source: ©Coldwater Creek Inc. www.coldwatercreek.com

1. **Familiarize.** We let $w =$ the length of the pearl on the pendant.

2. **Translate.** We see that this is a "how many more" situation. We can translate to an equation.

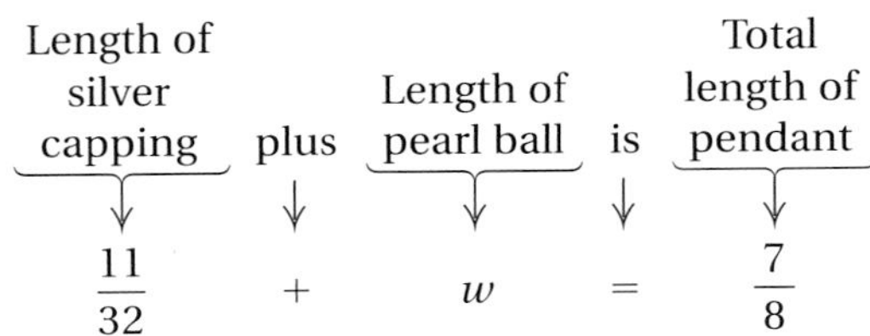

Length of silver capping
plus
Length of pearl ball
is
Total length of pendant
11/32
+
w
=
7/8

3. **Solve.** To solve the equation, we subtract $\frac{11}{32}$ on both sides:

$$\frac{11}{32} + w = \frac{7}{8}$$

$$\frac{11}{32} + w - \frac{11}{32} = \frac{7}{8} - \frac{11}{32}$$ Subtracting $\frac{11}{32}$ on both sides

$$w + 0 = \frac{7}{8} \cdot \frac{4}{4} - \frac{11}{32}$$ The LCD is 32. We multiply by 1 to obtain the LCD.

$$w = \frac{28}{32} - \frac{11}{32}$$

$$= \frac{17}{32}.$$

4. **Check.** To check, we return to the original problem and add:

$$\frac{11}{32} + \frac{17}{32} = \frac{28}{32} = \frac{7}{8} \cdot \frac{4}{4} = \frac{7}{8}.$$

5. **State.** The length of the pearl ball on the pendant is $\frac{17}{32}$ in.

Answer on page A-6

Do Exercise 15.

Translating for Success

Bubble Wrap. One-Stop Postal Center orders bubble wrap in 64-yd rolls. On the average, $\frac{3}{4}$ yd is used per small package. How many small packages can be prepared with 2 rolls of bubble wrap?

Distance from College. The post office is $\frac{7}{9}$ mi from the community college. The medical clinic is $\frac{2}{5}$ as far from the college as the post office is. How far is the clinic from the college?

Swimming. Andrew swims $\frac{7}{9}$ mi every day. One day he swims $\frac{2}{5}$ mi by 11:00 A.M. How much farther must Andrew swim to reach his daily goal?

Tuition. The average tuition at Waterside University is \$12,000. If a loan is obtained for $\frac{1}{3}$ of the tuition, how much is the loan?

Thermos Bottle Capacity. A thermos bottle holds $\frac{11}{12}$ gal. How much is in the bottle when it is $\frac{4}{7}$ full?

The goal of these matching questions is to practice step (2), *Translate,* of the five-step problem-solving process. Translate each word problem to an equation and select a correct translation from equations A–O.

A. $\frac{3}{4} \cdot 64 = x$

B. $\frac{1}{3} \cdot 12{,}000 = x$

C. $\frac{1}{3} + \frac{2}{5} = x$

D. $\frac{2}{5} + x = \frac{7}{9}$

E. $\frac{2}{5} \cdot \frac{7}{9} = x$

F. $\frac{3}{4} \cdot x = 64$

G. $\frac{4}{7} = x + \frac{11}{12}$

H. $\frac{2}{5} = x + \frac{7}{9}$

I. $\frac{4}{7} \cdot \frac{11}{12} = x$

J. $\frac{3}{4} \cdot x = 128$

K. $\frac{1}{3} \cdot x = 12{,}000$

L. $\frac{1}{3} + \frac{2}{5} + x = 1$

M. $\frac{4}{3} \cdot x = 64$

N. $\frac{4}{7} + x = \frac{11}{12}$

O. $\frac{1}{3} + x = \frac{2}{5}$

Answers on page A-6

6. *Cutting Rope.* A piece of rope $\frac{11}{12}$ yd long is cut into two pieces. One piece is $\frac{4}{7}$ yd long. How long is the other piece?

7. *Planting Corn.* Each year, Prairie State Farm plants 64 acres of corn. With good weather, $\frac{3}{4}$ of the planting can be completed by April 20. How many acres can be planted with good weather?

8. *Painting Trim.* A painter used $\frac{1}{3}$ gal of white paint for the trim in the library and $\frac{2}{5}$ gal in the family room. How much paint was used for the trim in the two rooms?

9. *Lottery Winnings.* Sally won \$12,000 in a state lottery and decides to give the net amount after taxes to three charities. One received $\frac{1}{3}$ of the money, and a second received $\frac{2}{5}$. What fractional part did the third charity receive?

10. *Reading Assignment.* Lowell read 64 pages of his political science assignment. This completed $\frac{3}{4}$ of his required reading. How many total pages were assigned?

3.3 EXERCISE SET

For Extra Help

MathXL | MyMathLab | InterAct Math | Math Tutor Center | Digital Video Tutor CD 2 Videotape 3 | Student's Solutions Manual

a Subtract and simplify.

1. $\frac{5}{6} - \frac{1}{6}$
2. $\frac{5}{8} - \frac{3}{8}$
3. $\frac{11}{12} - \frac{2}{12}$
4. $\frac{17}{18} - \frac{11}{18}$
5. $\frac{3}{4} - \frac{1}{8}$
6. $\frac{2}{3} - \frac{1}{9}$
7. $\frac{1}{8} - \frac{1}{12}$
8. $\frac{1}{6} - \frac{1}{8}$
9. $\frac{4}{3} - \frac{5}{6}$
10. $\frac{7}{8} - \frac{1}{16}$
11. $\frac{3}{4} - \frac{3}{28}$
12. $\frac{2}{5} - \frac{2}{15}$
13. $\frac{3}{4} - \frac{3}{20}$
14. $\frac{5}{6} - \frac{1}{2}$
15. $\frac{3}{4} - \frac{1}{20}$
16. $\frac{3}{4} - \frac{4}{16}$
17. $\frac{5}{12} - \frac{2}{15}$
18. $\frac{9}{10} - \frac{11}{16}$
19. $\frac{6}{10} - \frac{7}{100}$
20. $\frac{9}{10} - \frac{3}{100}$
21. $\frac{7}{15} - \frac{3}{25}$
22. $\frac{18}{25} - \frac{4}{35}$
23. $\frac{99}{100} - \frac{9}{10}$
24. $\frac{78}{100} - \frac{11}{20}$
25. $\frac{2}{3} - \frac{1}{8}$
26. $\frac{3}{4} - \frac{1}{2}$
27. $\frac{3}{5} - \frac{1}{2}$
28. $\frac{5}{6} - \frac{2}{3}$
29. $\frac{5}{12} - \frac{3}{8}$
30. $\frac{7}{12} - \frac{2}{9}$
31. $\frac{7}{8} - \frac{1}{16}$
32. $\frac{5}{12} - \frac{5}{16}$
33. $\frac{17}{25} - \frac{4}{15}$
34. $\frac{11}{18} - \frac{7}{24}$
35. $\frac{23}{25} - \frac{112}{150}$
36. $\frac{89}{90} - \frac{53}{120}$

b Use < or > for □ to write a true sentence.

37. $\frac{5}{8} \square \frac{6}{8}$
38. $\frac{7}{9} \square \frac{5}{9}$
39. $\frac{1}{3} \square \frac{1}{4}$
40. $\frac{1}{8} \square \frac{1}{6}$
41. $\frac{2}{3} \square \frac{5}{7}$
42. $\frac{3}{5} \square \frac{4}{7}$
43. $\frac{4}{5} \square \frac{5}{6}$
44. $\frac{3}{2} \square \frac{7}{5}$
45. $\frac{19}{20} \square \frac{4}{5}$
46. $\frac{5}{6} \square \frac{13}{16}$
47. $\frac{19}{20} \square \frac{9}{10}$
48. $\frac{3}{4} \square \frac{11}{15}$
49. $\frac{31}{21} \square \frac{41}{13}$
50. $\frac{12}{7} \square \frac{132}{49}$

c Solve.

51. $x + \frac{1}{30} = \frac{1}{10}$

52. $y + \frac{9}{12} = \frac{11}{12}$

53. $\frac{2}{3} + t = \frac{4}{5}$

54. $\frac{2}{3} + p = \frac{7}{8}$

55. $x + \frac{1}{3} = \frac{5}{6}$

56. $m + \frac{5}{6} = \frac{9}{10}$

d Solve.

57. Jaci spent $\frac{3}{4}$ hr listening to DVDs of Maroon 5 and U2. She spent $\frac{1}{3}$ hr listening to Maroon 5. How many hours were spent listening to U2?

58. As part of a fitness program, Deb swims $\frac{1}{2}$ mi every day. One day she had already swum $\frac{1}{5}$ mi. How much farther should Deb swim?

59. A server has a bowl containing $\frac{11}{12}$ cup of grated Parmesan cheese and puts $\frac{1}{8}$ cup on a customer's spaghetti and meatballs. How much remains in the bowl?

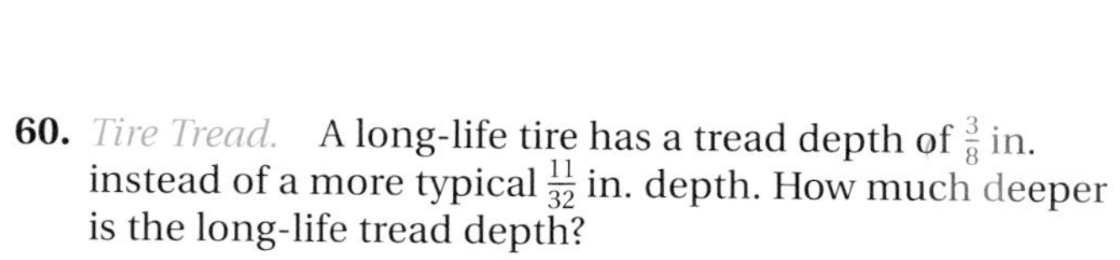

60. *Tire Tread.* A long-life tire has a tread depth of $\frac{3}{8}$ in. instead of a more typical $\frac{11}{32}$ in. depth. How much deeper is the long-life tread depth?

Source: *Popular Science*

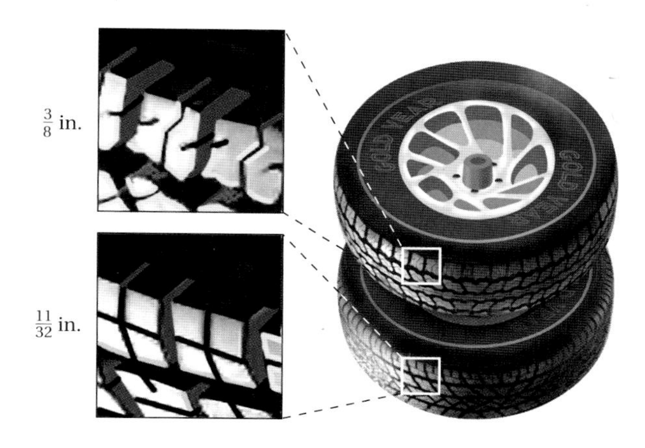

61. A Subway franchise is owned by Alicia, Erica, and Hannah. Alicia owns $\frac{7}{12}$ of the business and Erica owns $\frac{1}{6}$. What part of the business does Hannah own?

62. An estate was left to four children. One received $\frac{1}{4}$ of the estate, the second $\frac{1}{16}$, and the third $\frac{3}{8}$. How much did the fourth receive?

63. From a $\frac{4}{5}$-lb wheel of cheese, a $\frac{1}{4}$-lb piece was served. How much cheese remained on the wheel?

64. Jovan has an $\frac{11}{10}$-lb mixture of cashews and peanuts that includes $\frac{3}{5}$ lb of cashews. How many pounds of peanuts are in the mixture?

65. **D$_W$** A fellow student made the following error:

$$\frac{8}{5} - \frac{8}{2} = \frac{8}{3}.$$

Find at least two ways to convince him of the mistake.

66. **D$_W$** Explain how one could use pictures to convince someone that $\frac{7}{29}$ is larger than $\frac{13}{57}$.

SKILL MAINTENANCE

Divide, if possible. If not possible, write "not defined." [1.6b], [2.3b]

67. $\frac{38}{38}$

68. $\frac{38}{0}$

69. $\frac{124}{0}$

70. $\frac{124}{31}$

Divide and simplify. [2.7b]

71. $\frac{3}{7} \div \frac{9}{4}$

72. $\frac{9}{10} \div \frac{3}{5}$

73. $7 \div \frac{1}{3}$

74. $\frac{1}{4} \div 8$

Solve.

75. *Digital Tire Gauges.* A factory produces 3885 digital tire gauges per day. How long will it take to fill an order for 66,045 tire gauges? [1.8a]

76. A batch of fudge requires $\frac{3}{4}$ cup of sugar. How much sugar is needed to make 12 batches? [2.6b]

SYNTHESIS

Solve.

77. 🖩 $x + \frac{16}{323} = \frac{10}{187}$

78. 🖩 $x + \frac{7}{253} = \frac{12}{299}$

79. As part of a rehabilitation program, an athlete must swim and then walk a total of $\frac{9}{10}$ km each day. If one lap in the swimming pool is $\frac{3}{80}$ km, how far must the athlete walk after swimming 10 laps?

80. *Mountain Climbing.* A mountain climber, beginning at sea level, climbs $\frac{3}{5}$ km, descends $\frac{1}{4}$ km, climbs $\frac{1}{3}$ km, and then descends $\frac{1}{7}$ km. At what elevation does the climber finish?

Simplify. Use the rules for order of operations given in Section 1.9.

81. $\frac{7}{8} - \frac{1}{10} \times \frac{5}{6}$

82. $\frac{2}{5} + \frac{1}{6} \div 3$

83. $\left(\frac{2}{3}\right)^2 + \left(\frac{3}{4}\right)^2$

84. $5 \times \frac{3}{7} - \frac{1}{7} \times \frac{4}{5}$

Use <, >, or = for $\square$ to write a true sentence.

85. 🖩 $\frac{37}{157} + \frac{19}{107} \square \frac{6941}{16,799}$

86. 🖩 $\frac{12}{97} + \frac{67}{139} \square \frac{8167}{13,289}$

87. *Microsoft Interview.* The following is a question taken from an employment interview with Microsoft. Try to answer it. "Given a gold bar that can be cut exactly twice and a contractor who must be paid one-seventh of a gold bar every day for seven days, how should the bar be cut?"

Source: *Fortune Magazine*, January 22, 2001

3.4 MIXED NUMERALS

Objectives

a Convert between mixed numerals and fraction notation.

b Divide whole numbers, writing the quotient as a mixed numeral.

a Mixed Numerals

The following figure illustrates the use of a **mixed numeral.** The bolt shown is $2\frac{3}{8}$ in. long. The length is given as a whole-number part, 2, and a fractional part less than 1, $\frac{3}{8}$. We can represent the measurement of the bolt with fraction notation as $\frac{19}{8}$, but the meaning or interpretation of such a symbol is less understandable or visual.

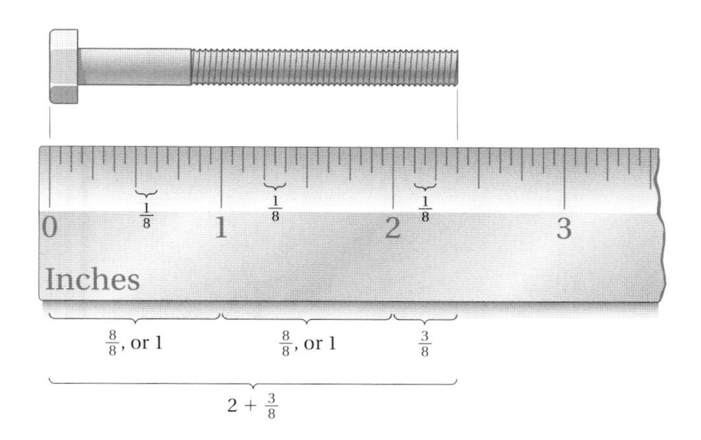

A mixed numeral $2\frac{3}{8}$ represents a sum:

$$2\frac{3}{8} \quad \text{means} \quad 2 + \frac{3}{8}$$

This is a whole number. This is a fraction less than 1.

EXAMPLES Convert to a mixed numeral.

1. $7 + \frac{2}{5} = 7\frac{2}{5}$

2. $4 + \frac{3}{10} = 4\frac{3}{10}$

Do Exercises 1–4.

The notation $2\frac{3}{4}$ has a plus sign left out. To aid in understanding, we sometimes write the missing plus sign.

EXAMPLES Convert to fraction notation.

3. $2\frac{3}{4} = 2 + \frac{3}{4}$ Inserting the missing plus sign

$= \frac{2}{1} + \frac{3}{4}$ $\quad 2 = \frac{2}{1}$

$= \frac{2}{1} \cdot \frac{4}{4} + \frac{3}{4}$ Finding a common denominator

$= \frac{8}{4} + \frac{3}{4} = \frac{11}{4}$

4. $4\frac{3}{10} = 4 + \frac{3}{10} = \frac{4}{1} + \frac{3}{10} = \frac{4}{1} \cdot \frac{10}{10} + \frac{3}{10} = \frac{40}{10} + \frac{3}{10} = \frac{43}{10}$

Convert to a mixed numeral.

1. $1 + \frac{2}{3} = \square\frac{\square}{\square}$

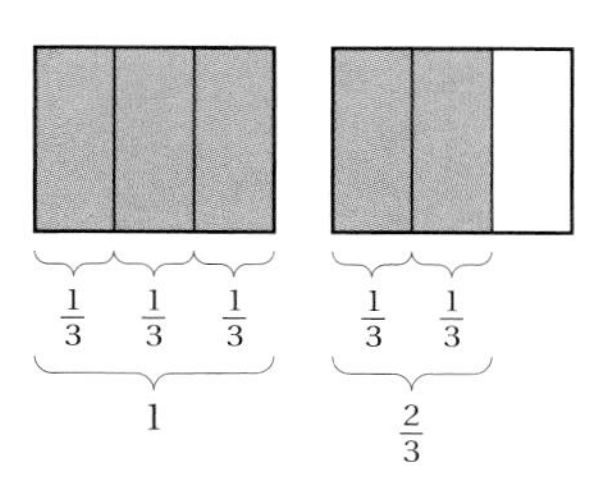

2. $2 + \frac{3}{4} = \square\frac{\square}{\square}$

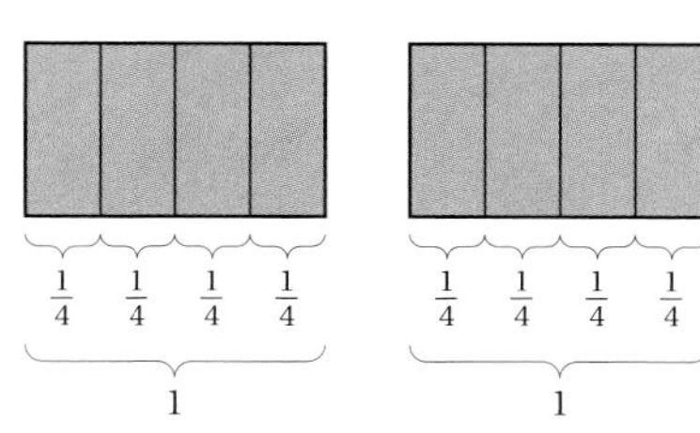

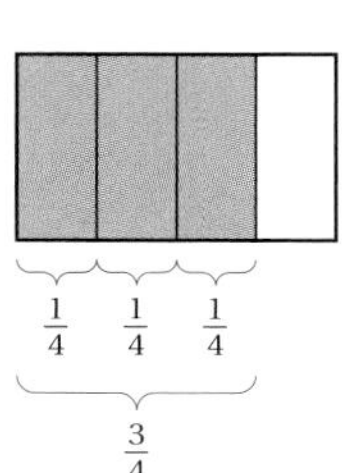

3. $8 + \frac{3}{4}$

4. $12 + \frac{2}{3}$

Answers on page A-7

Convert to fraction notation.

5. $4\frac{2}{5}$

6. $6\frac{1}{10}$

Convert to fraction notation. Use the faster method.

7. $4\frac{5}{6}$

8. $9\frac{1}{4}$

9. $20\frac{2}{3}$

Answers on page A-7

Do Exercises 5 and 6.

Let's now consider a faster method for converting a mixed numeral to fraction notation.

To convert from a mixed numeral to fraction notation:

ⓐ Multiply the whole number by the denominator: $4 \cdot 10 = 40$.
ⓑ Add the result to the numerator: $40 + 3 = 43$.
ⓒ Keep the denominator.

$$4\frac{3}{10} = \frac{43}{10}$$

EXAMPLES Convert to fraction notation.

5. $6\frac{2}{3} = \frac{20}{3}$ $\quad 6 \cdot 3 = 18,\ 18 + 2 = 20$

6. $8\frac{2}{9} = \frac{74}{9}$ $\quad 8 \cdot 9 = 72,\ 72 + 2 = 74$

7. $10\frac{7}{8} = \frac{87}{8}$ $\quad 10 \cdot 8 = 80,\ 80 + 7 = 87$

Do Exercises 7–9.

WRITING MIXED NUMERALS

We can find a mixed numeral for $\frac{5}{3}$ as follows:

$$\frac{5}{3} = \frac{3}{3} + \frac{2}{3} = 1 + \frac{2}{3} = 1\frac{2}{3}.$$

In terms of objects, we can think of $\frac{5}{3}$ as $\frac{3}{3}$, or 1, plus $\frac{2}{3}$, as shown below.

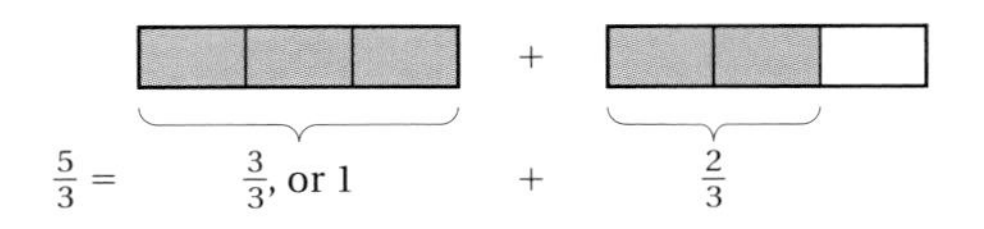

Fraction symbols like $\frac{5}{3}$ also indicate division; $\frac{5}{3}$ means $5 \div 3$. Let's divide the numerator by the denominator.

$$\begin{array}{r} 1 \\ 3\overline{)5} \\ \underline{3} \\ 2 \end{array} \leftarrow 2 \div 3 = \frac{2}{3}$$

Thus, $\frac{5}{3} = 1\frac{2}{3}$.

To convert from fraction notation to a mixed numeral, divide.

$\frac{13}{5}$

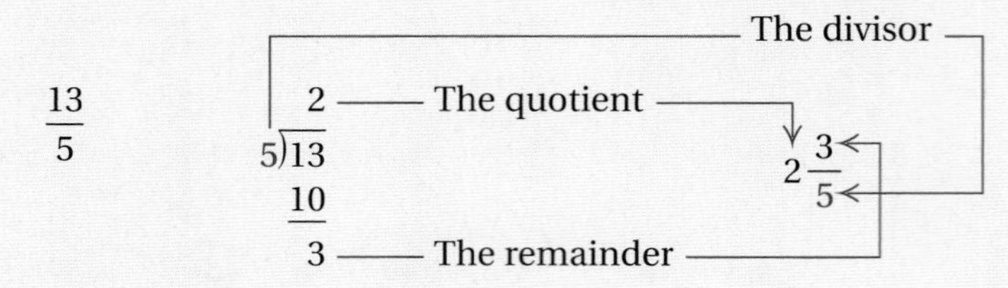

EXAMPLES Convert to a mixed numeral.

8. $\frac{69}{10}$

$$\begin{array}{r} 6 \\ 10\overline{)69} \\ \underline{60} \\ 9 \end{array} \qquad \frac{69}{10} = 6\frac{9}{10}$$

9. $\frac{122}{8}$

$$\begin{array}{r} 15 \\ 8\overline{)122} \\ \underline{80} \\ 42 \\ \underline{40} \\ 2 \end{array} \qquad \frac{122}{8} = 15\frac{2}{8} = 15\frac{1}{4}$$

Do Exercises 10–12.

b Writing Mixed Numerals for Quotients

It is quite common when dividing whole numbers to write the quotient using a mixed numeral. The remainder is the numerator of the fractional part of the mixed numeral.

EXAMPLE 10 Divide. Write a mixed numeral for the quotient.

$$7\overline{)6\,3\,4\,1}$$

We first divide as usual.

$$\begin{array}{r} 9\,0\,5 \\ 7\overline{)6\,3\,4\,1} \\ \underline{6\,3\,0\,0} \\ 4\,1 \\ \underline{3\,5} \\ 6 \end{array}$$

The answer is 905 R 6. We write a mixed numeral for the answer as follows:

$$905\frac{6}{7}.$$

The division $6341 \div 7$ can be expressed using fraction notation or a mixed numeral as follows:

$$\frac{6341}{7} = 905\frac{6}{7}.$$

Convert to a mixed numeral.

10. $\frac{7}{3}$

11. $\frac{11}{10}$

12. $\frac{110}{6}$

Answers on page A-7

Study Tips

FORMING A STUDY GROUP

Consider forming a study group with some of your fellow students. Exchange e-mail addresses, telephone numbers, and schedules so that you can coordinate study time for homework and tests.

Divide. Write a mixed numeral for the answer.

13. $6\overline{)4846}$

14. $45\overline{)6053}$

Answers on page A-7

EXAMPLE 11 Divide. Write a mixed numeral for the answer.

$42\overline{)8915}$

We first divide as usual.

$$\begin{array}{r} 212 \\ 42\overline{)8915} \\ 8400 \\ \hline 515 \\ 420 \\ \hline 95 \\ 84 \\ \hline 11 \end{array} \qquad \frac{8915}{42} = 212\frac{11}{42}$$

The answer is $212\frac{11}{42}$.

Do Exercises 13 and 14.

A fraction larger than 1, such as $\frac{27}{8}$, is sometimes referred to as an "improper" fraction. We will not use this terminology because notation such as $\frac{27}{8}$, $\frac{11}{9}$, and $\frac{89}{10}$ is quite "proper" and very common in algebra.

As we will see in subsequent sections, mixed numerals have many real-world applications, especially in the areas of garment manufacturing and carpentry, as well as many other fields.

CALCULATOR CORNER

Writing Quotients as Mixed Numerals When using a calculator to divide whole numbers, we can express the result using a mixed numeral. To do so, we first find the quotient and the remainder as shown in the Calculator Corner on p. 47. The quotient is the whole-number part of the mixed numeral, the remainder is the numerator of the fractional part, and the divisor is the denominator of the fractional part. For example, on p. 47, we saw that $453 \div 15 = 30 \text{ R } 3$. Then we can also write

$$453 \div 15 = 30\frac{3}{15} = 30\frac{1}{5}.$$

Exercises: Use a calculator to divide. Write the result as a mixed numeral.

1. $6\overline{)8857}$
2. $9\overline{)6088}$
3. $56\overline{)44{,}851}$
4. $18\overline{)234{,}567}$
5. $11\overline{)567{,}895}$
6. $32\overline{)234{,}567}$
7. $45\overline{)6033}$
8. $213\overline{)567{,}988}$
9. $112\overline{)400{,}003}$
10. $908\overline{)11{,}234}$

For Extra Help

MathXL · MyMathLab · InterAct Math · Math Tutor Center · Digital Video Tutor CD 2 Videotape 3 · Student's Solutions Manual

a

1. *Alarm Clock Dimensions.* A world time clock/radio with NOAA (National Oceanic and Atmospheric Administration) Weather Band measures $14\frac{1}{4}$ in. $\times$ $6\frac{3}{4}$ in. $\times$ $2\frac{1}{4}$ in. Convert $14\frac{1}{4}$, $6\frac{3}{4}$, and $2\frac{1}{4}$ to fraction notation.

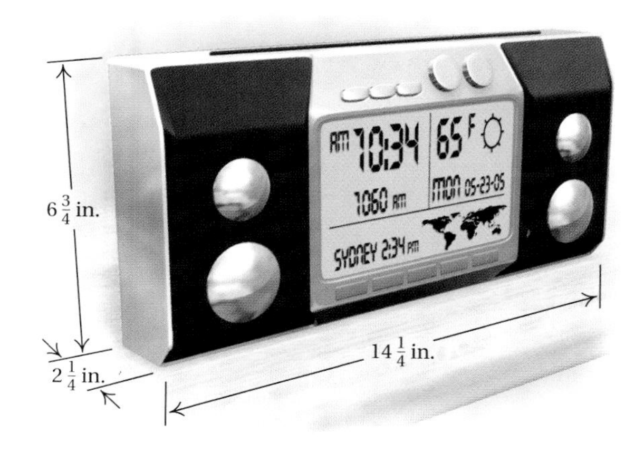

2. *Carpentry.* Dick Bonewitz, master carpenter, is making a jewelry box according to the design below. Convert each mixed numeral to fraction notation.

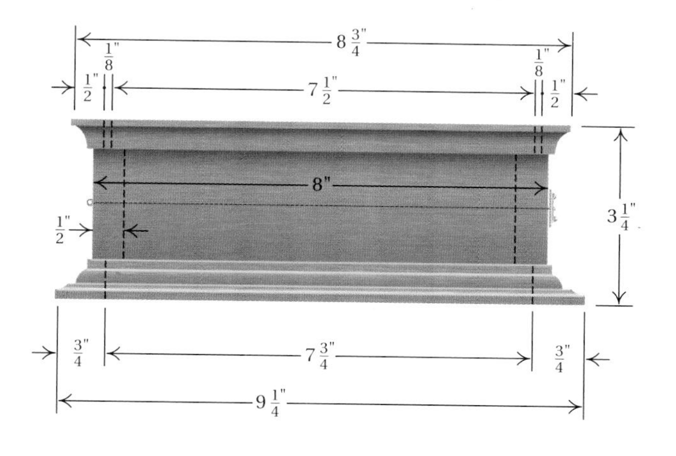

3. *Garment Manufacturing.* A tailoring shop determines that for a certain size dress, it must use $\frac{29}{8}$ yd of fabric that is 45 in. wide. To make the same dress with fabric that is 60 in. wide, it needs $\frac{11}{4}$ yd. Convert $\frac{29}{8}$ and $\frac{11}{4}$ to mixed numerals.

4. *Bake Sale* A fire department organized a bake sale for a charitable fund-raiser. Each pie was cut into 8 pieces and each cake into 12 pieces. They sold 73 pieces of pie, or $\frac{73}{8}$ pies, and 55 pieces of cake, or $\frac{55}{12}$ cakes. Convert $\frac{73}{8}$ and $\frac{55}{12}$ to mixed numerals.

Convert to fraction notation.

5. $5\frac{2}{3}$

6. $3\frac{4}{5}$

7. $3\frac{1}{4}$

8. $6\frac{1}{2}$

9. $10\frac{1}{8}$

10. $20\frac{1}{5}$

11. $5\frac{1}{10}$

12. $9\frac{1}{10}$

13. $20\frac{3}{5}$

14. $30\frac{4}{5}$

15. $9\frac{5}{6}$

16. $8\frac{7}{8}$

17. $7\frac{3}{10}$

18. $6\frac{9}{10}$

19. $1\frac{5}{8}$

20. $1\frac{3}{5}$

21. $12\frac{3}{4}$

22. $15\frac{2}{3}$

23. $4\frac{3}{10}$

24. $5\frac{7}{10}$

25. $2\frac{3}{100}$

26. $5\frac{7}{100}$

27. $66\frac{2}{3}$

28. $33\frac{1}{3}$

29. $5\frac{29}{50}$

30. $84\frac{3}{8}$

Convert to a mixed numeral.

31. $\frac{18}{5}$

32. $\frac{17}{4}$

33. $\frac{14}{3}$

34. $\frac{39}{8}$

35. $\frac{27}{6}$

36. $\frac{30}{9}$

37. $\frac{57}{10}$

38. $\frac{89}{10}$

39. $\frac{53}{7}$

40. $\frac{59}{8}$

41. $\frac{45}{6}$

42. $\frac{50}{8}$

43. $\frac{46}{4}$

44. $\frac{39}{9}$

45. $\frac{12}{8}$

46. $\frac{28}{6}$

47. $\frac{757}{100}$

48. $\frac{467}{100}$

49. $\frac{345}{8}$

50. $\frac{223}{4}$

b Divide. Write a mixed numeral for the answer.

51. $8\overline{)869}$

52. $3\overline{)2126}$

53. $5\overline{)3091}$

54. $9\overline{)9110}$

55. $21\overline{)852}$

56. $85\overline{)7672}$

57. $102\overline{)5612}$

58. $46\overline{)1081}$

59. $\mathbf{D_W}$ Describe in your own words a method for rewriting a fraction as a mixed numeral.

60. $\mathbf{D_W}$ Are the numbers $2\frac{1}{3}$ and $2 \cdot \frac{1}{3}$ equal? Why or why not?

SKILL MAINTENANCE

61. Round to the nearest hundred: 45,765. [1.4a]

62. Round to the nearest ten: 45,765. [1.4a]

Simplify. [2.5b]

63. $\frac{200}{375}$

64. $\frac{63}{75}$

65. $\frac{160}{270}$

66. $\frac{6996}{8028}$

Multiply and simplify. [2.4a], [2.6a]

67. $\frac{6}{5} \cdot 15$

68. $\frac{5}{12} \cdot 6$

69. $\frac{7}{10} \cdot \frac{5}{14}$

70. $\frac{1}{10} \cdot \frac{20}{5}$

Divide and simplify. [2.7b]

71. $\frac{2}{3} \div \frac{1}{36}$

72. $28 \div \frac{4}{7}$

73. $200 \div \frac{15}{64}$

74. $\frac{3}{4} \div \frac{9}{16}$

SYNTHESIS

Write a mixed numeral.

75. 🖩 $\frac{128{,}236}{541}$

76. 🖩 $\frac{103{,}676}{349}$

77. $\frac{56}{7} + \frac{2}{3}$

78. $\frac{72}{12} + \frac{5}{6}$

79. There are $\frac{366}{7}$ weeks in a leap year.

80. There are $\frac{365}{7}$ weeks in a year.

3.5 ADDITION AND SUBTRACTION USING MIXED NUMERALS; APPLICATIONS

Objectives

a Add using mixed numerals.

b Subtract using mixed numerals.

c Solve applied problems involving addition and subtraction with mixed numerals.

1. Add.

$$\begin{array}{r} 2\frac{3}{10} \\ +\,5\frac{1}{10} \\ \hline \end{array}$$

2. Add.

$$\begin{array}{r} 8\frac{2}{5} \\ +\,3\frac{7}{10} \\ \hline \end{array}$$

3. Add.

$$\begin{array}{r} 9\frac{3}{4} \\ +\,3\frac{5}{6} \\ \hline \end{array}$$

Answers on page A-7

a Addition Using Mixed Numerals

To find the sum $1\frac{5}{8} + 3\frac{1}{8}$, we first add the fractions. Then we add the whole numbers.

$$\begin{array}{rcl} 1\frac{5}{8} & = & 1\frac{5}{8} \\ +\,3\frac{1}{8} & = & +\,3\frac{1}{8} \\ \hline \frac{6}{8} & & 4\frac{6}{8} = 4\frac{3}{4} \end{array}$$

Add the fractions. Add the whole numbers. Simplifying

Do Exercise 1.

Sometimes we must write the fractional parts with a common denominator before we can add.

EXAMPLE 1 Add: $5\frac{2}{3} + 3\frac{5}{6}$. Write a mixed numeral for the answer.

The LCD is 6.

$$\begin{array}{rcl} 5\frac{2}{3}\cdot\frac{2}{2} & = & 5\frac{4}{6} \\ +\,3\frac{5}{6} & = & +\,3\frac{5}{6} \\ \hline & & 8\frac{9}{6} = 8 + \frac{9}{6} \\ & & = 8 + 1\frac{1}{2} \\ & & = 9\frac{1}{2} \end{array}$$

To find a mixed numeral for $\frac{9}{6}$, we divide:

$$\begin{array}{r} 1 \\ 6\overline{)9} \\ \underline{6} \\ 3 \end{array} \qquad \frac{9}{6} = 1\frac{3}{6} = 1\frac{1}{2}$$

Do Exercise 2.

EXAMPLE 2 Add: $10\frac{5}{6} + 7\frac{3}{8}$.

The LCD is 24.

$$\begin{array}{rcl} 10\frac{5}{6}\cdot\frac{4}{4} & = & 10\frac{20}{24} \\ +\,7\frac{3}{8}\cdot\frac{3}{3} & = & +\,7\frac{9}{24} \\ \hline & & 17\frac{29}{24} = 17 + \frac{29}{24} \\ & & = 17 + 1\frac{5}{24} \\ & & = 18\frac{5}{24} \end{array}$$

Writing $\frac{29}{24}$ as a mixed numeral, $1\frac{5}{24}$

Do Exercise 3.

b Subtraction Using Mixed Numerals

Subtraction is a lot like addition; we subtract the fractions and then the whole numbers.

EXAMPLE 3 Subtract: $7\frac{3}{4} - 2\frac{1}{4}$.

$$\begin{array}{r} 7\frac{3}{4} \\ -\,2\frac{1}{4} \\ \hline \frac{2}{4} \end{array} = \begin{array}{r} 7\frac{3}{4} \\ -\,2\frac{1}{4} \\ \hline 5\frac{2}{4} \end{array} = 5\frac{1}{2}$$

Subtract the fractions. Subtract the whole numbers. Simplifying

EXAMPLE 4 Subtract: $9\frac{4}{5} - 3\frac{1}{2}$.

The LCD is 10.

$$\begin{array}{rcr} 9\frac{4}{5}\cdot\frac{2}{2} & = & 9\frac{8}{10} \\ -\,3\frac{1}{2}\cdot\frac{5}{5} & = & -\,3\frac{5}{10} \\ \hline & & 6\frac{3}{10} \end{array}$$

Do Exercises 4 and 5.

EXAMPLE 5 Subtract: $7\frac{1}{6} - 2\frac{1}{4}$.

The LCD is 12.

$$\begin{array}{rcr} 7\frac{1}{6}\cdot\frac{2}{2} & = & 7\frac{2}{12} \\ -\,2\frac{1}{4}\cdot\frac{3}{3} & = & -\,2\frac{3}{12} \\ \hline \end{array}$$

We cannot subtract $\frac{3}{12}$ from $\frac{2}{12}$.
We borrow 1, or $\frac{12}{12}$, from 7:
$7\frac{2}{12} = 6 + 1 + \frac{2}{12} = 6 + \frac{12}{12} + \frac{2}{12} = 6\frac{14}{12}$.

We can write this as

$$\begin{array}{rcr} 7\frac{2}{12} & = & 6\frac{14}{12} \\ -\,2\frac{3}{12} & = & -\,2\frac{3}{12} \\ \hline & & 4\frac{11}{12} \end{array}$$

Do Exercise 6.

Subtract.

4. $\begin{array}{r} 10\frac{7}{8} \\ -\,9\frac{3}{8} \\ \hline \end{array}$

5. $\begin{array}{r} 8\frac{2}{3} \\ -\,5\frac{1}{2} \\ \hline \end{array}$

6. Subtract.

$\begin{array}{r} 8\frac{1}{9} \\ -\,4\frac{5}{6} \\ \hline \end{array}$

Answers on page A-7

7. Subtract.

$$\begin{array}{r} 5 \\ -\ 1\frac{1}{3} \\ \hline \end{array}$$

8. **Travel Distance.** Chrissy Jenkins is a college textbook sales representative for Addison-Wesley. On two business days, Chrissy drove $144\frac{9}{10}$ mi and $87\frac{1}{4}$ mi. What was the total distance Chrissy drove?

Answers on page A-7

EXAMPLE 6 Subtract: $12 - 9\frac{3}{8}$.

$$\begin{array}{rcr} 12 & = & 11\frac{8}{8} \\ -\ 9\frac{3}{8} & = & -\ 9\frac{3}{8} \\ \hline & & 2\frac{5}{8} \end{array} \qquad 12 = 11 + 1 = 11 + \frac{8}{8} = 11\frac{8}{8}$$

Do Exercise 7.

C Applications and Problem Solving

EXAMPLE 7 *Widening a Driveway.* Sherry and Woody are widening their existing $17\frac{1}{4}$-ft driveway by adding $5\frac{9}{10}$ ft on one side. What is the width of the new driveway?

1. **Familiarize.** We let w = the width of the new driveway.
2. **Translate.** We translate as follows.

Width of existing driveway	+	Width of additional driveway	=	Width of new driveway
↓	↓	↓	↓	↓
$17\frac{1}{4}$	+	$5\frac{9}{10}$	=	w

3. **Solve.** The translation tells us what to do. We add. The LCD is 20.

$$\begin{array}{rcrcr} 17\frac{1}{4} & = & 17\ \frac{1}{4}\cdot\frac{5}{5} & = & 17\frac{5}{20} \\ +\ 5\frac{9}{10} & = & +\ 5\ \frac{9}{10}\cdot\frac{2}{2} & = & +\ 5\frac{18}{20} \\ \hline & & & & 22\frac{23}{20} = 23\frac{3}{20} \end{array}$$

Thus, $w = 23\frac{3}{20}$.

4. **Check.** We check by repeating the calculation. We also note that the answer is larger than either of the widths, which means that the answer is reasonable.
5. **State.** The width of the new driveway is $23\frac{3}{20}$ ft.

Do Exercise 8.

EXAMPLE 8 *Communication Cable.* Celebrity Cable has two crews who install Internet communication cable. Crew A can install $38\frac{1}{8}$ ft per hour. Crew B can install $31\frac{2}{3}$ ft per hour. How many fewer feet can Crew B install per hour than Crew A?

1. **Familiarize.** The phrase "how many fewer" indicates subtraction. We let $c =$ the difference in the numbers of feet per hour.

2. **Translate.** We translate as follows.

Crew A: ft/hr − Crew B: ft/hr = Difference in ft/hr

$$38\frac{1}{8} - 31\frac{2}{3} = c$$

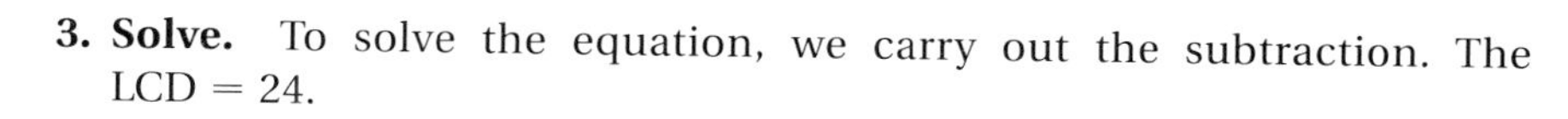

3. **Solve.** To solve the equation, we carry out the subtraction. The LCD = 24.

$$\begin{array}{rcrcrcr} 38\frac{1}{8} & = & 38\,\frac{1}{8}\cdot\frac{3}{3} & = & 38\frac{3}{24} & = & 37\frac{27}{24} \\ -31\frac{2}{3} & = & -31\,\frac{2}{3}\cdot\frac{8}{8} & = & -31\frac{16}{24} & = & -31\frac{16}{24} \\ \hline & & & & & & 6\frac{11}{24} \end{array}$$

Thus, $c = 6\frac{11}{24}$.

4. **Check.** To check, we add the difference, $6\frac{11}{24}$, to Crew B's rate:

$$\begin{aligned} 31\frac{2}{3} + 6\frac{11}{24} &= 31\frac{16}{24} + 6\frac{11}{24} \\ &= 37\frac{27}{24} \\ &= 38\frac{3}{24} \\ &= 38\frac{1}{8}. \end{aligned}$$

This checks.

5. **State.** Crew B installs $6\frac{11}{24}$ fewer feet per hour than Crew A.

Do Exercise 9.

9. **NCAA Football Goal Posts.** In college football, the distance between goal posts was reduced from $23\frac{1}{3}$ ft to $18\frac{1}{2}$ ft in 1991. By how much was it reduced?

Source: NCAA

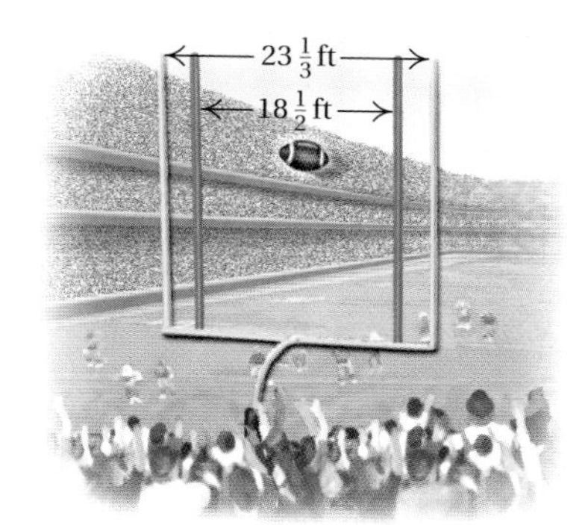

Answer on page A-7

10. Liquid Fertilizer. There are $283\frac{5}{8}$ gal of liquid fertilizer in a fertilizer application tank. After applying $178\frac{2}{3}$ gal to a soybean field, the farmer requests that Braden's Farm Supply deliver an additional 250 gal. How many gallons of fertilizer are in the tank after the delivery?

Answer on page A-7

MULTISTEP PROBLEMS

EXAMPLE 9 *Carpentry.* The following diagram shows the layout for the construction of a desk drawer. Find the missing length a.

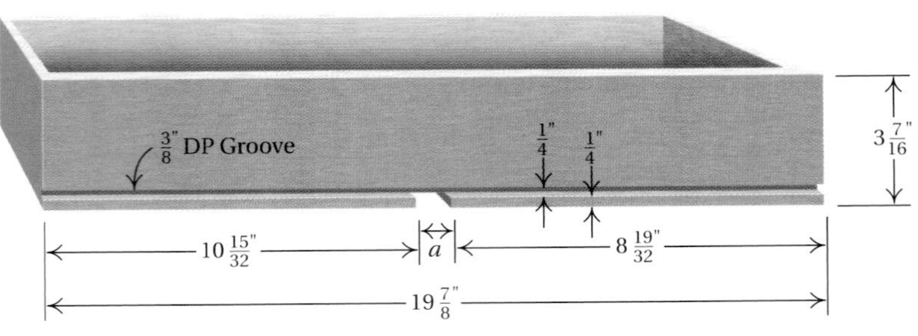

Middle Drawer / Back Layout

1. **Familiarize.** The length a is shown in the drawing.
2. **Translate.** From the drawing, we see that the length a is the full length of the drawer minus the sum of the lengths $10\frac{15}{32}$" and $8\frac{19}{32}$". Thus we have

$$a = 19\frac{7}{8} - \left(10\frac{15}{32} + 8\frac{19}{32}\right).$$

3. **Solve.** This is a two-step problem.

   a) We first add the two lengths, $10\frac{15}{32}$ and $8\frac{19}{32}$.

$$\begin{array}{r} 10\frac{15}{32} \\ +\ 8\frac{19}{32} \\ \hline 18\frac{34}{32} = 18\frac{17}{16} = 19\frac{1}{16} \end{array}$$

   b) Next, we subtract $19\frac{1}{16}$ from $19\frac{7}{8}$.

$$\begin{array}{rcr} 19\frac{7}{8} & = & 19\frac{14}{16} \\ -19\frac{1}{16} & = & -19\frac{1}{16} \\ \hline & & \frac{13}{16} \end{array}$$

   Thus, $a = \frac{13}{16}$.

4. **Check.** We check by repeating the calculation, or adding the three measurements:

$$10\frac{15}{32} + 8\frac{19}{32} + \frac{13}{16} = 10\frac{15}{32} + 8\frac{19}{32} + \frac{26}{32}$$
$$= 18\frac{60}{32} = 19\frac{28}{32} = 19\frac{7}{8}.$$

5. **State.** The length a in the diagram is $\frac{13}{16}$".

Do Exercise 10.

3.5 EXERCISE SET

For Extra Help

MathXL | MyMathLab | InterAct Math | Math Tutor Center | Digital Video Tutor CD 2 Videotape 3 | Student's Solutions Manual

a Add. Write a mixed numeral for the answer.

1. $20 + 8\frac{3}{4}$

2. $37 + 18\frac{2}{3}$

3. $129\frac{7}{8} + 56$

4. $2003\frac{4}{11} + 59$

5. $2\frac{7}{8} + 3\frac{5}{8}$

6. $4\frac{5}{6} + 3\frac{5}{6}$

7. $1\frac{1}{4} + 1\frac{2}{3}$

8. $4\frac{1}{3} + 5\frac{2}{9}$

9. $8\frac{3}{4} + 5\frac{5}{6}$

10. $4\frac{3}{8} + 6\frac{5}{12}$

11. $3\frac{2}{5} + 8\frac{7}{10}$

12. $5\frac{1}{2} + 3\frac{7}{10}$

13. $5\frac{3}{8} + 10\frac{5}{6}$

14. $\frac{5}{8} + 1\frac{5}{6}$

15. $12\frac{4}{5} + 8\frac{7}{10}$

16. $15\frac{5}{8} + 11\frac{3}{4}$

17. $14\frac{5}{8} + 13\frac{1}{4}$

18. $16\frac{1}{4} + 15\frac{7}{8}$

19. $7\frac{1}{8} + 9\frac{2}{3} + 10\frac{3}{4}$

20. $45\frac{2}{3} + 31\frac{3}{5} + 12\frac{1}{4}$

b Subtract. Write a mixed numeral for the answer.

21. $4\frac{1}{5} - 2\frac{3}{5}$

22. $5\frac{1}{8} - 2\frac{3}{8}$

23. $6\frac{3}{5} - 2\frac{1}{2}$

24. $7\frac{2}{3} - 6\frac{1}{2}$

25. $34\frac{1}{3} - 12\frac{5}{8}$

26. $23\frac{5}{16} - 16\frac{3}{4}$

27. $21 - 8\frac{3}{4}$

28. $42 - 3\frac{7}{8}$

29. $\begin{array}{r} 34 \\ -\ 18\frac{5}{8} \\ \hline \end{array}$

30. $\begin{array}{r} 23 \\ -\ 19\frac{3}{4} \\ \hline \end{array}$

31. $\begin{array}{r} 21\frac{1}{6} \\ -\ 13\frac{3}{4} \\ \hline \end{array}$

32. $\begin{array}{r} 42\frac{1}{10} \\ -\ 23\frac{7}{12} \\ \hline \end{array}$

33. $\begin{array}{r} 14\frac{1}{8} \\ -\ \ \frac{3}{4} \\ \hline \end{array}$

34. $\begin{array}{r} 28\frac{1}{6} \\ -\ 5 \\ \hline \end{array}$

35. $\begin{array}{r} 25\frac{1}{9} \\ -\ 13\frac{5}{6} \\ \hline \end{array}$

36. $\begin{array}{r} 23\frac{5}{16} \\ -\ 14\frac{7}{12} \\ \hline \end{array}$

Solve.

37. *Sewing from a Pattern.* Suppose you want to make an outfit in size 8. Using 45-in. fabric, you need $1\frac{3}{8}$ yd for the dress, $\frac{5}{8}$ yd of contrasting fabric for the band at the bottom, and $3\frac{3}{8}$ yd for the jacket. How many yards of 45-in. fabric are needed to make the outfit?

38. *Sewing from a Pattern.* Suppose you want to make an outfit in size 12. Using 45-in. fabric, you need $2\frac{3}{4}$ yd for the dress and $3\frac{1}{2}$ yd for the jacket. How many yards of 45-in. fabric are needed to make the outfit?

39. For a family barbecue, Cayla bought packages of hamburger weighing $1\frac{2}{3}$ lb and $5\frac{3}{4}$ lb. What was the total weight of the meat?

40. Marsha's Butcher Shop sold packages of sliced turkey breast weighing $1\frac{1}{3}$ lb and $4\frac{3}{5}$ lb. What was the total weight of the meat?

41. Tara is 66 in. tall and her son, Tom, is $59\frac{7}{12}$ in. tall. How much taller is Tara?

42. Nicholas is $73\frac{2}{3}$ in. tall and his daughter, Kendra, is $71\frac{5}{16}$ in. tall. How much shorter is Kendra?

43. *Cutco Knives.* The Kitchen Classics knife set sold by Cutco contains three basic knives: $7\frac{5}{8}$" Petite Chef, $4\frac{3}{4}$" Trimmer, and $2\frac{3}{4}$" Paring Knife. How much larger is the blade in the Petite Chef than in the Trimmer? than in the Paring Knife?

Source: Cutco Cutlery Corporation

44. *Upholstery Fabric.* Executive Car Care sells 45-in. upholstery fabric for car restoration. Art buys $9\frac{1}{4}$ yd and $10\frac{5}{6}$ yd for two car projects. How many total yards did Art buy?

45. A plumber uses pipes of lengths $10\frac{5}{16}$ in. and $8\frac{3}{4}$ in. in the installation of a sink. How much pipe was used?

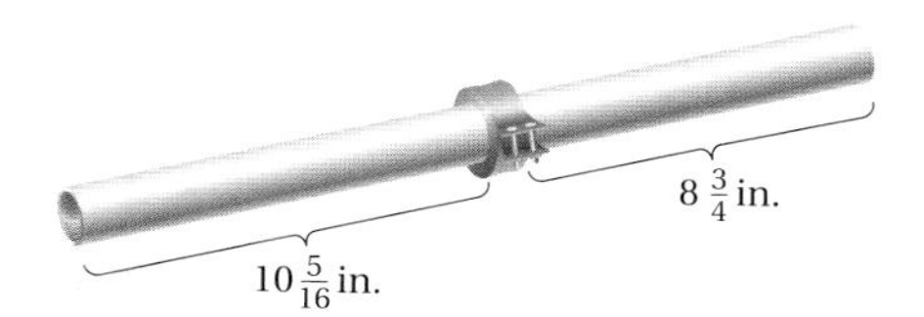

46. *Writing Supplies.* The standard pencil is $6\frac{7}{8}$ in. wood and $\frac{1}{2}$ in. eraser. What is the total length of the standard pencil?

Source: Eberhard Faber American

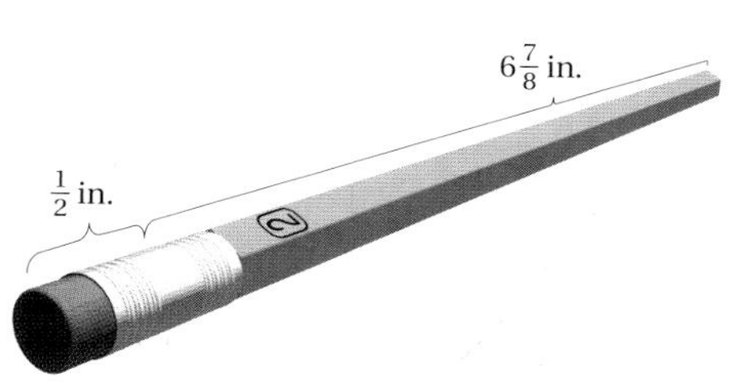

47. Kim Park is a computer technician. One day, she drove $180\frac{7}{10}$ mi away from Los Angeles for a service call. The next day, she drove $85\frac{1}{2}$ mi back toward Los Angeles for another service call. How far was she from Los Angeles?

48. Pilar is $4\frac{1}{2}$ in. taller than her daughter, Teresa. Teresa is $66\frac{2}{3}$ in. tall. How tall is Pilar?

49. *Book Size.* One standard book size is $8\frac{1}{2}$ in. by $9\frac{3}{4}$ in. What is the total distance around (perimeter of) the front cover of such a book?

50. *Copier Paper.* A standard sheet of copier paper is $8\frac{1}{2}$ in. by 11 in. What is the total distance around (perimeter of) the paper?

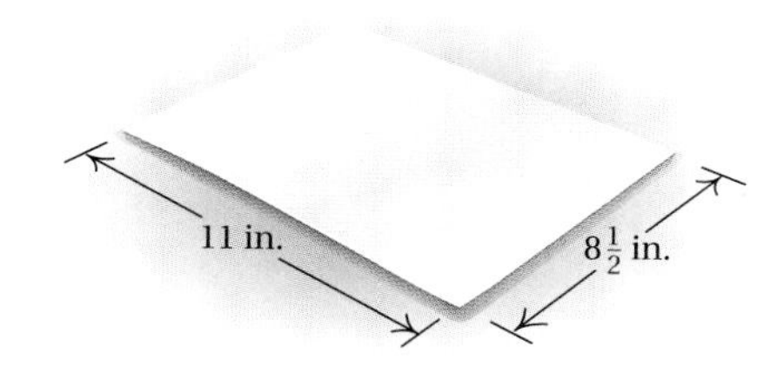

51. *Carpentry.* When cutting wood with a saw, a carpenter must take into account the thickness of the saw blade. Suppose that from a piece of wood 36 in. long, a carpenter cuts a $15\frac{3}{4}$-in. length with a saw blade that is $\frac{1}{8}$ in. thick. How long is the piece that remains?

52. *Painting.* A painter used $1\frac{3}{4}$ gal of paint for the Garcias' living room and $1\frac{1}{3}$ gal for their family room. How much paint was used in all?

53. Rene is $5\frac{1}{4}$ in. taller than his son, who is $72\frac{5}{6}$ in. tall. How tall is Rene?

54. A Boeing 767 flew 640 mi on a nonstop flight. On the return flight, it landed after having flown $320\frac{3}{10}$ mi. How far was the plane from its original point of departure?

55. *Interior Design.* Sue worked $10\frac{1}{2}$ hr over a three-day period on an interior design project. If she worked $2\frac{1}{2}$ hr on the first day and $4\frac{1}{5}$ hr on the second, how many hours did Sue work on the third day?

56. *Painting.* Geri had $3\frac{1}{2}$ gal of paint. It took $2\frac{3}{4}$ gal to paint the family room. She estimated that it would take $2\frac{1}{4}$ gal to paint the living room. How much more paint did Geri need?

Find the perimeter of (distance around) the figure.

57.

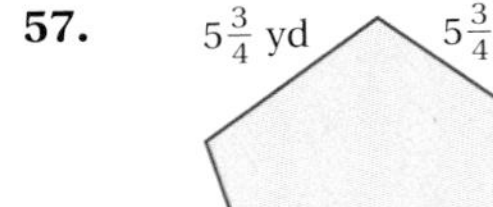

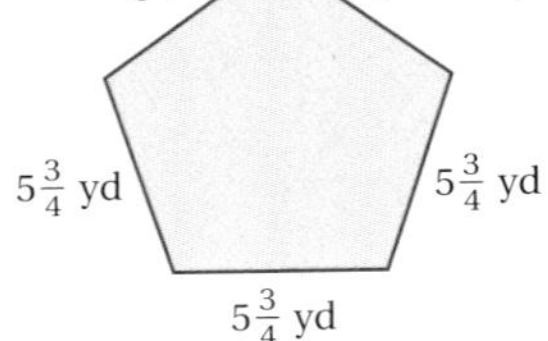

58.

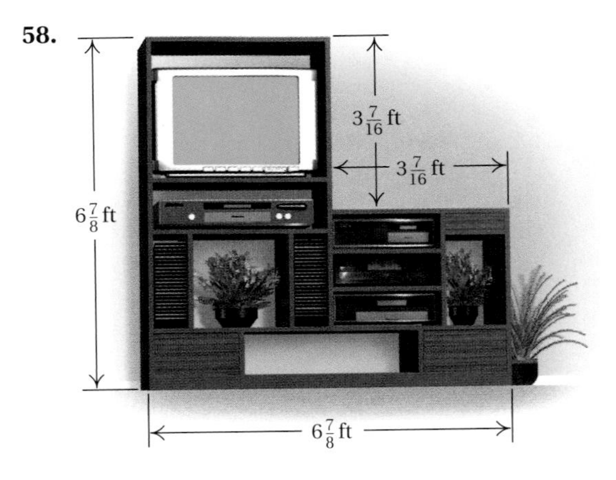

Find the length d in the figure.

59.

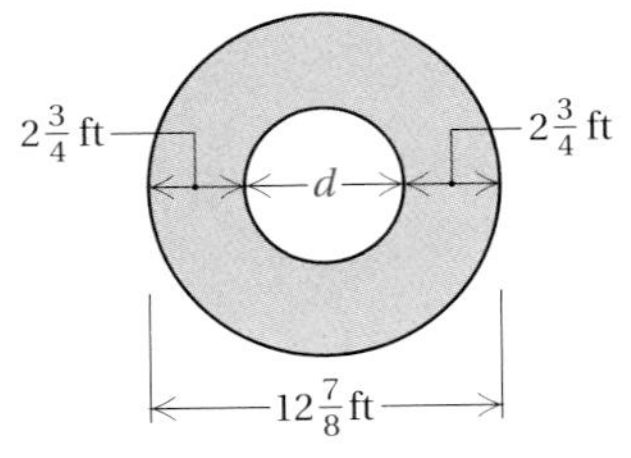

60.

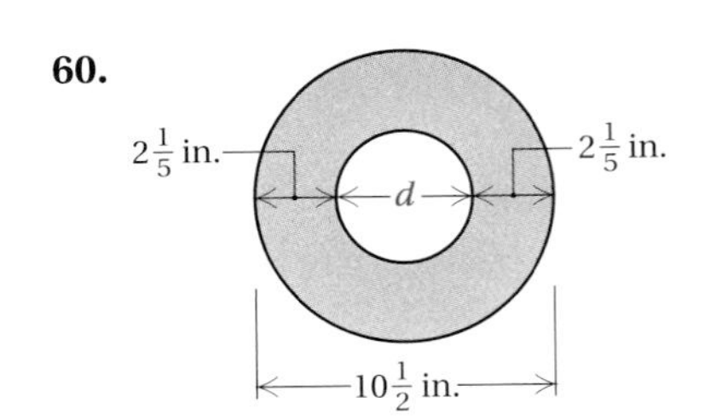

61. Find the smallest length of a bolt that will pass through a piece of tubing with an outside diameter of $\frac{1}{2}$ in., a washer $\frac{1}{16}$ in. thick, a piece of tubing with a $\frac{3}{4}$-in. outside diameter, another washer, and a nut $\frac{3}{16}$ in. thick.

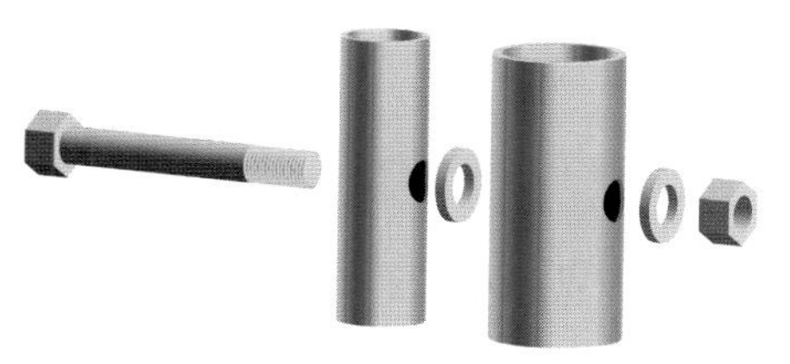

62. The front of the stage at the Lagrange Town Hall is $6\frac{1}{2}$ yd wide. If renovation work succeeds in adding $2\frac{3}{4}$ yd in width, how wide is the renovated stage?

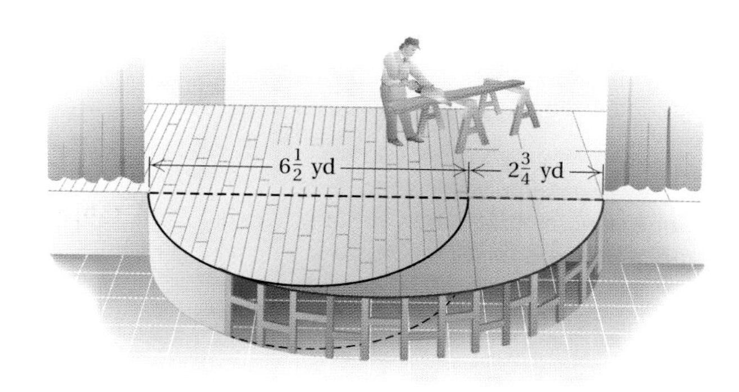

63. DW Write a problem for a classmate to solve. Design the problem so the solution is "The larger package holds $4\frac{1}{2}$ oz more than the smaller package."

64. DW Is the sum of two mixed numerals always a mixed numeral? Why or why not?

SKILL MAINTENANCE

Solve.

65. Rick's Market sells Swiss cheese in $\frac{3}{4}$-lb packages. How many packages can be made from a 12-lb slab of cheese? [2.7d]

66. Holstein's Dairy produced 4578 oz of milk one morning. How many 16-oz cartons could be filled? How much milk would be left over? [1.8a]

Determine whether the first number is divisible by the second. [2.2a]

67. 9993 by 3

68. 9993 by 9

69. 2345 by 9

70. 2345 by 5

71. 2335 by 10

72. 7764 by 6

73. 18,888 by 8

74. 18,888 by 4

75. Multiply and simplify: $\frac{15}{9} \cdot \frac{18}{39}$. [2.6a]

76. Divide and simplify: $\frac{12}{25} \div \frac{24}{5}$. [2.7b]

SYNTHESIS

Calculate each of the following. Write the result as a mixed numeral.

77. ▦ $3289\frac{1047}{1189} + 5278\frac{32}{41}$

78. ▦ $5798\frac{17}{53} - 3909\frac{1957}{2279}$

79. A post for a pier is 29 ft long. Half of the post extends above the water's surface and $8\frac{3}{4}$ ft of the post is buried in mud. How deep is the water at that point?

80. Solve: $47\frac{2}{3} + n = 56\frac{1}{4}$.

3.6 MULTIPLICATION AND DIVISION USING MIXED NUMERALS; APPLICATIONS

Objectives

a Multiply using mixed numerals.

b Divide using mixed numerals.

c Solve applied problems involving multiplication and division with mixed numerals.

1. Multiply: $6 \cdot 3\frac{1}{3}$.

2. Multiply: $2\frac{1}{2} \cdot \frac{3}{4}$.

3. Multiply: $2 \cdot 6\frac{2}{5}$.

4. Multiply: $3\frac{1}{3} \cdot 2\frac{1}{2}$.

Answers on page A-7

a Multiplication Using Mixed Numerals

Carrying out addition and subtraction with mixed numerals is usually easier if the numbers are left as mixed numerals. With multiplication and division, however, it is easier to convert the numbers to fraction notation first.

MULTIPLICATION USING MIXED NUMERALS

To multiply using mixed numerals, first convert to fraction notation and multiply. Then convert the answer to a mixed numeral, if appropriate.

EXAMPLE 1 Multiply: $6 \cdot 2\frac{1}{2}$.

$$6 \cdot 2\frac{1}{2} = \frac{6}{1} \cdot \frac{5}{2} = \frac{6 \cdot 5}{1 \cdot 2} = \frac{2 \cdot 3 \cdot 5}{2 \cdot 1} = \frac{2}{2} \cdot \frac{3 \cdot 5}{1} = 15$$

Note that fraction notation is needed to carry out the multiplication.

Do Exercise 1.

EXAMPLE 2 Multiply: $3\frac{1}{2} \cdot \frac{3}{4}$.

$$3\frac{1}{2} \cdot \frac{3}{4} = \frac{7}{2} \cdot \frac{3}{4} = \frac{21}{8} = 2\frac{5}{8}$$

Here we write fraction notation.

Do Exercise 2.

EXAMPLE 3 Multiply: $8 \cdot 4\frac{2}{3}$.

$$8 \cdot 4\frac{2}{3} = \frac{8}{1} \cdot \frac{14}{3} = \frac{112}{3} = 37\frac{1}{3}$$

Do Exercise 3.

EXAMPLE 4 Multiply: $2\frac{1}{4} \cdot 3\frac{2}{5}$.

$$2\frac{1}{4} \cdot 3\frac{2}{5} = \frac{9}{4} \cdot \frac{17}{5} = \frac{153}{20} = 7\frac{13}{20}$$

Do Exercise 4.

Caution!

$2\frac{1}{4} \cdot 3\frac{2}{5} \neq 6\frac{2}{20}$. A common error is to multiply the whole numbers and then the fractions. This does not give the correct answer, $7\frac{13}{20}$, which is found by converting to fraction notation first.

b Division Using Mixed Numerals

The division $1\frac{1}{2} \div \frac{1}{6}$ is shown here. *Think*: "How many $\frac{1}{6}$'s are in $1\frac{1}{2}$?" We see that the answer is 9.

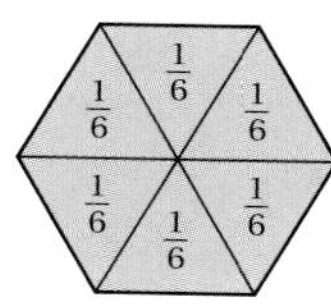

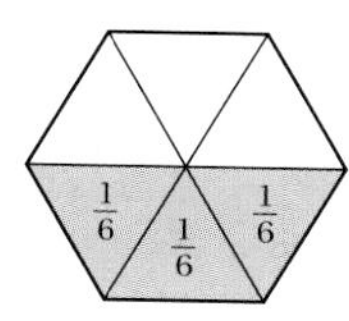

When we divide using mixed numerals, we convert to fraction notation first.

$$1\frac{1}{2} \div \frac{1}{6} = \frac{3}{2} \div \frac{1}{6} = \frac{3}{2} \cdot 6$$

$$= \frac{3 \cdot 6}{2} = \frac{3 \cdot 3 \cdot 2}{2 \cdot 1} = \frac{3 \cdot 3}{1} \cdot \frac{2}{2} = \frac{3 \cdot 3}{1} \cdot 1 = 9$$

DIVISION USING MIXED NUMERALS

To divide using mixed numerals, first write fraction notation and divide. Then convert the answer to a mixed numeral, if appropriate.

EXAMPLE 5 Divide: $32 \div 3\frac{1}{5}$.

$$32 \div 3\frac{1}{5} = \frac{32}{1} \div \frac{16}{5}$$ Writing the mixed numeral in fraction notation

$$= \frac{32}{1} \cdot \frac{5}{16} = \frac{32 \cdot 5}{1 \cdot 16} = \frac{2 \cdot 16 \cdot 5}{1 \cdot 16} = \frac{16}{16} \cdot \frac{2 \cdot 5}{1} = 1 \cdot \frac{2 \cdot 5}{1} = 10$$

Remember to multiply by the reciprocal.

Do Exercise 5.

EXAMPLE 6 Divide: $35 \div 4\frac{1}{3}$.

$$35 \div 4\frac{1}{3} = \frac{35}{1} \div \frac{13}{3} = \frac{35}{1} \cdot \frac{3}{13} = \frac{105}{13} = 8\frac{1}{13}$$

Do Exercise 6.

EXAMPLE 7 Divide: $2\frac{1}{3} \div 1\frac{3}{4}$.

$$2\frac{1}{3} \div 1\frac{3}{4} = \frac{7}{3} \div \frac{7}{4} = \frac{7}{3} \cdot \frac{4}{7} = \frac{7 \cdot 4}{7 \cdot 3} = \frac{7}{7} \cdot \frac{4}{3} = 1 \cdot \frac{4}{3} = \frac{4}{3} = 1\frac{1}{3}$$

Caution!

The reciprocal of $1\frac{3}{4}$ is not $1\frac{4}{3}$!

EXAMPLE 8 Divide: $1\frac{3}{5} \div 3\frac{1}{3}$.

$$1\frac{3}{5} \div 3\frac{1}{3} = \frac{8}{5} \div \frac{10}{3} = \frac{8}{5} \cdot \frac{3}{10} = \frac{8 \cdot 3}{5 \cdot 10} = \frac{2 \cdot 4 \cdot 3}{5 \cdot 2 \cdot 5} = \frac{2}{2} \cdot \frac{4 \cdot 3}{5 \cdot 5} = \frac{12}{25}$$

Do Exercises 7 and 8.

5. Divide: $84 \div 5\frac{1}{4}$.

6. Divide: $26 \div 3\frac{1}{2}$.

Divide.

7. $2\frac{1}{4} \div 1\frac{1}{5}$

8. $1\frac{3}{4} \div 2\frac{1}{2}$

Answers on page A-7

9. Kyle's pickup truck travels on an interstate highway at 65 mph for $3\frac{1}{2}$ hr. How far does it travel?

10. Holly's minivan travels 302 mi on $15\frac{1}{10}$ gal of gas. How many miles per gallon did it get?

Answers on page A-7

C Applications and Problem Solving

EXAMPLE 9 *Average Speed in Indianapolis 500.* Arie Luyendyk won the Indianapolis 500 in 1990 with the highest average speed of about 186 mph. This record high through 2004 is about $2\frac{12}{25}$ times the average speed of the first winner, Ray Harroun, in 1911. What was the average speed in the first Indianapolis 500?

Source: Indianapolis Motor Speedway

1. **Familiarize.** We ask the question, "186 is $2\frac{12}{25}$ times what number? We let $s =$ the average speed in 1911. Then the average speed in 1990 was $2\frac{12}{25} \cdot s$.
2. **Translate.** The problem can be translated to an equation as follows.

Average speed in 1990	is	$2\frac{12}{25}$	times	Average speed in 1911
↓	↓	↓	↓	↓
186	$=$	$2\frac{12}{25}$	$\cdot$	s

3. **Solve.** To solve the equation, we divide on both sides.

$$186 = \frac{62}{25} \cdot s \qquad \text{Converting } 2\frac{12}{25} \text{ to fraction notation}$$

$$\frac{186}{\frac{62}{25}} = \frac{\frac{62}{25} \cdot s}{\frac{62}{25}} \qquad \text{Dividing by } \frac{62}{25} \text{ on both sides}$$

$$\frac{186}{\frac{62}{25}} = s \qquad \text{Factoring and removing a factor of 1: } ((62/25)/(62/25)) = 1$$

$$186 \cdot \frac{25}{62} = s \qquad \text{Multiplying by the reciprocal}$$

$$75 = s \qquad \text{Simplifying: } 186 \cdot \frac{25}{62} = 3 \cdot 25 = 75$$

4. **Check.** If the average speed in 1911 was about 75 mph, we find the average speed in 1990 by multiplying 75 by $2\frac{12}{25}$:

$$2\frac{12}{25} \cdot 75 = \frac{62}{25} \cdot 75 = \frac{62 \cdot 75}{25} = \frac{62 \cdot 25 \cdot 3}{25} = 62 \cdot 3 = 186.$$

The answer checks.

5. **State.** The average speed in the first Indianapolis 500 was about 75 mph.

Do Exercises 9 and 10.

EXAMPLE 10 *Mirror Area.* The mirror-backed candle shelf, shown below with a carpenter's diagram, was designed and built by Harry Cooper. Such shelves were popular in Colonial times because the mirror provided extra lighting from the candle. A rectangular walnut board is used to make the back of the shelf. Find the area of the original board and the amount left over after the opening for the mirror has been cut out.

Source: Popular Science Woodworking Projects

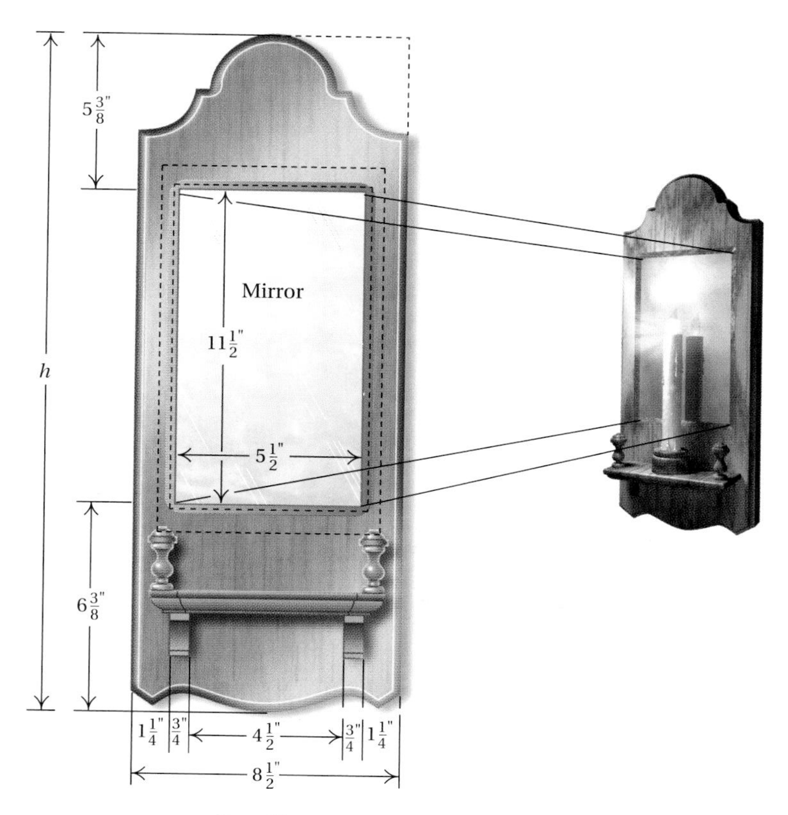

Front View

1. **Familiarize.** Refer to the figure above. We let h = the height of the back of the shelf and B = the area of the original board. We know the width of the original board, $8\frac{1}{2}$". (Remember, $8\frac{1}{2}$" means $8\frac{1}{2}$ in.) We let A = the area left over after the mirror has been cut out.

2. **Translate.** This is a multistep problem. To find B, which equals $8\frac{1}{2} \cdot h$, we first need to calculate h. We read the dimensions $5\frac{3}{8}$", $11\frac{1}{2}$", and $6\frac{3}{8}$" from the diagram and add them to find h:

$$h = 5\frac{3}{8} + 11\frac{1}{2} + 6\frac{3}{8}.$$

The dimensions of the mirror are $11\frac{1}{2}$" and $5\frac{1}{2}$". Then A is the area of the original board minus the area of the mirror. That is,

$$A = B - 11\frac{1}{2} \cdot 5\frac{1}{2}.$$

11. A room measures $22\frac{1}{2}$ ft by $15\frac{1}{2}$ ft. A 9-ft by 12-ft Oriental rug is placed in the center of the room. How much area is not covered by the rug?

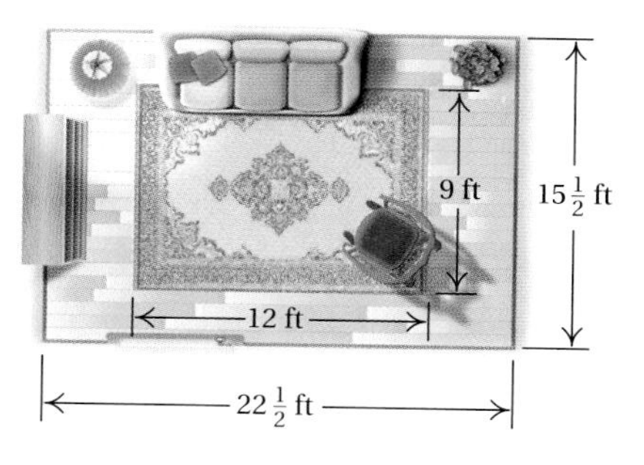

Answer on page A-7

3. **Solve.** We carry out each calculation as follows:

$$h = 5\frac{3}{8} + 11\frac{1}{2} + 6\frac{3}{8} \qquad\qquad B = 8\frac{1}{2} \cdot h$$

$$= 5\frac{3}{8} + 11\frac{4}{8} + 6\frac{3}{8} \qquad\qquad = 8\frac{1}{2} \cdot 23\frac{1}{4} = \frac{17}{2} \cdot \frac{93}{4}$$

$$= 22\frac{10}{8} = 22\frac{5}{4} = 23\frac{1}{4}; \qquad\qquad = \frac{1581}{8} = 197\frac{5}{8};$$

$$A = B - 11\frac{1}{2} \cdot 5\frac{1}{2}$$

$$= 197\frac{5}{8} - 11\frac{1}{2} \cdot 5\frac{1}{2}$$

$$= 197\frac{5}{8} - \frac{23}{2} \cdot \frac{11}{2} = 197\frac{5}{8} - \frac{253}{4}$$

$$= 197\frac{5}{8} - 63\frac{1}{4} = 197\frac{5}{8} - 63\frac{2}{8} = 134\frac{3}{8}.$$

4. **Check.** We perform a check by repeating the calculations.

5. **State.** The area of the original board is $197\frac{5}{8}$ in^2. The area left over is $134\frac{3}{8}$ in^2.

Do Exercise 11 on the preceding page.

CALCULATOR CORNER

Operations on Fractions and Mixed Numerals Fraction calculators can add, subtract, multiply, and divide fractions and mixed numerals. The [a b/c] key is used to enter fractions and mixed numerals. To find $\frac{3}{4} + \frac{1}{2}$, for example, we press [3] [a b/c] [4] [+] [1] [a b/c] [2] [=]. Note that 3/4 and 1/2 appear on the display as [3 ⌟ 4] and [1 ⌟ 2], respectively. The result is given as the mixed numeral $1\frac{1}{4}$ and is displayed as [1 ⌟ 1 ⌟ 4]. Fraction results that are greater than 1 are always displayed as mixed numerals. To express this result as a fraction, we press [SHIFT] [d/c]. We get [5 ⌟ 4], or 5/4.

To find $3\frac{2}{3} \cdot 4\frac{1}{5}$, we press [3] [a b/c] [2] [a b/c] [3] [×] [4] [a b/c] [1] [a b/c] [5] [=]. The calculator displays [15 ⌟ 2 ⌟ 5], so the product is $15\frac{2}{5}$.

Some calculators are capable of displaying mixed numerals in the way in which we write them, as shown below.

Exercises: Perform each calculation. Give the answer in fraction notation.

1. $\frac{1}{3} + \frac{1}{4}$
2. $\frac{7}{5} - \frac{3}{10}$
3. $\frac{15}{4} \cdot \frac{7}{12}$
4. $\frac{4}{5} \div \frac{8}{3}$

Perform each calculation. Give the answer as a mixed numeral.

5. $4\frac{1}{3} + 5\frac{4}{5}$
6. $9\frac{2}{7} - 8\frac{1}{4}$
7. $2\frac{1}{3} \cdot 4\frac{3}{5}$
8. $10\frac{7}{10} \div 3\frac{5}{6}$

Translating for Success

Raffle Tickets. At the Happy Hollow Camp Fall Festival, Rico and Becca, together, spent \$270 on raffle tickets that sell for $\$\frac{9}{20}$ each. How many tickets did they buy?

Irrigation Pipe. Jed uses two pipes, one of which measures $5\frac{1}{3}$ ft, to repair the irrigation system in the Buxtons' lawn. The total length of the two pipes is $8\frac{7}{12}$ ft. How long is the other pipe?

Vacation Days. Together, Helmut and Claire have 36 vacation days a year. Helmut has 22 vacation days per year. How many does Claire have?

Enrollment in Japanese Classes. Last year at the Lakeside Community College, 225 students enrolled in basic mathematics. This number is $4\frac{1}{2}$ times as many as the number who enrolled in Japanese. How many enrolled in Japanese?

Bicycling. Cole rode his bicycle $5\frac{1}{3}$ mi on Saturday and $8\frac{7}{12}$ mi on Sunday. How far did he ride on the weekend?

The goal of these matching questions is to practice step (2), *Translate,* of the five-step problem-solving process. Translate each word problem to an equation and select a correct translation from equations A–O.

A. $13\frac{11}{12} = x + 5\frac{1}{3}$

B. $\frac{3}{4} \cdot x = 1\frac{2}{3}$

C. $\frac{20}{9} \cdot x = 270$

D. $225 = 4\frac{1}{2} \cdot x$

E. $98 \div 2\frac{1}{3} = x$

F. $22 + x = 36$

G. $x = 4\frac{1}{2} \cdot 225$

H. $x = 5\frac{1}{3} + 8\frac{7}{12}$

I. $22 \cdot x = 36$

J. $x = \frac{3}{4} \cdot 1\frac{2}{3}$

K. $5\frac{1}{3} + x = 8\frac{7}{12}$

L. $\frac{9}{20} \cdot 270 = x$

M. $1\frac{2}{3} + \frac{3}{4} = x$

N. $98 - 2\frac{1}{3} = x$

O. $\frac{9}{20} \cdot x = 270$

Answers on page A-7

6. *Deli Order.* For a promotional open house for contractors last year, the Bayside Builders Association ordered 225 turkey sandwiches. Due to increased registrations this year, $4\frac{1}{2}$ times as many sandwiches will be needed. How many sandwiches are ordered?

7. *Dog Ownership.* In Sam's community, $\frac{9}{20}$ of the households own at least one dog. There are 270 households. How many own dogs?

8. *Magic Tricks.* A magic trick requires a piece of rope $2\frac{1}{3}$ ft long. Gerry, a magician, has 98 ft of rope and needs to divide it into $2\frac{1}{3}$-ft pieces. How many pieces can be cut from the rope?

9. *Painting.* Laura needs $1\frac{2}{3}$ gal of paint to paint the ceiling of the exercise room and $\frac{3}{4}$ gal of the same paint for the bathroom. How much paint does Laura need?

10. *Chocolate Fudge Bars.* A recipe for chocolate fudge bars that serves 16 includes $1\frac{2}{3}$ cups of sugar. How much sugar is needed for $\frac{3}{4}$ of this recipe?

3.6 EXERCISE SET

For Extra Help

MathXL | MyMathLab | InterAct Math | Math Tutor Center | Digital Video Tutor CD 2 Videotape 3 | Student's Solutions Manual

a Multiply. Write a mixed numeral for the answer.

1. $8 \cdot 2\frac{5}{6}$

2. $5 \cdot 3\frac{3}{4}$

3. $3\frac{5}{8} \cdot \frac{2}{3}$

4. $6\frac{2}{3} \cdot \frac{1}{4}$

5. $3\frac{1}{2} \cdot 2\frac{1}{3}$

6. $4\frac{1}{5} \cdot 5\frac{1}{4}$

7. $3\frac{2}{5} \cdot 2\frac{7}{8}$

8. $2\frac{3}{10} \cdot 4\frac{2}{5}$

9. $4\frac{7}{10} \cdot 5\frac{3}{10}$

10. $6\frac{3}{10} \cdot 5\frac{7}{10}$

11. $20\frac{1}{2} \cdot 10\frac{1}{5} \cdot 4\frac{2}{3}$

12. $21\frac{1}{3} \cdot 11\frac{1}{3} \cdot 3\frac{5}{8}$

b Divide. Write a mixed numeral for the answer.

13. $20 \div 3\frac{1}{5}$

14. $18 \div 2\frac{1}{4}$

15. $8\frac{2}{5} \div 7$

16. $3\frac{3}{8} \div 3$

17. $4\frac{3}{4} \div 1\frac{1}{3}$

18. $5\frac{4}{5} \div 2\frac{1}{2}$

19. $1\frac{7}{8} \div 1\frac{2}{3}$

20. $4\frac{3}{8} \div 2\frac{5}{6}$

21. $5\frac{1}{10} \div 4\frac{3}{10}$

22. $4\frac{1}{10} \div 2\frac{1}{10}$

23. $20\frac{1}{4} \div 90$

24. $12\frac{1}{2} \div 50$

C Solve.

25. *Beagles.* There are about 155,000 Labrador retrievers registered with The American Kennel Club. This is $3\frac{4}{9}$ times the number of beagles registered. How many beagles are registered?

Source: The American Kennel Club

26. *Exercise.* At one point during an aerobics class, Kea's bicycle wheel was completing $76\frac{2}{3}$ revolutions per minute. How many revolutions did the wheel complete in 6 min?

27. *Population.* The population of Louisiana is $2\frac{1}{2}$ times the population of West Virginia. The population of West Virginia is approximately 1,800,000. What is the population of Louisiana?

Source: U.S. Bureau of the Census

28. *Population.* The population of New York is $3\frac{1}{3}$ times that of Missouri. The population of New York is approximately 19,000,000. What is the population of Missouri?

Source: U.S. Bureau of the Census

29. *Mural.* Cecilia hired an artist to paint a mural on the wall in her twin sons' bedroom. Dimensions of the mural must be $6\frac{2}{3}$ ft by $9\frac{3}{8}$ ft. What is the area of the mural?

30. *Sidewalk.* A sidewalk alongside a garden at the conservatory is to be $14\frac{2}{5}$ yd long. Rectangular stone tiles that are each $1\frac{1}{8}$ yd long are used to form the sidewalk. How many tiles are used?

31. *Sodium Consumption.* The average American woman consumes $1\frac{1}{3}$ tsp of sodium each day. How much sodium do 10 average American women consume in one day?

Source: *Nutrition Action Health Letter*, March 1994, p. 6. 1875 Connecticut Ave., N.W., Washington, DC 20009-5728

32. *Aeronautics.* Most space shuttles orbit the earth once every $1\frac{1}{2}$ hr. How many orbits are made every 24 hr?

33. *Weight of Water.* The weight of water is $62\frac{1}{2}$ lb per cubic foot. What is the weight of $5\frac{1}{2}$ cubic feet of water?

34. *Weight of Water.* The weight of water is $62\frac{1}{2}$ lb per cubic foot. What is the weight of $2\frac{1}{4}$ cubic feet of water?

35. *Temperatures.* Fahrenheit temperature can be obtained from Celsius (centigrade) temperature by multiplying by $1\frac{4}{5}$ and adding 32°. What Fahrenheit temperature corresponds to a Celsius temperature of 20°?

36. *Temperature.* Fahrenheit temperature can be obtained from Celsius (centigrade) temperature by multiplying by $1\frac{4}{5}$ and adding 32°. What Fahrenheit temperature corresponds to the Celsius temperature of boiling water, which is 100°?

37. *Newspaper Circulation.* In 2002, daily circulation of the *Wall Street Journal* (New York, N.Y.) was about $5\frac{1}{4}$ times the daily circulation of the *Star Tribune* (Minneapolis). The circulation of the *Star Tribune* was about 343,000. What was the daily circulation of the *Wall Street Journal*?

Source: *Editor and Publisher International Year Book* 2003

38. *Language Enrollments.* In institutions of higher education in Fall 2002, the number of students who enrolled in Spanish was $14\frac{3}{10}$ times the number who enrolled in Japanese. About 746,270 students enrolled in Spanish. About how many enrolled in Japanese?

Source: Association of Departments of Foreign Languages at the Modern Language Association

39. *Creamy Peach Coffee Cake.* Listed below is the recipe for creamy peach coffee cake. What are the ingredients for $\frac{1}{2}$ cake? for 4 cakes?

Source: Reprinted with permission from *Taste of Home* magazine, www.tasteofhome.com

Creamy Peach Coffee Cake
$2\frac{1}{4}$ cups all-purpose flour
$\frac{3}{4}$ cup sugar
$\frac{3}{4}$ cup cold butter
$\frac{3}{4}$ cup sour cream
$\frac{1}{2}$ teaspoon baking powder
$\frac{1}{2}$ teaspoon baking soda
1 egg
1 teaspoon almond extract
FILLING:
1 package (8 ounces) cream cheese, softened
$\frac{1}{4}$ cup sugar
1 egg
$\frac{3}{4}$ cup peach preserves
$\frac{1}{2}$ cup sliced almonds

40. *Butterscotch Hard Candy.* Listed below is the recipe for butterscotch hard candy. What are the ingredients for $\frac{1}{2}$ batch? for 3 batches?

Source: Reprinted with permission from *Taste of Home* magazine, www.tasteofhome.com

Butterscotch Hard Candy
$2\frac{1}{2}$ cups sugar
$\frac{3}{4}$ cup water
$\frac{1}{2}$ cup light corn syrup
1 cup butter, cubed
$\frac{1}{4}$ cup honey
$\frac{1}{2}$ teaspoon salt
$\frac{1}{2}$ teaspoon rum extract

41. A car traveled 213 mi on $14\frac{2}{10}$ gal of gas. How many miles per gallon did it get?

42. A car traveled 385 mi on $15\frac{4}{10}$ gal of gas. How many miles per gallon did it get?

43. *Weight of Water.* The weight of water is $62\frac{1}{2}$ lb per cubic foot. How many cubic feet would be occupied by 250 lb of water?

44. *Weight of Water.* The weight of water is $62\frac{1}{2}$ lb per cubic foot. How many cubic feet would be occupied by 375 lb of water?

45. *Servings of Salmon.* A serving of filleted fish is generally considered to be about $\frac{1}{3}$ lb. How many servings can be prepared from $5\frac{1}{2}$ lb of salmon fillet?

46. *Servings of Tuna.* A serving of fish steak (cross section) is generally $\frac{1}{2}$ lb. How many servings can be prepared from a cleaned $18\frac{3}{4}$-lb tuna?

Find the area of the shaded region.

47.

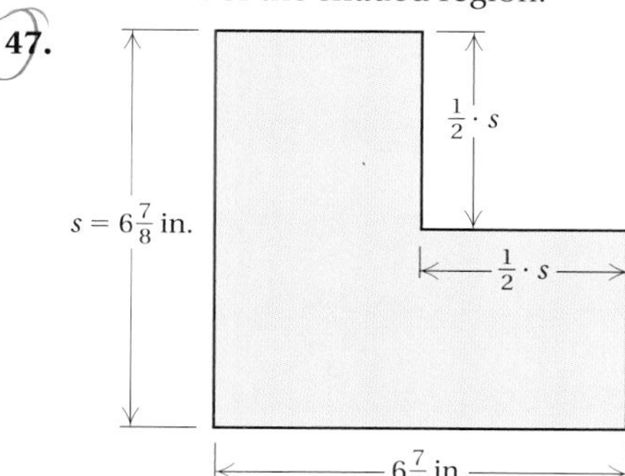

48.

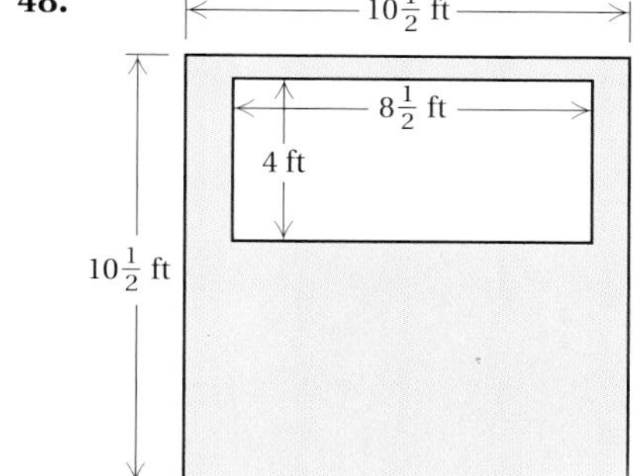

49. *Construction.* A rectangular lot has dimensions of $302\frac{1}{2}$ ft by $205\frac{1}{4}$ ft. A building with dimensions of 100 ft by $25\frac{1}{2}$ ft is built on the lot. How much area is left over?

50. *Word Processing.* Kelly wants to create a table using Microsoft® Word software for word processing. She needs to have two columns, each $1\frac{1}{2}$ in. wide, and five columns, each $\frac{3}{4}$ in. wide. Will this table fit on a piece of standard paper that is $8\frac{1}{2}$ in. wide? If so, how wide will each margin be if her margins on each side are to be of equal width?

51. $^{\mathbf{D}}\mathbf{W}$ Write a problem for a classmate to solve. Design the problem so that its solution is found by performing the multiplication $4\frac{1}{2} \cdot 33\frac{1}{3}$.

52. $^{\mathbf{D}}\mathbf{W}$ Under what circumstances is a pair of mixed numerals more easily added than multiplied?

SKILL MAINTENANCE

VOCABULARY REINFORCEMENT

In each of Exercises 53–60, fill in the blank with the correct term from the given list. Some of the choices may not be used and some may be used more than once.

53. In the equation $420 \div 60 = 7$, 60 is called the __________, 7 the __________, and 420 the __________. [1.6a]

54. When denominators are the same, we say that fractions have a __________ denominator. [2.5c]

55. The numbers 91, 95, and 111 are examples of __________ numbers. [2.1c]

56. The number 22,223,133 is __________ by 9 because the sum of its digits is __________ by 9. [2.2a]

57. When simplifying $24 \div 4 + 4 \times 12 - 6 \div 2$, do all __________ and __________ in order from left to right before doing all __________ and __________ in order from left to right. [1.9c]

58. In the equation $2 + 3 = 5$, 2 and 3 are called __________. [1.2a]

59. In the expression $\frac{c}{d}$, we call c the __________. [2.3a]

60. The number 0 has no __________. [2.7a]

identity
reciprocal
divisor
dividend
quotient
additions
subtractions
multiplications
divisions
numerator
denominator
equal
common
prime
composite
product
divisible
digits
factors
addends

SYNTHESIS

Multiply. Write the answer as a mixed numeral whenever possible.

61. 🖩 $15\frac{2}{11} \cdot 23\frac{31}{43}$

62. 🖩 $17\frac{23}{31} \cdot 19\frac{13}{15}$

Simplify.

63. $8 \div \frac{1}{2} + \frac{3}{4} + \left(5 - \frac{5}{8}\right)^2$

64. $\left(\frac{5}{9} - \frac{1}{4}\right) \times 12 + \left(4 - \frac{3}{4}\right)^2$

65. $\frac{1}{3} \div \left(\frac{1}{2} - \frac{1}{5}\right) \times \frac{1}{4} + \frac{1}{6}$

66. $\frac{7}{8} - 1\frac{1}{8} \times \frac{2}{3} + \frac{9}{10} \div \frac{3}{5}$

67. $4\frac{1}{2} \div 2\frac{1}{2} + 8 - 4 \div \frac{1}{2}$

68. $6 - 2\frac{1}{3} \times \frac{3}{4} + \frac{5}{8} \div \frac{2}{3}$

3.7 ORDER OF OPERATIONS; ESTIMATION

Objectives

a Simplify expressions using the rules for order of operations.

b Estimate with fraction and mixed-numeral notation.

a Order of Operations; Fraction Notation and Mixed Numerals

The rules for order of operations that we use with whole numbers (see Section 1.9) apply when we are simplifying expressions involving fraction notation and mixed numerals. For review, these rules are listed below.

RULES FOR ORDER OF OPERATIONS

1. Do all calculations within parentheses before operations outside.
2. Evaluate all exponential expressions.
3. Do all multiplications and divisions in order from left to right.
4. Do all additions and subtractions in order from left to right.

EXAMPLE 1 Simplify: $\frac{1}{6} + \frac{2}{3} \div \frac{1}{2} \cdot \frac{5}{8}$.

$\frac{1}{6} + \frac{2}{3} \div \frac{1}{2} \cdot \frac{5}{8} = \frac{1}{6} + \frac{2}{3} \cdot \frac{2}{1} \cdot \frac{5}{8}$ — Doing the division first by multiplying by the reciprocal of $\frac{1}{2}$

$= \frac{1}{6} + \frac{2 \cdot 2 \cdot 5}{3 \cdot 1 \cdot 8}$ — Doing the multiplications in order from left to right

$= \frac{1}{6} + \frac{\not{2} \cdot \not{2} \cdot 5}{3 \cdot 1 \cdot \not{2} \cdot \not{2} \cdot 2}$ — Factoring in order to simplify

$= \frac{1}{6} + \frac{5}{6}$ — Removing a factor of 1: $\frac{2 \cdot 2}{2 \cdot 2} = 1$; simplifying

$= \frac{6}{6}$, or 1 — Doing the addition

Do Exercises 1 and 2.

Simplify.

1. $\frac{2}{5} \cdot \frac{5}{8} + \frac{1}{4}$

2. $\frac{1}{3} \cdot \frac{3}{4} \div \frac{5}{8} - \frac{1}{10}$

EXAMPLE 2 Simplify: $\frac{2}{3} \cdot 24 - 11\frac{1}{2}$.

$\frac{2}{3} \cdot 24 - 11\frac{1}{2} = \frac{2 \cdot 24}{3} - 11\frac{1}{2}$ — Doing the multiplication first

$= \frac{2 \cdot \not{3} \cdot 8}{\not{3}} - 11\frac{1}{2}$ — Factoring the numerator

$= 2 \cdot 8 - 11\frac{1}{2}$ — Removing a factor of 1: $\frac{3}{3} = 1$

$= 16 - 11\frac{1}{2}$ — Completing the multiplication

$= 4\frac{1}{2}$, or $\frac{9}{2}$ — Doing the subtraction

3. Simplify: $\frac{3}{4} \cdot 16 + 8\frac{2}{3}$.

Answers on page A-7

4. After two weeks, Kurt's tomato seedlings measure $9\frac{1}{2}$ in., $10\frac{3}{4}$ in., $10\frac{1}{4}$ in., and 9 in. tall. Find their average height.

5. Find the average of

$$\frac{1}{2}, \frac{1}{3}, \text{ and } \frac{5}{6}.$$

6. Find the average of $\frac{3}{4}$ and $\frac{4}{5}$.

7. Simplify:

$$\left(\frac{2}{3}+\frac{3}{4}\right) \div 2\frac{1}{3} - \left(\frac{1}{2}\right)^3.$$

Answers on page A-7

Do Exercise 3 on the preceding page.

EXAMPLE 3 Melody has triplets. Their birth weights were $3\frac{1}{2}$ lb, $2\frac{3}{4}$ lb, and $3\frac{1}{8}$ lb. What was the average weight of her babies?

Recall that to compute an **average,** we add the numbers and then divide the sum by the number of addends (see Section 1.9). We have

$$\frac{3\frac{1}{2}+2\frac{3}{4}+3\frac{1}{8}}{3}.$$

We first add in the numerator:

$$\frac{3\frac{1}{2}+2\frac{3}{4}+3\frac{1}{8}}{3} = \frac{3\frac{4}{8}+2\frac{6}{8}+3\frac{1}{8}}{3}$$

$$= \frac{8\frac{11}{8}}{3} = \frac{\frac{75}{8}}{3}$$ Doing the additions; converting $8\frac{11}{8}$ to fraction notation, $\frac{75}{8}$

$$= \frac{75}{8}\cdot\frac{1}{3} = \frac{75}{24} = \frac{25}{8}$$ Multiplying by the reciprocal and simplifying

$$= 3\frac{1}{8}.$$ Converting to a mixed numeral

The average weight of the three babies is $3\frac{1}{8}$ lb.

Do Exercises 4–6.

EXAMPLE 4 Simplify: $\left(\frac{7}{8}-\frac{1}{3}\right)\times 48 + \left(13+\frac{4}{5}\right)^2.$

$$\left(\frac{7}{8}-\frac{1}{3}\right)\times 48 + \left(13+\frac{4}{5}\right)^2$$

$$= \left(\frac{7}{8}\cdot\frac{3}{3}-\frac{1}{3}\cdot\frac{8}{8}\right)\times 48 + \left(13\cdot\frac{5}{5}+\frac{4}{5}\right)^2$$ Carrying out operations inside parentheses first. To do so, we first multiply by 1 to obtain the LCD.

$$= \left(\frac{21}{24}-\frac{8}{24}\right)\times 48 + \left(\frac{65}{5}+\frac{4}{5}\right)^2$$

$$= \frac{13}{24}\times 48 + \left(\frac{69}{5}\right)^2$$ Completing the operations within parentheses

$$= \frac{13}{24}\times 48 + \frac{4761}{25}$$ Evaluating the exponential expression next

$$= 26 + \frac{4761}{25}$$ Doing the multiplication

$$= 26 + 190\frac{11}{25}$$ Converting to a mixed numeral

$$= 216\frac{11}{25}, \text{ or } \frac{5411}{25}$$ Adding

Answers can be given using either fraction notation or mixed numerals.

Do Exercise 7.

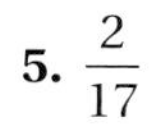

Estimation with Fraction Notation and Mixed Numerals

We now estimate with fraction notation and mixed numerals.

EXAMPLES Estimate each of the following as 0, $\frac{1}{2}$, or 1.

5. $\dfrac{2}{17}$

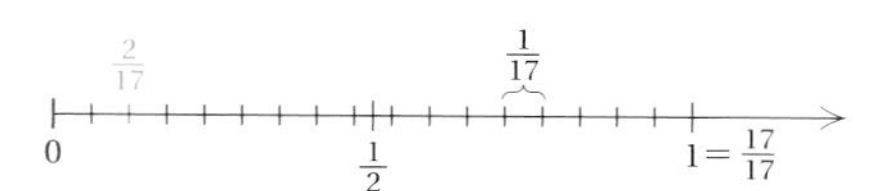

A fraction is close to 0 when the numerator is small in comparison to the denominator. Thus, 0 is an estimate for $\frac{2}{17}$ because 2 is small in comparison to 17. Thus, $\frac{2}{17} \approx 0$.

6. $\dfrac{11}{23}$

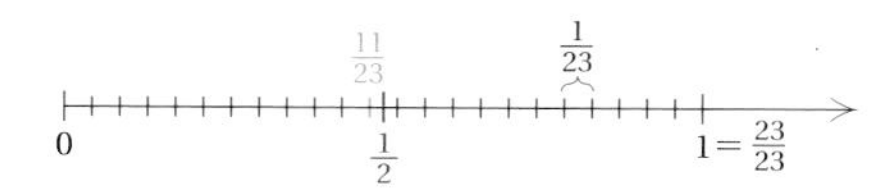

A fraction is close to $\frac{1}{2}$ when the denominator is about twice the numerator. Thus, $\frac{1}{2}$ is an estimate for $\frac{11}{23}$ because $2 \cdot 11 = 22$ and 22 is close to 23. Thus, $\frac{11}{23} \approx \frac{1}{2}$.

7. $\dfrac{37}{38}$

A fraction is close to 1 when the numerator is nearly equal to the denominator. Thus, 1 is an estimate for $\frac{37}{38}$ because 37 is nearly equal to 38. Thus, $\frac{37}{38} \approx 1$.

8. $\dfrac{43}{41}$

As in the preceding example, the numerator 43 is nearly equal to the denominator 41. Thus, $\frac{43}{41} \approx 1$.

Do Exercises 8–11.

EXAMPLE 9 Find a number for the blank so that $\dfrac{9}{\square}$ is close to but less than 1. Answers may vary.

If the number in the blank were 9, we would have 1, so we increase 9 to 10. An answer is 10; $\frac{9}{10}$ is close to 1. The number 11 would also be a correct answer; $\frac{9}{11}$ is close to 1.

Do Exercises 12 and 13.

Estimate each of the following as 0, $\frac{1}{2}$, or 1.

8. $\dfrac{3}{59}$

9. $\dfrac{61}{59}$

10. $\dfrac{29}{59}$

11. $\dfrac{57}{59}$

Find a number for the blank so that the fraction is close to but less than 1.

12. $\dfrac{11}{\square}$

13. $\dfrac{\square}{33}$

Answers on page A-7

Study Tips

SKILL MAINTENANCE EXERCISES

It is never too soon to begin reviewing for the final examination. The Skill Maintenance exercises found in each exercise set review and reinforce skills taught in earlier sections. Include all of these exercises in your weekly preparation. Answers to both odd-numbered and even-numbered exercises along with section references appear at the back of the book.

Find a number for the blank so that the fraction is close to but less than $\frac{1}{2}$.

14. $\dfrac{13}{\square}$

15. $\dfrac{\square}{31}$

Find a number for the blank so that the fraction is close to but greater than 0.

16. $\dfrac{\square}{37}$

17. $\dfrac{13}{\square}$

Estimate each part of the following as a whole number or as a mixed numeral where the fractional part is $\frac{1}{2}$.

18. $5\dfrac{9}{10} + 26\dfrac{1}{2} - 10\dfrac{3}{29}$

19. $10\dfrac{7}{8} \cdot \left(25\dfrac{11}{13} - 14\dfrac{1}{9}\right)$

20. $\left(10\dfrac{4}{5} + 7\dfrac{5}{9}\right) \div \dfrac{17}{30}$

Answers on page A-7

EXAMPLE 10 Find a number for the blank so that $\dfrac{9}{\square}$ is close to but less than $\frac{1}{2}$. Answers may vary.

If we double 9 to get 18 and use it for the blank, we have $\frac{1}{2}$. If we increase that denominator by 1, to get 19, and use it for the blank, we get a number less than $\frac{1}{2}$ but close to $\frac{1}{2}$. Thus, $\frac{9}{19} \approx \frac{1}{2}$.

Do Exercises 14 and 15.

EXAMPLE 11 Find a number for the blank so that $\dfrac{\square}{50}$ is close to but greater than 0.

Since 50 is rather large, any small number such as 1, 2, or 3 will make the fraction close to 0. For example, $\frac{1}{50} \approx 0$.

Do Exercises 16 and 17.

EXAMPLE 12 Estimate $16\frac{8}{9} + 11\frac{2}{13} - 4\frac{22}{43}$ as a whole number or as a mixed numeral where the fractional part is $\frac{1}{2}$.

We estimate each fraction as 0, $\frac{1}{2}$, or 1. Then we calculate:

$$16\frac{8}{9} + 11\frac{2}{13} - 4\frac{22}{43} \approx 17 + 11 - 4\frac{1}{2}$$
$$= 28 - 4\frac{1}{2}$$
$$= 23\frac{1}{2}.$$

Do Exercises 18–20.

3.7 EXERCISE SET

For Extra Help

MathXL | MyMathLab | InterAct Math | Math Tutor Center | Digital Video Tutor CD 2 Videotape 3 | Student's Solutions Manual

a Simplify.

1. $\frac{1}{2} \cdot \frac{1}{3} \cdot \frac{1}{4}$

2. $\frac{1}{3} \cdot \frac{1}{4} \cdot \frac{1}{5}$

3. $6 \div 3 \div 5$

4. $12 \div 4 \div 8$

5. $\frac{2}{3} \div \frac{4}{3} \div \frac{7}{8}$

6. $\frac{5}{6} \div \frac{3}{4} \div \frac{2}{5}$

7. $\frac{5}{8} \div \frac{1}{4} - \frac{2}{3} \cdot \frac{4}{5}$

8. $\frac{4}{7} \cdot \frac{7}{15} + \frac{2}{3} \div 8$

9. $\frac{3}{4} - \frac{2}{3} \cdot \left(\frac{1}{2} + \frac{2}{5}\right)$

10. $\frac{3}{4} \div \frac{1}{2} \cdot \left(\frac{8}{9} - \frac{2}{3}\right)$

11. $28\frac{1}{8} - 5\frac{1}{4} + 3\frac{1}{2}$

12. $10\frac{3}{5} - 4\frac{1}{10} - 1\frac{1}{2}$

13. $\frac{7}{8} \div \frac{1}{2} \cdot \frac{1}{4}$

14. $\frac{7}{10} \cdot \frac{4}{5} \div \frac{2}{3}$

15. $\left(\frac{2}{3}\right)^2 - \frac{1}{3} \cdot 1\frac{1}{4}$

16. $\left(\frac{3}{4}\right)^2 + 3\frac{1}{2} \div 1\frac{1}{4}$

17. $\frac{1}{2} - \left(\frac{1}{2}\right)^2 + \left(\frac{1}{2}\right)^3$

18. $1 + \frac{1}{4} + \left(\frac{1}{4}\right)^2 - \left(\frac{1}{4}\right)^3$

19. Find the average of $\frac{2}{3}$ and $\frac{7}{8}$.

20. Find the average of $\frac{1}{4}$ and $\frac{1}{5}$.

21. Find the average of $\frac{1}{6}$, $\frac{1}{8}$, and $\frac{3}{4}$.

22. Find the average of $\frac{4}{5}$, $\frac{1}{2}$, and $\frac{1}{10}$.

23. Find the average of $3\frac{1}{2}$ and $9\frac{3}{8}$.

24. Find the average of $10\frac{2}{3}$ and $24\frac{5}{6}$.

25. *Birth Weights.* The Piper quadruplets of Great Britain weighed $2\frac{9}{16}$ lb, $2\frac{9}{32}$ lb, $2\frac{1}{8}$ lb, and $2\frac{5}{16}$ lb at birth. Find their average birth weight.

Source: *The Guinness Book of Records,* 1998

26. *Vertical Leaps.* Eight-year-old Zachary registered vertical leaps of $12\frac{3}{4}$ in., $13\frac{3}{4}$ in., $13\frac{1}{2}$ in., and 14 in. Find his average vertical leap.

27. *Acceleration* The results of a *Motor Trend* road acceleration test for five cars are given in the graph below. The test measures the time in seconds required to go from 0 mph to 60 mph. What was the average time?

Source: *Motor Trend,* March 2005, pp. 134–142

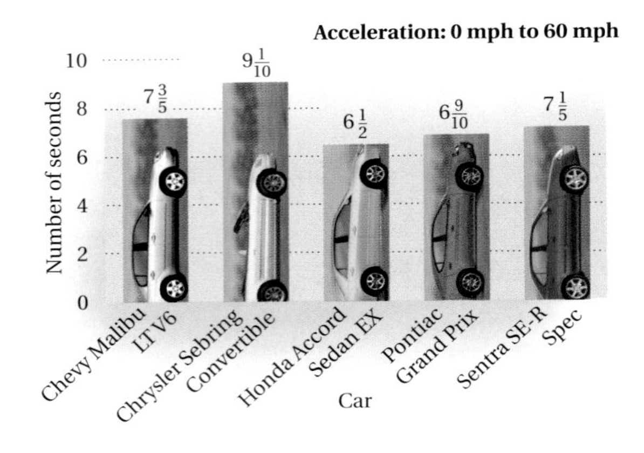

28. *Manufacturing.* A test of five light bulbs showed that they burned for the lengths of time given on the graph below. For how many days, on average, did the bulbs burn?

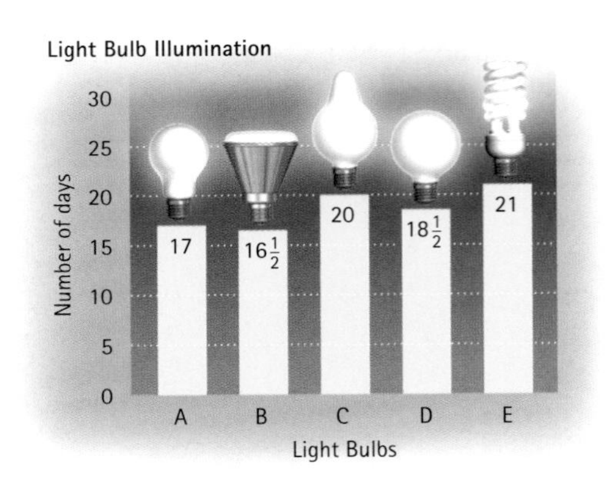

Simplify.

29. $\left(\frac{2}{3}+\frac{3}{4}\right) \div \left(\frac{5}{6}-\frac{1}{3}\right)$

30. $\left(\frac{3}{5}-\frac{1}{2}\right) \div \left(\frac{3}{4}-\frac{3}{10}\right)$

31. $\left(\frac{1}{2}+\frac{1}{3}\right)^2 \cdot 144 - \frac{5}{8} \div 10\frac{1}{2}$

32. $\left(3\frac{1}{2}-2\frac{1}{3}\right)^2 + 6 \cdot 2\frac{1}{2} \div 32$

b Estimate each of the following as 0, $\frac{1}{2}$, or 1.

33. $\frac{2}{47}$

34. $\frac{4}{5}$

35. $\frac{1}{13}$

36. $\frac{7}{8}$

37. $\frac{6}{11}$

38. $\frac{10}{13}$

39. $\frac{7}{15}$

40. $\frac{1}{16}$

41. $\frac{7}{100}$

42. $\frac{5}{9}$

43. $\frac{19}{20}$

44. $\frac{5}{12}$

Find a number for the blank so that the fraction is close to but greater than $\frac{1}{2}$. Answers may vary.

45. $\frac{\square}{11}$

46. $\frac{\square}{8}$

47. $\frac{\square}{23}$

48. $\frac{\square}{35}$

49. $\frac{10}{\square}$

50. $\frac{51}{\square}$

Find a number for the blank so that the fraction is close to but greater than 1. Answers may vary.

51. $\frac{7}{\square}$

52. $\frac{11}{\square}$

53. $\frac{13}{\square}$

54. $\frac{27}{\square}$

55. $\frac{\square}{15}$

56. $\frac{\square}{100}$

Estimate each part of the following as a whole number, $\frac{1}{2}$, or as a mixed numeral where the fractional part is $\frac{1}{2}$.

57. $2\frac{7}{8}$

58. $1\frac{1}{3}$

59. $12\frac{5}{6}$

60. $26\frac{6}{13}$

61. $\frac{4}{5} + \frac{7}{8}$

62. $\frac{1}{12} \cdot \frac{7}{15}$

63. $\frac{2}{3} + \frac{7}{13} + \frac{5}{9}$

64. $\frac{8}{9} + \frac{4}{5} + \frac{11}{12}$

65. $\frac{43}{100} + \frac{1}{10} - \frac{11}{1000}$

66. $\frac{23}{24} + \frac{37}{39} + \frac{51}{50}$

67. $7\frac{29}{60} + 10\frac{12}{13} \cdot 24\frac{2}{17}$

68. $5\frac{13}{14} - 1\frac{5}{8} + 1\frac{23}{28} \cdot 6\frac{35}{74}$

69. $24 \div 7\frac{8}{9}$

70. $43\frac{16}{17} \div 11\frac{2}{13}$

71. $76\frac{3}{14} + 23\frac{19}{20}$

72. $76\frac{13}{14} \cdot 23\frac{17}{20}$

73. $16\frac{1}{5} \div 2\frac{1}{11} + 25\frac{9}{10} - 4\frac{11}{23}$

74. $96\frac{2}{13} \div 5\frac{19}{20} + 3\frac{1}{7} \cdot 5\frac{18}{21}$

75. D_W A student insists that $3\frac{2}{5} \cdot 1\frac{3}{7} = 3\frac{6}{35}$. What mistake is he making and how should he have proceeded?

76. D_W A student insists that $5 \cdot 3\frac{2}{7} = (5 \cdot 3) \cdot \left(5 \cdot \frac{2}{7}\right)$. What mistake is she making and how should she have proceeded?

SKILL MAINTENANCE

77. Multiply: $27 \cdot 126$. [1.5a]

78. Multiply: $132 \cdot 7865$. [1.5a]

79. Divide: $7865 \div 132$. [1.6b]

Multiply and simplify. [2.4a], [2.6a]

80. $\frac{2}{3} \cdot 522$

81. $\frac{3}{2} \cdot 522$

Divide and simplify. [2.7b]

82. $\frac{4}{5} \div \frac{3}{10}$

83. $\frac{3}{10} \div \frac{4}{5}$

84. Classify the given numbers as prime, composite, or neither. [2.1c]

1, 5, 7, 9, 14, 23, 43

Solve.

85. *Luncheon Servings.* Ian purchased 6 lb of cold cuts for a luncheon. If Ian has allowed $\frac{3}{8}$ lb per person, how many people did he invite to the luncheon? [2.7d]

86. *Cholesterol.* A 3-oz serving of crabmeat contains 85 milligrams (mg) of cholesterol. A 3-oz serving of shrimp contains 128 mg of cholesterol. How much more cholesterol is in the shrimp? [1.8a]

SYNTHESIS

87. **a)** Find an expression for the sum of the areas of the two rectangles shown here.
b) Simplify the expression.
c) How is the computation in part (b) related to the rules for order of operations?

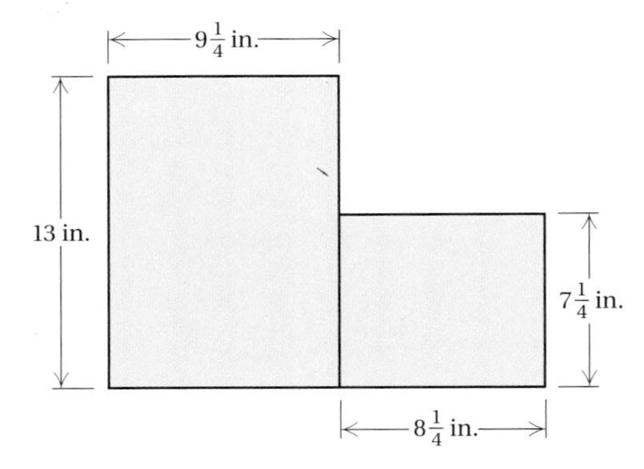

88. Find r if

$$\frac{1}{r} = \frac{1}{100} + \frac{1}{150} + \frac{1}{200}.$$

89. In the sum below, a and b are digits. Find a and b.

$$\frac{a}{17} + \frac{1b}{23} = \frac{35a}{391}$$

90. Consider only the numbers 3, 4, 5, and 6. Assume each can be placed in a blank in the following.

$$\square + \frac{\square}{\square} \cdot \square = ?$$

What placement of the numbers in the blanks yields the largest number?

91. Consider only the numbers 2, 3, 4, and 5. Assume each is placed in a blank in the following.

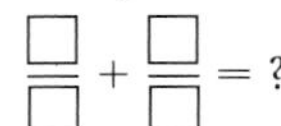

$$\frac{\square}{\square} + \frac{\square}{\square} = ?$$

What placement of the numbers in the blanks yields the largest sum?

92. Use a standard calculator. Arrange the following in order from smallest to largest.

$$\frac{3}{4}, \frac{17}{21}, \frac{13}{15}, \frac{7}{9}, \frac{15}{17}, \frac{13}{12}, \frac{19}{22}$$

3 Summary and Review

The review that follows is meant to prepare you for a chapter exam. It consists of two parts. The first part, Concept Reinforcement, is designed to increase understanding of the concepts through true/false exercises. The second part is the Review Exercises. These provide practice exercises for the exam, together with references to section objectives so you can go back and review. Before beginning, stop and look back over the skills you have obtained. What skills in mathematics do you have now that you did not have before studying this chapter?

CONCEPT REINFORCEMENT

Determine whether the statement is true or false. Answers are given at the back of the book.

_______ **1.** If $\frac{a}{b} > \frac{c}{b}$, $b \neq 0$, then $a > c$.

_______ **2.** All mixed numerals represent numbers larger than 1.

_______ **3.** The least common multiple of two natural numbers is the smallest number that is a factor of both.

_______ **4.** The mixed numeral $5\frac{2}{3}$ can be represented by the sum $5 \cdot \frac{3}{3} + \frac{2}{3}$.

_______ **5.** The least common multiple of two numbers is always larger than or equal to the larger number.

_______ **6.** To add fractions when denominators are the same, we keep the numerator and add the denominators.

Review Exercises

Find the LCM. [3.1a]

1. 12 and 18

2. 18 and 45

3. 3, 6, and 30

4. 26, 36, and 54

Add and simplify. [3.2a]

5. $\frac{6}{5} + \frac{3}{8}$

6. $\frac{5}{16} + \frac{1}{12}$

7. $\frac{6}{5} + \frac{11}{15} + \frac{3}{20}$

8. $\frac{1}{1000} + \frac{19}{100} + \frac{7}{10}$

Subtract and simplify. [3.3a]

9. $\frac{5}{9} - \frac{2}{9}$

10. $\frac{7}{8} - \frac{3}{4}$

11. $\frac{11}{27} - \frac{2}{9}$

12. $\frac{5}{6} - \frac{2}{9}$

Use $<$ or $>$ for $\square$ to write a true sentence. [3.3b]

13. $\frac{4}{7} \square \frac{5}{9}$

14. $\frac{8}{9} \square \frac{11}{13}$

Solve. [3.3c]

15. $x + \frac{2}{5} = \frac{7}{8}$

16. $\frac{1}{2} + y = \frac{9}{10}$

Convert to fraction notation. [3.4a]

17. $7\frac{1}{2}$

18. $8\frac{3}{8}$

19. $4\frac{1}{3}$

20. $10\frac{5}{7}$

Convert to a mixed numeral. [3.4a]

21. $\frac{7}{3}$

22. $\frac{27}{4}$

23. $\frac{63}{5}$

24. $\frac{7}{2}$

Divide. Write a mixed numeral for the answer. [3.4b]

25. $9\overline{)7896}$

26. $23\overline{)10{,}493}$

Add. Write a mixed numeral for the answer. [3.5a]

27. $5\frac{3}{5} + 4\frac{4}{5}$

28. $8\frac{1}{3} + 3\frac{2}{5}$

29. $5\frac{5}{6} + 4\frac{5}{6}$

30. $2\frac{3}{4} + 5\frac{1}{2}$

Subtract. Write a mixed numeral for the answer where appropriate. [3.5b]

31. $12 - 4\frac{2}{9}$

32. $9\frac{3}{5} - 4\frac{13}{15}$

33. $10\frac{1}{4} - 6\frac{1}{10}$

34. $24 - 10\frac{5}{8}$

Multiply. Write a mixed numeral for the answer where appropriate. [3.6a]

35. $6 \cdot 2\frac{2}{3}$

36. $5\frac{1}{4} \cdot \frac{2}{3}$

37. $2\frac{1}{5} \cdot 1\frac{1}{10}$

38. $2\frac{2}{5} \cdot 2\frac{1}{2}$

Divide. Write a mixed numeral for the answer where appropriate. [3.6b]

39. $27 \div 2\frac{1}{4}$

40. $2\frac{2}{5} \div 1\frac{7}{10}$

41. $3\frac{1}{4} \div 26$

42. $4\frac{1}{5} \div 4\frac{2}{3}$

Solve. [3.2b], [3.5c], [3.6c]

43. *Sewing.* Gloria wants to make a dress and jacket. She needs $1\frac{5}{8}$ yd of 60-in. fabric for the dress and $2\frac{5}{8}$ yd for the jacket. How many yards in all does Gloria need to make the outfit?

44. What is the sum of the areas in the figure below?

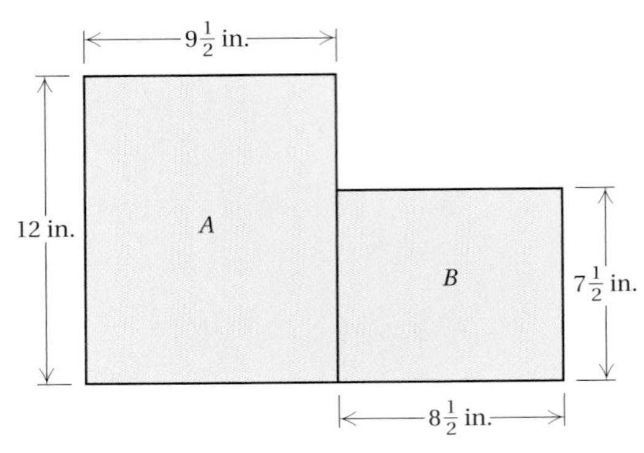

45. In the figure above, how much larger is the area of rectangle *A* than the area of rectangle *B*?

46. *Snapper Provençal.* Listed below is the recipe for Snapper Provençal. It serves 4. What are the quantities needed to prepare this dish when serving only 2? serving 12?

Source: Reprinted from: *The South Beach Diet: The Delicious, Doctor-Designed, Foolproof Plan for Fast and Healthy Weight Loss* by Arthur Agatston, M.D. ©2003 by Arthur Agatston, M.D. Permission granted by Rodale, Inc., Emmaus, PA 18098. Available wherever books are sold or directly from the publisher by calling (800) 848-4735 or visit their website at www.rodalestore.com. More information on the diet available at www.sbdiet.com.

Snapper Provençal

$\frac{1}{4}$ cup extra-virgin olive oil

$1\frac{1}{2}$ pounds fresh red snapper fillets

$\frac{1}{3}$ cup kalamata olives

$2\frac{1}{2}$ tablespoons capers

1 cup canned tomatoes

3 tablespoons chopped shallots

$\frac{1}{2}$ tablespoon fresh rosemary leaves

$\frac{1}{2}$ tablespoon minced garlic

$\frac{1}{3}$ cup white wine

47. *Carpentry.* A board $\frac{9}{10}$ in. thick is glued to a board $\frac{8}{10}$ in. thick. The glue is $\frac{3}{100}$ in. thick. How thick is the result?

48. *Turkey Servings.* Turkey contains $1\frac{1}{3}$ servings per pound. How many pounds are needed for 32 servings?

49. *Weightlifting.* In 1998, Sun Tianni of China snatched 111 kg. This amount was about $1\frac{3}{5}$ times her body weight. How much did Tianni weigh?

Source: *The Guinness Book of Records*, 2000

50. *Cake Recipe.* A wedding-cake recipe requires 12 cups of shortening. Being calorie-conscious, the wedding couple decides to reduce the shortening by $3\frac{5}{8}$ cups and replace it with prune purée. How many cups of shortening are used in their new recipe?

51. *Firefighters' Pie Sale.* Green River's Volunteer Fire Department recently hosted its annual ice cream social. Each of the 83 pies donated was cut into 6 pieces. At the end of the evening, they had sold 382 pieces of pie. How many pies did they sell? How many were left over? Express your answers in mixed numerals.

Simplify the expression using the rules for order of operations. [3.7a]

52. $\frac{1}{8} \div \frac{1}{4} + \frac{1}{2}$

53. $\frac{4}{5} - \frac{1}{2} \cdot \left(1 + \frac{1}{4}\right)$

54. $20\frac{3}{4} - 1\frac{1}{2} \times 12 + \left(\frac{1}{2}\right)^2$

55. Find the average of $\frac{1}{2}, \frac{1}{4}, \frac{1}{3}$, and $\frac{1}{5}$. [3.7a]

Estimate each of the following as 0, $\frac{1}{2}$, or 1. [3.7b]

56. $\frac{29}{59}$

57. $\frac{2}{59}$

58. $\frac{61}{59}$

Estimate each of the following as a whole number or as a mixed numeral where the fractional part is $\frac{1}{2}$. [3.7b]

59. $6\frac{7}{8}$

60. $10\frac{2}{17}$

61. $\frac{3}{10} + \frac{5}{6} + \frac{31}{29}$

62. $32\frac{14}{15} + 27\frac{3}{4} - 4\frac{25}{28} \cdot 6\frac{37}{76}$

63. $\mathbf{D_W}$ Discuss the role of least common multiples in adding and subtracting with fraction notation. [3.2a], [3.3a]

64. $\mathbf{D_W}$ Find a real-world situation that fits this equation: [3.5c], [3.6c]

$$2 \cdot 15\frac{3}{4} + 2 \cdot 28\frac{5}{8} = 88\frac{3}{4}.$$

SYNTHESIS

65. *Orangutan Circus Act.* Yuri and Olga are orangutans who perform in a circus by riding bicycles around a circular track. It takes Yuri 6 min and Olga 4 min to make one trip around the track. Suppose they start at the same point and then complete their act when they again reach the same point. How long is their act? [3.1a]

66. Place the numbers 3, 4, 5, and 6 in the boxes in order to make a true equation: [3.5a]

$$\frac{\square}{\square} + \frac{\square}{\square} = 3\frac{1}{4}.$$

3 Chapter Test

For Extra Help

Work It Out! Chapter Test Video on CD

Find the LCM.

1. 16 and 12

2. 15, 40, and 50

Add and simplify.

3. $\frac{1}{2} + \frac{5}{2}$

4. $\frac{7}{8} + \frac{2}{3}$

5. $\frac{7}{10} + \frac{19}{100} + \frac{31}{1000}$

Subtract and simplify.

6. $\frac{5}{6} - \frac{3}{6}$

7. $\frac{5}{6} - \frac{3}{4}$

8. $\frac{17}{24} - \frac{1}{15}$

Solve.

9. $\frac{1}{4} + y = 4$

10. $x + \frac{2}{3} = \frac{11}{12}$

11. Use $<$ or $>$ for $\square$ to write a true sentence:

$\frac{6}{7} \square \frac{21}{25}$.

Convert to fraction notation.

12. $3\frac{1}{2}$

13. $9\frac{7}{8}$

Convert to a mixed numeral.

14. $\frac{9}{2}$

15. $\frac{74}{9}$

Divide. Write a mixed numeral for the answer.

16. $11\overline{)1789}$

Add. Write a mixed numeral for the answer.

17. $6\frac{2}{5} + 7\frac{4}{5}$

18. $9\frac{1}{4} + 5\frac{1}{6}$

Subtract. Write a mixed numeral for the answer.

19. $10\frac{1}{6} - 5\frac{7}{8}$

20. $14 - 7\frac{5}{6}$

Multiply. Write a mixed numeral for the answer.

21. $9 \cdot 4\frac{1}{3}$

22. $6\frac{3}{4} \cdot \frac{2}{3}$

Divide. Write a mixed numeral for the answer.

23. $2\frac{1}{3} \div 1\frac{1}{6}$

24. $2\frac{1}{12} \div 75$

25. *Weightlifting.* In 2002, Hossein Rezazadeh of Iran did a clean and jerk of 263 kg. This amount was about $2\frac{1}{2}$ times his body weight. How much did Rezazadeh weigh?

Source: *The Guinness Book of Records*, 2005

26. *Book Order.* An order of books for a math course weighs 220 lb. Each book weighs $2\frac{3}{4}$ lb. How many books are in the order?

27. *Carpentry.* The following diagram shows a middle drawer support guide for a cabinet drawer. Find each of the following.

a) The short length a across the top
b) The length b across the bottom

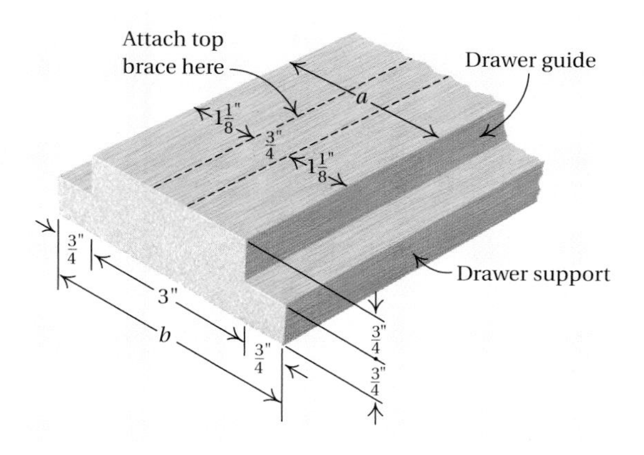

28. *Carpentry.* In carpentry, some pieces of plywood that are called "$\frac{3}{4}$-inch" plywood are actually $\frac{11}{16}$ in. thick. How much thinner is such a piece than its name indicates?

29. *Women's Dunks.* The first three women in the history of college basketball able to dunk a basketball are listed below. Their names, heights, and universities are:

Michelle Snow, $6\frac{5}{12}$ ft, Tennessee;
Charlotte Smith, $5\frac{11}{12}$ ft, North Carolina;
Georgeann Wells, $6\frac{7}{12}$ ft, West Virginia.

Find the average height of these women.

Source: *USA Today*, 11/30/00, p. 3C

Simplify.

30. $\frac{2}{3} + 1\frac{1}{3} \cdot 2\frac{1}{8}$

31. $1\frac{1}{2} - \frac{1}{2}\left(\frac{1}{2} \div \frac{1}{4}\right) + \left(\frac{1}{2}\right)^2$

Estimate each of the following as 0, $\frac{1}{2}$, or 1.

32. $\frac{3}{82}$

33. $\frac{93}{91}$

Estimate each of the following as a whole number or as a mixed numeral where the fractional part is $\frac{1}{2}$.

34. $3\frac{8}{9}$

35. $18\frac{9}{17}$

36. $256 \div 15\frac{19}{21}$

37. $43\frac{15}{31} \cdot 27\frac{5}{6} - 9\frac{15}{28} + 6\frac{5}{76}$

SYNTHESIS

38. The students in a math class can be organized into study groups of 8 each so that no students are left out. The same class of students can also be organized into groups of 6 so that no students are left out.

a) Find some class sizes for which this will work.
b) Find the smallest such class size.

39. Rebecca walks 17 laps at her health club. Trent walks 17 laps at his health club. If the track at Rebecca's health club is $\frac{1}{7}$ mi long, and the track at Trent's is $\frac{1}{8}$ mi long, who walks farther? How much farther?

CHAPTERS 1–3

Cumulative Review

Solve.

1. *Excelsior Made from Aspen.* Excelsior consists of slender, curved wood shavings and is often used for packing. Shown at right are examples of excelsior and the saw blades used to cut it, as made by Western Excelsior Corporation of Mancos, CO. The width of strips for craft decoration is either $\frac{1}{16}$ in. or $\frac{1}{8}$ in. The width for erosion control mats used for stabilizing soil and nourishing young crops is $\frac{1}{24}$ in.
   a) How much wider is the $\frac{1}{16}$-in. craft decoration excelsior than the erosion control excelsior?
   b) How much wider is the $\frac{1}{8}$-in. craft decoration excelsior than the erosion control excelsior?
   Source: Western Excelsior Corporation

2. *DVD Storage.* Gregory is making a home entertainment center. He is planning a 27-in. shelf that holds DVDs that are each $\frac{7}{16}$ in. thick. How many DVDs will the shelf hold?

3. *Cross-Country Skiing.* During a three-day holiday weekend trip, David and Sally Jean cross-country skied $3\frac{2}{3}$ mi on Friday, $6\frac{1}{8}$ mi on Saturday, and $4\frac{3}{4}$ mi on Sunday.
   a) Find the total miles they skied.
   b) Find the average miles they skied per day. Express your answer as a mixed numeral.

4. *Room Carpeting.* The Chandlers are carpeting an L-shaped family room consisting of a rectangle that is $8\frac{1}{2}$ ft by 11 ft and one that is $6\frac{1}{2}$ ft by $7\frac{1}{2}$ ft.
   a) Find the area of the carpet.
   b) Find the perimeter of the carpet.

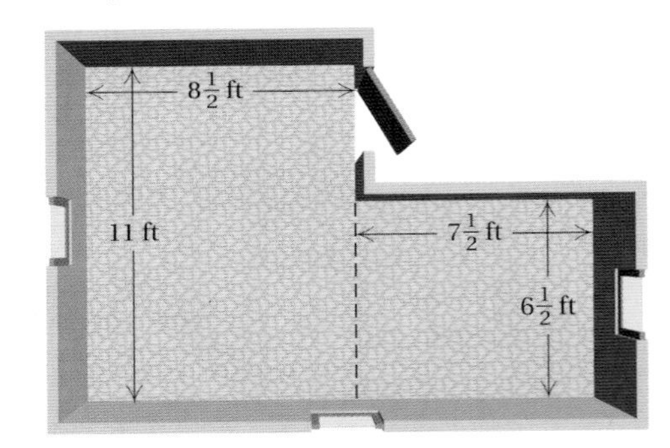

5. How many people can get equal \$16 shares from a total of \$496?

6. An emergency food pantry fund contains \$423. From this fund, \$148 and \$167 are withdrawn for expenses. How much is left in the fund?

7. A recipe calls for $\frac{4}{5}$ tsp of salt. How much salt should be used for $\frac{1}{2}$ recipe? for 5 recipes?

8. A book weighs $2\frac{3}{5}$ lb. How much do 15 books weigh?

9. How many pieces, each $2\frac{3}{8}$ ft long, can be cut from a piece of wire 38 ft long?

10. In a walkathon, Jermaine walked $\frac{9}{10}$ mi and Oleta walked $\frac{75}{100}$ mi. What was the total distance they walked?

11. In the number 2753, what digit names tens?

12. Write expanded notation for 6075.

13. Write a word name for the number in the following sentence: The diameter of Uranus is 29,500 miles.

14. What part is shaded?

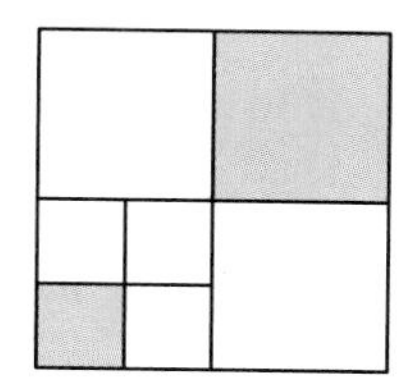

Calculate and simplify.

15. $\begin{array}{r} 628 \\ +\ 271 \\ \hline \end{array}$

16. $\begin{array}{r} 3704 \\ +\ 5278 \\ \hline \end{array}$

17. $\frac{3}{8} + \frac{1}{24}$

18. $\begin{array}{r} 2\frac{3}{4} \\ +\ 5\frac{1}{2} \\ \hline \end{array}$

19. $\begin{array}{r} 7469 \\ -\ 2345 \\ \hline \end{array}$

20. $\begin{array}{r} 7605 \\ -\ 3087 \\ \hline \end{array}$

21. $\frac{3}{4} - \frac{1}{3}$

22. $\begin{array}{r} 2\frac{1}{3} \\ -\ 1\frac{1}{6} \\ \hline \end{array}$

23. $\begin{array}{r} 278 \\ \times\ 18 \\ \hline \end{array}$

24. $\begin{array}{r} 894 \\ \times\ 328 \\ \hline \end{array}$

25. $\frac{9}{10} \cdot \frac{5}{3}$

26. $18 \cdot \frac{5}{6}$

27. $2\frac{1}{3} \cdot 3\frac{1}{7}$

Divide. Write the answer with the remainder in the form 34 R 7.

28. $6\overline{)4290}$

29. $45\overline{)2531}$

30. In Question 29, write a mixed numeral for the answer.

31. Simplify:

$$\left(\frac{1}{2} + \frac{2}{5}\right)^2 \div 3 + 6 \times \left(2 + \frac{1}{4}\right).$$

Divide and simplify, where appropriate.

32. $\frac{2}{5} \div \frac{7}{10}$

33. $2\frac{1}{5} \div \frac{3}{10}$

34. Round 38,478 to the nearest hundred.

35. Find the LCM of 18 and 24.

36. Determine whether 3718 is divisible by 8.

37. Find all factors of 16.

Use $<$, $>$, or $=$ for $\square$ to write a true sentence.

38. $\frac{4}{5} \square \frac{4}{6}$

39. $\frac{3}{13} \square \frac{9}{39}$

40. $\frac{5}{12} \square \frac{3}{7}$

Simplify.

41. $\frac{36}{45}$

42. $\frac{0}{27}$

43. $\frac{320}{10}$

44. Convert to fraction notation: $4\frac{5}{8}$.

45. Convert to a mixed numeral: $\frac{17}{3}$.

Solve.

46. $x + 24 = 117$

47. $x + \frac{7}{9} = \frac{4}{3}$

48. $\frac{7}{9} \cdot t = \frac{4}{3}$

49. $y = 32{,}580 \div 36$

Estimate each of the following as 0, $\frac{1}{2}$, or 1.

50. $\frac{29}{30}$

51. $\frac{15}{29}$

52. $\frac{2}{43}$

Estimate each of the following as a whole number or as a mixed numeral where the fractional part is $\frac{1}{2}$.

53. $30\frac{4}{53}$

54. $\frac{9}{10} - \frac{7}{8} + \frac{41}{39}$

55. $78\frac{14}{15} - 28\frac{7}{8} - 7\frac{25}{28} \div \frac{65}{66}$

56. *Matching.* Match each item in the first column with the appropriate item in the second column by drawing connecting lines.

Factors of 68	12, 54, 72, 300
Factorization of 68	2, 3, 17, 19, 23, 31, 47, 101
Prime factorization of 68	$2 \cdot 2 \cdot 17$
Numbers divisible by 6	$2 \cdot 34$
Numbers divisible by 8	8, 16, 24, 32, 40, 48, 64, 864
Numbers divisible by 5	1, 2, 4, 17, 34, 68
Prime numbers	70, 95, 215

For each of Exercises 57–60, choose the correct answer from the selections given.

57. In Arizona, people often install desert landscaping to conserve water. In a development, each home lot requires $\frac{4}{5}$ ton of gravel. How many tons of gravel are required for 40 lots?

a) $\frac{80}{100}$ **b)** 50 **c)** $\frac{80}{5}$
d) 32 **e)** None

58. A gasoline tank contains 20 gal when it is $\frac{3}{4}$ full. How many gallons can it hold when full?

a) $\frac{80}{3}$ **b)** $\frac{23}{4}$ **c)** 15
d) $\frac{83}{4}$ **e)** None

59. A gallon of ice cream provides $28\frac{1}{2}$ servings. How many gallons of ice cream are needed to serve 228 guests at a wedding reception?

a) 8 **b)** $8\frac{1}{2}$ **c)** 10
d) $10\frac{1}{2}$ **e)** None

60. For a certain type of load of dishes in a Kitchen-Aid® dishwasher, $1\frac{3}{4}$ oz of Cascade® detergent are required. How many ounces of detergent are needed for 8 loads?

a) 14 **b)** $9\frac{3}{4}$ **c)** $4\frac{4}{7}$
d) $7\frac{3}{32}$ **e)** None

SYNTHESIS

61. a) Simplify each of the following, using fraction notation for your answers.

$$\frac{1}{1 \cdot 2}$$

$$\frac{1}{1 \cdot 2} + \frac{1}{2 \cdot 3}$$

$$\frac{1}{1 \cdot 2} + \frac{1}{2 \cdot 3} + \frac{1}{3 \cdot 4}$$

$$\frac{1}{1 \cdot 2} + \frac{1}{2 \cdot 3} + \frac{1}{3 \cdot 4} + \frac{1}{4 \cdot 5}$$

b) Look for a pattern in your answers to part (a). Then find the following without carrying out the computations.

$$\frac{1}{1 \cdot 2} + \frac{1}{2 \cdot 3} + \frac{1}{3 \cdot 4} + \frac{1}{4 \cdot 5} + \frac{1}{5 \cdot 6} + \frac{1}{6 \cdot 7} + \frac{1}{7 \cdot 8} + \frac{1}{8 \cdot 9} + \frac{1}{9 \cdot 10}$$

62. Find the smallest prime number that is larger than 2000.

Decimal Notation

4

Real-World Application

The Panama Canal in Panama is 50.7 mi long. The Suez Canal in Egypt is 119.9 mi long. How much longer is the Suez Canal?

This problem appears as Example 1 in Section 4.7.

4.1 DECIMAL NOTATION, ORDER, AND ROUNDING

Objectives

a Given decimal notation, write a word name.

b Convert between fraction notation and decimal notation.

c Given a pair of numbers in decimal notation, tell which is larger.

d Round decimal notation to the nearest thousandth, hundredth, tenth, one, ten, hundred, or thousand.

Study Tips

QUIZ-TEST FOLLOW-UP

You may have just completed a chapter quiz or test. Immediately after each chapter quiz or test, write out a step-by-step solution of each question you missed. Visit your instructor or tutor for help with problems that are still giving you trouble. When the week of the final examination arrives, you will be glad to have the excellent study guide these corrected tests provide.

The set of **arithmetic numbers,** or **nonnegative rational numbers,** consists of the whole numbers 0, 1, 2, 3, 4, 5, 6, 7, 8, 9, 10, and so on, and fractions like $\frac{1}{2}, \frac{2}{3}, \frac{7}{8}, \frac{17}{10}$, and so on. Note that we can write the whole numbers using fraction notation. For example, 3 can be written as $\frac{3}{1}$. We studied the use of fraction notation for arithmetic numbers in Chapters 2 and 3.

In Chapter 4, we will study the use of *decimal notation.* The word *decimal* comes from the Latin word *decima,* meaning a tenth part. Although we are using different notation, we are still considering the nonnegative rational numbers. Using decimal notation, we can write 0.875 for $\frac{7}{8}$, for example, or 48.97 for $48\frac{97}{100}$.

a Decimal Notation and Word Names

A 5-quart stand mixer sells for \$249.98. The dot in \$249.98 is called a **decimal point.** Since \$0.98, or 98¢, is $\frac{98}{100}$ of a dollar, it follows that

$$\$249.98 = 249 + \frac{98}{100} \text{ dollars.}$$

Also, since \$0.98, or 98¢, has the same value as

9 dimes + 8 cents

and 1 dime is $\frac{1}{10}$ of a dollar and 1 cent is $\frac{1}{100}$ of a dollar, we can write

$$249.98 = 2 \cdot 100 + 4 \cdot 10 + 9 \cdot 1 + 9 \cdot \frac{1}{10} + 8 \cdot \frac{1}{100}.$$

This is an extension of the expanded notation for whole numbers that we used in Chapter 1. The place values are 100, 10, 1, $\frac{1}{10}, \frac{1}{100}$, and so on. We can see this on a **place-value chart.** The value of each place is $\frac{1}{10}$ as large as the one to its left.

Let's consider decimal notation using a place-value chart to represent 26.3385 min, the men's 10,000-meter run record held by Kenenisa Bekele from Ethiopia.

PLACE-VALUE CHART							
Hundreds	Tens	Ones	Ten*ths*	Hundred*ths*	Thousand*ths*	Ten-Thousand*ths*	Hundred-Thousand*ths*
100	10	1	$\frac{1}{10}$	$\frac{1}{100}$	$\frac{1}{1000}$	$\frac{1}{10,000}$	$\frac{1}{100,000}$
	2	6 .	3	3	8	5	

The decimal notation 26.3385 means

$$20 + 6 + \frac{3}{10} + \frac{3}{100} + \frac{8}{1000} + \frac{5}{10,000}, \text{ or } 26\frac{3385}{10,000}.$$

We read both 26.3385 and $26\frac{3385}{10,000}$ as

"Twenty-six and three thousand three hundred eighty-five ten-thousandths."

We can also read 26.3385 as

"Two six *point* three three eight five, or twenty-six *point* three three eight five."

To write a word name from decimal notation,

a) write a word name for the whole number (the number named to the left of the decimal point),

397.685 → Three hundred ninety-seven

b) write the word "and" for the decimal point, and

397.685 → Three hundred ninety-seven and

c) write a word name for the number named to the right of the decimal point, followed by the place value of the last digit.

397.685 → Three hundred ninety-seven and six hundred eighty-five *thousandths*

EXAMPLE 1 *Ice Cream.* Write a word name for the number in this sentence: The average person eats 26.3 servings of ice cream a year.

Source: NPD Group; J. M. Hirsch, Associated Press

Twenty-six and three tenths

EXAMPLE 2 Write a word name for 410.87.

Four hundred ten and eighty-seven hundredths

Write a word name for the numbers.

1. **Life Expectancy.** In 2004, the life expectancy at birth in Switzerland was 80.31. In the United States, it was 77.43.

Source: *CIA World Factbook*, 2004

2. **Kentucky Derby.** In 2004, racehorse Smarty Jones won the Kentucky Derby in a time of 2.06767 min.

Source: *Time Almanac*, 2005

Answers on page A-8

3. 245.89

4. 34.0064

5. 31,079.764

Answers on page A-8

EXAMPLE 3 *5000-m Run.* Write a word name for the number in this sentence: The world outdoor record in the women's 5000-m run is 14.4113 min, held by Elvan Abeylegesse of Turkey.

Fourteen and four thousand one hundred thirteen ten-thousandths

EXAMPLE 4 Write a word name for 1788.405.

One thousand, seven hundred eighty-eight and four hundred five thousandths

Do Exercises 1–5. (Exercises 1 and 2 are on the preceding page.)

b Converting Between Decimal Notation and Fraction Notation

Given decimal notation, we can convert to fraction notation as follows:

$$9.875 = 9 + \frac{8}{10} + \frac{7}{100} + \frac{5}{1000}$$

$$= 9 \cdot \frac{1000}{1000} + \frac{8}{10} \cdot \frac{100}{100} + \frac{7}{100} \cdot \frac{10}{10} + \frac{5}{1000}$$

$$= \frac{9000}{1000} + \frac{800}{1000} + \frac{70}{1000} + \frac{5}{1000} = \frac{9875}{1000}.$$

Decimal notation → 9.875 ← 3 decimal places

Fraction notation → $\frac{9875}{1000}$ ← 3 zeros

To convert from decimal to fraction notation,

a) count the number of decimal places, — 4.98 (2 places)

b) move the decimal point that many places to the right, and — 4.98. (Move 2 places.)

c) write the answer over a denominator with a 1 followed by that number of zeros. — $\frac{498}{100}$ (2 zeros)

EXAMPLE 5 Write fraction notation for 0.876. Do not simplify.

0.876   0.876. $0.876 = \frac{876}{1000}$

3 places   3 zeros

For a number like 0.876, we generally write a 0 before the decimal point to draw attention to the presence of the decimal point.

EXAMPLE 6 Write fraction notation for 56.23. Do not simplify.

56.23   56.23. $56.23 = \frac{5623}{100}$

2 places   2 zeros

EXAMPLE 7 Write fraction notation for 1.5018. Do not simplify.

1.5018   1.5018. $1.5018 = \frac{15,018}{10,000}$

4 places   4 zeros

Do Exercises 6–9.

If fraction notation has a denominator that is a power of ten, such as 10, 100, 1000, and so on, we reverse the procedure we used before.

To convert from fraction notation to decimal notation when the denominator is 10, 100, 1000, and so on,

a) count the number of zeros, and $\frac{8679}{1000}$ ← 3 zeros

b) move the decimal point that number of places to the left. Leave off the denominator.   8.679. Move 3 places. $\frac{8679}{1000} = 8.679$

EXAMPLE 8 Write decimal notation for $\frac{47}{10}$.

$\frac{47}{10}$ 4.7. $\frac{47}{10} = 4.7$

1 zero   1 place

Write fraction notation.

6. 0.896

7. 23.78

8. 5.6789

9. 1.9

Answers on page A-8

Write decimal notation.

10. $\frac{743}{100}$

11. $\frac{406}{1000}$

12. $\frac{67,089}{10,000}$

13. $\frac{9}{10}$

14. $\frac{57}{1000}$

15. $\frac{830}{10,000}$

Write decimal notation.

16. $4\frac{3}{10}$

17. $283\frac{71}{100}$

18. $456\frac{13}{1000}$

Answers on page A-8

EXAMPLE 9 Write decimal notation for $\frac{123,067}{10,000}$.

$\frac{123,067}{10,000}$ — 4 zeros; 12.3067. — 4 places; $\frac{123,067}{10,000} = 12.3067$

EXAMPLE 10 Write decimal notation for $\frac{13}{1000}$.

$\frac{13}{1000}$ — 3 zeros; 0.013. — 3 places; $\frac{13}{1000} = 0.013$

EXAMPLE 11 Write decimal notation for $\frac{570}{100,000}$.

$\frac{570}{100,000}$ — 5 zeros; 0.00570. — 5 places; $\frac{570}{100,000} = 0.0057$

Do Exercises 10–15.

When denominators are numbers other than 10, 100, and so on, we will use another method for conversion. It will be considered in Section 4.5.

If a mixed numeral has a fractional part with a denominator that is a power of ten, such as 10, 100, or 1000, and so on, we first write the mixed numeral as a sum of a whole number and a fraction. Then we convert to decimal notation.

EXAMPLE 12 Write decimal notation for $23\frac{59}{100}$.

$$23\frac{59}{100} = 23 + \frac{59}{100} = 23 \text{ and } \frac{59}{100} = 23.59$$

EXAMPLE 13 Write decimal notation for $772\frac{129}{10,000}$.

$$772\frac{129}{10,000} = 772 + \frac{129}{10,000} = 772 \text{ and } \frac{129}{10,000} = 772.0129$$

Do Exercises 16–18.

c Order

To understand how to compare numbers in decimal notation, consider 0.85 and 0.9. First note that $0.9 = 0.90$ because $\frac{9}{10} = \frac{90}{100}$. Then $0.85 = \frac{85}{100}$ and $0.90 = \frac{90}{100}$. Since $\frac{85}{100} < \frac{90}{100}$, it follows that $0.85 < 0.90$. This leads us to a quick way to compare two numbers in decimal notation.

COMPARING NUMBERS IN DECIMAL NOTATION

To compare two numbers in decimal notation, start at the left and compare corresponding digits moving from left to right. If two digits differ, the number with the larger digit is the larger of the two numbers. To ease the comparison, extra zeros can be written to the right of the last decimal place.

EXAMPLE 14 Which of 2.109 and 2.1 is larger?

		Think.
2.109	→	2.109
2.1		2.100

Same ↑ Different; $9 > 0$

Thus, 2.109 is larger than 2.1. That is, $2.109 > 2.1$.

EXAMPLE 15 Which of 0.09 and 0.108 is larger?

		Think.
0.09	→	0.090
0.108		0.108

Same ↑ Different; $1 > 0$

Thus, 0.108 is larger than 0.09. That is, $0.108 > 0.09$.

Do Exercises 19–24.

d Rounding

Rounding is done as for whole numbers. To understand, we first consider an example using a number line. It might help to review Section 1.4.

EXAMPLE 16 Round 0.37 to the nearest tenth.

Here is part of a number line.

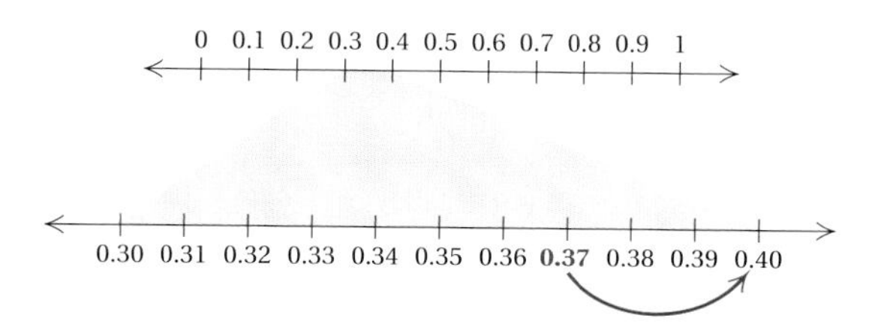

We see that 0.37 is closer to 0.40 than to 0.30. Thus, 0.37 rounded to the nearest tenth is 0.4.

Which number is larger?

19. 2.04, 2.039

2.040, 2.039

20. 0.06, 0.008

21. 0.5, 0.58

22. 1, 0.9999

23. 0.8989, 0.09898

24. 21.006, 21.05

Answers on page A-8

ROUNDING DECIMAL NOTATION

To round to a certain place:

a) Locate the digit in that place.
b) Consider the next digit to the right.
c) If the digit to the right is 5 or higher, round up; if the digit to the right is 4 or lower, round down.

EXAMPLE 17 Round 3872.2459 to the nearest tenth.

a) Locate the digit in the tenths place, 2.

3 8 7 2.2 4 5 9
↑

b) Consider the next digit to the right, 4.

3 8 7 2.2 4 5 9
↑

c) Since that digit, 4, is less than 5, round down.

3 8 7 2.2 ← This is the answer.

Caution!

3872.3 is not a correct answer to Example 17. It is *incorrect* to round from the ten-thousandths digit over to the tenths digit, as follows:

3872.246 → 3872.25 → 3872.3.

EXAMPLE 18 Round 3872.2459 to the nearest thousandth, hundredth, tenth, one, ten, hundred, and thousand.

Thousandth:	3872.246	Ten:	3870
Hundredth:	3872.25	Hundred:	3900
Tenth:	3872.2	Thousand:	4000
One:	3872		

EXAMPLE 19 Round 14.8973 to the nearest hundredth.

a) Locate the digit in the hundredths place, 9.

1 4.8 9 7 3
↑

b) Consider the next digit to the right, 7.

1 4.8 9 7 3
↑

c) Since that digit, 7, is 5 or higher, round up. When we make the hundredths digit a 10, we carry 1 to the tenths place.

The answer is 14.90. Note that the 0 in 14.90 indicates that the answer is correct to the nearest hundredth.

EXAMPLE 20 Round 0.008 to the nearest tenth.

a) Locate the digit in the tenths place, 0.

0.0 0 8
↑

b) Consider the next digit to the right, 0.

0.0 0 8
↑

c) Since that digit, 0, is less than 5, round down.

The answer is 0.0.

Do Exercises 25–43.

Round to the nearest tenth.

25. 2.76 **26.** 13.85

27. 234.448 **28.** 7.009

Round to the nearest hundredth.

29. 0.636 **30.** 7.834

31. 34.675 **32.** 0.025

Round to the nearest thousandth.

33. 0.9434 **34.** 8.0038

35. 43.1119 **36.** 37.4005

Round 7459.3548 to the nearest:

37. Thousandth.

38. Hundredth.

39. Tenth.

40. One.

41. Ten. (*Caution:* "Tens" are not "tenths.")

42. Hundred.

43. Thousand.

Answers on page A-9

EXERCISE SET

For Extra Help

MathXL | MyMathLab | InterAct Math | Math Tutor Center | Digital Video Tutor CD 2 Videotape 4 | Student's Solutions Manual

a Write a word name for the number in the sentence.

NYSE INDEX
MOST ACTIVE: SHARE VOLUME

	VOL. (000s)	LAST	CHANGE
Elan	97,202	6.65	−1.29
Lucent	46,527	3.08	−.08
NewsCpA n	32,126	17.28	+.48
BostonSci	24,330	30.90	−.66
Pfizer	21,024	26.59	−.02
ExxonMbl	20,149	63.05	+.37
WalMart	19,358	52.86	+.91
GenElec	17,824	35.55	−.05
Motorola	17,355	15.20	−.38
Texinst	17,234	26.58	−.39

1. *ExxonMbl.* Recently, the stock of Exxon Mobil sold for \$63.05 per share.

2. *Motorola.* Recently, the stock of Motorola sold for \$15.20 per share.

3. *Pfizer.* Recently, the stock of Pfizer sold for \$26.59 per share.

4. *Wal-Mart.* Recently, the stock of Wal-Mart sold for \$52.86 per share.

5. *Water Weight.* One gallon of water weighs 8.35 lb.

6. One gallon of paint is equal to 3.785 liters of paint.

7. *Tires.* A tire was on sale for \$86.89. This was the lowest price of the year.

8. *Digital Audio Jukebox.* The Apple iPod mini 4GB digital audio jukebox for Mac & Windows stores over 65 hours of music. Recently, it sold for \$249.99.

Write a word name.

9. 34.891

10. 27.1245

b Write fraction notation. Do not simplify.

11. 8.3

12. 0.17

13. 3.56

14. 203.6

15. 46.03

16. 1.509

17. 0.00013

18. 0.0109

19. 1.0008

20. 2.0114

21. 20.003

22. 4567.2

Write decimal notation.

23. $\frac{8}{10}$

24. $\frac{51}{10}$

25. $\frac{889}{100}$

26. $\frac{92}{100}$

27. $\frac{3798}{1000}$

28. $\frac{780}{1000}$

29. $\frac{78}{10,000}$

30. $\frac{56,788}{100,000}$

31. $\frac{19}{100,000}$

32. $\frac{2173}{100}$

33. $\frac{376,193}{1,000,000}$

34. $\frac{8,953,074}{1,000,000}$

35. $99\frac{44}{100}$

36. $4\frac{909}{1000}$

37. $3\frac{798}{1000}$

38. $67\frac{83}{100}$

39. $2\frac{1739}{10,000}$

40. $9243\frac{1}{10}$

41. $8\frac{953,073}{1,000,000}$

42. $2256\frac{3059}{10,000}$

c Which number is larger?

43. 0.06, 0.58

44. 0.008, 0.8

45. 0.905, 0.91

46. 42.06, 42.1

47. 0.0009, 0.001

48. 7.067, 7.054

49. 234.07, 235.07

50. 0.99999, 1

51. 0.004, $\frac{4}{100}$

52. $\frac{73}{10}$, 0.73

53. 0.432, 0.4325

54. 0.8437, 0.84384

d Round to the nearest tenth.

55. 0.11

56. 0.85

57. 0.49

58. 0.5794

59. 2.7449

60. 4.78

61. 123.65

62. 36.049

Round to the nearest hundredth.

63. 0.893

64. 0.675

65. 0.6666

66. 6.529

67. 0.995

68. 207.9976

69. 0.094

70. 11.4246

Round to the nearest thousandth.

71. 0.3246
72. 0.6666
73. 17.0015
74. 123.4562

75. 10.1011
76. 0.1161
77. 9.9989
78. 67.100602

Round 809.4732 to the nearest:

79. Hundred.
80. Tenth.
81. Thousandth.

82. Hundredth.
83. One.
84. Ten.

Round 34.54389 to the nearest:

85. Ten-thousandth.
86. Thousandth.
87. Hundredth.

88. Tenth.
89. One.
90. Ten.

91. DW Describe in your own words a procedure for converting from decimal notation to fraction notation.

92. DW A fellow student rounds 236.448 to the nearest one and gets 237. Explain the possible error.

SKILL MAINTENANCE

Round 6172 to the nearest: [1.4a]

93. Ten.
94. Hundred.
95. Thousand.

Find the prime factorization. [2.1d]

96. 2000
97. 1530
98. 2002
99. 4312

SYNTHESIS

100. Arrange the following numbers in order from smallest to largest.

0.99, 0.099, 1, 0.9999, 0.89999, 1.00009, 0.909, 0.9889

101. Arrange the following numbers in order from smallest to largest.

2.1, 2.109, 2.108, 2.018, 2.0119, 2.0302, 2.000001

Truncating. There are other methods of rounding decimal notation. To round using **truncating,** we drop off all decimal places past the rounding place, which is the same as changing all digits to the right to zeros. For example, rounding 6.78093456285102 to the ninth decimal place, using truncating, gives us 6.780934562. Use truncating to round each of the following to the fifth decimal place, that is, the hundred-thousandth.

102. 6.78346623
103. 6.783465902
104. 99.999999999
105. 0.030303030303

4.2 ADDITION AND SUBTRACTION

Objectives

a Add using decimal notation.

b Subtract using decimal notation.

c Solve equations of the type $x + a = b$ and $a + x = b$, where a and b may be in decimal notation.

d Balance a checkbook.

a Addition

Adding with decimal notation is similar to adding whole numbers. First we line up the decimal points so that we can add corresponding place-value digits. Then we add digits from the right. For example, we add the thousandths, then the hundredths, and so on, carrying if necessary. If desired, we can write extra zeros to the right of the decimal point so that the number of places is the same in all of the addends.

EXAMPLE 1 Add: 56.314 + 17.78.

$$\begin{array}{r} 56.314 \\ +\ 17.780 \\ \hline \end{array}$$

Lining up the decimal points in order to add
Writing an extra zero to the right of the decimal point

$$\begin{array}{r} 56.314 \\ +\ 17.780 \\ \hline 4 \end{array}$$

Adding thousandths

$$\begin{array}{r} 56.314 \\ +\ 17.780 \\ \hline 94 \end{array}$$

Adding hundredths

$$\begin{array}{r} \scriptstyle 1 \\ 56.314 \\ +\ 17.780 \\ \hline .094 \end{array}$$

Adding tenths
We get 10 tenths = 1 one + 0 tenths, so we carry the 1 to the ones column.
Write a decimal point in the answer.

$$\begin{array}{r} \scriptstyle 1\,1 \\ 56.314 \\ +\ 17.780 \\ \hline 4.094 \end{array}$$

Adding ones
We get 14 ones = 1 ten + 4 ones, so we carry the 1 to the tens column.

$$\begin{array}{r} \scriptstyle 1\,1 \\ 56.314 \\ +\ 17.780 \\ \hline 74.094 \end{array}$$

Adding tens

Do Exercises 1 and 2.

EXAMPLE 2 Add: 3.42 + 0.237 + 14.1.

$$\begin{array}{r} 3.420 \\ 0.237 \\ +\ 14.100 \\ \hline 17.757 \end{array}$$

Lining up the decimal points and writing extra zeros
Adding

Do Exercises 3–5.

Add.

1.
$$\begin{array}{r} 0.847 \\ +\ 10.07 \\ \hline \end{array}$$

2.
$$\begin{array}{r} 2.1 \\ 0.739 \\ +\ 31.3689 \\ \hline \end{array}$$

Add.

3. 0.02 + 4.3 + 0.649

4. 0.12 + 3.006 + 0.4357

5. 0.4591 + 0.2374 + 8.70894

Answers on page A-9

Consider the addition 3456 + 19.347. Keep in mind that any whole number has an "unwritten" decimal point at the right that can be followed by zeros. For example, 3456 can also be written 3456.000. When adding, we can always write in the decimal point and extra zeros if desired.

EXAMPLE 3 Add: 3456 + 19.347.

$$\begin{array}{rl} \overset{1}{3}\,4\,5\,6.0\,0\,0 & \text{Writing in the decimal point and extra zeros} \\ +\quad 1\,9.3\,4\,7 & \text{Lining up the decimal points} \\ \hline 3\,4\,7\,5.3\,4\,7 & \text{Adding} \end{array}$$

Do Exercises 6 and 7.

b Subtraction

Subtracting with decimal notation is similar to subtracting whole numbers. First we line up the decimal points so that we can subtract corresponding place-value digits. Then we subtract digits from the right. For example, we subtract the thousandths, then the hundredths, the tenths, and so on, borrowing if necessary.

EXAMPLE 4 Subtract: 56.314 − 17.78.

$$\begin{array}{rl} 5\,6.3\,1\,4 & \text{Lining up the decimal points in order to subtract} \\ -\,1\,7.7\,8\,0 & \text{Writing an extra 0} \\ \hline \end{array}$$

$$\begin{array}{rl} 5\,6.3\,1\,4 & \text{Subtracting thousandths} \\ -\,1\,7.7\,8\,0 & \\ \hline 4 & \end{array}$$

$$\begin{array}{rl} 5\,6.\overset{2}{\not 3}\,\overset{11}{\not 1}\,4 & \text{Borrowing tenths to subtract hundredths} \\ -\,1\,7.7\,8\,0 & \\ \hline 3\,4 & \end{array}$$

$$\begin{array}{rl} 5\,\overset{5}{\not 6}.\overset{12}{\not 2}\,\overset{11}{\not 1}\,4 & \text{Borrowing ones to subtract tenths} \\ -\,1\,7.7\,8\,0 & \\ \hline .5\,3\,4 & \text{Writing a decimal point} \end{array}$$

$$\begin{array}{rl} \overset{4}{\not 5}\,\overset{15}{\not 5}.\overset{12}{\not 2}\,\overset{11}{\not 1}\,4 & \text{Borrowing tens to subtract ones} \\ -\,1\,7.7\,8\,0 & \\ \hline 8.5\,3\,4 & \end{array}$$

$$\begin{array}{rl} \overset{4}{\not 5}\,\overset{15}{\not 5}.\overset{12}{\not 2}\,\overset{11}{\not 1}\,4 & \text{Subtracting tens} \\ -\,1\,7.7\,8\,0 & \\ \hline 3\,8.5\,3\,4 & \end{array}$$

Check:
$$\begin{array}{r} \overset{1}{3}\,\overset{1}{8}.\overset{1}{5}\,3\,4 \\ +\,1\,7.7\,8\,0 \\ \hline 5\,6.3\,1\,4 \end{array}$$

Do Exercises 8 and 9.

Add.

6. 789 + 123.67

7. 45.78 + 2467 + 1.993

Subtract.

8. 37.428 − 26.674

9.
$$\begin{array}{r} 0.3\,4\,7 \\ -\,0.0\,0\,8 \\ \hline \end{array}$$

Answers on page A-9

Subtract.

10. 1.2345 − 0.7

11. 0.9564 − 0.4392

12. 7.37 − 0.00008

Subtract.

13. 1277 − 82.78

14. 5 − 0.0089

Answers on page A-9

EXAMPLE 5 Subtract: 13.07 − 9.205.

```
   12
   2  10 6 10
 1 3 . 0 7 0     Writing an extra zero
 −   9 . 2 0 5
 ------------
     3 . 8 6 5   Subtracting
```

EXAMPLE 6 Subtract: 23.08 − 5.0053.

```
 1 13     7 9 10
 2 3 . 0 8 0 0     Writing two extra zeros
 −   5 . 0 0 5 3
 ---------------
 1 8 . 0 7 4 7     Subtracting
```

Check by addition:

```
 1      1 1
 1 8 . 0 7 4 7
 +   5 . 0 0 5 3
 ---------------
 2 3 . 0 8 0 0
```

The answer checks because this is the top number in the subtraction.

Do Exercises 10–12.

When subtraction involves a whole number, again keep in mind that there is an "unwritten" decimal point that can be written in if desired. Extra zeros can also be written in to the right of the decimal point.

EXAMPLE 7 Subtract: 456 − 2.467.

```
        5 9 9 10
 4 5 6 . 0 0 0     Writing in the decimal point and extra zeros
 −     2 . 4 6 7
 ---------------
 4 5 3 . 5 3 3     Subtracting
```

Do Exercises 13 and 14.

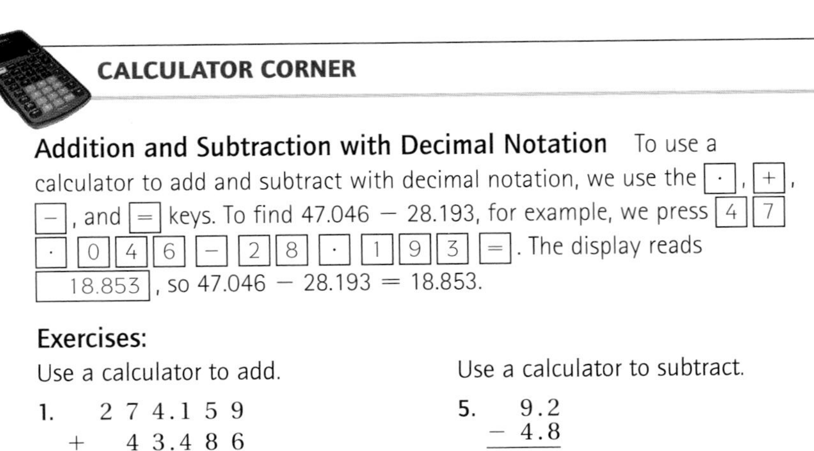

CALCULATOR CORNER

Addition and Subtraction with Decimal Notation To use a calculator to add and subtract with decimal notation, we use the [·], [+], [−], and [=] keys. To find 47.046 − 28.193, for example, we press [4] [7] [·] [0] [4] [6] [−] [2] [8] [·] [1] [9] [3] [=]. The display reads [18.853], so 47.046 − 28.193 = 18.853.

Exercises:

Use a calculator to add.

1. 274.159 + 43.486

2. 19.805 + 486.748

3. 1.7 + 14.56 + 0.89

4. 3.4 + 45 + 0.68

Use a calculator to subtract.

5. 9.2 − 4.8

6. 52.34 − 18.51

7. 489 − 34.26

8. 6.09 − 5.1

c Solving Equations

Now let's solve equations $x + a = b$ and $a + x = b$, where a and b may be in decimal notation. Proceeding as we have before, we subtract a on both sides.

EXAMPLE 8 Solve: $x + 28.89 = 74.567$.

We have

$$x + 28.89 - 28.89 = 74.567 - 28.89 \quad \text{Subtracting 28.89 on both sides}$$
$$x = 45.677.$$

$$\begin{array}{r} \overset{6}{\cancel{7}}\ \overset{13}{\cancel{4}}.\overset{14}{\cancel{5}}\overset{16}{\cancel{6}}\ 7 \\ -\ 2\ 8.8\ 9\ 0 \\ \hline 4\ 5.6\ 7\ 7 \end{array}$$

The solution is 45.677.

EXAMPLE 9 Solve: $0.8879 + y = 9.0026$.

We have

$$0.8879 + y - 0.8879 = 9.0026 - 0.8879 \quad \text{Subtracting 0.8879 on both sides}$$
$$y = 8.1147.$$

$$\begin{array}{r} \overset{8}{\cancel{9}}.\overset{9}{\cancel{0}}\ \overset{9}{\cancel{0}}\ \overset{11}{\cancel{2}}\ \overset{16}{\cancel{6}} \\ -\ 0.8\ 8\ 7\ 9 \\ \hline 8.1\ 1\ 4\ 7 \end{array}$$

The solution is 8.1147.

Do Exercises 15 and 16.

EXAMPLE 10 Solve: $120 + x = 4380.6$.

We have

$$120 + x - 120 = 4380.6 - 120 \quad \text{Subtracting 120 on both sides}$$
$$x = 4260.6$$

$$\begin{array}{r} 4\ 3\ 8\ 0.6 \\ -\quad 1\ 2\ 0.0 \\ \hline 4\ 2\ 6\ 0.6 \end{array}$$

The solution is 4260.6.

Do Exercise 17.

d Balancing a Checkbook

Let's use addition and subtraction with decimals to balance a checkbook.

EXAMPLE 11 Find the errors, if any, in the balances in this checkbook.

20___ RECORD ALL CHARGES OR CREDITS THAT AFFECT YOUR ACCOUNT

DATE	CHECK NUMBER	TRANSACTION DESCRIPTION	√ T	(−) PAYMENT/ DEBIT	(+ OR −) OTHER	(+) DEPOSIT/ CREDIT	BALANCE FORWARD
							8767 73
8/16	432	Burch Laundry		23 56			8744 16
8/19	433	Rogers TV		20 49			8764 65
8/20		Deposit				85 00	8848 65
8/21	434	Galaxy Records		48 60			8801 05
8/22	435	Electric Works		267 95			8533 09

Solve.

15. $x + 17.78 = 56.314$

16. $8.906 + t = 23.07$

17. Solve: $241 + y = 2374.5$.

Answers on page A-9

18. Find the errors, if any, in this checkbook.

20___ RECORD ALL CHARGES OR CREDITS THAT AFFECT YOUR ACCOUNT

DATE	CHECK NUMBER	TRANSACTION DESCRIPTION	√ T	(−) PAYMENT/ DEBIT	(+ OR −) OTHER	(+) DEPOSIT/ CREDIT	BALANCE FORWARD
							3078 92
12/1	888	H.H. Gregg Appliances		340 69			2738 23
12/3	889	Marie Callendar's Pies		78 56			2659 66
12/5		Deposit <Paycheck>				230 80	2890 46
12/6	890	Chili's Restaurant		13 14			2877 32
12/8	891	Stonycreek Golf Course		48 00			2829 32
12/8		Deposit <Molly>				39 58	2868 90
12/10	892	Galyan's Trading Post		102 87			2766 83
12/14	893	Goody's Music		68 59			2697 45
12/15	894	Salvation Army		100 00			2497 45

Answer on page A-9

There are two ways to determine whether there are errors. We assume that the amount \$8767.73 in the "Balance forward" column is correct. If we can determine that the ending balance is correct, we have some assurance that the checkbook is correct. But two errors could offset each other to give us that balance.

METHOD 1

a) We add the debits: $23.56 + 20.49 + 48.60 + 267.95 = 360.60$.

b) We add the deposits/credits. In this case, there is only one deposit, 85.00.

c) We add the total of the deposits to the balance brought forward:

$$8767.73 + 85.00 = 8852.73.$$

d) We subtract the total of the debits: $8852.73 - 360.60 = 8492.13$.

The result should be the ending balance, 8533.09. We see that $8492.13 \neq 8533.09$. Since the numbers are not equal, we proceed to method 2.

METHOD 2 We successively add or subtract deposit/credits and debits, and check the result in the "Balance forward" column.

$$8767.73 - 23.56 = 8744.17.$$

We have found our first error. The subtraction was incorrect. We correct it and continue, using 8744.17 as the corrected balance forward:

$$8744.17 - 20.49 = 8723.68.$$

It looks as though 20.49 was added instead of subtracted. Actually, we would have to correct this line even if it had been subtracted, because the error of 1¢ in the first step has been carried through successive calculations. We correct that balance line and continue, using 8723.68 as the balance and adding the deposit 85.00:

$$8723.68 + 85.00 = 8808.68.$$

We make the correction and continue subtracting the last two debits:

$$8808.68 - 48.60 = 8760.08.$$

Then

$$8760.08 - 267.95 = 8492.13.$$

The corrected checkbook is below.

20___ RECORD ALL CHARGES OR CREDITS THAT AFFECT YOUR ACCOUNT

DATE	CHECK NUMBER	TRANSACTION DESCRIPTION	√ T	(−) PAYMENT/ DEBIT	(+ OR −) OTHER	(+) DEPOSIT/ CREDIT	BALANCE FORWARD	
							8767 73	
8/16	432	Burch Laundry		23 56			~~8744 16~~	→ 8744.17
8/19	433	Rogers TV		20 49			~~8764 65~~	→ 8723.68
8/20		Deposit				85 00	~~8848 65~~	→ 8808.68
8/21	434	Galaxy Records		48 60			~~8801 05~~	→ 8760.08
8/22	435	Electric Works		267 95			~~8533 09~~	→ 8492.13

Do Exercise 18.

There are other ways in which errors can be made in checkbooks, such as forgetting to record a transaction or writing the amounts incorrectly, but we will not consider those here.

EXERCISE SET

For Extra Help

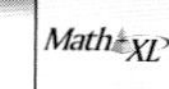

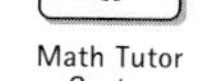

MathXL | MyMathLab | InterAct Math | Math Tutor Center | Digital Video Tutor CD 2 Videotape 4 | Student's Solutions Manual

a Add.

1. $316.25 + 18.12$

2. $641.803 + 14.935$

3. $659.403 + 916.812$

4. $4203.28 + 3.39$

5. $9.104 + 123.456$

6. $81.008 + 3.409$

7. 20.0124 + 30.0124

8. 0.687 + 0.9

9. 39 + 1.007

10. 2.3 + 0.729 + 23

11. $47.8 + 219.852 + 43.59 + 666.713$

12. $13.72 + 9.112 + 6542.7908 + 23.901$

13. 0.34 + 3.5 + 0.127 + 768

14. 17 + 3.24 + 0.256 + 0.3689

15. 99.6001 + 7285.18 + 500.042 + 870

16. 65.987 + 9.4703 + 6744.02 + 1.0003 + 200.895

b Subtract.

17. $51.31 - 2.29$

18. $44.345 - 3.105$

19. $92.341 - 6.42$

20. $97.01 - 3.15$

21. $2.5 - 0.0025$

22. $39.0 - 0.28$

23. $3.4 - 0.003$

24. $2.8 - 2.08$

25. 28.2 − 19.35

26. 100.16 − 0.118

27. 34.07 − 30.7

28. 36.2 − 16.28

29. $8.45 - 7.405$

30. $3.801 - 2.81$

31. $6.003 - 2.3$

32. $9.087 - 8.807$

33. $1 - 0.0098$

34. $2 - 1.0908$

35. $100 - 0.34$

36. $624 - 18.79$

37. $7.48 - 2.6$

38. $18.4 - 5.92$

39. $3 - 2.006$

40. $263.7 - 102.08$

41. $19 - 1.198$

42. $2548.98 - 2.007$

43. $65 - 13.87$

44. $45 - 0.999$

45. $\begin{array}{r} 32.7978 \\ -\ \ 0.0592 \\ \hline \end{array}$

46. $\begin{array}{r} 0.49634 \\ -\ 0.12678 \\ \hline \end{array}$

47. $\begin{array}{l} \ \ \ 6.07 \\ -\ 2.0078 \\ \hline \end{array}$

48. $\begin{array}{l} \ \ \ 1.0 \\ -\ 0.9999 \\ \hline \end{array}$

C Solve.

49. $x + 17.5 = 29.15$

50. $t + 50.7 = 54.07$

51. $17.95 + p = 402.63$

52. $w + 1.3004 = 47.8$

53. $13{,}083.3 = x + 12{,}500.33$

54. $100.23 = 67.8 + z$

55. $x + 2349 = 17{,}684.3$

56. $1830.4 + t = 23{,}067$

d Find the errors, if any, in each checkbook.

57.

20____ RECORD ALL CHARGES OR CREDITS THAT AFFECT YOUR ACCOUNT

DATE	CHECK NUMBER	TRANSACTION DESCRIPTION	√ T	(−) PAYMENT/ DEBIT	(+ OR −) OTHER	(+) DEPOSIT/ CREDIT	BALANCE FORWARD
							9704.56
8/8	342	Bill Rydman		27.44			9677.12
8/9		Deposit <Beauty Contest>				1000.00	10,677.12
8/12	343	Chuck Taylor		123.95			10,553.17
8/14	344	Jennifer Crum		124.02			10,677.19
8/22	345	Neon Johnny's Pizza		12.43			10,664.76
8/24		Deposit <Bowling Tournament>				2500.00	13,164.76
8/29	346	Border's Bookstore		137.78			13,302.54
9/2		Deposit <Bodybuilder Contest>				18.88	13,283.66
9/3	347	Fireman's Fund		2800.00			10,483.66

58.

20____ RECORD ALL CHARGES OR CREDITS THAT AFFECT YOUR ACCOUNT

DATE	CHECK NUMBER	TRANSACTION DESCRIPTION	√ T	(−) PAYMENT/ DEBIT	(+ OR −) OTHER	(+) DEPOSIT/ CREDIT	BALANCE FORWARD
							1876.43
4/1	500	Bart Kaufman		500.12			1376.31
4/3	501	Jim Lawler		28.56			1347.75
4/3		Deposit <State Lottery>				10,000.00	11,347.75
4/3	502	Victoria Montoya		464.00			10,883.75
4/3		Deposit <Jewelry Sale>				2500.00	8383.75
4/4	503	Baskin & Robbins		1600.00			6783.75
4/8	504	Golf Galaxy		1349.98			5433.77
4/12	505	Don Mitchell Pro Shops		658.97			4774.80
4/13		Deposit <Publisher's Clearing House>				100000.00	104,774.80
4/15	506	American Airlines		6885.58			98,889.22

59. DW Explain the error in the following:
Add.

$$13.07 + 9.205 = 10.512$$

60. DW Explain the error in the following:
Subtract.

$$73.089 - 5.0061 = 2.3028$$

SKILL MAINTENANCE

61. Round 34,567 to the nearest thousand. [1.4a]

62. Round 34,496 to the nearest thousand. [1.4a]

Subtract.

63. $\frac{13}{24} - \frac{3}{8}$ [3.3a]

64. $\frac{8}{9} - \frac{2}{15}$ [3.3a]

65. $8805 - 2639$ [1.3b]

66. $8005 - 2639$ [1.3b]

Solve.

67. A serving of filleted fish is generally considered to be about $\frac{1}{3}$ lb. How many servings can be prepared from $5\frac{1}{2}$ lb of flounder fillet? [3.6c]

68. A photocopier technician drove $125\frac{7}{10}$ mi away from Scottsdale for a repair call. The next day he drove $65\frac{1}{2}$ mi back toward Scottsdale for another service call. How far was the technician from Scottsdale? [3.5c]

SYNTHESIS

69. A student presses the wrong button when using a calculator and adds 235.7 instead of subtracting it. The incorrect answer is 817.2. What is the correct answer?

4.3 MULTIPLICATION

Objectives

a Multiply using decimal notation.

b Convert from notation like 45.7 million to standard notation, and between dollars and cents.

a Multiplication

Let's find the product

$2.3 \times 1.12.$

To understand how we find such a product, we first convert each factor to fraction notation. Next, we multiply the whole numbers 23 and 112, and then divide by 1000.

$$2.3 \times 1.12 = \frac{23}{10} \times \frac{112}{100} = \frac{23 \times 112}{10 \times 100} = \frac{2576}{1000} = 2.576$$

Note the number of decimal places.

$$\begin{array}{rl} 1.1\,2 & \text{(2 decimal places)} \\ \times \quad 2.3 & \text{(1 decimal place)} \\ \hline 2.5\,7\,6 & \text{(3 decimal places)} \end{array}$$

Now consider

$$0.011 \times 15.0002 = \frac{11}{1000} \times \frac{150{,}002}{10{,}000} = \frac{1{,}650{,}022}{10{,}000{,}000} = 0.1650022.$$

Note the number of decimal places.

$$\begin{array}{rl} 1\,5.0\,0\,0\,2 & \text{(4 decimal places)} \\ \times \quad 0.0\,1\,1 & \text{(3 decimal places)} \\ \hline 0.1\,6\,5\,0\,0\,2\,2 & \text{(7 decimal places)} \end{array}$$

To multiply using decimals:

a) Ignore the decimal points and multiply as though both factors were whole numbers.

b) Then place the decimal point in the result. The number of decimal places in the product is the sum of the numbers of places in the factors (count places from the right).

0.8×0.43

$$\begin{array}{rl} \overset{2}{0.4\,3} & \\ \times \quad 0.8 & \text{Ignore the decimal} \\ \hline 3\,4\,4 & \text{points for now.} \end{array}$$

$$\begin{array}{rl} 0.4\,3 & \text{(2 decimal places)} \\ \times \quad 0.8 & \text{(1 decimal place)} \\ \hline 0.3\,4\,4 & \text{(3 decimal places)} \end{array}$$

EXAMPLE 1 Multiply: 8.3×74.6.

a) Ignore the decimal points and multiply as though factors were whole numbers.

$$\begin{array}{r} {\scriptstyle 3\ 4} \\ {\scriptstyle 1\ 1} \\ 7\,4.6 \\ \times \quad 8.3 \\ \hline 2\,2\,3\,8 \\ 5\,9\,6\,8\,0 \\ \hline 6\,1\,9\,1\,8 \end{array}$$

b) Place the decimal point in the result. The number of decimal places in the product is the sum of the number of places in the factors, 1 + 1, or 2.

$$\begin{array}{rl} 7\,4.6 & \text{(1 decimal place)} \\ \times \quad 8.3 & \text{(1 decimal place)} \\ \hline 2\,2\,3\,8 & \\ 5\,9\,6\,8\,0 & \downarrow \\ \hline 6\,1\,9.1\,8 & \text{(2 decimal places)} \end{array}$$

Do Exercise 1.

EXAMPLE 2 Multiply: 0.0032 × 2148.

$$\begin{array}{rl} 2\,1\,4\,8 & \text{(0 decimal places)} \\ \times\ 0.0\,0\,3\,2 & \text{(4 decimal places)} \\ \hline 4\,2\,9\,6 & \\ 6\,4\,4\,4\,0 & \downarrow \\ \hline 6.8\,7\,3\,6 & \text{(4 decimal places)} \end{array}$$

EXAMPLE 3 Multiply: 0.14 × 0.867.

$$\begin{array}{rl} 0.8\,6\,7 & \text{(3 decimal places)} \\ \times \quad 0.1\,4 & \text{(2 decimal places)} \\ \hline 3\,4\,6\,8 & \\ 8\,6\,7\,0 & \downarrow \\ \hline 0.1\,2\,1\,3\,8 & \text{(5 decimal places)} \end{array}$$

Do Exercises 2 and 3.

MULTIPLYING BY 0.1, 0.01, 0.001, AND SO ON

Now let's consider some special kinds of products. The first involves multiplying by a tenth, hundredth, thousandth, or ten-thousandth. Let's look at those products.

$$0.1 \times 38 = \frac{1}{10} \times 38 = \frac{38}{10} = 3.8$$

$$0.01 \times 38 = \frac{1}{100} \times 38 = \frac{38}{100} = 0.38$$

$$0.001 \times 38 = \frac{1}{1000} \times 38 = \frac{38}{1000} = 0.038$$

$$0.0001 \times 38 = \frac{1}{10{,}000} \times 38 = \frac{38}{10{,}000} = 0.0038$$

Note in each case that the product is *smaller* than 38. That is, the decimal point in each product is farther to the left than the unwritten decimal point in 38.

1. Multiply.

$$\begin{array}{r} 8\,5.4 \\ \times \quad 6.2 \\ \hline \end{array}$$

Multiply.

2.
$$\begin{array}{r} 1\,2\,3\,4 \\ \times\ 0.0\,0\,4\,1 \\ \hline \end{array}$$

3.
$$\begin{array}{r} 4\,2.6\,5 \\ \times\ 0.8\,0\,4 \\ \hline \end{array}$$

Answers on page A-9

Multiply.

4. 0.1×3.48

5. 0.01×3.48

6. 0.001×3.48

7. 0.0001×3.48

To multiply any number by 0.1, 0.01, 0.001, and so on,

a) count the number of decimal places in the tenth, hundredth, or thousandth, and so on, and

b) move the decimal point that many places to the left.

0.001×34.45678 → 3 places

$0.001 \times 34.45678 = 0.034.45678$

Move 3 places to the left.

$0.001 \times 34.45678 = 0.03445678$

EXAMPLES Multiply.

4. $0.1 \times 14.605 = 1.4605$ 1.4.605

5. $0.01 \times 14.605 = 0.14605$

6. $0.001 \times 14.605 = 0.014605$

We write an extra zero.

7. $0.0001 \times 14.605 = 0.0014605$

We write two extra zeros.

Do Exercises 4–7.

MULTIPLYING BY 10, 100, 1000, AND SO ON

Next, let's consider multiplying by 10, 100, 1000, and so on. Let's look at those products.

$$10 \times 97.34 = 973.4$$
$$100 \times 97.34 = 9734$$
$$1000 \times 97.34 = 97{,}340$$
$$10{,}000 \times 97.34 = 973{,}400$$

Note in each case that the product is *larger* than 97.34. That is, the decimal point in each product is farther to the right than the decimal point in 97.34.

Multiply.

8. 10×3.48

9. 100×3.48

10. 1000×3.48

11. $10{,}000 \times 3.48$

To multiply any number by 10, 100, 1000, and so on,

a) count the number of zeros, and

b) move the decimal point that many places to the right.

1000×34.45678 → 3 zeros

$1000 \times 34.45678 = 34.456.78$

Move 3 places to the right.

$1000 \times 34.45678 = 34{,}456.78$

EXAMPLES Multiply.

8. $10 \times 14.605 = 146.05$ 14.6.05

9. $100 \times 14.605 = 1460.5$

10. $1000 \times 14.605 = 14{,}605$

11. $10{,}000 \times 14.605 = 146{,}050$ 14.6050. We write an extra zero.

Answers on page A-9

Do Exercises 8–11 on the preceding page.

b Applications Using Multiplication with Decimal Notation

NAMING LARGE NUMBERS

We often see notation like the following in newspapers and magazines and on television.

- The largest building in the world is the Pentagon, which has 3.7 million square feet of floor space.
- In 2003, consumers spent $51.7 billion on online retail products. Total spending in specific categories is listed in the following table.

ONLINE CONSUMER SPENDING
(in billions of dollars)

CATEGORY	2003	CATEGORY	2003
1. PCs	$8.8	7. Event Tickets	$2.7
2. Peripherals	$2.5	8. Apparel	$6.1
3. Software	$2.9	9. Jewelry	$1.3
4. Consumer Electronics	$2.5	10. Toys	$1.3
5. Books	$3.1	11. Housewares/ Small Appliances	$1.9
6. Music	$1.3	12. Office Products	$1.7

Source: Jupiter Media Metrix, Inc.

To understand such notation, consider the information in the following table.

NAMING LARGE NUMBERS

1 hundred = 100 = $10 \cdot 10 = 10^2$ (2 zeros)

1 thousand = 1000 = $10 \cdot 10 \cdot 10 = 10^3$ (3 zeros)

1 million = 1,000,000 = $10 \cdot 10 \cdot 10 \cdot 10 \cdot 10 \cdot 10 = 10^6$ (6 zeros)

1 billion = 1,000,000,000 = 10^9 (9 zeros)

1 trillion = 1,000,000,000,000 = 10^{12} (12 zeros)

CALCULATOR CORNER

Multiplication with Decimal Notation To use a calculator to multiply with decimal notation, we use the [·], [×], and [=] keys. To find 4.78 × 0.34, for example, we press [4] [·] [7] [8] [×] [·] [3] [4] [=]. The display reads [1.6252], so 4.78 × 0.34 = 1.6252.

Exercises: Use a calculator to multiply.

1. 5.4×9
2. 415×16.7
3. 17.63×8.1
4. 0.04×12.69
5. 586.4×13.5
6. 4.003×5.1

To convert a large number to standard notation, we proceed as follows.

EXAMPLE 12 Convert the number in this sentence to standard notation: In 1999, the U.S. Mint produced 11.6 billion pennies.

Source: U.S. Mint

$$\begin{aligned} 11.6 \text{ billion} &= 11.6 \times 1 \text{ billion} \\ &= 11.6 \times \underbrace{1{,}000{,}000{,}000}_{\text{9 zeros}} \\ &= 11{,}600{,}000{,}000 \quad \text{Moving the decimal point 9 places to the right} \end{aligned}$$

Do Exercises 12–14.

MONEY CONVERSION

Converting from dollars to cents is like multiplying by 100. To see why, consider $19.43.

$\$19.43 = 19.43 \times \$1$ We think of $19.43 as 19.43 × 1 dollar, or 19.43 × $1.

$= 19.43 \times 100¢$ Substituting 100¢ for $1: $1 = 100¢

$= 1943¢$ Multiplying

DOLLARS TO CENTS

To convert from dollars to cents, move the decimal point two places to the right and change the $ sign in front to a ¢ sign at the end.

EXAMPLES Convert from dollars to cents.

13. $189.64 = 18,964¢

14. $0.75 = 75¢

Do Exercises 15 and 16.

Converting from cents to dollars is like multiplying by 0.01. To see why, consider 65¢.

$65¢ = 65 \times 1¢$ We think of 65¢ as 65 × 1 cent, or 65 × 1¢.

$= 65 \times \$0.01$ Substituting $0.01 for 1¢: 1¢ = $0.01

$= \$0.65$ Multiplying

CENTS TO DOLLARS

To convert from cents to dollars, move the decimal point two places to the left and change the ¢ sign at the end to a $ sign in front.

EXAMPLES Convert from cents to dollars.

15. 395¢ = $3.95

16. 8503¢ = $85.03

Do Exercises 17 and 18.

Convert the number in the sentence to standard notation.

12. The largest building in the world is the Pentagon, which has 3.7 million square feet of floor space.

13. Online Apparel Sales. In 2003, consumers spent $6.1 billion for apparel.

Source: Jupiter Media Metrix, Inc.

14. Online Ticket Sales. In 2003, consumers spent $2.7 billion for event tickets.

Source: Jupiter Media Metrix, Inc.

Convert from dollars to cents.

15. $15.69

16. $0.17

Convert from cents to dollars.

17. 35¢

18. 577¢

Answers on page A-9

EXERCISE SET

For Extra Help

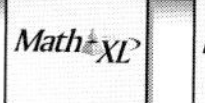
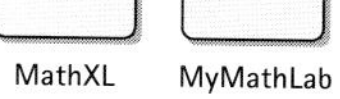

MathXL | MyMathLab | InterAct Math | Math Tutor Center | Digital Video Tutor CD 2 Videotape 4 | Student's Solutions Manual

a Multiply.

1. 8.6×7

2. 5.7×0.8

3. 0.84×8

4. 9.4×0.6

5. 6.3×0.04

6. 9.8×0.08

7. 87×0.006

8. 18.4×0.07

9. 10×23.76

10. 100×3.8798

11. 1000×583.686852

12. 0.34×1000

13. 7.8×100

14. 0.00238×10

15. 0.1×89.23

16. 0.01×789.235

17. 0.001×97.68

18. 8976.23×0.001

19. 78.2×0.01

20. 0.0235×0.1

21. 32.6×16

22. 9.28×8.6

23. 0.984×3.3

24. 8.489×7.4

25. 374×2.4

26. 865×1.08

27. 749×0.43

28. 978×20.5

29. 0.87×64

30. 7.25×60

31. 46.50×75

32. 8.24×703

33. $\begin{array}{r} 81.7 \\ \times\ 0.612 \\ \hline \end{array}$

34. $\begin{array}{r} 31.82 \\ \times\ 7.15 \\ \hline \end{array}$

35. $\begin{array}{r} 10.105 \\ \times\ 11.324 \\ \hline \end{array}$

36. $\begin{array}{r} 151.2 \\ \times\ 4.555 \\ \hline \end{array}$

37. $\begin{array}{r} 12.3 \\ \times\ 1.08 \\ \hline \end{array}$

38. $\begin{array}{r} 7.82 \\ \times\ 0.024 \\ \hline \end{array}$

39. $\begin{array}{r} 32.4 \\ \times\ 2.8 \\ \hline \end{array}$

40. $\begin{array}{r} 8.09 \\ \times\ 0.0075 \\ \hline \end{array}$

41. $\begin{array}{r} 0.00342 \\ \times\ 0.84 \\ \hline \end{array}$

42. $\begin{array}{r} 2.0056 \\ \times\ 3.8 \\ \hline \end{array}$

43. $\begin{array}{r} 0.347 \\ \times\ 2.09 \\ \hline \end{array}$

44. $\begin{array}{r} 2.532 \\ \times\ 1.067 \\ \hline \end{array}$

45. $\begin{array}{r} 3.005 \\ \times\ 0.623 \\ \hline \end{array}$

46. $\begin{array}{r} 16.34 \\ \times\ 0.000512 \\ \hline \end{array}$

47. 1000×45.678

48. 0.001×45.678

b Convert from dollars to cents.

49. $28.88

50. $67.43

51. $0.66

52. $1.78

Convert from cents to dollars.

53. 34¢

54. 95¢

55. 3445¢

56. 933¢

Convert the number in the sentence to standard notation.

57. *Wal-Mart.* In 2003, Wal-Mart stores had revenues of about $258.7 billion.

Source: Fortune 500

58. *Magazine Circulation.* The total paid circulation in 2003 of the magazine *Better Homes and Gardens* was about 7.6 million copies.

Source: Audit Bureau of Circulations, tabulated by Magazine Publishers of America

59. *Broadway.* Broadway productions grossed $748.9 million in 2004.

Source: League of American Theatres and Producers

60. *Rail Transportation.* People in the United States traveled 13.6 billion miles on city rail transportation in 2003.

Source: American Public Transportation Association, NASA

61. **DW** If two rectangles have the same perimeter, will they also have the same area? Experiment with different dimensions. Be sure to use decimals. Explain your answer.

62. **DW** A student insists that $346.708 \times 0.1 = 3467.08$. How could you convince him that a mistake had been made without checking on a calculator?

SKILL MAINTENANCE

Calculate.

63. $2\frac{1}{3} \cdot 4\frac{4}{5}$ [3.6a]

64. $2\frac{1}{3} \div 4\frac{4}{5}$ [3.6b]

65. $4\frac{4}{5} - 2\frac{1}{3}$ [3.5b]

66. $4\frac{4}{5} + 2\frac{1}{3}$ [3.5a]

Divide. [1.6b]

67. $24\overline{)8208}$

68. $4\overline{)348}$

69. $7\overline{)31{,}962}$

70. $18\overline{)22{,}626}$

71. $40\overline{)3480}$

72. $17\overline{)20{,}006}$

SYNTHESIS

Consider the following names for large numbers in addition to those already discussed in this section:

$$1 \text{ quadrillion} = 1{,}000{,}000{,}000{,}000{,}000 = 10^{15};$$
$$1 \text{ quintillion} = 1{,}000{,}000{,}000{,}000{,}000{,}000 = 10^{18};$$
$$1 \text{ sextillion} = 1{,}000{,}000{,}000{,}000{,}000{,}000{,}000 = 10^{21};$$
$$1 \text{ septillion} = 1{,}000{,}000{,}000{,}000{,}000{,}000{,}000{,}000 = 10^{24}.$$

Find each of the following. Express the answer with a name that is a power of 10.

73. (1 trillion) · (1 billion)

74. (1 million) · (1 billion)

75. (1 trillion) · (1 trillion)

76. Is a billion millions the same as a million billions? Explain.

4.4 DIVISION

Objectives

a Divide using decimal notation.

b Solve equations of the type $a \cdot x = b$, where a and b may be in decimal notation.

c Simplify expressions using the rules for order of operations.

Divide.

1. $9 \overline{)5.4}$

2. $15 \overline{)22.5}$

3. $82 \overline{)38.54}$

Answers on page A-9

a Division

WHOLE-NUMBER DIVISORS

We use the following method when we divide a decimal quantity by a whole number.

To divide by a whole number,

a) place the decimal point directly above the decimal point in the dividend, and

b) divide as though dividing whole numbers.

```
                0.8 4 <- Quotient
Divisor -> 7 ) 5.8 8 <- Dividend
               5 6 0
               -----
                 2 8
                 2 8
                 ---
                   0 <- Remainder
```

EXAMPLE 1 Divide: $379.2 \div 8$.

```
          4 7.4     <- Place the decimal point.
    8 ) 3 7 9.2
        3 2 0 0
        -------
          5 9 2      Divide as though dividing whole numbers.
          5 6 0
          -----
            3 2
            3 2
            ---
              0
```

EXAMPLE 2 Divide: $82.08 \div 24$.

```
            3.4 2     <- Place the decimal point.
    2 4 ) 8 2.0 8
          7 2 0 0
          -------
          1 0 0 8      Divide as though dividing whole numbers.
            9 6 0
            -----
              4 8
              4 8
              ---
                0
```

Do Exercises 1–3.

Suppose the dividend is a whole number. We can think of it as having a decimal point at the end with as many 0's as we wish after the decimal point. For example,

$$12 = 12. = 12.0 = 12.00 = 12.000, \text{ and so on.}$$

EXAMPLE 3 Divide: $30 \div 8$.

$$\begin{array}{r} 3. \\ 8\overline{)\,30.} \\ \underline{24} \\ 6 \end{array}$$

Place the decimal point and divide to find how many ones.

$$\begin{array}{r} 3. \\ 8\overline{)\,30.0} \\ \underline{24}\downarrow \\ 60 \end{array}$$

Write an extra zero.

$$\begin{array}{r} 3.7 \\ 8\overline{)\,30.0} \\ \underline{24} \\ 60 \\ \underline{56} \\ 4 \end{array}$$

Divide to find how many tenths.

$$\begin{array}{r} 3.7 \\ 8\overline{)\,30.00} \\ \underline{24} \\ 60 \\ \underline{56}\downarrow \\ 40 \end{array}$$

Write another zero.

$$\begin{array}{r} 3.75 \\ 8\overline{)\,30.00} \\ \underline{24} \\ 60 \\ \underline{56} \\ 40 \\ \underline{40} \\ 0 \end{array}$$

Divide to find how many hundredths.

Check:
$$\begin{array}{r} \scriptstyle 6\ 4 \\ 3.75 \\ \times\ 8 \\ \hline 30.00 \end{array}$$

EXAMPLE 4 Divide: $4 \div 25$.

$$\begin{array}{r} 0.16 \\ 25\overline{)\,4.00} \\ \underline{25} \\ 150 \\ \underline{150} \\ 0 \end{array}$$

Check:
$$\begin{array}{r} \scriptstyle 1 \\ \scriptstyle 3 \\ 0.16 \\ \times25 \\ \hline 80 \\ 320 \\ \hline 4.00 \end{array}$$

Do Exercises 4–6.

Divide.

4. $25\overline{)\,8}$

5. $4\overline{)\,15}$

6. $86\overline{)\,21.5}$

Answers on page A-9

7. a) Complete.

$$\frac{3.75}{0.25} = \frac{3.75}{0.25} \times \frac{100}{100}$$
$$= \frac{(\ )}{25}$$

b) Divide.

$0.25\overline{)3.75}$

Divide.

8. $0.83\overline{)4.067}$

9. $3.5\overline{)44.8}$

Answers on page A-9

DIVISORS THAT ARE NOT WHOLE NUMBERS

Consider the division

$0.24\overline{)8.208}$

We write the division as $\frac{8.208}{0.24}$. Then we multiply by 1 to change to a whole-number divisor:

$$\frac{8.208}{0.24} = \frac{8.208}{0.24} \times \frac{100}{100} = \frac{820.8}{24}.$$

The division $0.24\overline{)8.208}$ is the same as $24\overline{)820.8}$.

The divisor is now a whole number.

To divide when the divisor is not a whole number,

a) move the decimal point (multiply by 10, 100, and so on) to make the divisor a whole number;

$0.24\overline{)8.208}$ Move 2 places to the right.

b) move the decimal point (multiply the same way) in the dividend the same number of places; and

$0.24\overline{)8.208}$ Move 2 places to the right.

c) place the decimal point directly above the new decimal point in the dividend and divide as though dividing whole numbers.

```
          3 4.2
0.2 4 ) 8.2 0^8
        7 2 0 0
        1 0 0 8
          9 6 0
            4 8
            4 8
              0
```

(The new decimal point in the dividend is indicated by a caret.)

EXAMPLE 5 Divide: $5.848 \div 8.6$.

$8.6\overline{)5.848}$ Multiply the divisor by 10 (move the decimal point 1 place). Multiply the same way in the dividend (move 1 place).

```
         0.6 8
8.6 ) 5.8^4 8
      5 1 6 0
        6 8 8
        6 8 8
            0
```

Place a decimal point above the new decimal point and then divide.

Note: $\frac{5.848}{8.6} = \frac{5.848}{8.6} \cdot \frac{10}{10} = \frac{58.48}{86}$.

Do Exercises 7–9.

Sometimes it helps to write extra zeros to the right of the decimal point in the dividend.

EXAMPLE 6 Divide: $12 \div 0.64$.

$0.64 \overline{)12.}$ Place a decimal point at the end of the whole number.

$0.64 \overline{)12.00}$ Multiply the divisor by 100 (move the decimal point 2 places). Multiply the same way in the dividend (move 2 places).

Place a decimal point above and then divide.

$$\begin{array}{r} 18.75 \\ 0.64_\wedge \overline{)12.00_\wedge 00} \\ \underline{640} \\ 560 \\ \underline{512} \\ 480 \\ \underline{448} \\ 320 \\ \underline{320} \\ 0 \end{array}$$

Do Exercise 10.

DIVIDING BY 10, 100, 1000, AND SO ON

It is often helpful to be able to divide quickly by a ten, hundred, or thousand, or by a tenth, hundredth, or thousandth. Each procedure we use is based on multiplying by 1. Consider the following example:

$$\frac{23.789}{1000} = \frac{23.789}{1000} \cdot \frac{1000}{1000} = \frac{23{,}789}{1{,}000{,}000} = 0.023789.$$

We are dividing by a number greater than 1: The result is *smaller* than 23.789.

To divide by 10, 100, 1000, and so on,

a) count the number of zeros in the divisor, and

$\frac{713.49}{100}$ → 2 zeros

b) write the quotient by moving the decimal point in the dividend that number of places to the left.

$\frac{713.49}{100}$, 7.13.49, $\frac{713.49}{100} = 7.1349$

2 places to the left

EXAMPLE 7 Divide: $\frac{0.0104}{10}$.

$\frac{0.0104}{10}$, 0.0.0104, $\frac{0.0104}{10} = 0.00104$

1 zero — 1 place to the left

10. Divide.

$1.6 \overline{)25}$

CALCULATOR CORNER

Division with Decimal Notation To use a calculator to divide with decimal notation, we use the [·], [÷], and [=] keys. To find $237.12 \div 5.2$, for example, we press [2] [3] [7] [·] [1] [2] [÷] [5] [·] [2] [=]. The display reads [45.6], so $237.12 \div 5.2 = 45.6$.

Exercises: Use a calculator to divide.

1. $12.4 \overline{)177.32}$
2. $49 \overline{)125.44}$
3. $3.2 \overline{)640}$
4. $16 \overline{)12}$
5. $14 \div 0.7$
6. $1.6 \div 25$
7. $474.14 \div 30.2$
8. $518.472 \div 6.84$

Answer on page A-9

Study Tips

HOMEWORK TIPS

Prepare for your homework assignment by reading the explanations of concepts and following the step-by-step solutions of examples in the text. The time you spend preparing will save valuable time when you do your assignment.

Divide.

11. $\dfrac{0.1278}{0.01}$

12. $\dfrac{0.1278}{100}$

13. $\dfrac{98.47}{1000}$

14. $\dfrac{6.7832}{0.1}$

Answers on page A-9

DIVIDING BY 0.1, 0.01, 0.001, AND SO ON

Now consider the following example:

$$\frac{23.789}{0.01} = \frac{23.789}{0.01} \cdot \frac{100}{100} = \frac{2378.9}{1} = 2378.9.$$

We are dividing by a number less than 1: The result is *larger* than 23.789. We use the following procedure.

To divide by 0.1, 0.01, 0.001, and so on,

a) count the number of decimal places in the divisor, and

$\dfrac{713.49}{0.001}$ → 3 places

b) write the quotient by moving the decimal point in the dividend that number of places to the right.

$$\frac{713.49}{0.001}, \quad 713.490. \quad \frac{713.49}{0.001} = 713{,}490$$

3 places to the right

EXAMPLE 8 Divide: $\dfrac{23.738}{0.001}$.

$$\frac{23.738}{0.001}, \quad 23.738. \quad \frac{23.738}{0.001} = 23{,}738$$

3 places   3 places to the right to change 0.001 to 1

Do Exercises 11–14.

b Solving Equations

Now let's solve equations of the type $a \cdot x = b$, where a and b may be in decimal notation. Proceeding as before, we divide by a on both sides.

EXAMPLE 9 Solve: $8 \cdot x = 27.2$.

We have

$$\frac{8 \cdot x}{8} = \frac{27.2}{8} \qquad \text{Dividing by 8 on both sides}$$

$$x = 3.4.$$

$$\begin{array}{r} 3.4 \\ 8\,\overline{)\,27.2} \\ 24\,0 \\ \hline 3\,2 \\ 3\,2 \\ \hline 0 \end{array}$$

The solution is 3.4.

EXAMPLE 10 Solve: $2.9 \cdot t = 0.14616$.

We have

$$\frac{2.9 \cdot t}{2.9} = \frac{0.14616}{2.9} \quad \text{Dividing by 2.9 on both sides}$$

$$t = 0.0504.$$

```
         0.0 5 0 4
2.9 ) 0.1^4 6 1 6
        1 4 5 0 0
        ---------
            1 1 6
            1 1 6
            -----
                0
```

The solution is 0.0504.

Do Exercises 15 and 16.

C Order of Operations: Decimal Notation

The same rules for order of operations used with whole numbers and fraction notation apply when simplifying expressions with decimal notation.

RULES FOR ORDER OF OPERATIONS

1. Do all calculations within grouping symbols before operations outside.
2. Evaluate all exponential expressions.
3. Do all multiplications and divisions in order from left to right.
4. Do all additions and subtractions in order from left to right.

EXAMPLE 11 Simplify: $2.56 \times 25.6 \div 25{,}600 \times 256$.

There are no exponents or parentheses, so we multiply and divide from left to right:

$$\begin{aligned} 2.56 \times 25.6 \div 25{,}600 \times 256 &= 65.536 \div 25{,}600 \times 256 \\ &= 0.00256 \times 256 \\ &= 0.65536. \end{aligned}$$

Doing all multiplications and divisions in order from left to right

EXAMPLE 12 Simplify: $(5 - 0.06) \div 2 + 3.42 \times 0.1$.

$$(5 - 0.06) \div 2 + 3.42 \times 0.1 = 4.94 \div 2 + 3.42 \times 0.1$$ Carrying out the operation inside parentheses

$$= 2.47 + 0.342$$

$$= 2.812$$ Doing all multiplications and divisions in order from left to right

Solve.

15. $100 \cdot x = 78.314$

16. $0.25 \cdot y = 276.4$

Answers on page A-9

Simplify.

17. $625 \div 62.5 \times 25 \div 6250$

18. $0.25 \cdot (1 + 0.08) - 0.0274$

19. $20^2 - 3.4^2 + \{2.5[20(9.2 - 5.6)] + 5(10 - 5)\}$

20. Mountains in Peru. Refer to the figure in Example 14. Find the average height of the mountains, in meters.

Answers on page A-9

EXAMPLE 13 Simplify: $10^2 \times \{[(3 - 0.24) \div 2.4] - (0.21 - 0.092)\}$.

$10^2 \times \{[(3 - 0.24) \div 2.4] - (0.21 - 0.092)\}$

$= 10^2 \times \{[2.76 \div 2.4] - 0.118\}$ Doing the calculations in the innermost parentheses first

$= 10^2 \times \{1.15 - 0.118\}$ Again, doing the calculations in the innermost parentheses

$= 10^2 \times 1.032$ Subtracting inside the parentheses

$= 100 \times 1.032$ Evaluating the exponential expression

$= 103.2$

Do Exercises 17–19.

EXAMPLE 14 *Mountains in Peru.* The following figure shows a range of very high mountains in Peru, together with their altitudes, given both in feet and in meters. Find the average height of these mountains, in feet.

Source: *National Geographic,* July 1968, p. 130

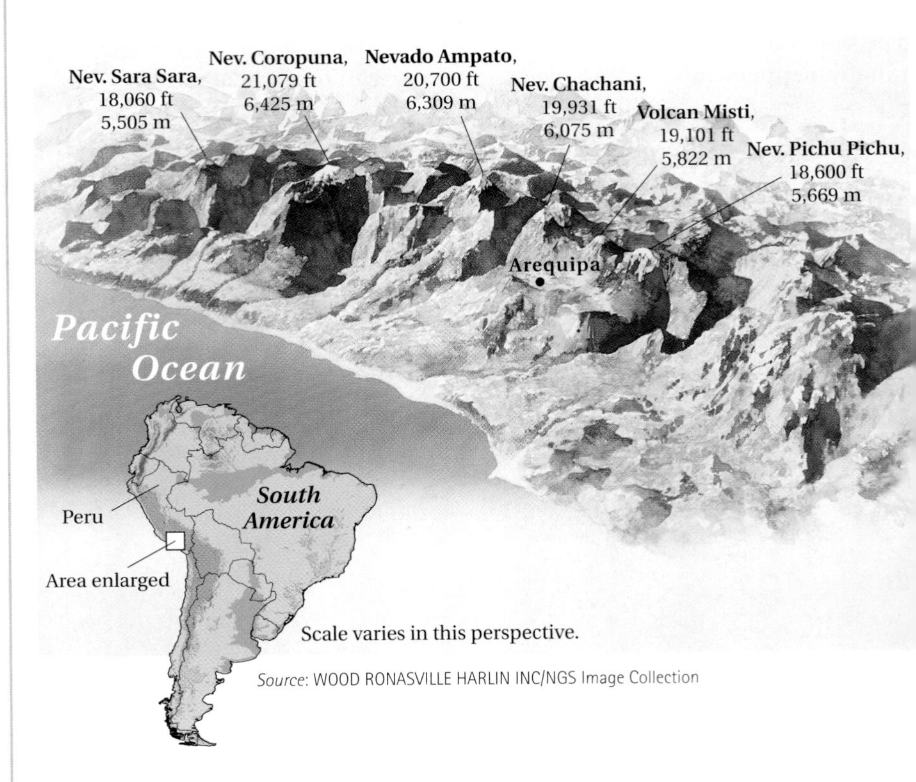

Source: WOOD RONASVILLE HARLIN INC/NGS Image Collection

The **average** of a set of numbers is the sum of the numbers divided by the number of addends. (See Section 1.9.) We find the sum of the heights divided by the number of addends, 6:

$$\frac{18{,}060 + 21{,}079 + 20{,}700 + 19{,}931 + 19{,}101 + 18{,}600}{6} = \frac{117{,}471}{6} = 19{,}578.5.$$

Thus the average height of these mountains is 19,578.5 ft.

Do Exercise 20.

4.4 EXERCISE SET

For Extra Help

MathXL

MyMathLab

InterAct Math

Math Tutor Center

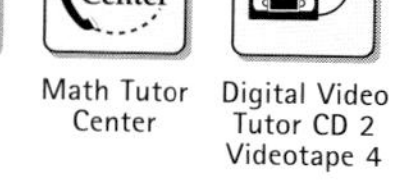

Digital Video Tutor CD 2 Videotape 4

Student's Solutions Manual

a Divide.

1. $2\overline{)5.98}$

2. $5\overline{)18}$

3. $4\overline{)95.12}$

4. $8\overline{)25.92}$

5. $12\overline{)89.76}$

6. $23\overline{)25.07}$

7. $33\overline{)237.6}$

8. $12.4 \div 4$

9. $9.144 \div 8$

10. $4.5 \div 9$

11. $12.123 \div 3$

12. $7\overline{)5.6}$

13. $5\overline{)0.35}$

14. $0.04\overline{)1.68}$

15. $0.12\overline{)8.4}$

16. $0.36\overline{)2.88}$

17. $3.4\overline{)68}$

18. $0.25\overline{)5}$

19. $15\overline{)6}$

20. $12\overline{)1.8}$

21. $36\overline{)14.76}$

22. $52\overline{)119.6}$

23. $3.2\overline{)27.2}$

24. $8.5\overline{)27.2}$

25. $4.2\overline{)39.06}$

26. $4.8\overline{)0.1104}$

27. $8\overline{)5}$

28. $8\overline{)3}$

29. $0.47\overline{)0.1222}$

30. $1.08\overline{)0.54}$

31. $4.8\overline{)75}$

32. $0.28\overline{)63}$

33. $0.032\overline{)0.07488}$

34. $0.017\overline{)1.581}$

35. $82\overline{)38.54}$

36. $34\overline{)0.1462}$

37. $\dfrac{213.4567}{1000}$

38. $\dfrac{213.4567}{100}$

39. $\dfrac{213.4567}{10}$

40. $\dfrac{100.7604}{0.1}$

41. $\dfrac{1.0237}{0.001}$

42. $\dfrac{1.0237}{0.01}$

b Solve.

43. $4.2 \cdot x = 39.06$

44. $36 \cdot y = 14.76$

45. $1000 \cdot y = 9.0678$

46. $789.23 = 0.25 \cdot q$

47. $1048.8 = 23 \cdot t$

48. $28.2 \cdot x = 423$

c Simplify.

49. $14 \times (82.6 + 67.9)$

50. $(26.2 - 14.8) \times 12$

51. $0.003 + 3.03 \div 0.01$

52. $9.94 + 4.26 \div (6.02 - 4.6) - 0.9$

53. $42 \times (10.6 + 0.024)$

54. $(18.6 - 4.9) \times 13$

55. $4.2 \times 5.7 + 0.7 \div 3.5$

56. $123.3 - 4.24 \times 1.01$

57. $9.0072 + 0.04 \div 0.1^2$

58. $12 \div 0.03 - 12 \times 0.03^2$

59. $(8 - 0.04)^2 \div 4 + 8.7 \times 0.4$

60. $(5 - 2.5)^2 \div 100 + 0.1 \times 6.5$

61. $86.7 + 4.22 \times (9.6 - 0.03)^2$

62. $2.48 \div (1 - 0.504) + 24.3 - 11 \times 2$

63. $4 \div 0.4 + 0.1 \times 5 - 0.1^2$

64. $6 \times 0.9 + 0.1 \div 4 - 0.2^3$

65. $5.5^2 \times [(6 - 4.2) \div 0.06 + 0.12]$

66. $12^2 \div (12 + 2.4) - [(2 - 1.6) \div 0.8]$

67. $200 \times \{[(4 - 0.25) \div 2.5] - (4.5 - 4.025)\}$

68. $0.03 \times \{1 \times 50.2 - [(8 - 7.5) \div 0.05]\}$

69. Find the average of $1276.59, $1350.49, $1123.78, and $1402.58.

70. Find the average weight of two wrestlers who weigh 308 lb and 296.4 lb.

71. *Income Taxes.* The amount paid in United States individual income taxes in a 5-year period is shown in the bar graph below. Find the average amount paid per year in individual income tax over the 5-year period.

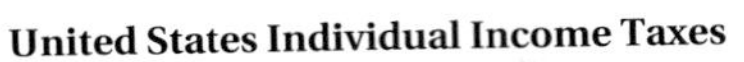

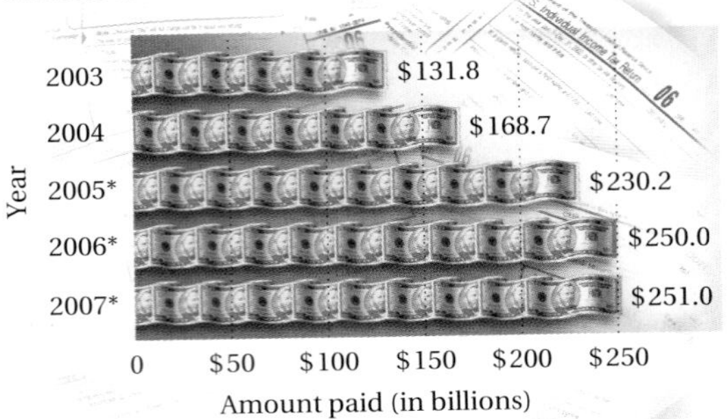

*Figures are estimates.

Sources: Department of the Treasury; Office of Management and Budget

72. *Herbal Supplement.* Biologically based therapies in complementary and alternative medicine use substances found in nature such as herbs, food, and vitamins. The graph below shows spending on the herbal supplement echinacea, also known as purple coneflower, which is thought to strengthen the immune system. Find the average amount spent per year from 1999 to 2003.

Spending on Herbal Supplement Echinacea

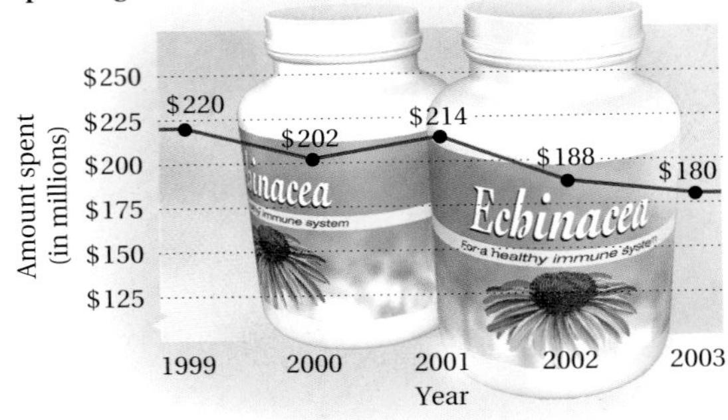

Source: *Nutrition Business Journal*

73. $\mathbf{D_W}$ How is division with decimal notation similar to division of whole numbers? How is it different?

74. $\mathbf{D_W}$ Kayla made these two computational mistakes:

$0.247 \div 0.1 = 0.0247$; $\quad 0.247 \div 10 = 2.47$.

In each case, how could you convince her that a mistake has been made?

SKILL MAINTENANCE

Simplify. [2.5b]

75. $\dfrac{36}{42}$

76. $\dfrac{56}{64}$

77. $\dfrac{38}{146}$

78. $\dfrac{114}{438}$

Find the prime factorization. [2.1d]

79. 684

80. 162

81. 2007

82. 2005

83. Add: $10\frac{1}{2} + 4\frac{5}{8}$. [3.5a]

84. Subtract: $10\frac{1}{2} - 4\frac{5}{8}$. [3.5b]

SYNTHESIS

Simplify.

85. 🖩 $9.0534 - 2.041^2 \times 0.731 \div 1.043^2$

86. 🖩 $23.042(7 - 4.037 \times 1.46 - 0.932^2)$

In Exercises 87–90, find the missing value.

87. $439.57 \times 0.01 \div 1000 \times \square = 4.3957$

88. $5.2738 \div 0.01 \times 1000 \div \square = 52.738$

89. $0.0329 \div 0.001 \times 10^4 \div \square = 3290$

90. $0.0047 \times 0.01 \div 10^4 \times \square = 4.7$

4.5 CONVERTING FROM FRACTION NOTATION TO DECIMAL NOTATION

Objectives

a Convert from fraction notation to decimal notation.

b Round numbers named by repeating decimals in problem solving.

c Calculate using fraction and decimal notation together.

a Fraction Notation to Decimal Notation

When a denominator has no prime factors other than 2's and 5's, we can find decimal notation by multiplying by 1. We multiply to get a denominator that is a power of ten, like 10, 100, or 1000.

EXAMPLE 1 Find decimal notation for $\frac{3}{5}$.

$$\frac{3}{5} = \frac{3}{5} \cdot \frac{2}{2} = \frac{6}{10} = 0.6$$

We use $\frac{2}{2}$ for 1 to get a denominator of 10.

EXAMPLE 2 Find decimal notation for $\frac{7}{20}$.

$$\frac{7}{20} = \frac{7}{20} \cdot \frac{5}{5} = \frac{35}{100} = 0.35$$

We use $\frac{5}{5}$ for 1 to get a denominator of 100.

EXAMPLE 3 Find decimal notation for $\frac{87}{25}$.

$$\frac{87}{25} = \frac{87}{25} \cdot \frac{4}{4} = \frac{348}{100} = 3.48$$

We use $\frac{4}{4}$ for 1 to get a denominator of 100.

EXAMPLE 4 Find decimal notation for $\frac{9}{40}$.

$$\frac{9}{40} = \frac{9}{40} \cdot \frac{25}{25} = \frac{225}{1000} = 0.225$$

We use $\frac{25}{25}$ for 1 to get a denominator of 1000.

Do Exercises 1–4.

We can always divide to find decimal notation.

EXAMPLE 5 Find decimal notation for $\frac{3}{5}$.

$$\frac{3}{5} = 3 \div 5$$

$$\begin{array}{r} 0.6 \\ 5\,\overline{)\,3.0} \\ \underline{3\,0} \\ 0 \end{array}$$

$$\frac{3}{5} = 0.6$$

EXAMPLE 6 Find decimal notation for $\frac{7}{8}$.

$$\frac{7}{8} = 7 \div 8$$

$$\begin{array}{r} 0.8\,7\,5 \\ 8\,\overline{)\,7.0\,0\,0} \\ \underline{6\,4} \\ 6\,0 \\ \underline{5\,6} \\ 4\,0 \\ \underline{4\,0} \\ 0 \end{array}$$

$$\frac{7}{8} = 0.875$$

Do Exercises 5 and 6.

Find decimal notation. Use multiplying by 1.

1. $\frac{4}{5}$

2. $\frac{9}{20}$

3. $\frac{11}{40}$

4. $\frac{33}{25}$

Find decimal notation.

5. $\frac{2}{5}$

6. $\frac{3}{8}$

Answers on page A-10

Find decimal notation.

7. $\frac{1}{6}$

8. $\frac{2}{3}$

Find decimal notation.

9. $\frac{5}{11}$

10. $\frac{12}{11}$

Answers on page A-10

In Examples 5 and 6, the division *terminated,* meaning that eventually we got a remainder of 0. A **terminating decimal** occurs when the denominator has only 2's or 5's, or both, as factors, as in $\frac{17}{25}$, $\frac{5}{8}$, or $\frac{83}{100}$. This assumes that the fraction notation has been simplified.

Consider a different situation:

$$\frac{5}{6}, \quad \text{or} \quad \frac{5}{2 \cdot 3}.$$

Since 6 has a 3 as a factor, the division will not terminate. Although we can still use division to get decimal notation, the answer will be a **repeating decimal,** as follows.

EXAMPLE 7 Find decimal notation for $\frac{5}{6}$.

$$\frac{5}{6} = 5 \div 6 \qquad 6\overline{)5.000}$$

```
      0.8 3 3
  6 ) 5.0 0 0
      4 8
        2 0
        1 8
          2 0
          1 8
            2
```

Since 2 keeps reappearing as a remainder, the digits repeat and will continue to do so; therefore,

$$\frac{5}{6} = 0.83333\ldots.$$

The red dots indicate an endless sequence of digits in the quotient. When there is a repeating pattern, the dots are often replaced by a bar to indicate the repeating part—in this case, only the 3:

$$\frac{5}{6} = 0.8\overline{3}.$$

Do Exercises 7 and 8.

EXAMPLE 8 Find decimal notation for $\frac{4}{11}$.

$$\frac{4}{11} = 4 \div 11$$

```
        0.3 6 3 6
  1 1 ) 4.0 0 0 0
        3 3
          7 0
          6 6
            4 0
            3 3
              7 0
              6 6
                4
```

Since 7 and 4 keep repeating as remainders, the sequence of digits "36" repeats in the quotient, and

$$\frac{4}{11} = 0.363636\ldots, \quad \text{or} \quad 0.\overline{36}.$$

Do Exercises 9 and 10.

EXAMPLE 9 Find decimal notation for $\frac{5}{7}$.

$$\begin{array}{r} 0.714285 \\ 7\,\overline{)\,5.000000} \\ \underline{4\,9} \\ 10 \\ \underline{7} \\ 30 \\ \underline{28} \\ 20 \\ \underline{14} \\ 60 \\ \underline{56} \\ 40 \\ \underline{35} \\ 5 \end{array}$$

Since 5 appears as a remainder, the sequence of digits "714285" repeats in the quotient, and

$$\frac{5}{7} = 0.714285714285\ldots, \quad \text{or} \quad 0.\overline{714285}.$$

The length of a repeating part can be very long—too long to find on a calculator. An example is $\frac{5}{97}$, which has a repeating part of 96 digits.

Do Exercise 11.

11. Find decimal notation for $\frac{3}{7}$.

b Rounding in Problem Solving

In applied problems, repeating decimals are rounded to get approximate answers. To round a repeating decimal, we can extend the decimal notation at least one place past the rounding digit, and then round as before.

EXAMPLES Round each of the following to the nearest tenth, hundredth, and thousandth.

	Nearest tenth	*Nearest hundredth*	*Nearest thousandth*
10. $0.8\overline{3} = 0.83333\ldots$	0.8	0.83	0.833
11. $0.\overline{09} = 0.090909\ldots$	0.1	0.09	0.091
12. $0.\overline{714285} = 0.714285714285\ldots$	0.7	0.71	0.714

Do Exercises 12–14.

Round each to the nearest tenth, hundredth, and thousandth.

12. $0.\overline{6}$

13. $0.\overline{80}$

14. $6.\overline{245}$

Answers on page A-10

CONVERTING RATIOS TO DECIMAL NOTATION

When solving applied problems, we often convert ratios to decimal notation.

EXAMPLE 13 *Forest Fires.* In October 2003, 15 devastating forest fires burned in Southern California for two weeks, burning 800,000 acres. Find the ratio of number of acres burned to number of fires and convert it to decimal notation. Round to the nearest thousandth.

Source: National Forest Service

15. Coin Tossing. A coin is tossed 51 times. It lands heads 26 times. Find the ratio of heads to tosses and convert it to decimal notation. Round to the nearest thousandth. (This is also the experimental probability of getting heads.)

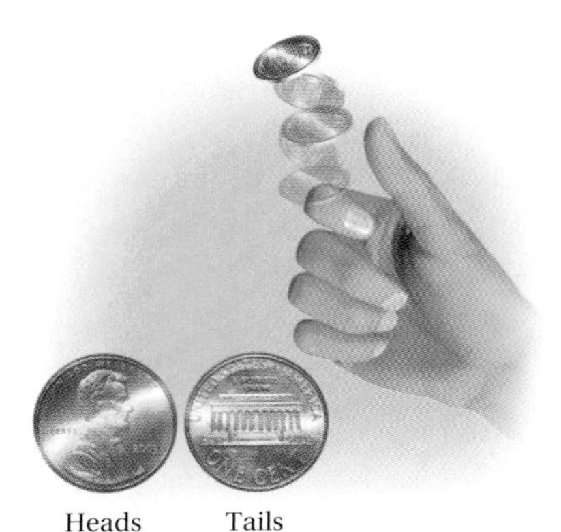

Heads Tails

16. Gas Mileage. A car goes 380 mi on 15.7 gal of gasoline. Find the gasoline mileage and convert the ratio to decimal notation rounded to the nearest tenth.

17. Camera Sales. Refer to the data on sales of digital cameras in Example 15. Find the average sales of these cameras for the 6-yr period.

Source: Photo Marketing Association International

Answers on page A-10

We have

$$\frac{\text{Acres burned}}{\text{Number of fires}} = \frac{800{,}000 \text{ acres}}{15 \text{ fires}} \approx 53{,}333.\overline{3}.$$

There were about 53,333.333 acres burned per fire.

EXAMPLE 14 *Gas Mileage.* A car goes 457 mi on 16.4 gal of gasoline. The ratio of number of miles driven to amount of gasoline used is *gas mileage.* Find the gas mileage and convert the ratio to decimal notation rounded to the nearest tenth.

$$\frac{\text{Miles driven}}{\text{Gasoline used}} = \frac{457}{16.4} \approx 27.86 \qquad \text{Dividing to 2 decimal places}$$

$$\approx 27.9 \qquad \text{Rounding to 1 decimal place}$$

The gas mileage is 27.9 miles to the gallon.

Do Exercises 15 and 16.

AVERAGES

When finding an average, we may at times need to round an answer.

EXAMPLE 15 *Camera Sales.* As sales of digital cameras have increased, analog camera sales have decreased. The bar graph below shows the sales of both types of cameras from 2000 to 2005, in millions of cameras. Find the average sales of analog cameras for this period.

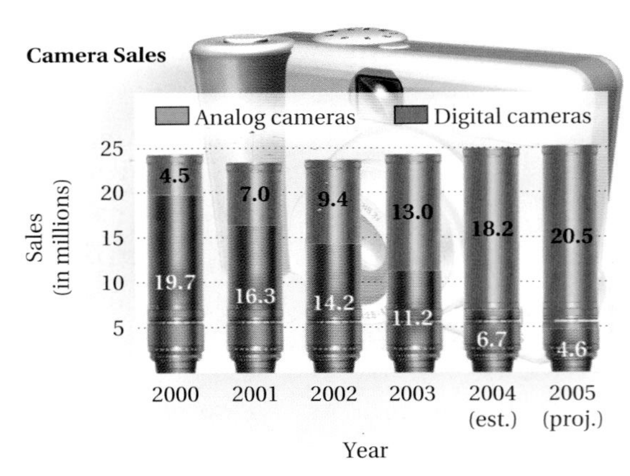

Source: Photo Marketing Association International

We add the sales totals shown on the bar graph and divide by the number of addends, 6. Since all the units are in millions, we need not convert them to standard notation. The average is

$$\frac{19.7 + 16.3 + 14.2 + 11.2 + 6.7 + 4.6}{6} = \frac{72.7}{6} \approx 12.12.$$

The average number of analog cameras sold per year for the 6-yr period is about 12.12 million.

Do Exercise 17.

C Calculations with Fraction and Decimal Notation Together

In certain kinds of calculations, fraction and decimal notation might occur together. In such cases, there are at least three ways in which we might proceed.

EXAMPLE 16 Calculate: $\frac{2}{3} \times 0.576$.

METHOD 1 One way to do this calculation is to convert the fraction notation to decimal notation so that both numbers are in decimal notation. Since $\frac{2}{3}$ converts to repeating decimal notation, it is first rounded to some chosen decimal place. We choose three decimal places. Then, using decimal notation, we multiply.

$$\frac{2}{3} \times 0.576 = 0.\overline{6} \times 0.576 \approx 0.667 \times 0.576 = 0.384192$$

METHOD 2 A second way to do this calculation is to convert the decimal notation to fraction notation so that both numbers are in fraction notation. The answer can be left in fraction notation and simplified, or we can convert to decimal notation and round, if appropriate.

$$\begin{aligned}\frac{2}{3} \times 0.576 &= \frac{2}{3} \cdot \frac{576}{1000} = \frac{2 \cdot 576}{3 \cdot 1000}\\ &= \frac{2 \cdot 2 \cdot 2 \cdot 2 \cdot 2 \cdot 2 \cdot 2 \cdot 3 \cdot 3}{2 \cdot 2 \cdot 2 \cdot 3 \cdot 5 \cdot 5 \cdot 5}\\ &= \frac{2 \cdot 2 \cdot 2 \cdot 3}{2 \cdot 2 \cdot 2 \cdot 3} \cdot \frac{2 \cdot 2 \cdot 2 \cdot 2 \cdot 3}{5 \cdot 5 \cdot 5}\\ &= 1 \cdot \frac{2 \cdot 2 \cdot 2 \cdot 2 \cdot 3}{5 \cdot 5 \cdot 5}\\ &= \frac{2 \cdot 2 \cdot 2 \cdot 2 \cdot 3}{5 \cdot 5 \cdot 5} = \frac{48}{125}, \text{ or } 0.384\end{aligned}$$

METHOD 3 A third way to do this calculation is to treat 0.576 as $\frac{0.576}{1}$. Then we multiply 0.576 by 2, and divide the result by 3.

$$\frac{2}{3} \times 0.576 = \frac{2}{3} \times \frac{0.576}{1} = \frac{2 \times 0.576}{3} = \frac{1.152}{3} = 0.384$$

Note that we get an exact answer with methods 2 and 3, but method 1 gives an approximation since we rounded decimal notation for $\frac{2}{3}$.

Do Exercise 18.

EXAMPLE 17 Calculate: $\frac{2}{3} \times 0.576 + 3.287 \div \frac{4}{5}$.

We use the rules for order of operations, doing first the multiplication and then the division. Then we add.

$$\begin{aligned}\frac{2}{3} \times 0.576 + 3.287 \div \frac{4}{5} &= 0.384 + 3.287 \cdot \frac{5}{4}\\ &= 0.384 + 4.10875\\ &= 4.49275\end{aligned}$$

Method 3:
$\frac{2}{3} \times \frac{0.576}{1} = 0.384;$
$\frac{3.287}{1} \times \frac{5}{4} = 4.10875$

Do Exercises 19 and 20.

18. Calculate: $\frac{5}{6} \times 0.864$.

Calculate.

19. $\frac{1}{3} \times 0.384 + \frac{5}{8} \times 0.6784$

20. $\frac{5}{6} \times 0.864 + 14.3 \div \frac{8}{5}$

Answers on page A-10

4.5 EXERCISE SET

For Extra Help

MathXL | MyMathLab | InterAct Math | Math Tutor Center | Digital Video Tutor CD 2 Videotape 4 | Student's Solutions Manual

a Find decimal notation.

1. $\frac{23}{100}$
2. $\frac{9}{100}$
3. $\frac{3}{5}$
4. $\frac{19}{20}$
5. $\frac{13}{40}$
6. $\frac{3}{16}$
7. $\frac{1}{5}$
8. $\frac{4}{5}$
9. $\frac{17}{20}$
10. $\frac{11}{20}$
11. $\frac{3}{8}$
12. $\frac{7}{8}$
13. $\frac{39}{40}$
14. $\frac{31}{40}$
15. $\frac{13}{25}$
16. $\frac{61}{125}$
17. $\frac{2502}{125}$
18. $\frac{181}{200}$
19. $\frac{1}{4}$
20. $\frac{1}{2}$
21. $\frac{29}{25}$
22. $\frac{37}{25}$
23. $\frac{19}{16}$
24. $\frac{5}{8}$
25. $\frac{4}{15}$
26. $\frac{7}{9}$
27. $\frac{1}{3}$
28. $\frac{1}{9}$
29. $\frac{4}{3}$
30. $\frac{8}{9}$
31. $\frac{7}{6}$
32. $\frac{7}{11}$
33. $\frac{4}{7}$
34. $\frac{14}{11}$
35. $\frac{11}{12}$
36. $\frac{5}{12}$

b

37.–47. Odds. Round each answer of the odd-numbered Exercises 25–35 to the nearest tenth, hundredth, and thousandth.

38.–48. Evens. Round each answer of the even-numbered Exercises 26–36 to the nearest tenth, hundredth, and thousandth.

Round each to the nearest tenth, hundredth, and thousandth.

49. $0.\overline{18}$
50. $0.\overline{83}$
51. $0.2\overline{7}$
52. $3.5\overline{4}$

53. For this set of people, what is the ratio, in decimal notation rounded to the nearest thousandth, where appropriate, of:

a) women to the total number of people?
b) women to men?
c) men to the total number of people?
d) men to women?

54. For this set of nuts and bolts, what is the ratio, in decimal notation rounded to the nearest thousandth, where appropriate, of:

a) nuts to bolts?
b) bolts to nuts?
c) nuts to the total?
d) total number to nuts?

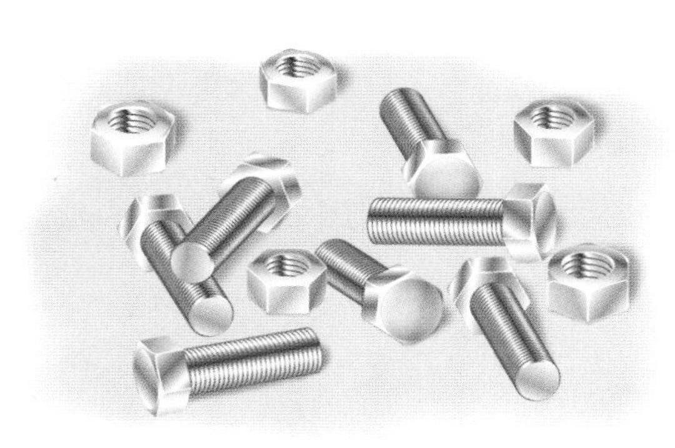

Gas Mileage. In each of Exercises 55–58, find the gas mileage rounded to the nearest tenth.

55. 285 mi; 18 gal

56. 396 mi; 17 gal

57. 324.8 mi; 18.2 gal

58. 264.8 mi; 12.7 gal

59. *Windy Cities.* Although nicknamed the Windy City, Chicago is not the windiest city in the United States. Listed in the table below are the six windiest cities and their average wind speeds. Find the average of these wind speeds and round your answer to the nearest tenth.

Source: *The Handy Geography Answer Book*

CITY	AVERAGE WIND SPEED (in miles per hour)
Mt. Washington, NH	35.3
Boston, MA	12.5
Honolulu, HI	11.3
Dallas TX	10.7
Kansas City, MO	10.7
Chicago, IL	10.4

60. *Areas of the New England States.* The table below lists the areas of the New England states. Find the average area and round your answer to the nearest tenth.

Source: *The New York Times Almanac*

STATE	TOTAL AREA (in square miles)
Maine	33,265
New Hampshire	9,279
Vermont	9,614
Massachusetts	8,284
Connecticut	5,018
Rhode Island	1,211

Stock Prices. At one time stock prices were given using mixed numerals involving halfs, fourths, eighths, and, more recently, sixteenths. The Securities and Exchange Commission has mandated the use of decimal notation. Thus a price of $\$23\frac{13}{16}$ is now converted to decimal notation rounded to the nearest hundredth, that is, \$23.81. Complete the following table.

Sources: *The Indianapolis Star,* 1/30/01; www.yahoo.com

	STOCK	PRICE PER SHARE	DECIMAL NOTATION	ROUNDED TO NEAREST HUNDREDTH
61.	Pfizer	$\$24\frac{9}{16}$		______
62.	Lucent	$\$3\frac{31}{64}$		______
63.	Qwest	$\$3\frac{47}{64}$		______
64.	Merck	$\$32\frac{5}{8}$		______
65.	Exxon Mobil	$\$59\frac{7}{8}$		______
66.	Hewlett Packard	$\$21\frac{7}{16}$		______

Source: New York Stock Exchange

C Calculate.

67. $\frac{7}{8} \times 12.64$

68. $\frac{4}{5} \times 384.8$

69. $2\frac{3}{4} + 5.65$

70. $4\frac{4}{5} + 3.25$

71. $\frac{47}{9} \times 79.95$

72. $\frac{7}{11} \times 2.7873$

73. $\frac{1}{2} - 0.5$

74. $3\frac{1}{8} - 2.75$

75. $4.875 - 2\frac{1}{16}$

76. $55\frac{3}{5} - 12.22$

77. $\frac{5}{6} \times 0.0765 + \frac{5}{4} \times 0.1124$

78. $\frac{3}{5} \times 6384.1 - \frac{3}{8} \times 156.56$

79. $\frac{4}{5} \times 384.8 + 24.8 \div \frac{8}{3}$

80. $102.4 \div \frac{2}{5} - 12 \times \frac{5}{6}$

81. $\frac{7}{8} \times 0.86 - 0.76 \times \frac{3}{4}$

82. $17.95 \div \frac{5}{8} + \frac{3}{4} \times 16.2$

83. $3.375 \times 5\frac{1}{3}$

84. $2.5 \times 3\frac{5}{8}$

85. $6.84 \div 2\frac{1}{2}$

86. $8\frac{1}{2} \div 2.125$

87. D_W When is long division *not* the fastest way to convert from fraction notation to decimal notation?

88. D_W Examine Example 16 of this section. How could the problem be changed so that method 1 would give a result that is completely accurate?

SKILL MAINTENANCE

Multiply. [3.6a]

89. $9 \cdot 2\frac{1}{3}$

90. $10\frac{1}{2} \cdot 22\frac{3}{4}$

Divide. [3.6b]

91. $84 \div 8\frac{2}{5}$

92. $8\frac{3}{5} \div 10\frac{2}{5}$

Add. [3.5a]

93. $17\frac{5}{6} + 32\frac{3}{8}$

94. $14\frac{3}{5} + 16\frac{1}{10}$

Subtract. [3.5b]

95. $16\frac{1}{10} - 14\frac{3}{5}$

96. $32\frac{3}{8} - 17\frac{5}{6}$

Solve. [3.5c]

97. A recipe for bread calls for $\frac{2}{3}$ cup of water, $\frac{1}{4}$ cup of milk, and $\frac{1}{8}$ cup of oil. How many cups of liquid ingredients does the recipe call for?

98. A board $\frac{7}{10}$ in. thick is glued to a board $\frac{3}{5}$ in. thick. The glue is $\frac{3}{100}$ in. thick. How thick is the result?

SYNTHESIS

Find decimal notation.

99. $\frac{1}{7}$

100. $\frac{2}{7}$

101. $\frac{3}{7}$

102. $\frac{4}{7}$

103. $\frac{5}{7}$

104. From the pattern of Exercises 99–103, guess the decimal notation for $\frac{6}{7}$. Check on your calculator.

Find decimal notation.

105. $\frac{1}{9}$

106. $\frac{1}{99}$

107. $\frac{1}{999}$

108. From the pattern of Exercises 105–107, guess the decimal notation for $\frac{1}{9999}$. Check on your calculator.

4.6 ESTIMATING

Objective

a Estimate sums, differences, products, and quotients.

a Estimating Sums, Differences, Products, and Quotients

Estimating has many uses. It can be done before a problem is even attempted in order to get an idea of the answer. It can be done afterward as a check, even when we are using a calculator. In many situations, an estimate is all we need. We usually estimate by rounding the numbers so that there are one or two nonzero digits, depending on how accurate we want our estimate. Consider the following advertisements for Examples 1–3.

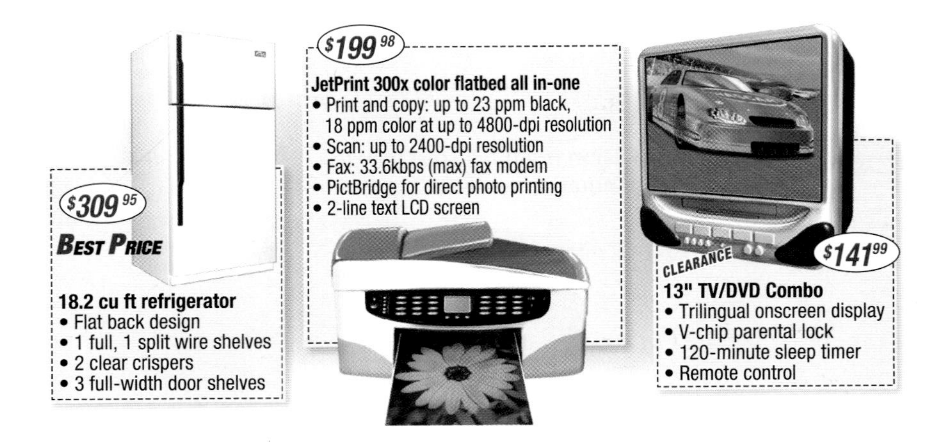

EXAMPLE 1 Estimate by rounding to the nearest ten the total cost of one all-in-one printer/copier and one TV/DVD combo.

We are estimating the sum

$\$199.98 + \$141.99 = \text{Total cost}.$

The estimate found by rounding the addends to the nearest ten is

$\$200 + \$140 = \$340.$ (Estimated total cost)

Do Exercise 1.

EXAMPLE 2 About how much more does the refrigerator cost than the TV/DVD combo? Estimate by rounding to the nearest ten.

We are estimating the difference

$\$309.95 - \$141.99 = \text{Price difference}.$

The estimate to the nearest ten is

$\$310 - \$140 = \$170.$ (Estimated price difference)

Do Exercise 2.

---

1. Estimate by rounding to the nearest ten the total cost of one refrigerator and one printer/copier. Which of the following is an appropriate estimate?

   a) \$500 b) \$510
   c) \$5100 d) \$51

2. About how much more does the printer/copier cost than the TV/DVD combo? Estimate by rounding to the nearest ten. Which of the following is an appropriate estimate?

   a) \$60 b) \$600
   c) \$340 d) \$6000

Answers on page A-10

Study Tips

ASKING QUESTIONS

Don't be afraid to ask questions in class. Most instructors welcome this and encourage students to ask them. Other students probably have the same questions you do.

"Better to ask twice than lose your way once."

Danish Proverb

EXAMPLE 3 Estimate the total cost of 4 TV/DVD combos.

We are estimating the product

$4 \times \$141.99 = \text{Total cost}.$

The estimate is found by rounding 141.99 to the nearest ten:

$4 \times \$140 = \$560.$

Do Exercise 3.

EXAMPLE 4 About how many Xbox™ Online Ready Video Game Systems at $149.99 each can be purchased for $1480?

We estimate the quotient

$\$1480 \div \$149.99.$

Since we want a whole-number estimate, we choose our rounding appropriately. Rounding $149.99 to the nearest one, we get $150. Since $1480 is close to $1500, which is a multiple of 150, we estimate

$\$1500 \div \$150,$

so the answer is 10.

Do Exercise 4.

EXAMPLE 5 Estimate: 4.8×52. Do not find the actual product. Which of the following is an appropriate estimate?

a) 25 **b)** 250 **c)** 2500 **d)** 360

We round 4.8 to the nearest one and 52 to the nearest ten.

$5 \times 50 = 250.$ (Estimated product)

Thus an approximate estimate is (b).

Other estimates we might have used in Example 5 are

$5 \times 52 = 260$ or $4.8 \times 50 = 240.$

The estimate in Example 5, $5 \times 50 = 250$, is the easiest to do because it has the fewest nonzero digits. You could probably do it mentally. In general, we try to round so that a computation has as few nonzero digits as possible while still keeping the estimated value close to the original value.

Do Exercises 5–10.

3. Estimate the total cost of 6 refrigerators. Which of the following is an appropriate estimate?

a) $1200 b) $186
c) $18,600 d) $1860

4. About how many Xbox™ Game Systems can be purchased for $4530? Which of the following is an appropriate estimate?

a) 90 b) 30
c) 300 d) 45

Estimate the product. Do not find the actual product. Which of the following is an appropriate estimate?

5. 2.4×8

a) 16 b) 34
c) 125 d) 5

6. 24×0.6

a) 200 b) 5
c) 110 d) 20

7. 0.86×0.432

a) 0.04 b) 0.4
c) 1.1 d) 4

8. 0.82×0.1

a) 800 b) 8
c) 0.08 d) 80

9. 0.12×18.248

a) 180 b) 1.8
c) 0.018 d) 18

10. 24.234×5.2

a) 200 b) 120
c) 12.5 d) 234

Answers on page A-10

EXAMPLE 6 Estimate: $82.08 \div 24$. Which of the following is an appropriate estimate?

a) 400 b) 16 c) 40 d) 4

This is about $80 \div 20$, so the answer is about 4. Thus an appropriate estimate is (d).

EXAMPLE 7 Estimate: $94.18 \div 3.2$. Which of the following is an appropriate estimate?

a) 30 b) 300 c) 3 d) 60

This is about $90 \div 3$, so the answer is about 30. Thus an appropriate estimate is (a).

EXAMPLE 8 Estimate: $0.0156 \div 1.3$. Which of the following is an appropriate estimate?

a) 0.2 b) 0.002 c) 0.02 d) 20

This is about $0.02 \div 1$, so the answer is about 0.02. Thus an appropriate estimate is (c).

Do Exercises 11–13.

In some cases, it is easier to estimate a quotient directly rather than by rounding the divisor and the dividend.

EXAMPLE 9 Estimate: $0.0074 \div 0.23$. Which of the following is an appropriate estimate?

a) 0.3 b) 0.03 c) 300 d) 3

We estimate 3 for a quotient. We check by multiplying.

$$0.23 \times 3 = 0.69$$

We make the estimate smaller. We estimate 0.3 and check by multiplying.

$$0.23 \times 0.3 = 0.069$$

We make the estimate smaller. We estimate 0.03 and check by multiplying.

$$0.23 \times 0.03 = 0.0069$$

This is about 0.0074, so the quotient is about 0.03. Thus an appropriate estimate is (b).

Do Exercise 14.

Estimate the quotient. Which of the following is an appropriate estimate?

11. $59.78 \div 29.1$

a) 200 b) 20
c) 2 d) 0.2

12. $82.08 \div 2.4$

a) 40 b) 4.0
c) 400 d) 0.4

13. $0.1768 \div 0.08$

a) 8 b) 10
c) 2 d) 20

14. Estimate: $0.0069 \div 0.15$. Which of the following is an appropriate estimate?

a) 0.5 b) 50
c) 0.05 d) 0.004

Answers on page A-10

EXERCISE SET

For Extra Help

MathXL MyMathLab InterAct Math Math Tutor Center Digital Video Tutor CD 2 Videotape 4 Student's Solutions Manual

a Consider the following advertisements for Exercises 1–8. Estimate the sums, differences, products, or quotients involved in these problems. Indicate which of the choices is an appropriate estimate.

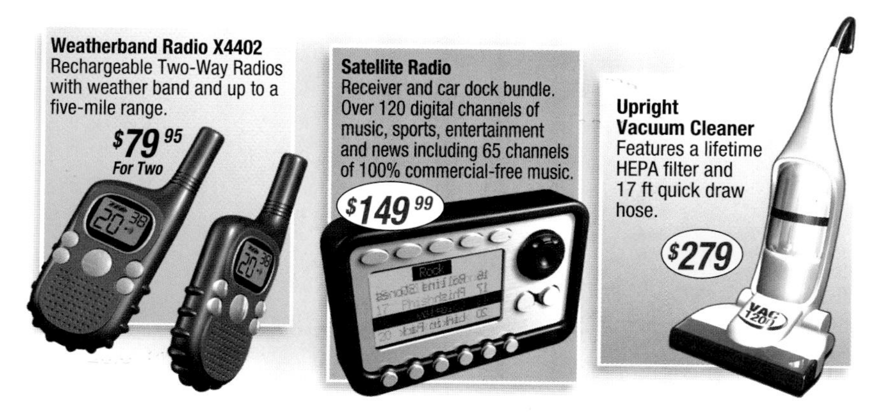

Weatherband Radio X4402
Rechargeable Two-Way Radios with weather band and up to a five-mile range.
$79.95
For Two
Satellite Radio
Receiver and car dock bundle. Over 120 digital channels of music, sports, entertainment and news including 65 channels of 100% commercial-free music.
$149.99
Upright Vacuum Cleaner
Features a lifetime HEPA filter and 17 ft quick draw hose.
$279

1. Estimate the total cost of one vacuum cleaner and one satellite radio.
a) \$43 **b)** \$4300 **c)** \$360 **d)** \$430

2. Estimate the total cost of one satellite radio and one set of two-way radios.
a) \$230 **b)** \$23 **c)** \$2300 **d)** \$400

3. About how much more does the vacuum cleaner cost than the satellite radio?
a) \$1300 **b)** \$200 **c)** \$130 **d)** \$13

4. About how much more does the satellite radio cost than the two-way radios?
a) \$7000 **b)** \$70 **c)** \$130 **d)** \$700

5. Estimate the total cost of 6 sets of two-way radios.
a) \$480 **b)** \$48 **c)** \$240 **d)** \$4800

6. Estimate the total cost of 4 vacuum cleaners.
a) \$1200 **b)** \$1120 **c)** \$11,200 **d)** \$600

7. About how many sets of two-way radios can be purchased for \$830?
a) 120 **b)** 100 **c)** 10 **d)** 1000

8. About how many vacuum cleaners can be purchased for \$5627?
a) 200 **b)** 20 **c)** 1800 **d)** 2000

Estimate by rounding as directed.

9. $0.02 + 1.31 + 0.34$; nearest tenth

10. $0.88 + 2.07 + 1.54$; nearest one

11. $6.03 + 0.007 + 0.214$; nearest one

12. $1.11 + 8.888 + 99.94$; nearest one

13. $52.367 + 1.307 + 7.324$; nearest one

14. $12.9882 + 1.0115$; nearest tenth

15. 2.678 − 0.445; nearest tenth

16. 12.9882 − 1.0115; nearest one

17. 198.67432 − 24.5007; nearest ten

Estimate. Choose a rounding digit that gives one or two nonzero digits. Indicate which of the choices is an appropriate estimate.

18. 234.12321 − 200.3223
a) 600 b) 60
c) 300 d) 30

19. 49 × 7.89
a) 400 b) 40
c) 4 d) 0.4

20. 7.4 × 8.9
a) 95 b) 63
c) 124 d) 6

21. 98.4 × 0.083
a) 80 b) 12
c) 8 d) 0.8

22. 78 × 5.3
a) 400 b) 800
c) 40 d) 8

23. 3.6 ÷ 4
a) 10 b) 1
c) 0.1 d) 0.01

24. 0.0713 ÷ 1.94
a) 3.5 b) 0.35
c) 0.035 d) 35

25. 74.68 ÷ 24.7
a) 9 b) 3
c) 12 d) 120

26. 914 ÷ 0.921
a) 10 b) 100
c) 1000 d) 1

27. *Fence Posts* A zoo plans to construct a fence around its proposed African wildlife exhibit. The perimeter of the area to be fenced is 1760 ft. Estimate the number of wooden fence posts needed if the posts are placed 8.625 ft apart?

28. *Ticketmaster.* Recently, Ticketmaster stock sold for $8.63 per share. Estimate how many shares can be purchased for $27,000.

29. D_W Describe a situation in which an estimation is made by rounding to the nearest 10,000 and then multiplying.

30. D_W A roll of fiberglass insulation costs $21.95. Describe two situations involving estimating and the cost of fiberglass insulation. Devise one situation so that $21.95 is rounded to $22. Devise the other situation so that $21.95 is rounded to $20.

SKILL MAINTENANCE

VOCABULARY REINFORCEMENT

In each of Exercises 31–38, fill in the blank with the correct term from the given list. Some of the choices may not be used and some may be used more than once.

31. The decimal $0.57\overline{3}$ is an example of a ______________ decimal. [4.5a]

32. The least common ______________ of two natural numbers is the smallest number that is a multiple of both. [3.1a]

33. The sentence $5(3 + 8) = 5 \cdot 3 + 5 \cdot 8$ illustrates the ______________ law. [1.5a]

34. A ______________ of an equation is a replacement for the variable that makes the equation true. [1.7a]

35. The number 1 is the ______________ identity. [1.5a]

36. The sentence $13 + 7 = 7 + 13$ illustrates the ______________ law of addition. [1.2a]

37. The least common ______________ of two or more fractions is the least common ______________ of their denominators. [3.2a]

38. The number 3728 is ______________ by 4 if the number named by the last two digits is ______________ by 4. [2.2a]

additive
multiplicative
numerator
denominator
commutative
associative
distributive
solution
divisible
terminating
repeating
multiple
factor

SYNTHESIS

The following were done on a calculator. Estimate to determine whether the decimal point was placed correctly.

39. $178.9462 \times 61.78 = 11{,}055.29624$

40. $14{,}973.35 \div 298.75 = 501.2$

41. $19.7236 - 1.4738 \times 4.1097 = 1.366672414$

42. $28.46901 \div 4.9187 - 2.5081 = 3.279813473$

43. Use one of $+$, $-$, $\times$, and $\div$ in each blank to make a true sentence.

a) $(0.37 \square 18.78) \square 2^{13} = 156{,}876.8$

b) $2.56 \square 6.4 \square 51.2 \square 17.4 = 312.84$

4.7 APPLICATIONS AND PROBLEM SOLVING

Objective

a Solve applied problems involving decimals.

1. **Photofinishing Market.** Because an increasing number of people are printing their own digital photographs at home, the commercial photofinishing industry's revenue is declining. Between 1994 and 2005, the total revenue from film processing ranged in value from a low of \$3.7 billion to a high of \$6.2 billion. By how much did the high value differ from the low value?

Photofinishing Market

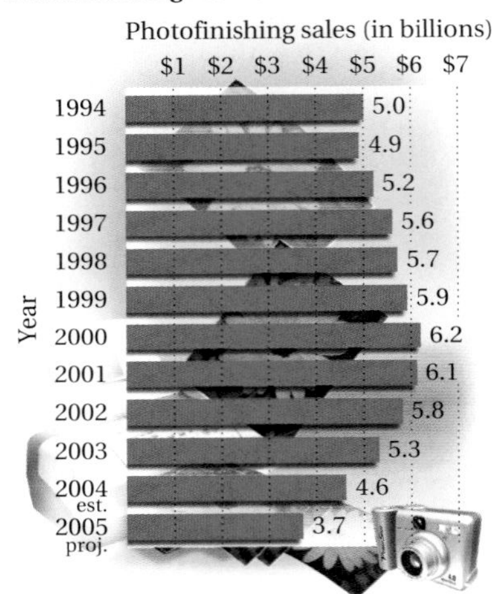

Source: PMA Marketing Research

Answer on page A-10

a Solving Applied Problems

Solving applied problems with decimals is like solving applied problems with whole numbers. We translate first to an equation that corresponds to the situation. Then we solve the equation.

EXAMPLE 1 *Canals.* The Panama Canal in Panama is 50.7 mi long. The Suez Canal in Egypt is 119.9 mi long. How much longer is the Suez Canal?

1. **Familiarize.** This is a "how much more" situation. We let l = the distance in miles that the length of the longer canal differs from the length of the shorter canal.
2. **Translate.** We translate as follows, using the given information.

Length of Panama Canal, the shorter canal	plus	Additional length	is	Length of Suez Canal, the longer canal
↓	↓	↓	↓	↓
50.7 mi	$+$	l	$=$	119.9 mi

3. **Solve.** We solve the equation by subtracting 50.7 mi on both sides:

$$50.7 + l = 119.9$$
$$50.7 + l - 50.7 = 119.9 - 50.7$$
$$l = 69.2.$$

4. **Check.** We can check by adding 69.2 to 50.7 to get 119.9.
5. **State.** The Suez Canal is 69.2 mi longer than the Panama Canal.

Do Exercise 1.

EXAMPLE 2 *Injections of Medication.* A patient was given injections of 2.8 mL, 1.35 mL, 2.0 mL, and 1.88 mL over a 24-hr period. What was the total amount of the injections?

1. **Familiarize.** We make a drawing or at least visualize the situation. We let $t =$ the amount of the injections.

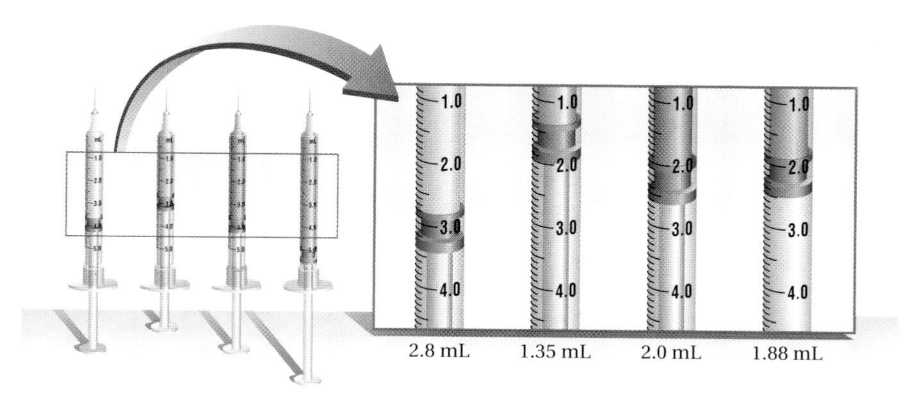

2. **Translate.** Amounts are being combined. We translate to an equation.

First	plus	Second	plus	Third	plus	Fourth	is	Total
↓	↓	↓	↓	↓	↓	↓	↓	↓
2.8	+	1.35	+	2.0	+	1.88	=	t

3. **Solve.** To solve, we carry out the addition.

$$\begin{array}{r} \scriptstyle 2\ \ 1\ \ \\ 2.8\,0 \\ 1.3\,5 \\ 2.0\,0 \\ +\ 1.8\,8 \\ \hline 8.0\,3 \end{array}$$

Thus, $t = 8.03$.

4. **Check.** We can check by repeating our addition. We can also see whether our answer is reasonable by first noting that it is indeed larger than any of the numbers being added. We can also partially check by rounding:

$$\begin{aligned} 2.8 + 1.35 + 2.0 + 1.88 &\approx 3 + 1 + 2 + 2 \\ &= 8 \approx 8.03. \end{aligned}$$

If we had gotten an answer like 80.3 or 0.803, then our estimate, 8, would have told us that we did something wrong, like not lining up the decimal points.

5. **State.** The total amount of the injections was 8.03 mL.

Do Exercise 2.

2. **Meat and Seafood Consumption.** In 2002, the average American consumed about 64.5 lb of beef, 48.2 lb of pork, 56.8 lb of chicken, 14.0 lb of turkey, and 15.6 lb of seafood. What is the total amount of meat and seafood that the average American consumed?

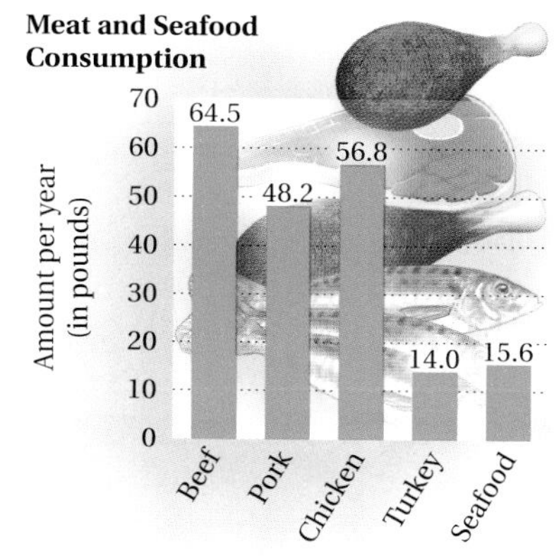

Source: *Statistical Abstract of the United States*, 2004

Answer on page A-10

3. **Printing Costs.** At a printing company, the cost of copying is 11 cents per page. How much, in dollars, would it cost to make 466 copies?

1D=100C.

EXAMPLE 3 *IRS Driving Allowance.* In 2004, the Internal Revenue Service allowed a tax deduction of 37.5¢ per mile for mileage driven for business purposes. What deduction, in dollars, would be allowed for driving 9143 mi?

Source: Internal Revenue Service

1. **Familiarize.** We first make a drawing or at least visualize the situation. Repeated addition fits this situation. We let $d =$ the deduction, in dollars, allowed for driving 9143 mi.

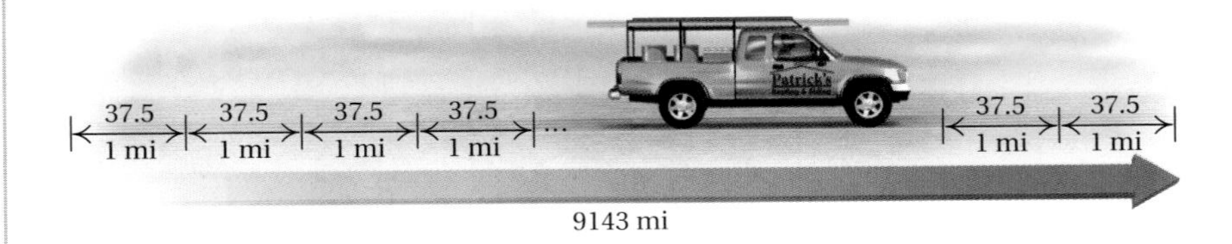

2. **Translate.** We translate as follows.

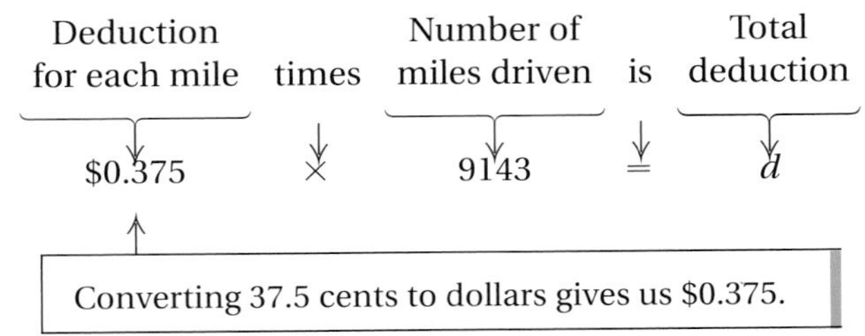

Deduction for each mile
times
Number of miles driven
is
Total deduction
$0.375
×
9143
=
d
Converting 37.5 cents to dollars gives us $0.375.

3. **Solve.** To solve the equation, we carry out the multiplication.

```
      9 1 4 3
×     0.3 7 5
-------------
    4 5 7 1 5
  6 4 0 0 1 0
2 7 4 2 9 0 0
-------------
3 4 2 8.6 2 5
```

Thus, $d = 3428.625 \approx \$3428.63$.

4. **Check.** We can obtain a partial check by rounding and estimating:

$$9143 \times 0.375 \approx 9000 \times 0.4 = 3600 \approx 3428.63.$$

5. **State.** The total allowable deduction would be $3428.63.

Do Exercise 3.

EXAMPLE 4 *Loan Payments.* A car loan of $7382.52 is to be paid off in 36 monthly payments. How much is each payment?

1. **Familiarize.** We first make a drawing. We let $n =$ the amount of each payment.

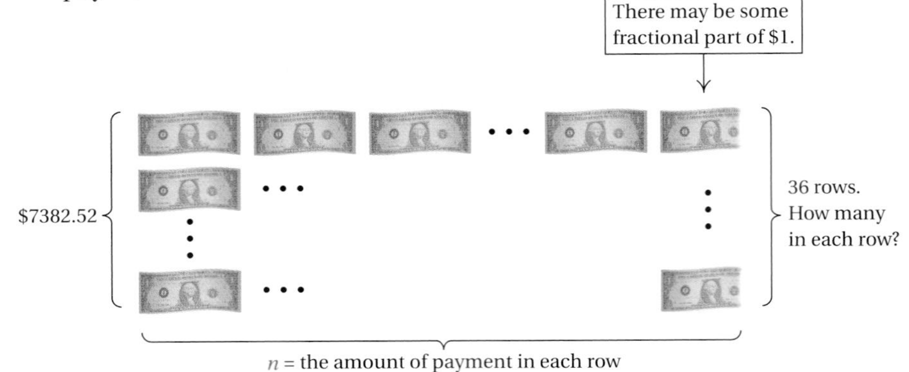

There may be some fractional part of $1.

36 rows. How many in each row?

Answer on page A-10

2. **Translate.** The problem can be translated to the following equation, thinking that

(Total loan) ÷ (Number of payments) = Amount of each payment

$$\$7382.52 \div 36 = n.$$

3. **Solve.** To solve the equation, we carry out the division.

$$\begin{array}{r} 205.07 \\ 36\,)\overline{7382.52} \\ \underline{720000} \\ 18252 \\ \underline{18000} \\ 252 \\ \underline{252} \\ 0 \end{array}$$

Thus, $n = 205.07$.

4. **Check.** A partial check can be obtained by estimating the quotient: $\$7382.56 \div 36 \approx 8000 \div 40 = 200 \approx 205.07$. The estimate checks.

5. **State.** Each payment is \$205.07.

Do Exercise 4.

4. **Loan Payments.** A loan of \$4425 is to be paid off in 12 monthly payments. How much is each payment?

Answer on page A-10

EXAMPLE 5 *Jackie Robinson Poster.* A limited-edition poster by sports artist Leroy Neiman commemorates Jackie Robinson's entrance into Major League Baseball in 1947. Robinson was the first African-American player in the major leagues. The dimensions of the poster are 19.3 in. by 27.4 in. Find the area.

Source: Barton L. Kaufman, private collection

1. **Familiarize.** We first make a drawing. We let A = the area.

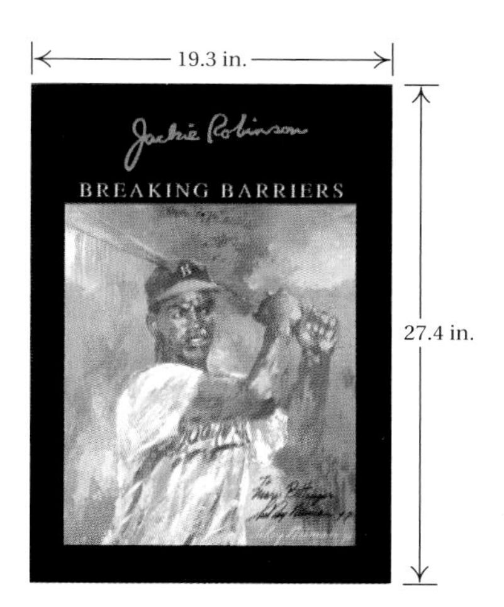

Study Tips

FIVE STEPS FOR PROBLEM SOLVING

Are you using the five steps for problem solving that were developed in Section 1.8?

1. **Familiarize** yourself with the situation.
   a) Carefully read and reread until you understand what you are being asked to find.
   b) Draw a diagram or see if there is a formula that applies.
   c) Assign a letter, or *variable,* to the unknown.
2. **Translate** the problem to an equation using the letter or variable.
3. **Solve** the equation.
4. **Check** the answer in the original wording of the problem.
5. **State** the answer to the problem clearly with appropriate units.

5. **Index Cards.** A standard-size index card measures 12.7 cm by 7.6 cm. Find its area.

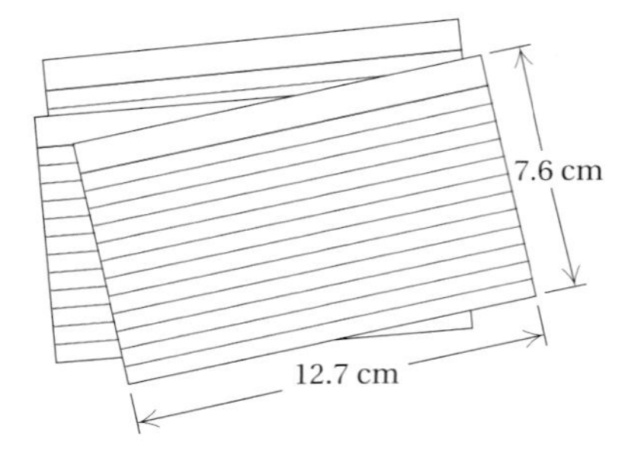

2. **Translate.** We use the formula $A = l \cdot w$ and substitute.

$$A = l \cdot w$$
$$A = 27.4 \times 19.3.$$

3. **Solve.** We solve by carrying out the multiplication.

```
      2 7.4
  ×   1 9.3
  ---------
      8 2 2
  2 4 6 6 0
  2 7 4 0 0
  ---------
  5 2 8.8 2
```

4. **Check.** We obtain a partial check by estimating the product:

$$A = 27.4 \times 19.3 \approx 25 \times 20 = 500.$$

This estimate is close to 528.82 so it's a good check.

5. **State.** The area of the Jackie Robinson poster is 528.82 in^2.

Do Exercise 5.

EXAMPLE 6 *Digital Camera Purchase.* Dawson Real Estate spent $10,399.74 on a set of 26 Nikon COOLPIX 5200 cameras, so that its realtors can place photos of properties on the firm's Web site. How much did each camera cost?

1. **Familiarize.** We let c = the cost of each camera.

2. **Translate.** We translate as follows.

Cost of each camera	is	Total cost of purchase	divided by	Number of cameras purchased
↓	↓	↓	↓	↓
c	$=$	$10,399.74	$\div$	26

3. **Solve.** To solve, we carry out the division.

```
           3 9 9.9 9
 2 6 ) 1 0,3 9 9.7 4
         7 8
         -----
         2 5 9
         2 3 4
         -----
           2 5 9
           2 3 4
           -----
             2 5 7
             2 3 4
             -----
               2 3 4
               2 3 4
               -----
                   0
```

4. **Check.** We check by estimating

$$10{,}399.74 \div 26 \approx 10{,}400 \div 25 = 416.$$

Since 416 is close to 399.99, the answer is probably correct.

5. **State.** The cost of each camera was $399.99.

Do Exercise 6.

6. One pound of lean boneless ham contains 4.5 servings. It costs $5.99 per pound. What is the cost per serving? Round to the nearest cent.

Answers on page A-10

MULTISTEP PROBLEMS

EXAMPLE 7 *Gas Mileage.* A driver filled the gasoline tank and noted that the odometer read 67,507.8. After the next filling, the odometer read 68,006.1. It took 16.5 gal to fill the tank. How many miles per gallon did the driver get?

1. **Familiarize.** We first make a drawing.

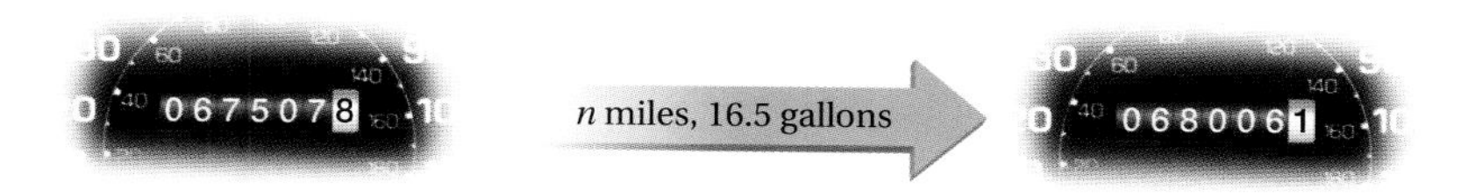

This is a two-step problem. First, we find the number of miles that have been driven between fillups. We let n = the number of miles driven.

2., 3. Translate and **Solve.** This is a "how many more" situation. We translate and solve as follows.

First odometer reading	plus	Number of miles driven	is	Second odometer reading
↓	↓	↓	↓	↓
67,507.8	+	n	=	68,006.1

To solve the equation, we subtract 67,507.8 on both sides:

$$n = 68,006.1 - 67,507.8 = 498.3.$$

$$\begin{array}{r} 68,006.1 \\ -\ 67,507.8 \\ \hline 498.3 \end{array}$$

Second, we divide the total number of miles driven by the number of gallons. This gives us m = the number of miles per gallon—that is, the mileage. The division that corresponds to the situation is

$$498.3 \div 16.5 = m.$$

To find the number m, we divide.

$$\begin{array}{r} 30.2 \\ 16.5\overline{)\,498.3_{\wedge}0} \\ \underline{495\,0} \\ 33\,0 \\ \underline{33\,0} \\ 0 \end{array}$$

Thus, $m = 30.2$.

4. **Check.** To check, we first multiply the number of miles per gallon times the number of gallons:

$$16.5 \times 30.2 = 498.3.$$

Then we add 498.3 to 67,507.8:

$$67,507.8 + 498.3 = 68,006.1.$$

The mileage 30.2 checks.

5. **State.** The driver got 30.2 miles per gallon.

Do Exercise 7.

7. **Gas Mileage.** A driver filled the gasoline tank and noted that the odometer read 38,320.8. After the next filling, the odometer read 38,735.5. It took 14.5 gal to fill the tank. How many miles per gallon did the driver get?

Answer on page A-10

EXAMPLE 8 *Home-Cost Comparison.* Suppose you own a home like the one shown here and it is valued at $225,000 in Tulsa, Oklahoma. What would it cost to buy a similar (replacement) home in Boston, Massachusetts? To find out, we can use an index table prepared by Coldwell Banker Real Estate Corporation. (For a complete index table, contact your local representative.) We use the following formula:

$$\begin{pmatrix}\text{Cost of your}\\ \text{home in new city}\end{pmatrix} = \frac{\begin{pmatrix}\text{Value of}\\ \text{your home}\end{pmatrix} \times \begin{pmatrix}\text{Index of}\\ \text{new city}\end{pmatrix}}{\begin{pmatrix}\text{Index of}\\ \text{your city}\end{pmatrix}}.$$

Find the cost of your Tulsa home in Boston. Round to the nearest one.

STATE	CITY	INDEX	STATE	CITY	INDEX
Massachusetts	Boston	297	Texas	Dallas	67
Illinois	Chicago	215	Oklahoma	Tulsa	39
California	Palo Alto Sacramento San Diego	342 98 171	Florida	Miami/Coral Gables Orlando Key West	143 69 214
Alaska	Juneau	119	Colorado	Boulder	125
Kentucky	Louisville	63	North Carolina	Charlotte	54

Source: Coldwell Banker Real Estate Corporation

Refer to the table in Example 8 to answer Margin Exercises 8 and 9.

8. **Home-Cost Comparison.** Find the cost of a $180,000 home in Charlotte if you were to try to replace it when moving to Chicago. Round to the nearest one.

9. Find the cost of a $315,000 home in Boulder if you were to try to replace it when moving to Louisville. Round to the nearest one.

1. **Familiarize.** We let $C =$ the cost of the home in Boston. We use the table and look up the indexes of the city in which you now live and the city to which you are moving.
2. **Translate.** Using the formula, we translate to the following equation:

$$C = \frac{\$225{,}000 \times 297}{39}.$$

3. **Solve.** To solve, we carry out the computations using the rules for order of operations (see Section 4.4).

$$\begin{aligned} C &= \frac{\$225{,}000 \times 297}{39} \\ &= \frac{66{,}825{,}000}{39} && \text{Carrying out the multiplication first} \\ &\approx \$1{,}713{,}462. && \text{Carrying out the division and rounding to the nearest one} \end{aligned}$$

On a calculator, the computation could be done in one step.

4. **Check.** We can repeat our computations.
5. **State.** A home that sells for $225,000 in Tulsa would cost about $1,713,462 in Boston.

Do Exercises 8 and 9.

Answers on page A-10

Translating for Success

. *Gas Mileage.* Art filled his SUV's gas tank and noted that the odometer read 38,271.8. At the next filling, the odometer read 38,677.92. It took 28.4 gal to fill the tank. How many miles per gallon did the SUV get?

. *Dimensions of a Parking Lot.* Seals' parking lot is a rectangle that measures 85.2 ft by 52.3 ft. What is the area of the parking lot?

Game Snacks. Three students pay $18.40 for snacks at a football game. What is each person's share?

Electrical Wiring. An electrician needs 1314 ft of wiring cut into $2\frac{1}{2}$-ft pieces. How many pieces will she have?

College Tuition. Wayne needs $4638 for the fall semester's tuition. On the day of registration, he had only $3092. How much does he need to borrow?

The goal of these matching questions is to practice step (2), *Translate,* of the five-step problem-solving process. Translate each word problem to an equation and select a correct translation from equations A–O.

A. $2\frac{1}{2} \cdot n = 1314$

B. $18.4 \times 1.87 = n$

C. $n = 85.2 \times 52.3$

D. $1314.28 - 437 = n$

E. $3 \times 18.40 = n$

F. $2\frac{1}{2} \cdot 1314 = n$

G. $3092 + n = 4638$

H. $18.4 \cdot n = 1.87$

I. $\frac{406.12}{28.4} = n$

J. $52.3 \cdot n = 85.2$

K. $n = 1314.28 + 437$

L. $52.3 + n = 85.2$

M. $3092 + 4638 = n$

N. $3 \cdot n = 18.40$

O. $85.2 + 52.3 = n$

Answers on page A-10

6. *Cost of Gasoline.* What is the cost, in dollars, of 18.4 gal of gasoline at $1.87 per gallon?

7. *Savings Account Balance.* Margaret has $1314.28 in her savings account. Before using her debit card to buy an office chair, she transferred $437 to her checking account. How much was left in her savings account?

8. *Acres Planted.* This season Sam planted 85.2 acres of corn and 52.3 acres of soybeans. Find the total number of acres that he planted.

9. *Amount Inherited.* Tara inherited $2\frac{1}{2}$ times as much as her cousin. Her cousin received $1314. How much did Tara receive?

10. *Travel Funds.* The athletic department needs travel funds of $4638 for the tennis team and $3092 for the golf team. What is the total amount needed for travel?

4.7 EXERCISE SET

For Extra Help

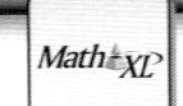

MathXL | MyMathLab | InterAct Math | Math Tutor Center | Digital Video Tutor CD 2 Videotape 4 | Student's Solutions Manual

a Solve.

1. *Hurricanes.* The five most costly hurricanes in the United States are listed in the following chart. The costs, adjusted to 2003 dollars by the Insurance Information Institute, ranged from a low of \$2.5 billion to a high of \$20.3 billion. By how much did the high value differ from the low value?

COSTLY HURRICANES

RANK	HURRICANE	DATE	IN 2003 DOLLARS (in billions)
1	Andrew	Aug. 23–26, 1992	\$20.3
2	Charley	Aug. 13–15, 2004	\$6.8
3	Hugo	Sept. 17–22, 1989	\$6.2
4	Georges	Sept. 21–28, 1998	\$3.3
5	Opal	Oct. 4, 1995	\$2.5

Sources: Insurance Services Office; Insurance Information Institute

2. *Compact Discs* Because of increased sales of DVDs, the sales of recorded music in compact disc (CD) format are declining. Between 1996 and 2003, the total sales of recorded music on CDs ranged from a high of 942.5 million to a low of 745.9 million. By how much did the high value differ from the low value?

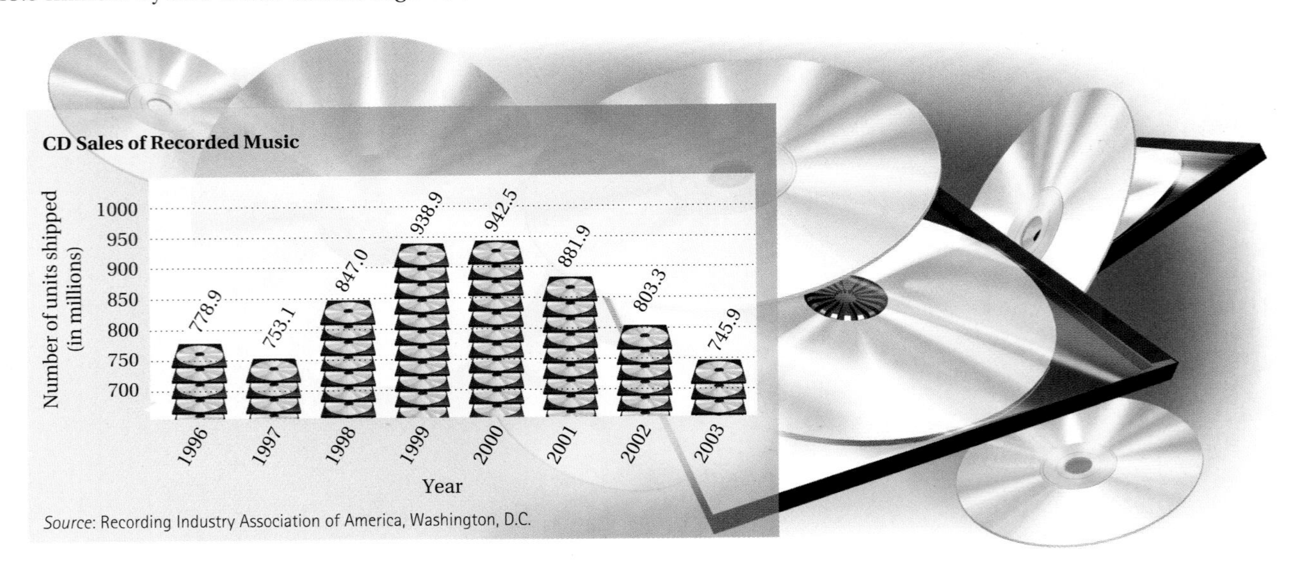

Source: Recording Industry Association of America, Washington, D.C.

3. Andrew bought a DVD of the complete ninth season of the TV show "Friends" for \$29.24 plus \$1.61 sales tax. He paid for it with a \$50 bill. How much change did he receive?

4. Jaden bought the CD "Feels Like Home" by Norah Jones for \$13.49 plus \$0.81 sales tax. She paid for it with a \$20 bill. How much change did she receive?

5. *Body Temperature.* Normal body temperature is 98.6°F. During an illness, a patient's temperature rose 4.2°. What was the new temperature?

6. *Gasoline Cost.* What is the cost, in dollars, of 20.4 gal of gasoline at $2.13 per gallon? Round the answer to the nearest cent.

7. *Lottery Winnings.* The largest lotto jackpot ever won in California totaled $193,000,000 and was shared equally by 3 winners. How much was each winner's share? Round to the cent.

Source: California State Lottery

8. *Lunch Costs.* A group of 4 students pays $47.84 for lunch. What is each person's share?

9. *Stamp.* Find the area and the perimeter of the stamp shown here.

10. *Pole Vault Pit.* Find the area and the perimeter of the landing area of the pole vault pit shown here.

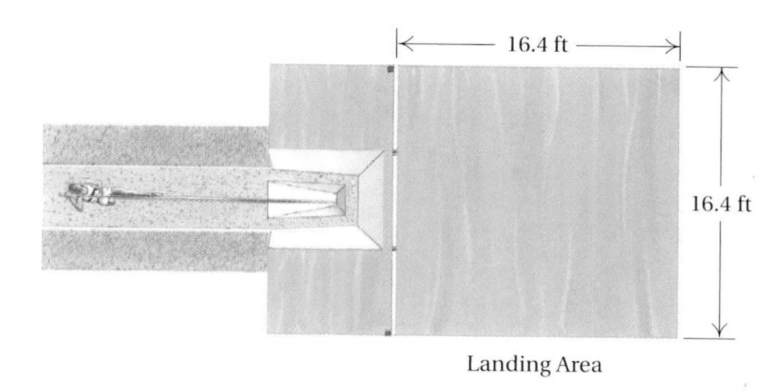

11. *Odometer Reading.* A family checked the odometer before starting a trip. It read 22,456.8 and they know that they will be driving 234.7 mi. What will the odometer read at the end of the trip?

12. *Miles Driven.* Petra bought gasoline when the odometer read 14,296.3. At the next gasoline purchase, the odometer read 14,515.8. How many miles had been driven?

13. *Gas Mileage.* Peggy filled her van's gas tank and noted that the odometer read 26,342.8. After the next filling, the odometer read 26,736.7. It took 19.5 gal to fill the tank. How many miles per gallon did the van get?

14. *Gas Mileage.* Henry filled his Honda's gas tank and noted that the odometer read 18,943.2. After the next filling, the odometer read 19,306.2. It took 13.2 gal to fill the tank. How many miles per gallon did the car get?

15. *Cost of Video Game.* A certain video game costs 75 cents and runs for 1.5 min. Assuming a player does not win any free games and plays continuously, how much money, in dollars, does it cost to play the video game for 1 hr?

1 h = 60 m
1 m = 75

16. *Property Taxes.* The Colavitos own a house with an assessed value of $184,500. For every $1000 of assessed value, they pay $7.68 in taxes. How much do they pay in taxes?

17. *Chemistry.* The water in a filled tank weighs 748.45 lb. One cubic foot of water weighs 62.5 lb. How many cubic feet of water does the tank hold?

18. *Highway Routes.* You can drive from home to work using either of two routes:

Route A: Via interstate highway, 7.6 mi, with a speed limit of 65 mph.

Route B: Via a country road, 5.6 mi, with a speed limit of 50 mph.

Assuming you drive at the posted speed limit, which route takes less time? (Use the formula *Distance* = *Speed* × *Time*.)

Find the perimeter of the figure.

19.

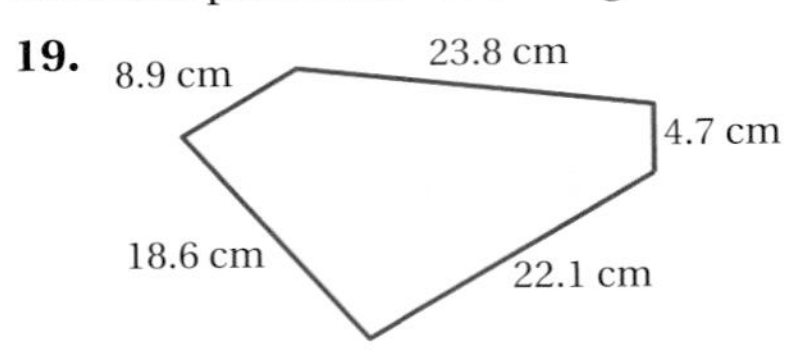

20.

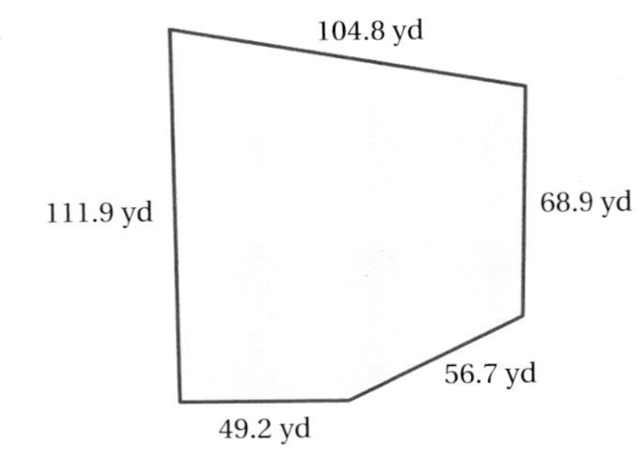

21.

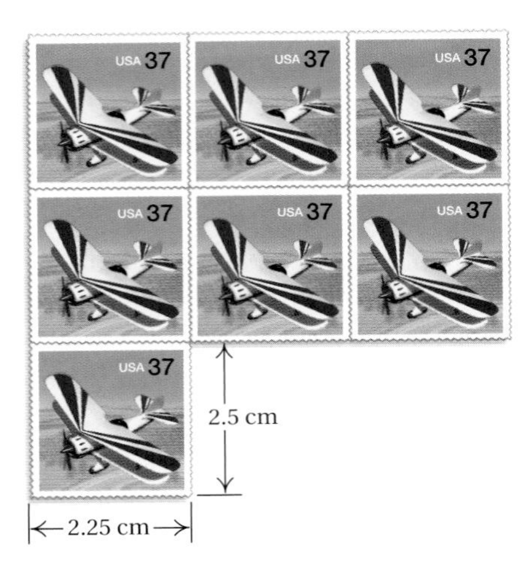

22.

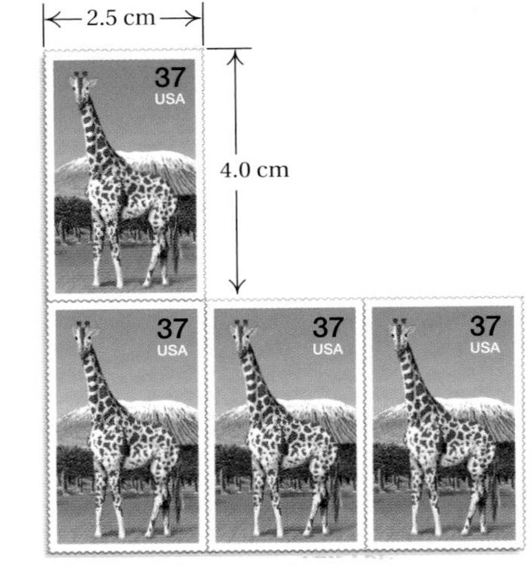

Find the length *d* in the figure.

23.

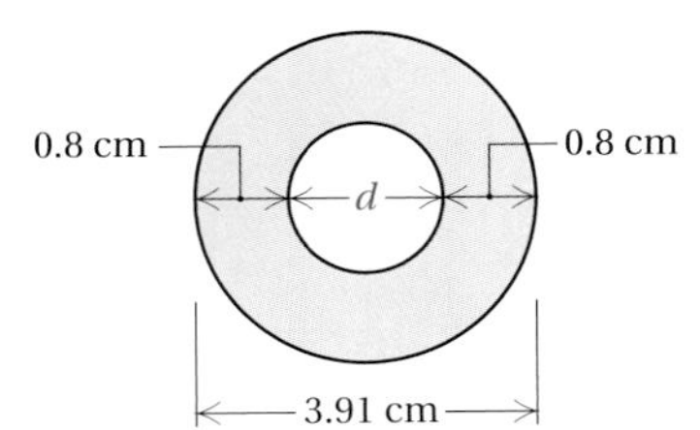

24.

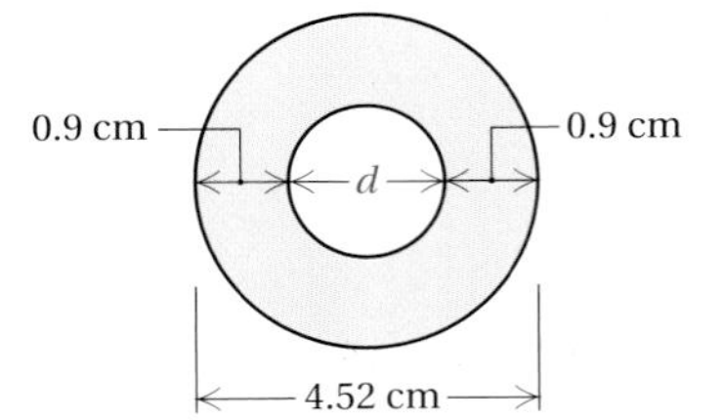

25. *Calories Burned Mowing.* A person weighing 150 lb burns 7.3 calories per minute while mowing a lawn with a power lawnmower. How many calories would be burned in 2 hr of mowing?
Source: *The Handy Science Answer Book*

26. Lot A measures 250.1 ft by 302.7 ft. Lot B measures 389.4 ft by 566.2 ft. What is the total area of the two lots?

27. Holly had $1123.56 in her checking account. She used her debit card to pay bills of $23.82, $507.88, and $98.32. She then deposited a bonus check of $678.20. How much is in her account after these changes?

28. Natalie had $185.00 to spend for fall clothes: $44.95 was spent on shoes, $71.95 for a jacket, and $55.35 for pants. How much was left?

29. A rectangular yard is 20 ft by 15 ft. The yard is covered with grass except for an 8.5-ft square flower garden. How much grass is in the yard?

30. Rita earns a gross paycheck (before deductions) of $495.72. Her deductions are $59.60 for federal income tax, $29.00 for FICA, and $29.00 for medical insurance. What is her take-home paycheck?

31. *Batting Averages.* For the 2004 season, Barry Bonds of the San Francisco Giants won the National League batting title with 135 hits in 373 times at bat. What part of his at-bats were hits? Give decimal notation to the nearest thousandth. (This is a player's *batting average.*)
Source: Major League Baseball

32. *Batting Averages.* For the 2004 season, Ichiro Suzuki of the Seattle Mariners won the American League batting title with 262 hits in 704 times at bat. What part of his at-bats were hits? Give decimal notation to the nearest thousandth.
Source: Major League Baseball

33. *Verizon Wireless.* In 2005, Verizon Wireless offered its America's Choice cellular phone plan with 450 anytime minutes per month for a monthly access fee of $39.99. Minutes in excess of 450 were charged at the rate of $0.45 per minute. In June, Leila used her cellphone for 479 minutes. What was she charged?
Source: www.vzwshop.com

34. *Verizon Wireless.* In 2005, Verizon Wireless offered its America's Choice cellular phone plan with 900 anytime minutes per month for a monthly access fee of $59.99. Minutes in excess of 900 were charged at the rate of $0.40 per minute. One month Jeff used his cellphone for 946 minutes. What was the charge?
Source: www.vzwshop.com

35. *Field Dimensions.* The dimensions of a World Cup soccer field are 114.9 yd by 74.4 yd. The dimensions of a standard football field are 120 yd by 53.3 yd. How much greater is the area of a World Cup soccer field?

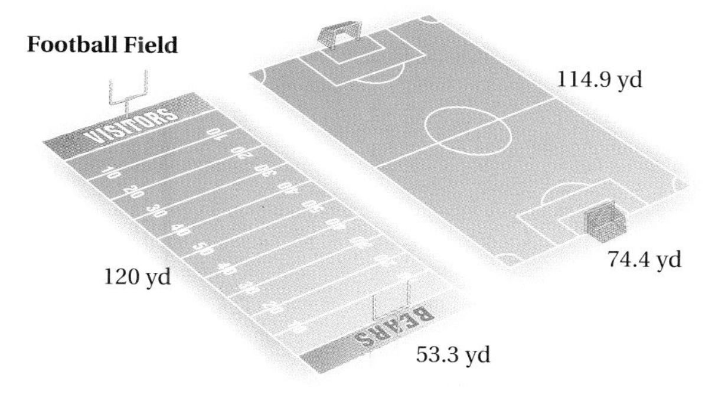

36. *Loan Payment.* In order to make money on loans, financial institutions are paid back more money than they loan. You borrow $120,000 to buy a house and agree to make monthly payments of $880.52 for 30 yr. How much do you pay back altogether? How much more do you pay back than the amount of the loan?

37. *Egg Costs.* A restaurant owner bought 20 dozen eggs for $25.80. Find the cost of each egg to the nearest tenth of a cent (thousandth of a dollar).

38. *Weight Loss.* A person weighing 170 lb burns 8.6 calories per minute while mowing a lawn. One must burn about 3500 calories in order to lose 1 lb. How many pounds would be lost by mowing for 2 hr? Round to the nearest tenth.

39. *Construction Pay.* A construction worker is paid $18.50 per hour for the first 40 hr of work, and time and a half, or $27.75 per hour, for any overtime exceeding 40 hr per week. One week she works 46 hr. How much is her pay?

40. *Summer Work.* Zachary worked 53 hr during a week one summer. He earned $6.50 per hour for the first 40 hr and $9.75 per hour for overtime (hours exceeding 40). How much did Zachary earn during the week?

41. *Projected World Population.* Using the information in the following bar graph, determine the average population of the world for the years 1950 through 2040.

World Population (historical and projected)

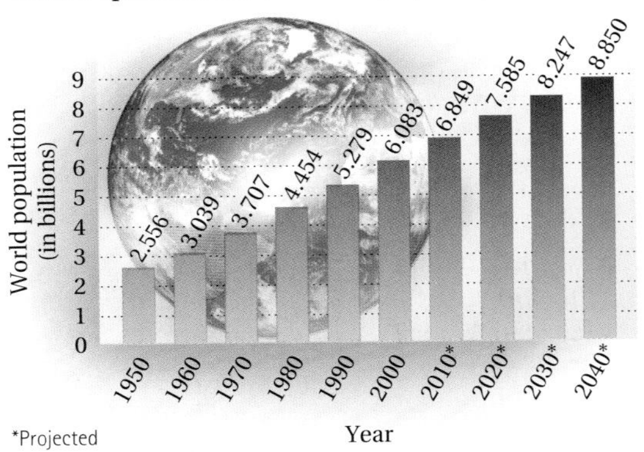

*Projected

Source: U.S. Bureau of the Census

42. *Volunteer Hours.* The chart at right lists the average number of hours per month that volunteers of specific age groups spend doing volunteer work. What is the average number of volunteer hours of all age groups?

Source: *Statistical Abstract of the United States*

HOURS SPENT VOLUNTEERING

AGE	AVERAGE HOURS PER MONTH
21–24 yrs	12.1
25–34 yrs	15.9
35–44 yrs	16.1
45–54 yrs	14.7
55–64 yrs	12.1
65–74 yrs	14.1
75 yrs and older	19.5

43. *Body Temperature.* Normal body temperature is 98.6°F. A baby's bath water should be 100°F. How many degrees above normal body temperature is this?

44. *Body Temperature.* Normal body temperature is 98.6°F. The lowest temperature at which a patient has survived is 69°F. How many degrees below normal is this?

Home-Cost Comparison. Use the table and formula from Example 8. In each of the following cases, find the value of the house in the new location.

Source: Coldwell Banker Real Estate Corporation

	VALUE	PRESENT LOCATION	NEW LOCATION	NEW VALUE
45.	$125,000	Dallas	Miami/Coral Gables	
46.	$180,000	San Diego	Key West	
47.	$96,000	Orlando	Juneau	
48.	$300,000	Palo Alto	Dallas	
49.	$240,000	Louisville	Boston	
50.	$160,000	Charlotte	Sacramento	

Powells.com logo used with permission.

Comparison Shopping. The Internet now provides many sites to shop for a product rather than shopping in a local store. The following lists various Web sites for purchasing the recent best-selling novel *The Da Vinci Code* by Dan Brown.

51. ^{D}W Complete the total costs in the table below for each merchant. (Assume you live in a state in which sales taxes are not charged on Internet purchases.) Then decide which site you would use to make a purchase. Discuss possible ways in which answers might vary.

MERCHANT	PRICE	SHIPPING AND HANDLING	SALES TAX	TOTAL COST
Amazon.com www.amazon.com	$14.97	$3.00 per order, plus $0.99 per item	$0	?
Barnes & Noble www.bn.com	$14.97	$3.00 per order, plus $0.99 per item	$0	?
Powell's Books www.powells.com	$24.95	$2.50 per order, plus $1.00 per item	$0	?
Costco Wholesale www.costco.com	$21.99	$5.00	$0	?
Overstock.com www.overstock.com	$13.47	$1.40	$0	?
Borders Bookstore, local store	$17.46	$0	$1.05	?

52. $\mathbf{D_W}$ *Internet Project.* Consider using the Internet to buy a copy of the book *The Broker* by John Grisham. Then compare your costs with buying it at a local bookstore. Decide which way you would make a purchase. Discuss possible ways in which answers might vary.

SKILL MAINTENANCE

Add.

53. 4569 + 1766 [1.2a]

54. $\frac{2}{3} + \frac{5}{8}$ [3.2a]

55. $4\frac{1}{3} + 2\frac{1}{2}$ [3.5a]

56. $\frac{5}{6} + \frac{7}{10}$ [3.2a]

Subtract.

57. $\frac{2}{3} - \frac{5}{8}$ [3.3a]

58. 4569 − 1766 [1.3b]

59. $\frac{5}{6} - \frac{7}{10}$ [3.3a]

60. $4\frac{1}{3} - 2\frac{1}{2}$ [3.5b]

Simplify. [2.5b]

61. $\frac{3225}{6275}$

62. $\frac{125}{400}$

63. $\frac{325}{625}$

64. $\frac{625}{475}$

Solve. [3.6c]

65. If a water wheel made 469 revolutions at a rate of $16\frac{3}{4}$ revolutions per minute, how long did it rotate?

66. If a bicycle wheel made 480 revolutions at a rate of $66\frac{2}{3}$ revolutions per minute, how long did it rotate?

67. *Calories in Pie.* A piece of pecan pie ($\frac{1}{8}$ of a 9-in. pie) has 502 calories. A piece of pumpkin pie has 316 calories. How many more calories does a piece of pecan pie have than a piece of pumpkin pie? [1.8a]

68. *Calories in Turkey.* Dark meat turkey contains 187 calories per 3.5-oz serving. White meat turkey contains 157 calories per 3.5-oz serving. How many more calories per 3.5-oz serving does the dark turkey contain than the white turkey? [1.8a]

SYNTHESIS

69. You buy a half-dozen packs of basketball cards with a dozen cards in each pack. The cost is twelve dozen cents for each half-dozen cards. How much do you pay for the cards?

4 Summary and Review

The review that follows is meant to prepare you for a chapter exam. It consists of two parts. The first part, Concept Reinforcement, is designed to increase understanding of the concepts through true/false exercises. The second part is the Review Exercises. These provide practice exercises for the exam, together with references to section objectives so you can go back and review. Before beginning, stop and look back over the skills you have obtained. What skills in mathematics do you have now that you did not have before studying this chapter?

CONCEPT REINFORCEMENT

Determine whether the statement is true or false. Answers are given at the back of the book.

______ **1.** In the number 308.00567, the digit 6 names the hundreds place.

______ **2.** To multiply any number by a multiple of 10, count the number of zeros and move the decimal point that many places to the right.

______ **3.** One thousand billions is one trillion.

______ **4.** The number of decimal places in the product of two numbers is the product of the number of places in the factors.

______ **5.** When writing a word name for decimal notation, we write the word "and" for the decimal point.

Review Exercises

Convert the number in the sentence to standard notation.

1. Russia has the largest total area of any country in the world, at 6.59 million square miles. [4.3b]

2. The total weight of the turkeys consumed by Americans during the Thanksgiving holidays is about 6.9 million pounds. [4.3b]

Write a word name. [4.1a]

3. 3.47

4. 0.031

5. 27.00011

6. 0.000007

Write fraction notation. [4.1b]

7. 0.09

8. 4.561

9. 0.089

10. 3.0227

Write decimal notation. [4.1b]

11. $\frac{34}{1000}$

12. $\frac{42,603}{10,000}$

13. $27\frac{91}{100}$

14. $867\frac{6}{1000}$

Which number is larger? [4.1c]

15. 0.034, 0.0185

16. 0.91, 0.19

17. 0.741, 0.6943

18. 1.038, 1.041

Round 17.4287 to the nearest: [4.1d]

19. Tenth.

20. Hundredth.

21. Thousandth.

22. One.

Add. [4.2a]

23. $\begin{array}{r} 2.048 \\ 65.371 \\ +\ 507.1 \\ \hline \end{array}$

24. $\begin{array}{r} 0.6 \\ 0.004 \\ 0.07 \\ +0.0098 \\ \hline \end{array}$

25. $219.3 + 2.8 + 7$

26. $0.41 + 4.1 + 41 + 0.041$

Subtract. [4.2b]

27. $\begin{array}{r} 30.0 \\ -\ 0.7908 \\ \hline \end{array}$

28. $\begin{array}{r} 845.08 \\ -\ 54.79 \\ \hline \end{array}$

29. $37.645 - 8.497$

30. $70.8 - 0.0109$

Multiply. [4.3a]

31. $\begin{array}{r} 48 \\ \times\ 0.27 \\ \hline \end{array}$

32. $\begin{array}{r} 0.174 \\ \times\ 0.83 \\ \hline \end{array}$

33. 100×0.043

34. 0.001×24.68

Divide. [4.4a]

35. $8\overline{)\,60}$

36. $52\overline{)\,23.4}$

37. $2.6\overline{)\,117.52}$

38. $2.14\overline{)\,2.18708}$

39. $\dfrac{276.3}{1000}$

40. $\dfrac{13.892}{0.01}$

Solve. [4.2c], [4.4b]

41. $x + 51.748 = 548.0275$

42. $3 \cdot x = 20.85$

43. $10 \cdot y = 425.4$

44. $0.0089 + y = 5$

Solve. [4.7a]

45. Stacia, a coronary intensive care nurse, earned $620.74 during a recent 40-hr week. What was her hourly wage? Round to the nearest cent.

46. *Nutrition.* The average person eats 683.6 lb of fruits and vegetables in a year. What is the average consumption in one day (Use 1 year = 365 days.) Round to the nearest tenth of a pound.

47. Derek had $1283.67 in his checking account. He used $370.99 to buy a Sony home theater system with his debit card. How much was left in his account?

48. *Verizon Wireless.* In 2005, Verizon Wireless offered its America's Choice cellular phone plan with 1350 anytime minutes per month for a monthly access fee of $79.99. Minutes in excess of 1350 were charged $0.35 per minute. One month Maria used her cellphone for 2000 min. What was the charge?

Source: www.vzwshop.com

49. *Gas Mileage.* A driver wants to estimate gas mileage per gallon. At 36,057.1 mi, the tank is filled with 10.7 gal. At 36,217.6 mi, the tank is filled with 11.1 gal. Find the mileage per gallon. Round to the nearest tenth.

50. *Seafood Consumption.* The following graph shows the annual consumption, in pounds, of seafood per person in the United States in recent years. [4.4c]

a) Find the total per capita consumption for the seven years.

b) Find the average per capita consumption.

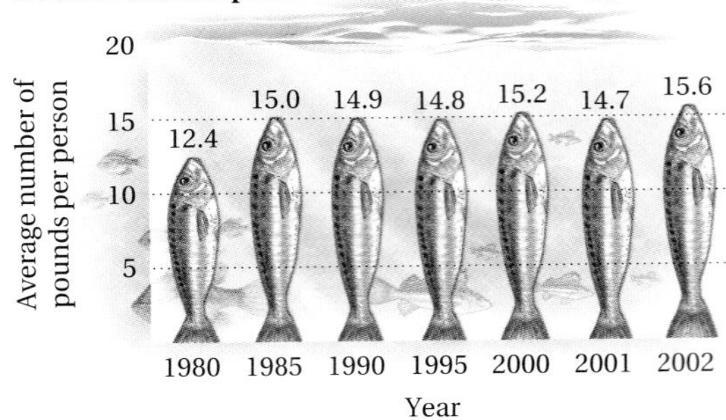

Estimate each of the following. [4.6a]

51. The product 7.82×34.487 by rounding to the nearest one

52. The difference $219.875 - 4.478$ by rounding to the nearest one

53. The quotient $82.304 \div 17.287$ by rounding to the nearest ten

54. The sum \$45.78 + \$78.99 by rounding to the nearest one

Find decimal notation. Use multiplying by 1. [4.5a]

55. $\frac{13}{5}$ **56.** $\frac{32}{25}$ **57.** $\frac{11}{4}$

Find decimal notation. Use division. [4.5a]

58. $\frac{13}{4}$ **59.** $\frac{7}{6}$ **60.** $\frac{17}{11}$

Round the answer to Exercise 60 to the nearest: [4.5b]

61. Tenth. **62.** Hundredth. **63.** Thousandth.

Convert from cents to dollars. [4.3b]

64. 8273 cents **65.** 487 cents

Convert from dollars to cents. [4.3b]

66. \$24.93 **67.** \$9.86

Calculate. [4.4c], [4.5c]

68. $(8 - 1.23) \div 4 + 5.6 \times 0.02$

69. $(1 + 0.07)^2 + 10^3 \div 10^2 + [4(10.1 - 5.6) + 8(11.3 - 7.8)]$

70. $\frac{3}{4} \times 20.85$

71. $\frac{1}{3} \times 123.7 + \frac{4}{9} \times 0.684$

72. D_W Consider finding decimal notation for $\frac{44}{125}$. Discuss as many ways as you can for finding such notation and give the answer. [4.5a]

73. D_W Explain how we can use fraction notation to understand why we count decimal places when multiplying with decimal notation. [4.3a]

SYNTHESIS

74. In each of the following, use one of $+$, $-$, $\times$, and $\div$ in each blank to make a true sentence. [4.4c]

a) $2.56 \square 6.4 \square 51.2 \square 17.4 \square 89.7 = 72.62$
b) $(11.12 \square 0.29) \square 3^4 = 877.23$

75. Find repeating decimal notation for 1 and explain. Use the following hints. [4.5a]

$$\frac{1}{3} = 0.33333333\ldots,$$

$$\frac{2}{3} = 0.66666666\ldots$$

76. Find repeating decimal notation for 2. [4.5a]

77. In the subtraction below, a and b are digits. Find a and b. [4.2b]

$$\begin{array}{r} b876.a4321 \\ -1234.a678b \\ \hline 8641.b7a32 \end{array}$$

4 Chapter Test

For Extra Help

Work It Out! Chapter Test Video on CD

Convert the number in the sentence to standard notation.

1. The annual sales of antibiotics in the United States is \$8.9 billion.
   Source: IMS Health

2. There are 3.756 million people enrolled in bowling organizations in the United States.
   Source: *Bowler's Journal International*, December 2000

Write a word name.

3. 2.34

4. 105.0005

Write fraction notation.

5. 0.91

6. 2.769

Write decimal notation.

7. $\frac{74}{1000}$

8. $\frac{37,047}{10,000}$

9. $756\frac{9}{100}$

10. $91\frac{703}{1000}$

Which number is larger?

11. 0.07, 0.162

12. 0.078, 0.06

13. 0.09, 0.9

Round 5.6783 to the nearest:

14. One.

15. Hundredth.

16. Thousandth.

17. Tenth.

Calculate.

18. $\begin{array}{r} 0.7 \\ 0.08 \\ 0.009 \\ +\ 0.0012 \\ \hline \end{array}$

19. 102.4 + 6.1 + 78

20. 0.93 + 9.3 + 93 + 930

21. $\begin{array}{r} 52.678 \\ -\ \ 4.321 \\ \hline \end{array}$

22. $\begin{array}{r} 20.0 \\ -\ \ 0.9099 \\ \hline \end{array}$

23. 234.6788 − 81.7854

24. $\begin{array}{r} 0.125 \\ \times\ \ 0.24 \\ \hline \end{array}$

25. 0.001 × 213.45

26. 1000 × 73.962

27. $4\overline{)19}$

28. $3.3\overline{)100.32}$

29. $82\overline{)15.58}$

30. $\dfrac{346.89}{1000}$

31. $\dfrac{346.89}{0.01}$

Solve.

32. $4.8 \cdot y = 404.448$

33. $x + 0.018 = 9$

34. *Verizon Wireless.* In 2005, Verizon Wireless offered its America's Choice cellular phone plan with 2000 anytime minutes per month for a monthly access fee of $99.99. Minutes in excess of 2000 were charged $0.25 per minute. One month Trey used his cell phone for 2860 min. What was the charge?

Source: www.vzwshop.com

35. *Gas Mileage.* Tina wants to estimate the gas mileage per gallon in her economy car. At 76,843 mi, the tank is filled with 14.3 gal of gasoline. At 77,310 mi, the tank is filled with 16.5 gal of gasoline. Find the mileage per gallon. Round to the nearest tenth.

36. *Checking Account Balance.* Nicholas has a balance of $10,200 in his checking account before making purchases of $123.89, $56.68, and $3446.98 with his debit card. What was the balance after making the purchases?

37. The Drake, Smith, and Nicholas law firm buys 7 cases of copy paper at $25.99 per case. What is the total cost?

38. *Airport Passengers.* The following graph shows the number of passengers in 2003 who traveled through the country's busiest airports. Find the average number of passengers through these airports.

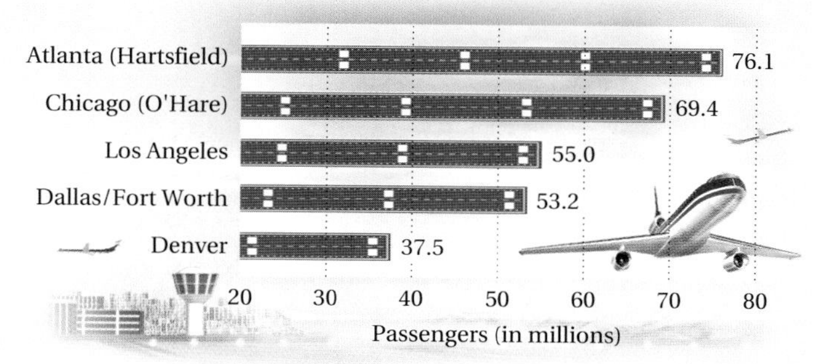

Source: Airports Council International, www.airports.org

Estimate each of the following.

39. The product 8.91×22.457 by rounding to the nearest one

40. The quotient $78.2209 \div 16.09$ by rounding to the nearest ten

Find decimal notation. Use multiplying by 1.

41. $\frac{8}{5}$

42. $\frac{22}{25}$

43. $\frac{21}{4}$

Find decimal notation. Use division.

44. $\frac{3}{4}$

45. $\frac{11}{9}$

46. $\frac{15}{7}$

Round the answer to Question 46 to the nearest:

47. Tenth.

48. Hundredth.

49. Thousandth.

50. Convert from cents to dollars: 949 cents.

Calculate.

51. $256 \div 3.2 \div 2 - 1.56 + 78.325 \times 0.02$

52. $(1 - 0.08)^2 + 6[5(12.1 - 8.7) + 10(14.3 - 9.6)]$

53. $\frac{7}{8} \times 345.6$

54. $\frac{2}{3} \times 79.95 - \frac{7}{9} \times 1.235$

SYNTHESIS

55. The Silver's Health Club generally charges a $79 membership fee and $42.50 a month. Allise has a coupon that will allow her to join the club for $299 for six months. How much will Allise save if she uses the coupon?

56. Arrange from smallest to largest.

$\frac{2}{3}, \frac{15}{19}, \frac{11}{13}, \frac{5}{7}, \frac{13}{15}, \frac{17}{20}$

5

Ratio and Proportion

Real-World Application

The number of women attending Purdue University's veterinary school of medicine has grown to surpass the number of men in the past three decades. In 1971, 53 men and 12 women were enrolled. In 1979, 36 men and 36 women were enrolled, and in 2004, there were 58 women and 12 men. What was the ratio of women to men in 1971, in 1979, and in 2004? What was the ratio of men to total enrollment in 2004?

Source: Purdue University School of Veterinary Medicine, *Indianapolis Star*

This problem appears as Example 7 in Section 5.1.

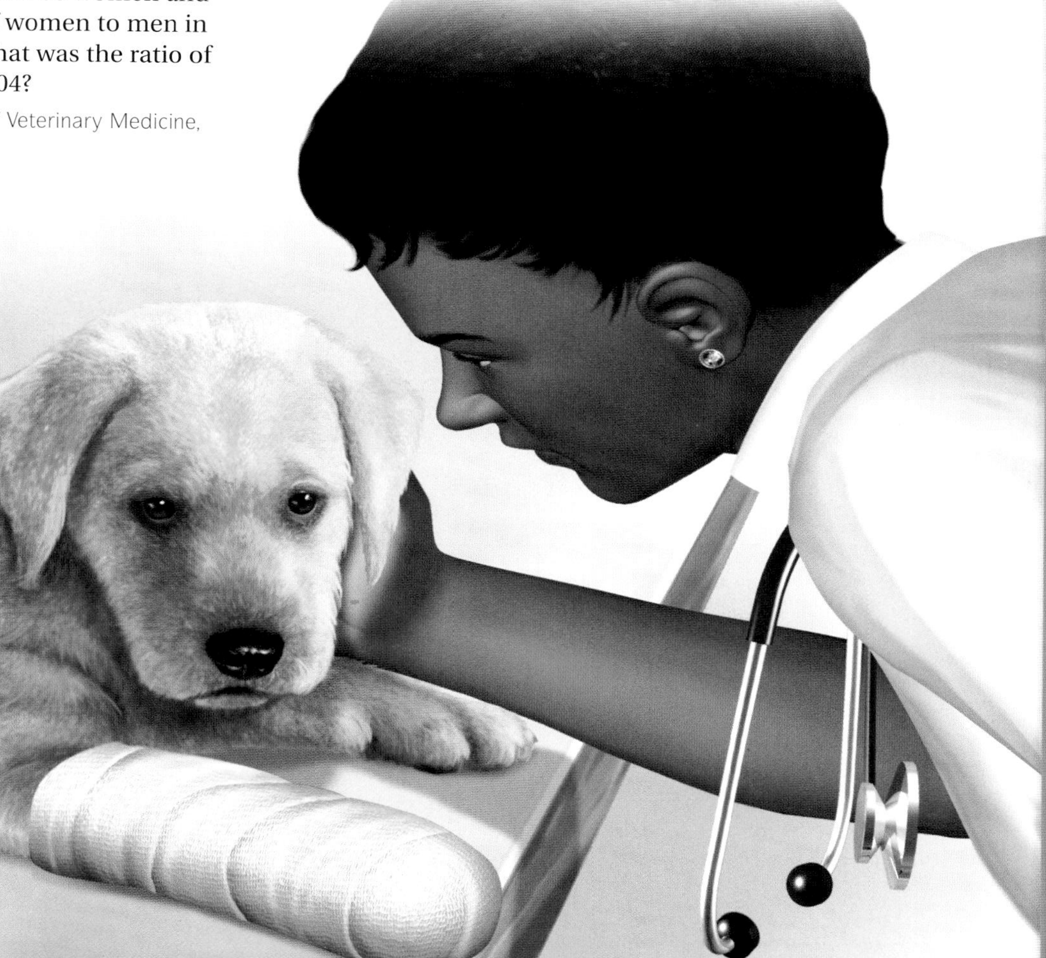

5.1 INTRODUCTION TO RATIOS

Objectives

a Find fraction notation for ratios.

b Simplify ratios.

1. Find the ratio of 5 to 11.

2. Find the ratio of 57.3 to 86.1.

3. Find the ratio of $6\frac{3}{4}$ to $7\frac{2}{5}$.

4. **Rainfall.** The greatest amount of rainfall ever recorded for a 12-month period was 739 in. in Kukui, Maui, Hawaii, from December 1981 to December 1982. Find the ratio of rainfall to time in months.
Source: *The Handy Science Answer Book*

Answers on page A-11

a Ratios

RATIO

A **ratio** is the quotient of two quantities.

In the 2004–2005 season, the Detroit Pistons basketball team averaged 93.3 points per game and allowed their opponents an average of 89.5 points per game. The *ratio* of points earned to points allowed is given by the fraction notation

$$\frac{\text{Points earned} \rightarrow 93.3}{\text{Points allowed} \rightarrow 89.5} \quad \text{or by the colon notation} \quad 93.3:89.5.$$

We read both forms of notation as "the ratio of 93.3 to 89.5," listing the numerator first and the denominator second.

RATIO NOTATION

The **ratio** of a to b is given by the fraction notation $\frac{a}{b}$, where a is the numerator and b is the denominator, or by the colon notation $a:b$.

EXAMPLE 1 Find the ratio of 7 to 8.

The ratio is $\frac{7}{8}$, or $7:8$.

EXAMPLE 2 Find the ratio of 31.4 to 100.

The ratio is $\frac{31.4}{100}$, or $31.4:100$.

EXAMPLE 3 Find the ratio of $4\frac{2}{3}$ to $5\frac{7}{8}$. You need not simplify.

The ratio is $\frac{4\frac{2}{3}}{5\frac{7}{8}}$, or $4\frac{2}{3}:5\frac{7}{8}$.

Do Exercises 1–3.

In most of our work, we will use fraction notation for ratios.

EXAMPLE 4 *Wind Speeds.* The average wind speed in Chicago is 10.4 mph. The average wind speed in Boston is 12.5 mph. Find the ratio of the wind speed in Chicago to the wind speed in Boston.
Source: *The Handy Geography Answer Book*

The ratio is $\frac{10.4}{12.5}$.

EXAMPLE 5 *Batting.* In the 2004 season, Vladimir Guerrero of the Anaheim Angels got 206 hits in 612 at-bats. What was the ratio of hits to at-bats? of at-bats to hits?

Source: Major League Baseball

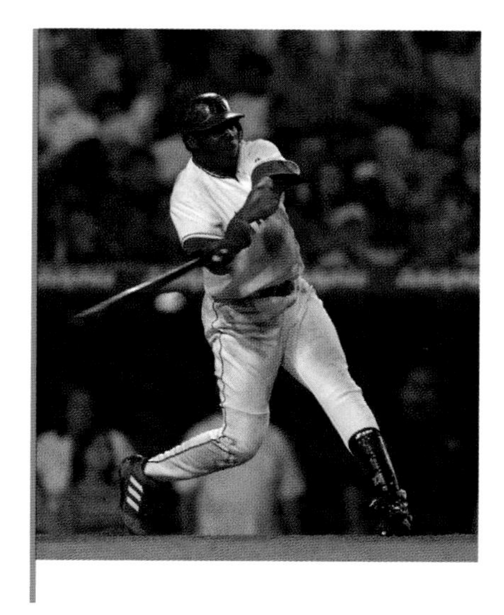

The ratio of hits to at-bats is

$$\frac{206}{612}.$$

The ratio of at-bats to hits is

$$\frac{612}{206}.$$

Do Exercises 4–6. (Exercise 4 is on the preceding page.)

EXAMPLE 6 Refer to the triangle below.

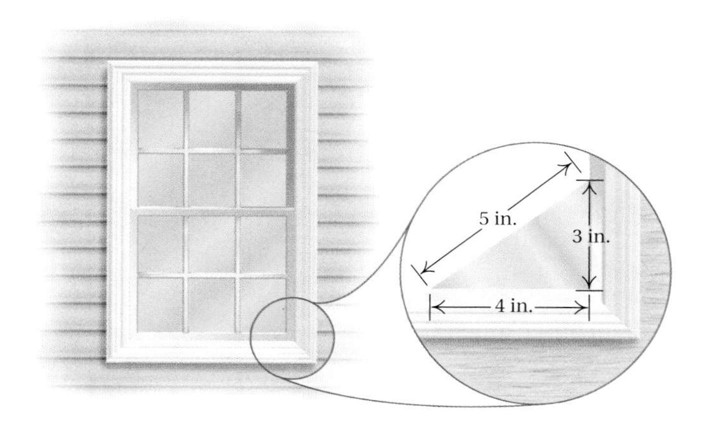

a) What is the ratio of the length of the longest side to the length of the shortest side?

$$\frac{5}{3}$$

b) What is the ratio of the length of the shortest side to the length of the longest side?

$$\frac{3}{5}$$

Do Exercise 7.

5. Fat Grams. In one serving $\left(\frac{1}{2}\text{-cup}\right)$ of fried scallops, there are 12 g of fat. In one serving $\left(\frac{1}{2}\text{-cup}\right)$ of fried oysters, there are 14 g of fat. What is the ratio of grams of fat in one serving of scallops to grams of fat in one serving of oysters?

Source: *Better Homes and Gardens: A New Cook Book*

6. Earned Runs. In the 2004 season, Roger Clemens of the Houston Astros gave up 71 earned runs in 214.1 innings pitched. What was the ratio of earned runs to innings pitched? of innings pitched to earned runs?

Source: Major League Baseball

7. In the triangle below, what is the ratio of the length of the shortest side to the length of the longest side?

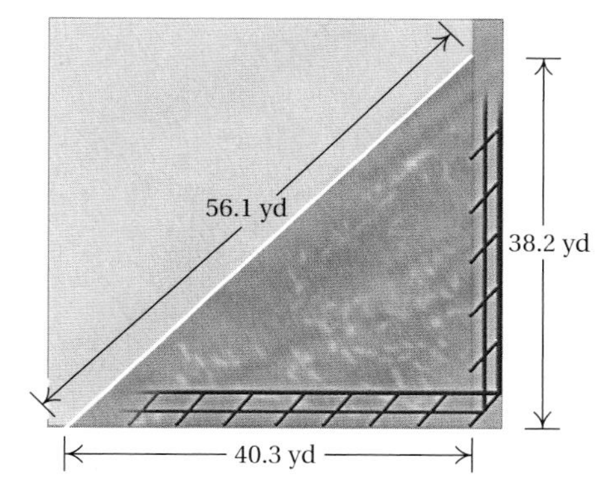

Answers on page A-11

8. Soap Box Derby. Participation in the All-American Soap Box Derby World Championship has increased by more than 300 competitors since 1985. In 2004, there were 483 participants, 278 boys and 205 girls. What was the ratio of girls to boys? of boys to girls? of boys to total number of participants?

Source: All-American Soap Box Derby

EXAMPLE 7 *Veterinary Medicine.* The number of women attending Purdue University's veterinary school of medicine has grown to surpass the number of men in the past three decades.

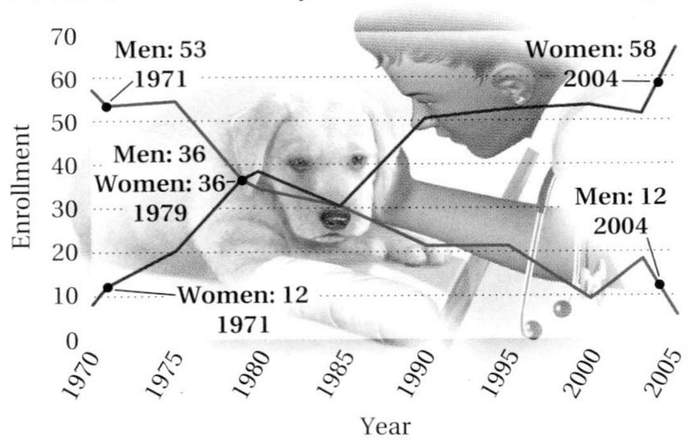

Sources: Purdue University School of Veterinary Medicine; *Indianapolis Star*

a) What was the ratio of women to men in 1971? in 1979? in 2004?

b) What was the ratio of men to women in 1971? in 1979? in 2004?

c) What was the ratio of women to total enrollment in 2004?

d) What was the ratio of men to total enrollment in 2004?

a) The ratio of women to men

in 1971: $\frac{12}{53}$; in 1979: $\frac{36}{36}$; in 2004: $\frac{58}{12}$.

b) The ratio of men to women

in 1971: $\frac{53}{12}$; in 1979: $\frac{36}{36}$; in 2004: $\frac{12}{58}$.

c) The ratio of women to total enrollment

in 2004: $\frac{58}{70}$.

d) The ratio of men to total enrollment

in 2004: $\frac{12}{70}$.

Do Exercise 8.

b Simplifying Notation for Ratios

Sometimes a ratio can be simplified. This provides a means of finding other numbers with the same ratio.

EXAMPLE 8 Find the ratio of 6 to 8. Then simplify and find two other numbers in the same ratio.

Answer on page A-11

We write the ratio in fraction notation and then simplify:

$$\frac{6}{8} = \frac{2 \cdot 3}{2 \cdot 4} = \frac{2}{2} \cdot \frac{3}{4} = 1 \cdot \frac{3}{4} = \frac{3}{4}.$$

Thus, 3 and 4 have the same ratio as 6 and 8. We can express this by saying "6 is to 8" as "3 is to 4."

Do Exercise 9.

EXAMPLE 9 Find the ratio of 2.4 to 10. Then simplify and find two other numbers in the same ratio.

We first write the ratio in fraction notation. Next, we multiply by 1 to clear the decimal from the numerator. Then we simplify.

$$\frac{2.4}{10} = \frac{2.4}{10} \cdot \frac{10}{10} = \frac{24}{100} = \frac{4 \cdot 6}{4 \cdot 25} = \frac{4}{4} \cdot \frac{6}{25} = \frac{6}{25}$$

Thus, 2.4 is to 10 as 6 is to 25.

Do Exercises 10 and 11.

EXAMPLE 10 A standard HDTV screen with a width of 40 in. has a height of $22\frac{1}{2}$ in. Find the ratio of width to height and simplify.

$$\text{The ratio is } \frac{40}{22\frac{1}{2}} = \frac{40}{22.5} = \frac{400}{225}$$
$$= \frac{25 \cdot 16}{25 \cdot 9} = \frac{25}{25} \cdot \frac{16}{9}$$
$$= \frac{16}{9}.$$

Thus we can say the ratio of width to height is 16 to 9, which can also be expressed as 16:9.

Do Exercise 12.

9. Find the ratio of 18 to 27. Then simplify and find two other numbers in the same ratio.

10. Find the ratio of 3.6 to 12. Then simplify and find two other numbers in the same ratio.

11. Find the ratio of 1.2 to 1.5. Then simplify and find two other numbers in the same ratio.

12. In Example 10, find the ratio of the height of the HDTV screen to the width and simplify.

Answers on page A-11

Study Tips — TIME MANAGEMENT (PART 2)

Here are some additional tips to help you with time management. (See also the Study Tips on time management in Sections 1.5 and 5.5.)

- **Avoid "time killers."** We live in a media age, and the Internet, e-mail, television, and movies all are time killers. Allow yourself a break to enjoy some college and outside activities. But keep track of the time you spend on such activities and compare it to the time you spend studying.
- **Prioritize your tasks.** Be careful about taking on too many college activities that fall outside of academics. Examples of such activities are decorating a homecoming float, joining a fraternity or sorority, and participating on a student council committee. Any of these is important but keep your involvement to a minimum to be sure that you have enough time for your studies.
- **Be aggressive about your study tasks.** Instead of worrying over your math homework or test preparation, do something to get yourself started. Work a problem here and a problem there, and before long you will accomplish the task at hand. If the task is large, break it down into smaller parts, and do one at a time. You will be surprised at how quickly the large task can then be completed.

5.1 EXERCISE SET

For Extra Help

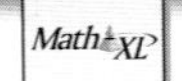

MathXL | MyMathLab | InterAct Math | Math Tutor Center | Digital Video Tutor CD 3 Videotape 5 | Student's Solutions Manual

a Find fraction notation for the ratio. You need not simplify.

1. 4 to 5

2. 3 to 2

3. 178 to 572

4. 329 to 967

5. 0.4 to 12

6. 2.3 to 22

7. 3.8 to 7.4

8. 0.6 to 0.7

9. 56.78 to 98.35

10. 456.2 to 333.1

11. $8\frac{3}{4}$ to $9\frac{5}{6}$

12. $10\frac{1}{2}$ to $43\frac{1}{4}$

13. *Corvette Accidents.* Of every 5 fatal accidents involving a Corvette, 4 do not involve another vehicle. Find the ratio of fatal accidents involving just a Corvette to those involving a Corvette and at least one other vehicle.
Source: *Harper's Magazine*

14. *Price of a Book.* The recent paperback book *A Short History of Nearly Everything* by Bill Bryson had a list price of $15.95 but was sold by Amazon.com for $10.85. What was the ratio of the sale price to the list price? of the list price to the sale price?
Source: www.amazon.com

15. *Physicians.* In 2001, there were 356 physicians in Connecticut per 100,000 residents. In Wyoming, there were 173 physicians per 100,000 residents. Find the ratio of the number of physicians to residents in Connecticut and in Wyoming.
Source: U.S. Census Bureau, American Medical Association

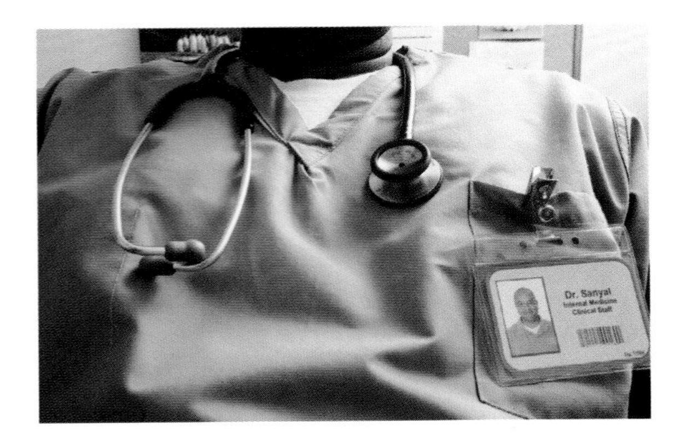

16. *Silicon in the Earth's Crust.* Of every 100 tons of the earth's crust, there will be about 28 tons of silicon in its content. What is the ratio of silicon to the weight of crust? of the weight of crust to the weight of silicon?
Source: *The Handy Science Answer Book*

17. *Heart Disease.* In the state of Minnesota, of every 1000 people, 93.2 will die of heart disease. Find the ratio of those who die of heart disease to all people.
Source: "Reforming the Health Care System; State Profiles 1999," AARP

18. *Cancer Deaths.* In the state of Texas, of every 1000 people, 122.8 will die of cancer. Find the ratio of those who die of cancer to all people.
Source: "Reforming the Health Care System; State Profiles 1999," AARP

19. *Batting.* In the 2004 season, Todd Helton of the Colorado Rockies got 190 hits in 547 at-bats. What was the ratio of hits to at-bats? of at-bats to hits?

Source: Major League Baseball

20. *Batting.* In the 2004 season, Manny Ramirez of the Boston Red Sox got 175 hits in 568 at-bats. What was the ratio of hits to at-bats? of at-bats to hits?

Source: Major League Baseball

21. *Field Hockey.* A diagram of the playing area for field hockey is shown below. What is the ratio of width to length? of length to width?

Source: *Sports: The Complete Visual Reference*

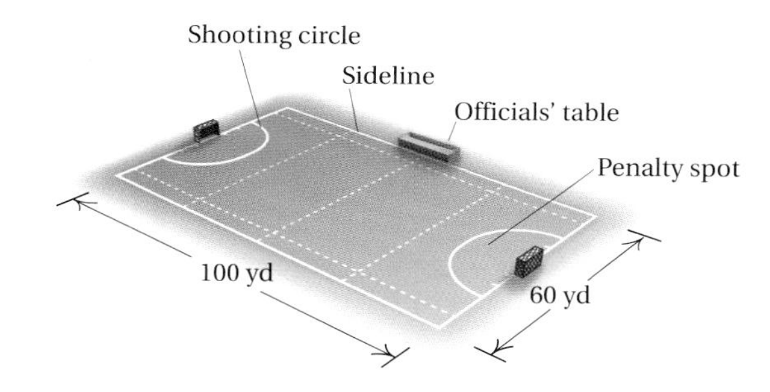

22. *The Leaning Tower of Pisa.* At the time of this writing, the Leaning Tower of Pisa is still standing. It is 184.5 ft tall but leans about 17 ft out from its base. What is the ratio of the distance it leans to its height? its height to the distance it leans?

Source: *The Handy Science Answer Book*

b Find the ratio of the first number to the second and simplify.

23. 4 to 6

24. 6 to 10

25. 18 to 24

26. 28 to 36

27. 4.8 to 10

28. 5.6 to 10

29. 2.8 to 3.6

30. 4.8 to 6.4

31. 20 to 30

32. 40 to 60

33. 56 to 100

34. 42 to 100

35. 128 to 256

36. 232 to 116

37. 0.48 to 0.64

38. 0.32 to 0.96

39. In this rectangle, find the ratios of length to width and of width to length.

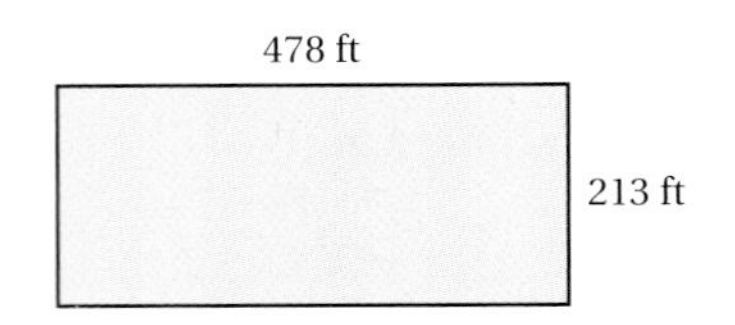

40. In this right triangle, find the ratios of shortest length to longest length and of longest length to shortest length.

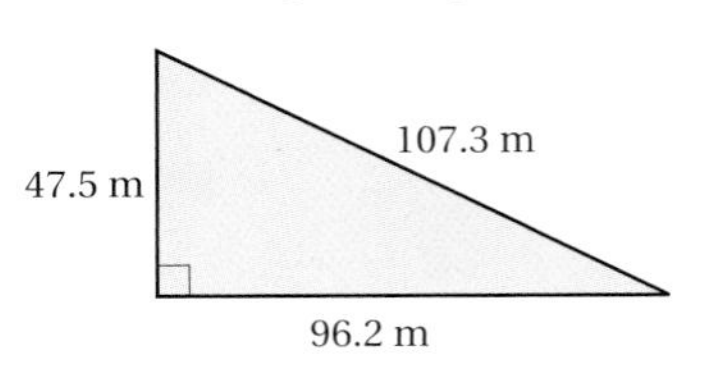

41. DW Can every ratio be written as the ratio of some number to 1? Why or why not?

42. DW What can be concluded about a rectangle's width if the ratio of length to perimeter is 1 to 3? Make some sketches and explain your reasoning.

SKILL MAINTENANCE

Use = or ≠ for □ to write a true sentence. [2.5c]

43. $\frac{12}{8}$ □ $\frac{6}{4}$

44. $\frac{4}{7}$ □ $\frac{5}{9}$

45. $\frac{7}{2}$ □ $\frac{31}{9}$

46. $\frac{17}{25}$ □ $\frac{68}{100}$

Divide. Write decimal notation for the answer. [4.4a]

47. $200 \div 4$

48. $95 \div 10$

49. $232 \div 16$

50. $342 \div 2.25$

Solve. [3.5c]

51. Rocky is $187\frac{1}{10}$ cm tall and his daughter is $180\frac{3}{4}$ cm tall. How much taller is Rocky?

52. Aunt Louise is $168\frac{1}{4}$ cm tall and her son is $150\frac{7}{10}$ cm tall. How much taller is Aunt Louise?

SYNTHESIS

53. Find the ratio of $3\frac{3}{4}$ to $5\frac{7}{8}$ and simplify.

Fertilizer. Exercises 54 and 55 refer to a common lawn fertilizer known as "5, 10, 15." This mixture contains 5 parts of potassium for every 10 parts of phosphorus and 15 parts of nitrogen (this is often denoted 5:10:15).

54. Find the ratio of potassium to nitrogen and of nitrogen to phosphorus.

55. Simplify the ratio 5:10:15.

5.2 RATES AND UNIT PRICES

Objectives

a Give the ratio of two different measures as a rate.

b Find unit prices and use them to compare purchases.

a Rates

A 2005 Kia Sportage EX 4WD can go 414 miles on 18 gallons of gasoline. Let's consider the ratio of miles to gallons:

Source: Kia Motors America, Inc.

$$\frac{414 \text{ mi}}{18 \text{ gal}} = \frac{414}{18}\,\frac{\text{miles}}{\text{gallon}} = \frac{23}{1}\,\frac{\text{miles}}{\text{gallon}}$$

$$= 23 \text{ miles per gallon} = 23 \text{ mpg.}$$

"per" means "division," or "for each."

The ratio

$$\frac{414 \text{ mi}}{18 \text{ gal}}, \quad \text{or} \quad \frac{414}{18}\,\frac{\text{mi}}{\text{gal}}, \quad \text{or } 23 \text{ mpg}$$

is called a **rate.**

RATE

When a ratio is used to compare two different kinds of measure, we call it a **rate.**

Suppose David says his car goes 392.4 mi on 16.8 gal of gasoline. Is the mpg (mileage) of his car better than that of the Kia Sportage above? To determine this, it helps to convert the ratio to decimal notation and perhaps round. Then we have

$$\frac{392.4 \text{ miles}}{16.8 \text{ gallons}} = \frac{392.4}{16.8} \text{ mpg} \approx 23.357 \text{ mpg.}$$

Since $23.357 > 23$, David's car gets better mileage than the Kia Sportage does.

EXAMPLE 1 It takes 60 oz of grass seed to seed 3000 sq ft of lawn. What is the rate in ounces per square foot?

$$\frac{60 \text{ oz}}{3000 \text{ sq ft}} = \frac{1}{50}\,\frac{\text{oz}}{\text{sq ft}}, \quad \text{or} \quad 0.02\,\frac{\text{oz}}{\text{sq ft}}$$

EXAMPLE 2 A cook buys 10 lb of potatoes for \$3.69. What is the rate in cents per pound?

$$\frac{\$3.69}{10 \text{ lb}} = \frac{369 \text{ cents}}{10 \text{ lb}}, \quad \text{or} \quad 36.9\,\frac{\text{cents}}{\text{lb}}$$

Study Tips

RECORDING YOUR LECTURES

Consider recording your lectures and playing them back when convenient, say, while commuting to campus. (Be sure to get permission from your instructor before doing so, however.) Important points can be emphasized verbally. We consider this idea so worthwhile that we provide a series of audio recordings that accompany the book. (See the Preface for more information.)

A ratio of distance traveled to time is called *speed*. What is the rate, or speed, in miles per hour?

1. 45 mi, 9 hr

2. 120 mi, 10 hr

3. 89 km, 13 hr (Round to the nearest hundredth.)

What is the rate, or speed, in feet per second?

4. 2200 ft, 2 sec

5. 52 ft, 13 sec

6. 242 ft, 16 sec

7. Ratio of Home Runs to Strikeouts. Referring to Example 4, determine Rolen's home-run to strikeout rate.

Source: Major League Baseball

Answers on page A-11

EXAMPLE 3 A pharmacy student working as a pharmacist's assistant earned \$3690 for working 3 months one summer. What was the rate of pay per month?

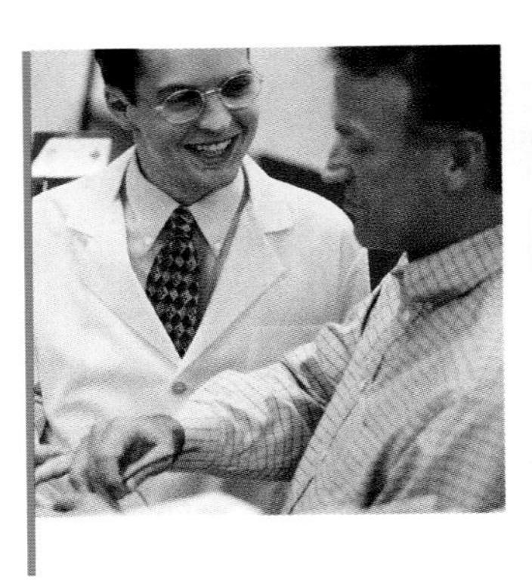

The rate of pay is the ratio of money earned per length of time worked, or

$$\frac{\$3690}{3\text{ mo}} = 1230\,\frac{\text{dollars}}{\text{month}}, \quad \text{or}$$

\$1230 per month.

EXAMPLE 4 *Ratio of Strikeouts to Home Runs.* In the 2004 season, Scott Rolen of the St. Louis Cardinals had 92 strikeouts and 34 home runs. What was his strikeout to home-run rate?

Source: Major League Baseball

$$\frac{92\text{ strikeouts}}{34\text{ home runs}} = \frac{92}{34}\,\frac{\text{strikeouts}}{\text{home runs}} = \frac{92}{34}\text{ strikeouts per home run}$$

$$\approx 2.71\text{ strikeouts per home run}$$

Do Exercises 1–8. (Exercise 8 is on the following page.)

b Unit Pricing

UNIT PRICE

A **unit price,** or **unit rate,** is the ratio of price to the number of units.

EXAMPLE 5 *Unit Price of Pears.* A consumer bought a $15\frac{1}{4}$-oz can of pears for \$1.07. What is the unit price in cents per ounce?

Often it is helpful to change the cost to cents so we can compare unit prices more easily:

$$\text{Unit price} = \frac{\text{Price}}{\text{Number of units}}$$

$$= \frac{\$1.07}{15\frac{1}{4}\text{ oz}} = \frac{107\text{ cents}}{15.25\text{ oz}} = \frac{107}{15.25}\,\frac{\text{cents}}{\text{oz}}$$

$$\approx 7.016\text{ cents per ounce.}$$

Do Exercise 9 on the following page.

To do comparison shopping, it helps to compare unit prices.

EXAMPLE 6 *Unit price of Heinz Ketchup.* Many factors can contribute to determining unit pricing in food, such as variations in store pricing and special discounts. Heinz produces ketchup in containers of various sizes. The table below lists several examples of pricing for these packages from a Meijer store. Starting with the price given for each package, compute the unit prices and decide which is the best purchase on the basis of unit price per ounce alone.

Source: Meijer Stores

PACKAGE	PRICE	UNIT PRICE
14 oz	\$1.29	9.214¢/oz
24 oz	\$1.47	6.125¢/oz
36 oz	\$2.49	6.917¢/oz
64 oz	\$4.45	6.953¢/oz
101-oz twin pack (two $50\frac{1}{2}$-oz packages)	\$5.69	5.634¢/oz Lowest unit price

We compute the unit price for the 24-oz package and leave the remaining prices to the student to check. The unit price for the 24-oz, \$1.47 package is given by

$$\frac{\$1.47}{24\text{ oz}} = \frac{147\text{ cents}}{24\text{ oz}} = \frac{147}{24}\,\frac{\text{cents}}{\text{oz}} = 6.125\text{ cents per ounce} = 6.125\text{ ¢/oz}.$$

On the basis of unit price alone, we see that the 101-oz twin pack is the best buy.

Sometimes, as you will see in Margin Exercise 10, a larger size may not have the lower unit price. It is also worth noting that "bigger" is not always "cheaper." (For example, you may not have room for larger packages or the food may go to waste before it is used.)

Do Exercise 10.

8. Babe Ruth. In his entire career, Babe Ruth had 1330 strikeouts and 714 home runs. What was his home-run to strikeout rate? How does it compare to Rolen's?

Source: Major League Baseball

9. Unit Price of Mustard. A consumer bought a 20-oz container of French's yellow mustard for \$1.49. What is the unit price in cents per ounce?

10. Meijer Brand Olives. Complete the following table of unit prices for Meijer Brand olives. Which package has the better unit price?

Source: Meijer Stores

PACKAGE	PRICE	UNIT PRICE
7 oz	\$1.69	
10 oz	\$2.59	
$5\frac{3}{4}$ oz	\$1.39	

Answers on page A-11

EXERCISE SET

For Extra Help

MathXL | MyMathLab | InterAct Math | Math Tutor Center | Digital Video Tutor CD 3 Videotape 5 | Student's Solutions Manual

a In Exercises 1–4, find the rate, or speed, as a ratio of distance to time. Round to the nearest hundredth where appropriate.

1. 120 km, 3 hr

2. 18 mi, 9 hr

3. 217 mi, 29 sec

4. 443 m, 48 sec

5. *Chevrolet Cobalt LS—City Driving.* A 2005 Chevrolet Cobalt LS will go 300 mi on 12.5 gallons of gasoline in city driving. What is the rate in miles per gallon?
Source: *Car and Driver,* April 2005, p. 104

6. *Audi A6 4.2 Quattro—City Driving.* A 2005 Audi A6 4.2 Quattro will go 246.5 miles on 14.5 gallons of gasoline in city driving. What is the rate in miles per gallon?
Source: *Car and Driver,* May 2005, p. 52

7. *Audi A6 4.2 Quattro—Highway Driving.* A 2005 Audi A6 4.2 Quattro will go 448.5 miles on 19.5 gallons of gasoline in highway driving. What is the rate in miles per gallon?
Source: *Car and Driver,* May 2005, p. 52

8. *Chevrolet Cobalt LS—Highway Driving.* A 2005 Chevrolet Cobalt LS will go 432 miles on 13.5 gallons of gasoline in highway driving. What is the rate in miles per gallon?
Source: *Car and Driver,* April 2005, p. 104

9. *Population Density of Monaco.* Monaco is a tiny country on the Mediterranean coast of France. It has an area of 0.75 square mile and a population of 32,270 people. What is the rate of number of people per square mile? The rate per square mile is called the *population density.* Monaco has the highest population density in the world.
Source: *Time Almanac, 2005*

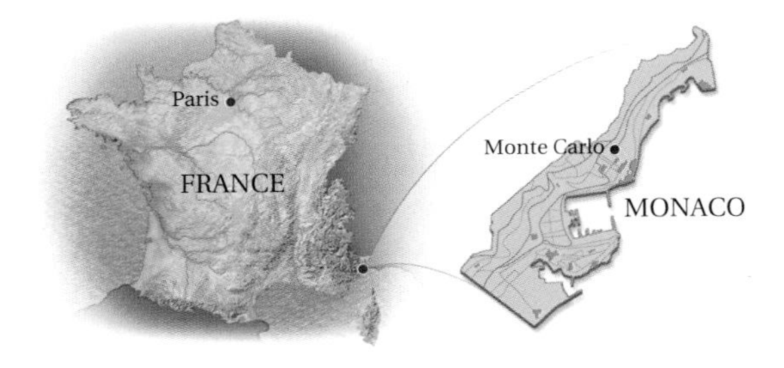

10. *Population Density of Australia.* The continent of Australia, with the island state of Tasmania, has an area of 2,967,893 sq mi and a population of 19,913,144 people. What is the rate of number of people per square mile? The rate per square mile is called the *population density.* Australia has one of the lowest population densities in the world.
Source: *Time Almanac, 2005*

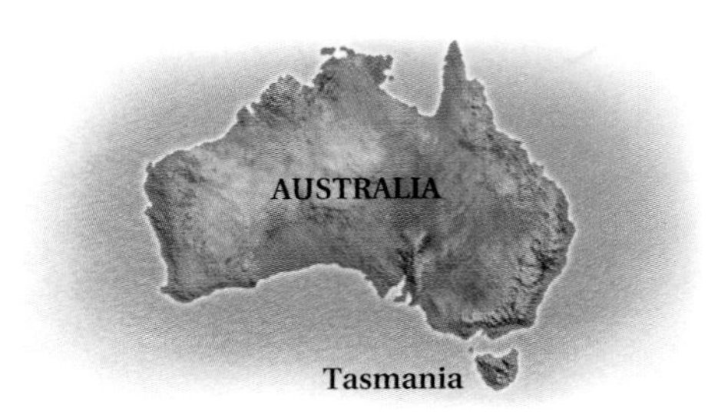

11. A car is driven 500 mi in 20 hr. What is the rate in miles per hour? in hours per mile?

12. A student eats 3 hamburgers in 15 min. What is the rate in hamburgers per minute? in minutes per hamburger?

13. *Points per Game.* In the 2004–2005 season, Yao Ming of the Houston Rockets scored 1465 points in 80 games. What was the rate in points per game?
Source: National Basketball Association

14. *Points per Game.* In the 2004–2005 season, Tayshaun Prince of the Detroit Pistons scored 1206 points in 82 games. What was the rate in points per game?
Source: National Basketball Association

15. *Lawn Watering.* To water a lawn adequately requires 623 gal of water for every 1000 ft^2. What is the rate in gallons per square foot?

16. A car is driven 200 km on 40 L of gasoline. What is the rate in kilometers per liter?

17. *Speed of Light.* Light travels 186,000 mi in 1 sec. What is its rate, or speed, in miles per second?
Source: *The Handy Science Answer Book*

18. *Speed of Sound.* Sound travels 1100 ft in 1 sec. What is its rate, or speed, in feet per second?
Source: *The Handy Science Answer Book*

19. Impulses in nerve fibers travel 310 km in 2.5 hr. What is the rate, or speed, in kilometers per hour?

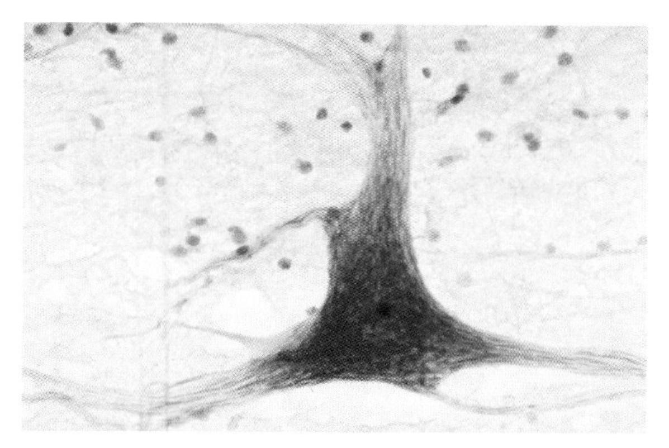

20. A black racer snake can travel 4.6 km in 2 hr. What is its rate, or speed, in kilometers per hour?

21. *Elephant Heartbeat.* The heart of an elephant, at rest, will beat an average of 1500 beats in 60 min. What is the rate in beats per minute?
Source: *The Handy Science Answer Book*

22. *Human Heartbeat.* The heart of a human, at rest, will beat an average of 4200 beats in 60 min. What is the rate in beats per minute?
Source: *The Handy Science Answer Book*

b Find each unit price in each of Exercises 23–32. Then determine which size has the lowest unit price.

23. *Pert Plus Shampoo.*

PACKAGE	PRICE	UNIT PRICE
13.5 oz	$2.59	
25.4 oz	$3.99	

24. *Roll-on Deodorant.*

PACKAGE	PRICE	UNIT PRICE
2.25 oz	$2.49	
3.5 oz	$3.98	

25. *Miracle Whip.*

PACKAGE	PRICE	UNIT PRICE
16-oz jar	$1.84	
18-oz squeezable	$2.49	

26. *Bush's Homestyle Baked Beans.*

PACKAGE	PRICE	UNIT PRICE
16 oz	$0.99	
28 oz	$1.44	

27. *Meijer Coffee.*

PACKAGE	PRICE	UNIT PRICE
11.5 oz	$2.09	
34.5 oz	$5.27	

28. *Maxwell House Coffee.*

PACKAGE	PRICE	UNIT PRICE
13 oz	$3.74	
34.5 oz	$6.36	

29. *Jif Creamy Peanut Butter.*

PACKAGE	PRICE	UNIT PRICE
18 oz	$1.89	
28 oz	$3.25	
40 oz	$4.99	
64 oz	$7.99	

30. *Downy Fabric Softener.*

PACKAGE	PRICE	UNIT PRICE
40 oz	$3.23	
60 oz	$4.79	
80 oz	$7.99	
120 oz	$10.69	

31. *Tide Liquid Laundry Detergent.*

PACKAGE	PRICE	UNIT PRICE
50 fl oz	$4.29	
100 fl oz	$5.29	
200 fl oz	$10.49	
300 fl oz	$15.79	

32. *Del Monte Green Beans.*

PACKAGE	PRICE	UNIT PRICE
8 oz	$0.59	
14.5 oz	$0.69	
28 oz	$1.19	

Use the unit prices listed in Exercises 23–32 when doing Exercises 33 and 34.

33. $\mathbf{D_W}$ Look over the unit prices for each size package of Downy fabric softener. What seems to violate common sense about these unit prices? Why do you think the products are sold this way?

34. $\mathbf{D_W}$ Compare the prices and unit prices for the 16-oz jar of Miracle Whip and the 18-oz squeezable container of Miracle Whip. What seems unusual about these prices? Explain why you think this has happened.

SKILL MAINTENANCE

Solve.

35. There are 20.6 million people in this country who play the piano and 18.9 million who play the guitar. How many more play the piano than the guitar? [4.7a]

36. A serving of fish steak (cross section) is generally $\frac{1}{2}$ lb. How many servings can be prepared from a cleaned $18\frac{3}{4}$-lb tuna? [3.6c]

37. *Surf Expo.* In a swimwear showing at Surf Expo, a trade show for retailers of beach supplies, each swimsuit test takes 8 minutes (min). If the show runs for 240 min, how many tests can be scheduled? [1.8a]

38. *Eating Habits.* Each year, Americans eat 24.8 billion hamburgers and 15.9 billion hot dogs. How many more hamburgers than hot dogs do Americans eat? [4.7a]

Multiply. [4.3a]

39. $\begin{array}{r} 45.67 \\ \times \quad 2.4 \\ \hline \end{array}$

40. $\begin{array}{r} 678.19 \\ \times \quad 100 \\ \hline \end{array}$

41. 84.3×69.2

42. 1002.56×465

SYNTHESIS

43. Recently, certain manufacturers have been changing the size of their containers in such a way that the consumer thinks the price of a product has been lowered when, in reality, a higher unit price is being charged.

Some aluminum juice cans are now concave (curved in) on the bottom. Suppose the volume of the can in the figure has been reduced from a fluid capacity of 6 oz to 5.5 oz, and the price of each can has been reduced from 65¢ to 60¢. Find the unit price of each container in cents per ounce.

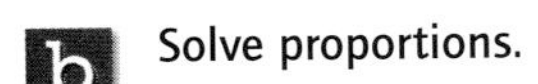

5.3 PROPORTIONS

Objectives

a Determine whether two pairs of numbers are proportional.

b Solve proportions.

During the 2004 season, Peyton Manning of the Indianapolis Colts completed 336 passes out of 497 attempts. His pass-completion rate was

$$\text{Completion rate} = \frac{336 \text{ completions}}{497 \text{ attempts}} = \frac{336}{497} \frac{\text{completion}}{\text{attempt}}$$

$$\approx 0.676 \frac{\text{completion}}{\text{attempt}}.$$

The rate was 0.676 completion per attempt.

Ben Roethlisberger of the Pittsburgh Steelers completed 196 passes out of 295 attempts. His pass-completion rate was

$$\text{Completion rate} = \frac{196 \text{ completions}}{295 \text{ attempts}} = \frac{196}{295} \frac{\text{completion}}{\text{attempt}}$$

$$\approx 0.664 \frac{\text{completion}}{\text{attempt}}.$$

The rate was 0.664 completion per attempt. We can see that the rates are not equal.

Source: National Football League

Instead of comparing the rates in decimal notation, we can compare the ratios

$$\frac{336}{497} \quad \text{and} \quad \frac{196}{295}$$

using the test for equality considered in Section 2.5. We compare cross products.

$$336 \cdot 295 = 99{,}120 \qquad \frac{336}{497} \stackrel{?}{=} \frac{196}{295} \qquad 497 \cdot 196 = 97{,}412$$

Since

$99{,}120 \neq 97{,}412$, we know that

$$\frac{336}{497} \neq \frac{196}{295}.$$

Thus the ratios are not equal. If the ratios had been equal, we would say they are proportional.

a Proportions

When two pairs of numbers (such as 3, 2 and 6, 4) have the same ratio, we say that they are **proportional.** The equation

$$\frac{3}{2} = \frac{6}{4}$$

states that the pairs 3, 2 and 6, 4 are proportional. Such an equation is called a **proportion.** We sometimes read $\frac{3}{2} = \frac{6}{4}$ as "3 is to 2 as 6 is to 4."

EXAMPLE 1 Determine whether 1, 2, and 3, 6 are proportional.

We can use cross products:

$$1 \cdot 6 = 6 \qquad \frac{1}{2} \stackrel{?}{=} \frac{3}{6} \qquad 2 \cdot 3 = 6.$$

Since the cross products are the same, $6 = 6$, we know that $\frac{1}{2} = \frac{3}{6}$, so the numbers are proportional.

EXAMPLE 2 Determine whether 2, 5 and 4, 7 are proportional.

We can use cross products:

$$2 \cdot 7 = 14 \qquad \frac{2}{5} \stackrel{?}{=} \frac{4}{7} \qquad 5 \cdot 4 = 20.$$

Since the cross products are not the same, $14 \neq 20$, we know that $\frac{2}{5} \neq \frac{4}{7}$, so the numbers are not proportional.

Do Exercises 1–3.

EXAMPLE 3 Determine whether 3.2, 4.8 and 0.16, 0.24 are proportional.

We can use cross products:

$$3.2 \times 0.24 = 0.768 \qquad \frac{3.2}{4.8} \stackrel{?}{=} \frac{0.16}{0.24} \qquad 4.8 \times 0.16 = 0.768.$$

Since the cross products are the same, $0.768 = 0.768$, we know that $\frac{3.2}{4.8} = \frac{0.16}{0.24}$, so the numbers are proportional.

Do Exercises 4 and 5.

EXAMPLE 4 Determine whether $4\frac{2}{3}$, $5\frac{1}{2}$ and $8\frac{7}{8}$, $16\frac{1}{3}$ are proportional.

We can use cross products:

$$\frac{4\frac{2}{3}}{5\frac{1}{2}} \stackrel{?}{=} \frac{8\frac{7}{8}}{16\frac{1}{3}}$$

$$4\frac{2}{3} \cdot 16\frac{1}{3} = \frac{14}{3} \cdot \frac{49}{3} = \frac{686}{9} = 76\frac{2}{9}$$

$$5\frac{1}{2} \cdot 8\frac{7}{8} = \frac{11}{2} \cdot \frac{71}{8} = \frac{781}{16} = 48\frac{13}{16}.$$

Since the cross products are not the same, $76\frac{2}{9} \neq 48\frac{13}{16}$, we know that the numbers are not proportional.

Do Exercise 6.

b Solving Proportions

Let's now look at solving proportions. Consider the proportion

$$\frac{x}{3} = \frac{4}{6}.$$

One way to solve a proportion is to use cross products. Then we can divide on

Determine whether the two pairs of numbers are proportional.

1. 3, 4 and 6, 8

2. 1, 4 and 10, 39

3. 1, 2 and 20, 39

Determine whether the two pairs of numbers are proportional.

4. 6.4, 12.8 and 5.3, 10.6

5. 6.8, 7.4 and 3.4, 4.2

6. Determine whether $4\frac{2}{3}$, $5\frac{1}{2}$ and 14, $16\frac{1}{2}$ are proportional.

Answers on page A-11

7. Solve: $\frac{x}{63} = \frac{2}{9}$.

8. Solve: $\frac{x}{9} = \frac{5}{4}$.

Answers on page A-11

Study Tips

WRITING ALL THE STEPS

Take the time to include all the steps when working your homework problems. Doing so will help you organize your thinking and avoid computational errors. If you find a wrong answer, having all the steps allows easier checking of your work. It will also give you complete, step-by-step solutions of the exercises that can be used to study for an exam.

Writing down all the steps and keeping your work organized may also give you a better chance of getting partial credit.

"Success comes before work only in the dictionary."

Anonymous

both sides to get the variable alone:

$x \cdot 6 = 3 \cdot 4$ Equating cross products (finding cross products and setting them equal)

$\frac{x \cdot 6}{6} = \frac{3 \cdot 4}{6}$ Dividing by 6 on both sides

$$x = \frac{3 \cdot 4}{6} = \frac{12}{6} = 2.$$

We can check that 2 is the solution by replacing x with 2 and using cross products:

$$2 \cdot 6 = 12 \quad \frac{2}{3} \stackrel{?}{=} \frac{4}{6} \quad 3 \cdot 4 = 12.$$

Since the cross products are the same, it follows that $\frac{2}{3} = \frac{4}{6}$; so the numbers 2, 3 and 4, 6 are proportional, and 2 is the solution of the equation.

SOLVING PROPORTIONS

To solve $\frac{x}{a} = \frac{c}{d}$, equate *cross products* and divide on both sides to get x alone.

Do Exercise 7.

EXAMPLE 5 Solve: $\frac{x}{7} = \frac{5}{3}$. Write a mixed numeral for the answer.

We have

$\frac{x}{7} = \frac{5}{3}$

$x \cdot 3 = 7 \cdot 5$ Equating cross products

$\frac{x \cdot 3}{3} = \frac{7 \cdot 5}{3}$ Dividing by 3

$x = \frac{7 \cdot 5}{3}$

$x = \frac{35}{3}$, or $11\frac{2}{3}$.

The solution is $11\frac{2}{3}$.

Do Exercise 8.

EXAMPLE 6 Solve: $\frac{7.7}{15.4} = \frac{y}{2.2}$.

We have

$\frac{7.7}{15.4} = \frac{y}{2.2}$

$7.7 \times 2.2 = 15.4 \times y$ Equating cross products

$\frac{7.7 \times 2.2}{15.4} = \frac{15.4 \times y}{15.4}$. Dividing by 15.4

Then

$$\frac{7.7 \times 2.2}{15.4} = y$$

$$\frac{16.94}{15.4} = y \quad \text{Multiplying}$$

$$1.1 = y. \quad \text{Dividing: } 15.4 \overline{)16.94} = 1.1$$

$$\begin{array}{r} 1.1 \\ 15.4\overline{)16.9_\wedge4} \\ \underline{1540} \\ 154 \\ \underline{154} \\ 0 \end{array}$$

The solution is 1.1.

Do Exercise 9.

9. Solve: $\frac{21}{5} = \frac{n}{2.5}$.

EXAMPLE 7 Solve: $\frac{8}{x} = \frac{5}{3}$. Write decimal notation for the answer.

We have

$$\frac{8}{x} = \frac{5}{3}$$

$$8 \cdot 3 = x \cdot 5 \quad \text{Equating cross products}$$

$$\frac{8 \cdot 3}{5} = \frac{x \cdot 5}{5} \quad \text{Dividing by 5}$$

$$\frac{8 \cdot 3}{5} = x$$

$$\frac{24}{5} = x \quad \text{Multiplying}$$

$$4.8 = x. \quad \text{Simplifying}$$

The solution is 4.8.

Do Exercise 10.

10. Solve: $\frac{6}{x} = \frac{25}{11}$.

EXAMPLE 8 Solve: $\frac{3.4}{4.93} = \frac{10}{n}$.

We have

$$\frac{3.4}{4.93} = \frac{10}{n}$$

$$3.4 \times n = 4.93 \times 10 \quad \text{Equating cross products}$$

$$\frac{3.4 \times n}{3.4} = \frac{4.93 \times 10}{3.4} \quad \text{Dividing by 3.4}$$

$$n = \frac{4.93 \times 10}{3.4}$$

$$n = \frac{49.3}{3.4} \quad \text{Multiplying}$$

$$n = 14.5. \quad \text{Dividing}$$

The solution is 14.5.

Do Exercise 11.

11. Solve: $\frac{0.4}{0.9} = \frac{4.8}{t}$.

Answers on page A-11

12. Solve:

$$\frac{8\frac{1}{3}}{x} = \frac{10\frac{1}{2}}{3\frac{3}{4}}.$$

Answer on page A-11

EXAMPLE 9 Solve: $\frac{4\frac{2}{3}}{5\frac{1}{2}} = \frac{14}{x}$. Write a mixed numeral for the answer.

We have

$$\frac{4\frac{2}{3}}{5\frac{1}{2}} = \frac{14}{x}$$

$$4\frac{2}{3} \cdot x = 14 \cdot 5\frac{1}{2} \qquad \text{Equating cross products}$$

$$\frac{14}{3} \cdot x = 14 \cdot \frac{11}{2} \qquad \text{Converting to fraction notation}$$

$$\frac{\frac{14}{3} \cdot x}{\frac{14}{3}} = \frac{14 \cdot \frac{11}{2}}{\frac{14}{3}} \qquad \text{Dividing by } \frac{14}{3}$$

$$x = 14 \cdot \frac{11}{2} \div \frac{14}{3}$$

$$x = \cancel{14} \cdot \frac{11}{2} \cdot \frac{3}{\cancel{14}} \qquad \text{Multiplying by the reciprocal of the divisor}$$

$$x = \frac{11 \cdot 3}{2} \qquad \text{Simplifying by removing a factor of 1: } \frac{14}{14} = 1$$

$$x = \frac{33}{2}, \text{ or } 16\frac{1}{2}.$$

The solution is $16\frac{1}{2}$.

Do Exercise 12.

CALCULATOR CORNER

Solving Proportions Note in Examples 5–9 that when we solve a proportion, we equate cross products and then we divide on both sides to isolate the variable on one side of the equation. We can use a calculator to do the calculations in this situation. In Example 8, for instance, after equating cross products and dividing by 3.4 on both sides, we have

$$n = \frac{4.93 \times 10}{3.4}.$$

To find n on a calculator, we can press [4] [.] [9] [3] [×] [1] [0] [÷] [3] [.] [4] [=]. The result is 14.5, so $n = 14.5$.

Exercises

1. Use a calculator to solve each of the proportions in Examples 5–7.
2. Use a calculator to solve each of the proportions in Margin Exercises 7–11.

Solve each proportion.

3. $\frac{15.75}{20} = \frac{a}{35}$
4. $\frac{32}{x} = \frac{25}{20}$
5. $\frac{t}{57} = \frac{17}{64}$
6. $\frac{71.2}{a} = \frac{42.5}{23.9}$
7. $\frac{29.6}{3.15} = \frac{x}{4.23}$
8. $\frac{a}{3.01} = \frac{1.7}{0.043}$

5.3 EXERCISE SET

For Extra Help

MathXL | MyMathLab | InterAct Math | Math Tutor Center | Digital Video Tutor CD 3 Videotape 5 | Student's Solutions Manual

a Determine whether the two pairs of numbers are proportional.

1. 5, 6 and 7, 9

2. 7, 5 and 6, 4

3. 1, 2 and 10, 20

4. 7, 3 and 21, 9

5. 2.4, 3.6 and 1.8, 2.7

6. 4.5, 3.8 and 6.7, 5.2

7. $5\frac{1}{3}, 8\frac{1}{4}$ and $2\frac{1}{5}, 9\frac{1}{2}$

8. $2\frac{1}{3}, 3\frac{1}{2}$ and 14, 21

Pass-Completion Rates. The table below lists the records of four NFL quarterbacks from the 2000 season.

PLAYER	TEAM	NUMBER OF PASSES COMPLETED	NUMBER OF PASSES ATTEMPTED	NUMBER OF COMPLETIONS PER ATTEMPT (completion rate)
m Brady	New England Patriots	288	474	
ew Brees	San Diego Chargers	262	400	
unte Culpepper	Minnesota Vikings	379	548	
n Roethlisberger	Pittsburgh Steelers	196	295	

rce: National Football League

9. Find each pass-completion rate rounded to the nearest hundredth. Are any the same?

10. Use cross products to determine whether any quarterback completion rates are the same.

b Solve.

11. $\frac{18}{4} = \frac{x}{10}$

12. $\frac{x}{45} = \frac{20}{25}$

13. $\frac{x}{8} = \frac{9}{6}$

14. $\frac{8}{10} = \frac{n}{5}$

15. $\frac{t}{12} = \frac{5}{6}$

16. $\frac{12}{4} = \frac{x}{3}$

17. $\frac{2}{5} = \frac{8}{n}$

18. $\frac{10}{6} = \frac{5}{x}$

19. $\frac{n}{15} = \frac{10}{30}$

20. $\frac{2}{24} = \frac{x}{36}$

21. $\frac{16}{12} = \frac{24}{x}$

22. $\frac{7}{11} = \frac{2}{x}$

23. $\frac{6}{11} = \frac{12}{x}$

24. $\frac{8}{9} = \frac{32}{n}$

25. $\frac{20}{7} = \frac{80}{x}$

26. $\frac{5}{x} = \frac{4}{10}$

27. $\frac{12}{9} = \frac{x}{7}$

28. $\frac{x}{20} = \frac{16}{15}$

29. $\frac{x}{13} = \frac{2}{9}$

30. $\frac{1.2}{4} = \frac{x}{9}$

31. $\frac{t}{0.16} = \frac{0.15}{0.40}$

32. $\frac{x}{11} = \frac{7.1}{2}$

33. $\frac{100}{25} = \frac{20}{n}$

34. $\frac{35}{125} = \frac{7}{m}$

35. $\frac{7}{\frac{1}{4}} = \frac{28}{x}$

36. $\frac{x}{6} = \frac{1}{6}$

37. $\frac{\frac{1}{4}}{\frac{1}{2}} = \frac{\frac{1}{2}}{x}$

38. $\frac{1}{7} = \frac{x}{4\frac{1}{2}}$

39. $\frac{1}{2} = \frac{7}{x}$

40. $\frac{x}{3} = \frac{0}{9}$

41. $\frac{\frac{2}{7}}{\frac{3}{4}} = \frac{\frac{5}{6}}{y}$

42. $\frac{\frac{5}{4}}{\frac{5}{8}} = \frac{\frac{3}{2}}{Q}$

43. $\frac{2\frac{1}{2}}{3\frac{1}{3}} = \frac{x}{4\frac{1}{4}}$

44. $\frac{5\frac{1}{5}}{6\frac{1}{6}} = \frac{y}{3\frac{1}{2}}$

45. $\frac{1.28}{3.76} = \frac{4.28}{y}$

46. $\frac{10.4}{12.4} = \frac{6.76}{t}$

47. $\frac{10\frac{3}{8}}{12\frac{2}{3}} = \frac{5\frac{3}{4}}{y}$

48. $\frac{12\frac{7}{8}}{20\frac{3}{4}} = \frac{5\frac{2}{3}}{y}$

49. D_W Instead of equating cross products, a student solves $\frac{x}{7} = \frac{5}{3}$ (see Example 5) by multiplying on both sides by the least common denominator, 21. Is his approach a good one? Why or why not?

50. D_W An instructor predicts that a student's test grade will be proportional to the amount of time the student spends studying. What is meant by this? Write an example of a proportion that involves the grades of two students and their study times.

SKILL MAINTENANCE

VOCABULARY REINFORCEMENT

In each of Exercises 51–58, fill in the blank with the correct term from the given list. Some of the choices may not be used.

51. A ratio is the ______________ of two quantities. [5.1a]

52. A number is divisible by 9 if the ______________ of the digits is divisible by 9. [2.2a]

53. To compute a(n) ______________ of a set of numbers, we add the numbers and then divide by the number of addends. [3.7a]

54. To convert from ______________ to ______________, move the decimal point two places to the right and change the \$ sign in front to the ¢ sign at the end. [4.3b]

55. In the equation $103 - 13 = 90$, the ______________ is 13. [1.3a]

56. The decimal 0.125 is an example of a(n) ______________ decimal. [4.5a]

57. The sentence $\frac{2}{5} \cdot \frac{4}{9} = \frac{4}{9} \cdot \frac{2}{5}$ illustrates the ______________ law of multiplication. [1.5a]

58. To solve $\frac{x}{a} = \frac{c}{d}$, equate the ______________ and divide on both sides to get x alone. [5.3b]

cross products
cents
dollars
terminating
repeating
sum
difference
minuend
subtrahend
associative
commutative
average
product
quotient

SYNTHESIS

 Solve.

59. $\frac{1728}{5643} = \frac{836.4}{x}$

60. $\frac{328.56}{627.48} = \frac{y}{127.66}$

61. *Strikeouts per Home Run.* Baseball Hall-of-Famer Babe Ruth had 1330 strikeouts and 714 home runs in his career. Hall-of-Famer Mike Schmidt had 1883 strikeouts and 548 home runs in his career.

a) Find the unit rate of each player in terms of strikeouts per home run. (These rates were considered among the highest in the history of the game and yet each made the Hall of Fame.)

b) Which player had the higher rate?

5.4 APPLICATIONS OF PROPORTIONS

Objective

a Solve applied problems involving proportions.

a Applications and Problem Solving

Proportions have applications in such diverse fields as business, chemistry, health sciences, and home economics, as well as to many areas of daily life. Proportions are useful in making predictions.

EXAMPLE 1 *Predicting Total Distance.* Donna drives her delivery van 800 mi in 3 days. At this rate, how far will she drive in 15 days?

1. **Familiarize.** We let $d =$ the distance traveled in 15 days.
2. **Translate.** We translate to a proportion. We make each side the ratio of distance to time, with distance in the numerator and time in the denominator.

$$\text{Distance in 15 days} \rightarrow \frac{d}{15} = \frac{800}{3} \leftarrow \text{Distance in 3 days}$$
$$\text{Time} \rightarrow 15 \qquad 3 \leftarrow \text{Time}$$

It may help to verbalize the proportion above as "the unknown distance d is to 15 days as the known distance 800 miles is to 3 days."

3. **Solve.** Next, we solve the proportion:

$$3 \cdot d = 15 \cdot 800 \quad \text{Equating cross products}$$
$$\frac{3 \cdot d}{3} = \frac{15 \cdot 800}{3} \quad \text{Dividing by 3 on both sides}$$
$$d = \frac{15 \cdot 800}{3}$$
$$d = 4000. \quad \text{Multiplying and dividing}$$

4. **Check.** We substitute into the proportion and check cross products:

$$\frac{4000}{15} = \frac{800}{3};$$
$$4000 \cdot 3 = 12{,}000; \qquad 15 \cdot 800 = 12{,}000.$$

The cross products are the same.

5. **State.** Donna will drive 4000 mi in 15 days.

Do Exercise 1.

Problems involving proportion can be translated in more than one way. In Example 1, any one of the following is a correct translation:

$$\frac{800}{3} = \frac{d}{15}, \quad \frac{15}{d} = \frac{3}{800}, \quad \frac{15}{3} = \frac{d}{800}, \quad \frac{800}{d} = \frac{3}{15}.$$

Equating the cross products in each equation gives us the equation $3 \cdot d = 15 \cdot 800$.

1. **Calories Burned.** Your author generally exercises for 2 hr each day. The readout on an exercise machine tells him that if he exercises for 24 min, he will burn 356 calories. How many calories will he burn if he exercises for 30 min?

Source: Star Trac Treadmill

Answer on page A-12

EXAMPLE 2 *Recommended Dosage.* To control a fever, a doctor suggests that a child who weighs 28 kg be given 320 mg of Tylenol. If the dosage is proportional to the child's weight, how much Tylenol is recommended for a child who weighs 35 kg?

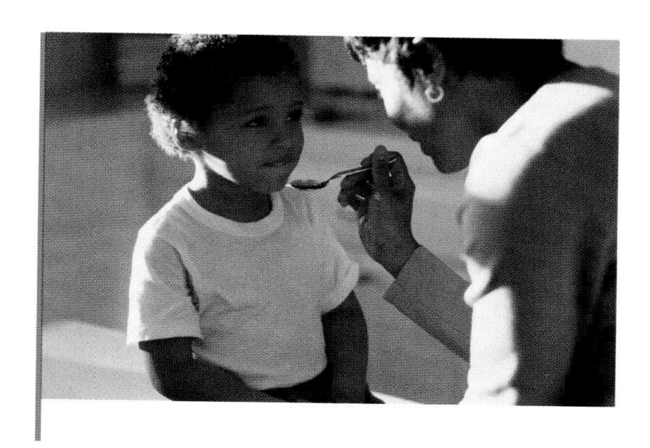

1. **Familiarize.** We let $t =$ the number of milligrams of Tylenol.
2. **Translate.** We translate to a proportion, keeping the amount of Tylenol in the numerators.

$$\text{Tylenol suggested} \rightarrow \frac{320}{28} = \frac{t}{35} \leftarrow \text{Tylenol suggested}$$
$$\text{Child's weight} \rightarrow \qquad \qquad \leftarrow \text{Child's weight}$$

3. **Solve.** Next, we solve the proportion:

$$320 \cdot 35 = 28 \cdot t \qquad \text{Equating cross products}$$
$$\frac{320 \cdot 35}{28} = \frac{28 \cdot t}{28} \qquad \text{Dividing by 28 on both sides}$$
$$\frac{320 \cdot 35}{28} = t$$
$$400 = t. \qquad \text{Multiplying and dividing}$$

4. **Check.** We substitute into the proportion and check cross products:

$$\frac{320}{28} = \frac{400}{35};$$
$$320 \cdot 35 = 11{,}200; \qquad 28 \cdot 400 = 11{,}200.$$

The cross products are the same.

5. **State.** The dosage for a child who weighs 35 kg is 400 mg.

Do Exercise 2.

EXAMPLE 3 *Purchasing Tickets.* Carey bought 8 tickets to an international food festival for $52. How many tickets could she purchase with $90?

1. **Familiarize.** We let $n =$ the number of tickets that can be purchased with $90.
2. **Translate.** We translate to a proportion, keeping the number of tickets in the numerators.

$$\text{Tickets} \rightarrow \frac{8}{52} = \frac{n}{90} \leftarrow \text{Tickets}$$
$$\text{Cost} \rightarrow \qquad \qquad \leftarrow \text{Cost}$$

2. **Determining Paint Needs.** Lowell and Chris run a summer painting company to support their college expenses. They can paint 1600 ft^2 of clapboard with 4 gal of paint. How much paint would be needed for a building with 6000 ft^2 of clapboard?

Answer on page A-12

3. Purchasing Shirts. If 2 shirts can be bought for \$47, how many shirts can be bought with \$200?

4. Men's Hip Measurements. It is recommended that a man's waist-to-hip ratio be 0.95 (or lower). Malcolm's hip measurement is 40 in. To meet the recommendation, what should Malcolm's waist measurement be?

Source: David Schmidt, "Lifting Weight Myths," *Nutrition Action Newsletter* 20, no. 4, October 1993

Answers on page A-12

3. Solve. Next, we solve the proportion:

$$52 \cdot n = 8 \cdot 90 \quad \text{Equating cross products}$$

$$\frac{52 \cdot n}{52} = \frac{8 \cdot 90}{52} \quad \text{Dividing by 52 on both sides}$$

$$n = \frac{8 \cdot 90}{52}$$

$$n \approx 13.8. \quad \text{Multiplying and dividing}$$

Because it is impossible to buy a fractional part of a ticket, we must round our answer *down* to 13.

4. Check. As a check, we use a different approach: We find the cost per ticket and then divide \$90 by that price. Since $52 \div 8 = 6.50$ and $90 \div 6.50 \approx 13.8$, we have a check.

5. State. Carey could purchase 13 tickets with \$90.

Do Exercise 3.

EXAMPLE 4 *Women's Hip Measurements.* For improved health, it is recommended that a woman's waist-to-hip ratio be 0.85 (or lower). Marta's hip measurement is 40 in. To meet the recommendation, what should Marta's waist measurement be?

Source: David Schmidt, "Lifting Weight Myths," *Nutrition Action Newsletter* 20, no. 4, October 1993

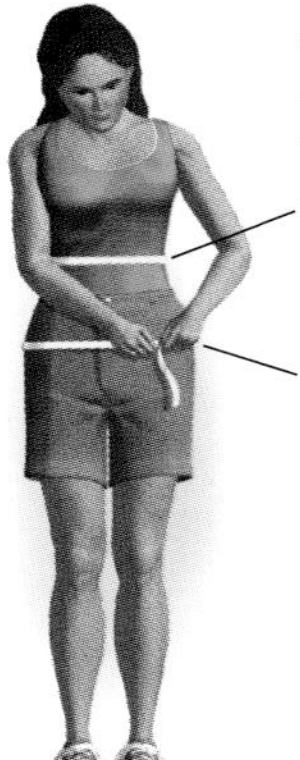
Waist measurement is the smallest measurement below the ribs but above the navel.
Hip measurement is the largest measurement around the widest part of the buttocks.

1. Familiarize. Note that $0.85 = \frac{85}{100}$. We let w = Marta's waist measurement.

2. Translate. We translate to a proportion as follows:

$$\frac{w}{40} = \frac{85}{100}.$$

Waist measurement → w; Hip measurement → 40; $\frac{85}{100}$: Recommended waist-to-hip ratio

3. Solve. Next, we solve the proportion:

$$100 \cdot w = 40 \cdot 85 \quad \text{Equating cross products}$$

$$\frac{100 \cdot w}{100} = \frac{40 \cdot 85}{100} \quad \text{Dividing by 100 on both sides}$$

$$w = \frac{40 \cdot 85}{100}$$

$$w = 34. \quad \text{Multiplying and dividing}$$

4. **Check.** As a check, we divide 34 by 40: $34 \div 40 = 0.85$. This is the desired ratio.
5. **State.** Marta's recommended waist measurement is 34 in. (or less).

Do Exercise 4 on the preceding page.

EXAMPLE 5 *Construction Plans.* Architects make blueprints of projects being constructed. These are scale drawings in which lengths are in proportion to actual sizes. The Hennesseys are constructing a rectangular deck just outside their house. The architectural blueprints are rendered such that $\frac{3}{4}$ in. on the drawing is actually 2.25 ft on the deck. The width of the deck on the drawing is 4.3 in. How wide is the deck in reality?

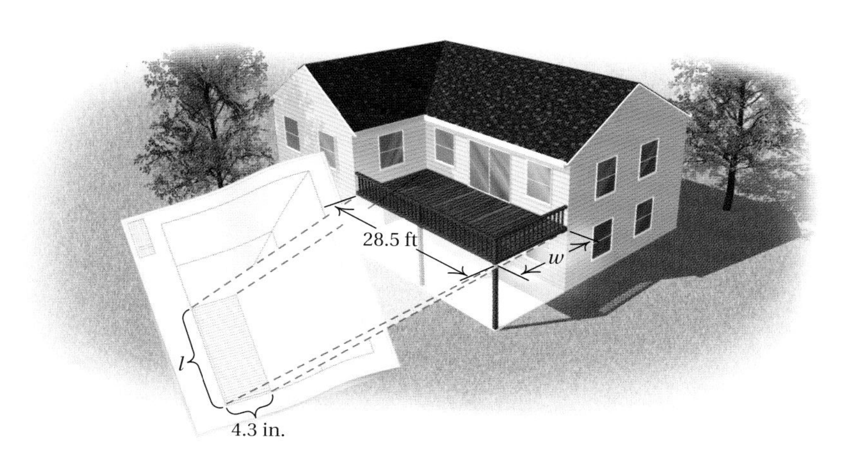

1. **Familiarize.** We let $w =$ the width of the deck.
2. **Translate.** Then we translate to a proportion, using 0.75 for $\frac{3}{4}$ in.

$$\text{Measure on drawing} \rightarrow \frac{0.75}{2.25} = \frac{4.3}{w} \leftarrow \text{Width of drawing}$$
$$\text{Measure on deck} \rightarrow \qquad \leftarrow \text{Width of deck}$$

3. **Solve.** Next, we solve the proportion:

$$0.75 \times w = 2.25 \times 4.3 \qquad \text{Equating cross products}$$
$$\frac{0.75 \times w}{0.75} = \frac{2.25 \times 4.3}{0.75} \qquad \text{Dividing by 0.75 on both sides}$$
$$w = \frac{2.25 \times 4.3}{0.75}$$
$$w = 12.9.$$

4. **Check.** We substitute into the proportion and check cross products:

$$\frac{0.75}{2.25} = \frac{4.3}{12.9};$$
$$0.75 \times 12.9 = 9.675; \qquad 2.25 \times 4.3 = 9.675.$$

The cross products are the same.

5. **State.** The width of the deck is 12.9 ft.

Do Exercise 5.

5. **Construction Plans.** In Example 5, the length of the actual deck is 28.5 ft. What is the length of the deck on the blueprints?

Answer on page A-12

6. Estimating a Deer Population. To determine the number of deer in a forest, a conservationist catches 612 deer, tags them, and releases them. Later, 244 deer are caught, and it is found that 72 of them are tagged. Estimate how many deer are in the forest.

Answer on page A-12

EXAMPLE 6 *Estimating a Wildlife Population.* To determine the number of fish in a lake, a conservationist catches 225 fish, tags them, and throws them back into the lake. Later, 108 fish are caught, and it is found that 15 of them are tagged. Estimate how many fish are in the lake.

1. **Familiarize.** We let $F =$ the number of fish in the lake.
2. **Translate.** We translate to a proportion as follows:

$$\begin{array}{rcl}\text{Fish tagged originally} \rightarrow & \dfrac{225}{F} = \dfrac{15}{108}. & \leftarrow \text{Tagged fish caught later} \\ \text{Fish in lake} \rightarrow & & \leftarrow \text{Fish caught later}\end{array}$$

3. **Solve.** Next, we solve the proportion:

$$225 \cdot 108 = F \cdot 15 \qquad \text{Equating cross products}$$

$$\frac{225 \cdot 108}{15} = \frac{F \cdot 15}{15} \qquad \text{Dividing by 15 on both sides}$$

$$\frac{225 \cdot 108}{15} = F$$

$$1620 = F. \qquad \text{Multiplying and dividing}$$

4. **Check.** We substitute into the proportion and check cross products:

$$\frac{225}{1620} = \frac{15}{108};$$

$$225 \cdot 108 = 24{,}300; \qquad 1620 \cdot 15 = 24{,}300.$$

The cross products are the same.

5. **State.** We estimate that there are 1620 fish in the lake.

Do Exercise 6.

Translating for Success

The goal of these matching questions is to practice step (2), *Translate*, of the five-step problem-solving process. Translate each word problem to an equation and select a correct translation from equations A–O.

1. *Calories in Cereal.* There are 140 calories in a $1\frac{1}{2}$-cup serving of Brand A cereal. How many calories are there in 6 cups of the cereal?

2. *Calories in Cereal.* There are 140 calories in 6 cups of Brand B cereal. How many calories are there in a $1\frac{1}{2}$-cup serving of the cereal?

3. *Gallons of Gasoline.* Jared's SUV traveled 310 miles on 15.5 gallons of gasoline. At this rate, how many gallons would be needed to travel 465 miles?

4. *Gallons of Gasoline.* Elizabeth's new fuel-efficient car traveled 465 miles on 15.5 gallons of gasoline. At this rate, how many gallons will be needed to travel 310 miles?

5. *Perimeter.* Find the perimeter of a rectangular field that measures 83.7 m by 62.4 m.

A. $\frac{310}{15.5} = \frac{465}{x}$

B. $180 = 1\frac{1}{2} \cdot x$

C. $x = 71\frac{1}{8} - 76\frac{1}{2}$

D. $71\frac{1}{8} \cdot x = 74$

E. $74 \cdot 71\frac{1}{8} = x$

F. $x = 83.7 + 62.4$

G. $71\frac{1}{8} + x = 76\frac{1}{2}$

H. $x = 1\frac{2}{3} \cdot 180$

I. $\frac{140}{6} = \frac{x}{1\frac{1}{2}}$

J. $x = 2(83.7 + 62.4)$

K. $\frac{465}{15.5} = \frac{310}{x}$

L. $x = 83.7 \cdot 62.4$

M. $x = 180 \div 1\frac{2}{3}$

N. $\frac{140}{1\frac{1}{2}} = \frac{x}{6}$

O. $x = 1\frac{2}{3} \div 180$

Answers on page A-12

6. *Electric Bill.* Last month Todd's electric bills for his two rentals were \$83.70 and \$62.40. What was the total electric bill for the two properties?

7. *Package Tape.* A postal service center uses rolls of package tape that each contain 180 feet of tape. If it takes an average of $1\frac{2}{3}$ ft per package, how many packages can be taped with one roll?

8. *Online Price.* Jane spent \$180 for an area rug in a department store. Later she saw the same rug for sale online and realized she had paid $1\frac{1}{2}$ times the online price. What was the online price?

9. *Heights of Sons.* Henry's three sons play basket-ball on three different college teams. Jeff, Jason, and Jared's heights are 74 in., $71\frac{1}{8}$ in., and $76\frac{1}{2}$ in., respectively. How much taller is Jared than Jason?

10. *Total Investment.* An investor bought 74 shares of stock at $\$71\frac{1}{8}$ per share. What was the total investment?

5.4 EXERCISE SET

For Extra Help

MathXL

MyMathLab

InterAct Math

Math Tutor Center

Digital Video Tutor CD 3 Videotape 5

Student's Solutions Manual

a Solve.

1. *Study Time and Test Grades.* An English instructor asserted that students' test grades are directly proportional to the amount of time spent studying. Lisa studies 9 hr for a particular test and gets a score of 75. At this rate, how many hours would she have had to study to get a score of 92?

2. *Study Time and Test Grades.* A mathematics instructor asserted that students' test grades are directly proportional to the amount of time spent studying. Brent studies 15 hr for a particular test and gets a score of 85. At this rate, what score would he have received if he had studied 16 hr?

3. *Cap'n Crunch's Peanut Butter Crunch® Cereal.* The nutritional chart on the side of a box of Quaker Cap'n Crunch's Peanut Butter Crunch® Cereal states that there are 110 calories in a $\frac{3}{4}$-cup serving. How many calories are there in 6 cups of the cereal?

Nutrition Facts

Serving Size 3/4 Cup (27g)
Servings Per Container about 15

Amount Per Serving	Cereal Alone	with 1/2 Cup Vitamin A&D Fortified Skim Milk
Calories	110	150
Calories from Fat	25	25
	% Daily Value	
Total Fat 2.5g	4%	4%
Saturated Fat 0.5g	3%	3%
Polyunsaturated Fat 0.5g		
Monounsaturated Fat 1g		
Cholesterol 0mg	0%	0%
Sodium 200mg	8%	11%
Potassium 65mg	2%	8%
Total Carbohydrate 21g	7%	9%
Dietary Fiber 1g	3%	3%
Sugars 9g		
Other Carbohydrate 12g		
Protein 2g		

4. *Rice Krispies® Cereal.* The nutritional chart on the side of a box of Kellogg's Rice Krispies® Cereal states that there are 120 calories in a $1\frac{1}{4}$-cup serving. How many calories are there in 5 cups of the cereal?

Nutrition Facts

Serving Size $1\frac{1}{4}$ Cups (33g/1.2oz)
Servings Per Container about 12

Amount Per Serving	Cereal	Cereal with 1/2 Cup Vitamins A&D Fat Free Milk
Calories	120	160
Calories from Fat	0	0
	% Daily Value	
Total Fat 0g	0%	0%
Saturated Fat 0g	0%	0%
Trans Fat 0g		
Cholesterol 0mg	0%	0%
Sodium 320mg	13%	16%
Potassium 40mg	1%	7%
Total Carbohydrate 29g	10%	11%
Dietary Fiber 0g	0%	0%
Sugars 3g		
Other Carbohydrate 26g		
Protein 2g		

5. *Overweight Americans.* A study recently confirmed that of every 100 Americans, 60 are considered overweight. There were 295 million Americans in 2005. How many would be considered overweight?

Source: U.S. Centers for Disease Control

6. *Cancer Death Rate in Illinois.* It is predicted that for every 1000 people in the state of Illinois, 130.9 will die of cancer. The population of Chicago is about 2,721,547. How many of these people will die of cancer?

Source: *2001 New York Times Almanac*

7. *Gasoline Mileage.* Nancy's van traveled 84 mi on 6.5 gal of gasoline. At this rate, how many gallons would be needed to travel 126 mi?

8. *Bicycling.* Roy bicycled 234 mi in 14 days. At this rate, how far would Roy travel in 42 days?

9. *Quality Control.* A quality-control inspector examined 100 lightbulbs and found 7 of them to be defective. At this rate, how many defective bulbs will there be in a lot of 2500?

10. *Grading.* A professor must grade 32 essays in a literature class. She can grade 5 essays in 40 min. At this rate, how long will it take her to grade all 32 essays?

11. *Painting.* Fred uses 3 gal of paint to cover 1275 ft^2 of siding. How much siding can Fred paint with 7 gal of paint?

12. *Waterproofing.* Bonnie can waterproof 450 ft^2 of decking with 2 gal of sealant. How many gallons should Bonnie buy for a 1200-ft^2 deck?

13. *Publishing.* Every 6 pages of an author's manuscript corresponds to 5 published pages. How many published pages will a 540-page manuscript become?

14. *Turkey Servings.* An 8-lb turkey breast contains 36 servings of meat. How many pounds of turkey breast would be needed for 54 servings?

15. *Exchanging Money.* On 26 April 2005, 1 U.S. dollar was worth about 0.52521 British pound.

a) How much would 45 U.S. dollars be worth in British pounds?

b) How much would a car cost in U.S. dollars that costs 8640 British pounds?

16. *Exchanging Money.* On 26 April 2005, 1 U.S. dollar was worth about 1.1894 Swiss francs.

a) How much would 360 U.S. dollars be worth in Swiss francs?

b) How much would a pair of jeans cost in U.S. dollars that costs 80 Swiss francs?

17. *Exchanging Money.* On 26 April 2005, 1 U.S. dollar was worth about 106.03 Japanese yen.

a) How much would 200 U.S. dollars be worth in Japanese yen?

b) Dan was traveling in Japan and bought a skateboard that cost 3180 Japanese yen. How much would it cost in U.S. dollars?

18. *Exchanging Money.* On 26 April 2005, 1 U.S. dollar was worth about 11.059 Mexican pesos.

a) How much would 120 U.S. dollars be worth in Mexican pesos?

b) Jackie was traveling in Mexico and bought a watch that cost 3600 Mexican pesos. How much would it cost in U.S. dollars?

19. *Gas Mileage.* A 2005 Ford Mustang GT Convertible will go 372 mi on 15.5 gal of gasoline in highway driving.

a) How many gallons of gasoline will it take to drive 2690 mi from Boston to Phoenix?

b) How far can the car be driven on 140 gal of gasoline?

Source: Ford

20. *Gas Mileage.* A 2005 Volkswagen Passat will go 462 mi on 16.5 gal of gasoline in highway driving.

a) How many gallons of gasoline will it take to drive 1650 mi from Pittsburgh to Albuquerque?

b) How far can the car be driven on 130 gal of gasoline?

Source: Volkswagen of America, Inc.

21. *Lefties.* In a class of 40 students, on average, 6 will be left-handed. If a class includes 9 "lefties," how many students would you estimate are in the class?

22. *Sugaring.* When 38 gal of maple sap are boiled down, the result is 2 gal of maple syrup. How much sap is needed to produce 9 gal of syrup?

23. *Mileage.* Jean bought a new car. In the first 8 months, it was driven 9000 mi. At this rate, how many miles will the car be driven in 1 yr?

24. *Coffee Production.* Coffee beans from 14 trees are required to produce the 17 lb of coffee that the average person in the United States drinks each year. How many trees are required to produce 375 lb of coffee?

25. *Metallurgy.* In a metal alloy, the ratio of zinc to copper is 3 to 13. If there are 520 lb of copper, how many pounds of zinc are there?

26. *Class Size.* A college advertises that its student-to-faculty ratio is 14 to 1. If 56 students register for Introductory Spanish, how many sections of the course would you expect to see offered?

27. *Painting.* Helen can paint 950 ft^2 with 2 gal of paint. How many 1-gal cans does she need in order to paint a 30,000-ft^2 wall?

28. *Snow to Water.* Under typical conditions, $1\frac{1}{2}$ ft of snow will melt to 2 in. of water. To how many inches of water will $5\frac{1}{2}$ ft of snow melt?

29. *Grass-Seed Coverage.* It takes 60 oz of grass seed to seed 3000 ft^2 of lawn. At this rate, how much would be needed for 5000 ft^2 of lawn?

30. *Grass-Seed Coverage.* In Exercise 29, how much seed would be needed for 7000 ft^2 of lawn?

31. *Estimating a Deer Population.* To determine the number of deer in a game preserve, a forest ranger catches 318 deer, tags them, and releases them. Later, 168 deer are caught, and it is found that 56 of them are tagged. Estimate how many deer are in the game preserve.

32. *Estimating a Trout Population.* To determine the number of trout in a lake, a conservationist catches 112 trout, tags them, and throws them back into the lake. Later, 82 trout are caught, and it is found that 32 of them are tagged. Estimate how many trout there are in the lake.

33. *Map Scaling.* On a road atlas map, 1 in. represents 16.6 mi. If two cities are 3.5-in. apart on the map, how far apart are they in reality?

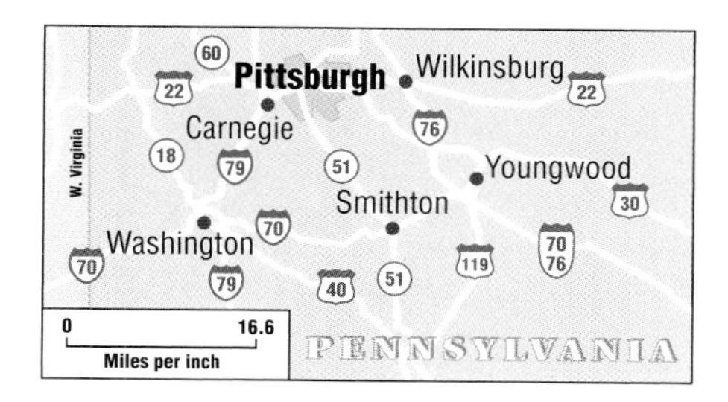

34. *Map Scaling.* On a map, $\frac{1}{4}$ in. represents 50 mi. If two cities are $3\frac{1}{4}$-in. apart on the map, how far apart are they in reality?

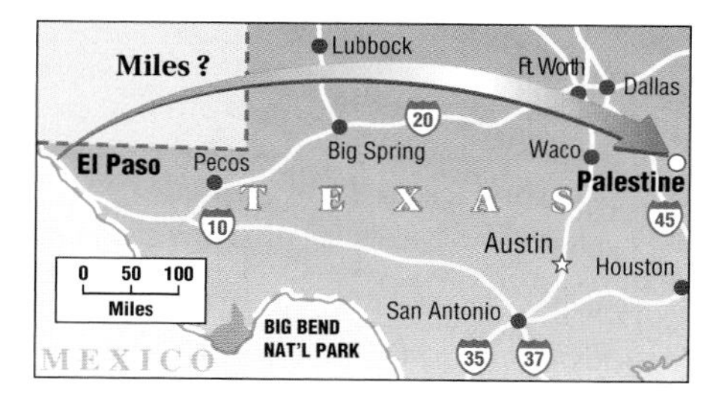

35. *Points per Game.* At one point in the 2000–2001 NBA season, Allen Iverson of the Philadelphia 76ers had scored 884 points in 33 games.

a) At this rate, how many games would it take him to score 1500 points?

b) There are 82 games in an entire NBA season. At this rate, how many points would Iverson score in the entire season?

Source: National Basketball Association

36. *Points per Game.* At one point in the 2000–2001 NBA season, Shaquille O'Neal, then of the Los Angeles Lakers, had scored 826 points in 32 games.

a) At this rate, how many games would it take him to score 2000 points?

b) There are 82 games in an entire NBA season. At this rate, how many points would O'Neal score in the entire season?

37. **DW** Can unit prices be used to solve proportions that involve money? Explain why or why not.

38. **DW** *Earned Run Average.* In baseball, the average number of runs given up by a pitcher in nine innings is his *earned run average,* or *ERA*. Set up a formula for determining a player's ERA. Then verify it using the fact that in the 2000 season, Daryl Kile of the St. Louis Cardinals gave up 101 earned runs in $232\frac{1}{3}$ innings to compile an ERA of 3.91. Then use your formula to determine the ERA of Randy Johnson of the Arizona Diamondbacks, who gave up 73 earned runs in $248\frac{2}{3}$ innings. Is a low ERA considered good or bad?

Source: Major League Baseball

SKILL MAINTENANCE

Determine whether each number is prime, composite, or neither. [2.1c]

39. 1 **40.** 28 **41.** 83 **42.** 93 **43.** 47

Find the prime factorization of each number. [2.1d]

44. 808 **45.** 28 **46.** 866 **47.** 93 **48.** 2020

SYNTHESIS

49. Carney College is expanding from 850 to 1050 students. To avoid any rise in the student-to-faculty ratio, the faculty of 69 professors must also increase. How many new faculty positions should be created?

50. In recognition of her outstanding work, Sheri's salary has been increased from $26,000 to $29,380. Tim is earning $23,000 and is requesting a proportional raise. How much more should he ask for?

51. *Baseball Statistics.* Cy Young, one of the greatest baseball pitchers of all time, gave up an average of 2.63 earned runs every 9 innings. Young pitched 7356 innings, more than anyone in the history of baseball. How many earned runs did he give up?

52. *Real-Estate Values.* According to Coldwell Banker Real Estate Corporation, a home selling for $189,000 in Austin, Texas, would sell for $665,795 in San Francisco. How much would a $450,000 home in San Francisco sell for in Austin? Round to the nearest $1000.

Source: Coldwell Banker Real Estate Corporation

53. The ratio 1:3:2 is used to estimate the relative costs of a CD player, receiver, and speakers when shopping for a stereo. That is, the receiver should cost three times the amount spent on the CD player and the speakers should cost twice as much as the amount spent on the CD player. If you had $900 to spend, how would you allocate the money, using this ratio?

5.5 GEOMETRIC APPLICATIONS

Objectives

a Find lengths of sides of similar triangles using proportions.

b Use proportions to find lengths in pairs of figures that differ only in size.

a Proportions and Similar Triangles

Look at the pair of triangles below. Note that they appear to have the same shape, but their sizes are different. These are examples of **similar triangles.** By using a magnifying glass, you could imagine enlarging the smaller triangle to get the larger. This process works because the corresponding sides of each triangle have the same ratio. That is, the following proportion is true.

$$\frac{a}{d} = \frac{b}{e} = \frac{c}{f}$$

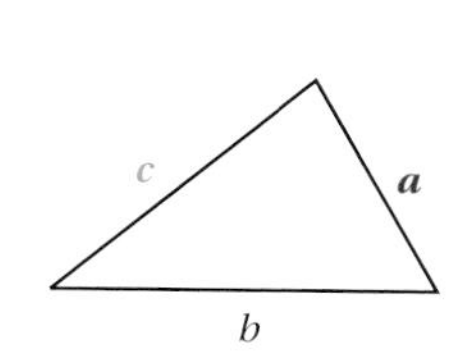

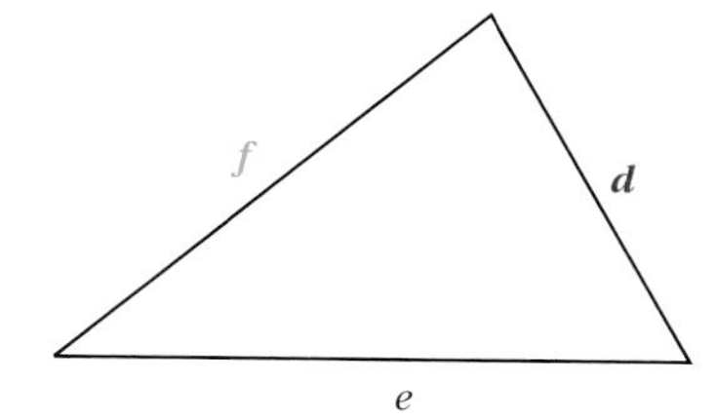

SIMILAR TRIANGLES

Similar triangles have the same shape. The lengths of their corresponding sides have the same ratio—that is, they are proportional.

EXAMPLE 1 The triangles below are similar triangles. Find the missing length x.

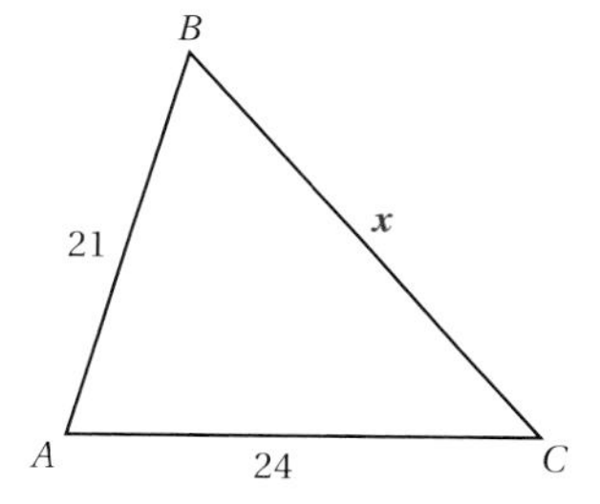

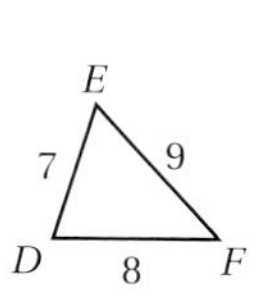

The ratio of x to 9 is the same as the ratio of 24 to 8 or 21 to 7. We get the proportions

$$\frac{x}{9} = \frac{24}{8} \quad \text{and} \quad \frac{x}{9} = \frac{21}{7}.$$

We can solve either one of these proportions. We use the first:

$$\frac{x}{9} = \frac{24}{8}$$

$x \cdot 8 = 24 \cdot 9$ Equating cross products

$\dfrac{x \cdot 8}{8} = \dfrac{24 \cdot 9}{8}$ Dividing by 8 on both sides

$x = 27.$ Simplifying

The missing length x is 27. Other proportions could also be used.

Do Exercise 1.

1. This pair of triangles is similar. Find the missing length x.

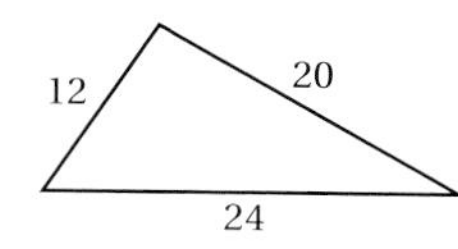

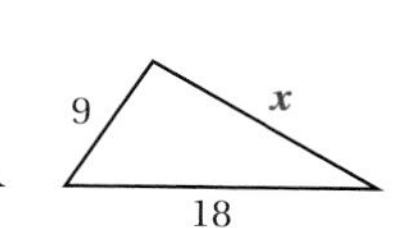

Answer on page A-12

2. How high is a flagpole that casts a 45-ft shadow at the same time that a 5.5-ft woman casts a 10-ft shadow?

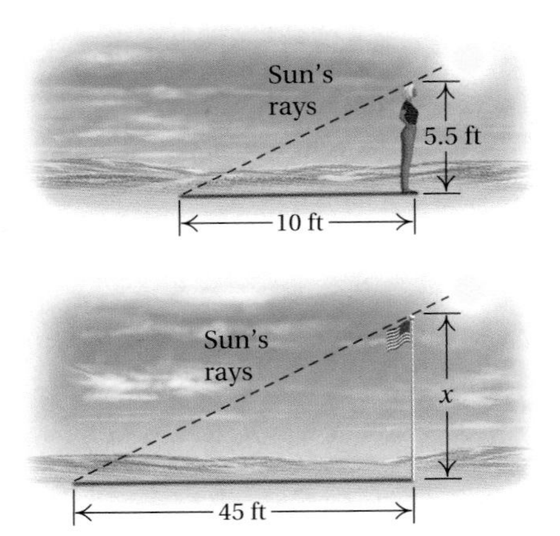

Similar triangles and proportions can often be used to find lengths that would ordinarily be difficult to measure. For example, we could find the height of a flagpole without climbing it or the distance across a river without crossing it.

EXAMPLE 2 How high is a flagpole that casts a 56-ft shadow at the same time that a 6-ft man casts a 5-ft shadow?

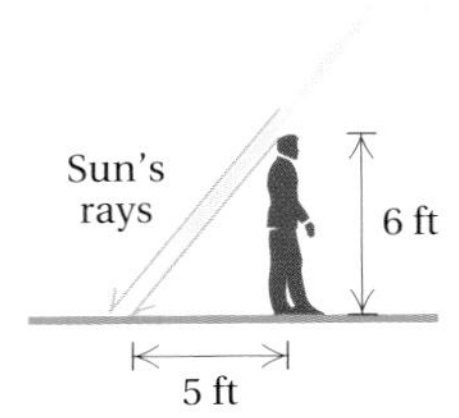

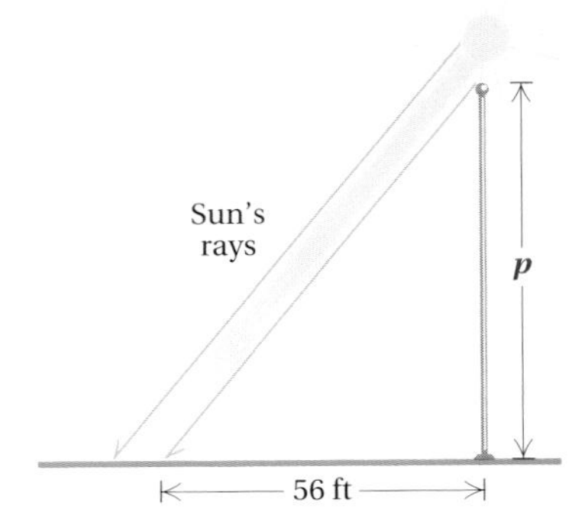

If we use the sun's rays to represent the third side of the triangle in our drawing of the situation, we see that we have similar triangles. Let p = the height of the flagpole. The ratio of 6 to p is the same as the ratio of 5 to 56. Thus we have the proportion

$$\begin{array}{r} \text{Height of man} \rightarrow \\ \text{Height of pole} \rightarrow \end{array} \frac{6}{p} = \frac{5}{56}. \begin{array}{l} \leftarrow \text{Length of shadow of man} \\ \leftarrow \text{Length of shadow of pole} \end{array}$$

Solve: $6 \cdot 56 = 5 \cdot p$ — Equating cross products

$$\frac{6 \cdot 56}{5} = \frac{5 \cdot p}{5}$$ Dividing by 5 on both sides

$$\frac{6 \cdot 56}{5} = p$$ Simplifying

$$67.2 = p$$

The height of the flagpole is 67.2 ft.

Do Exercise 2.

EXAMPLE 3 *Rafters of a House.* Carpenters use similar triangles to determine the lengths of rafters for a house. They first choose the pitch of the roof, or the ratio of the rise over the run. Then using a triangle with that ratio, they calculate the length of the rafter needed for the house. Loren is constructing rafters for a roof with a 6/12 pitch on a house that is 30 ft wide. Using a rafter guide, Loren knows that the rafter length corresponding to the 6/12 pitch is 13.4. Find the length x of the rafter of the house to the nearest tenth of a foot.

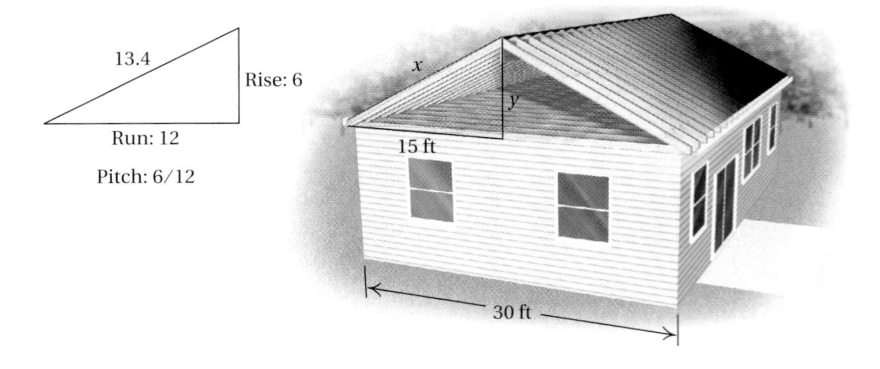

Answer on page A-12

Thus we have the proportion

$$\begin{matrix}\text{Length of rafter in 6/12 triangle} \rightarrow \\ \text{Length of rafter on the house} \rightarrow\end{matrix} \frac{13.4}{x} = \frac{12}{15} \begin{matrix}\leftarrow \text{Run in 6/12 triangle} \\ \leftarrow \text{Run in similar triangle on the house}\end{matrix}$$

Solve: $13.4 \cdot 15 = x \cdot 12$ Equating cross products

$\frac{13.4 \cdot 15}{12} = \frac{x \cdot 12}{12}$ Dividing by 12 on both sides

$\frac{13.4 \cdot 15}{12} = x$

$16.8 \text{ ft} \approx x$ Rounding to the nearest tenth of a foot

The length of the rafter x of the house is about 16.8 ft.

Do Exercise 3.

3. **Rafters of a House.** Referring to Example 3, find the length y in the rafter of the house to the nearest tenth of a foot.

b Proportions and Other Geometric Shapes

When one geometric figure is a magnification of another, the figures are similar. Thus the corresponding lengths are proportional.

EXAMPLE 4 The sides in the negative and photograph below are proportional. Find the width of the photograph.

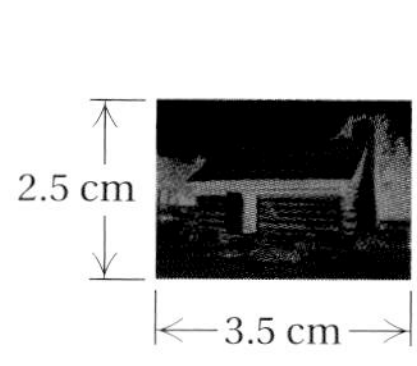

We let x = the width of the photograph. Then we translate to a proportion.

$$\begin{matrix}\text{Photo width} \rightarrow \\ \text{Negative width} \rightarrow\end{matrix} \frac{x}{2.5} = \frac{10.5}{3.5} \begin{matrix}\leftarrow \text{Photo length} \\ \leftarrow \text{Negative length}\end{matrix}$$

Solve: $3.5 \times x = 2.5 \times 10.5$ Equating cross products

$\frac{3.5 \times x}{3.5} = \frac{2.5 \times 10.5}{3.5}$ Dividing by 3.5 on both sides

$x = \frac{2.5 \times 10.5}{3.5}$ Simplifying

$x = 7.5$

Thus the width of the photograph is 7.5 cm.

Do Exercise 4.

4. The sides in the photographs below are proportional. Find the width of the larger photograph.

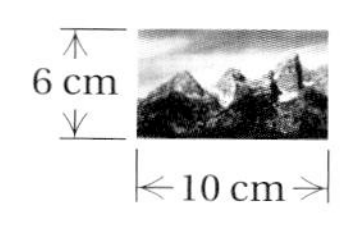

Answers on page A-12

5. Refer to the figures in Example 5. If a model skylight is 3 cm wide, how wide will the actual skylight be?

Answer on page A-12

EXAMPLE 5 A scale model of an addition to an athletic facility is 12 cm wide at the base and rises to a height of 15 cm. If the actual base is to be 116 ft, what will be the actual height of the addition?

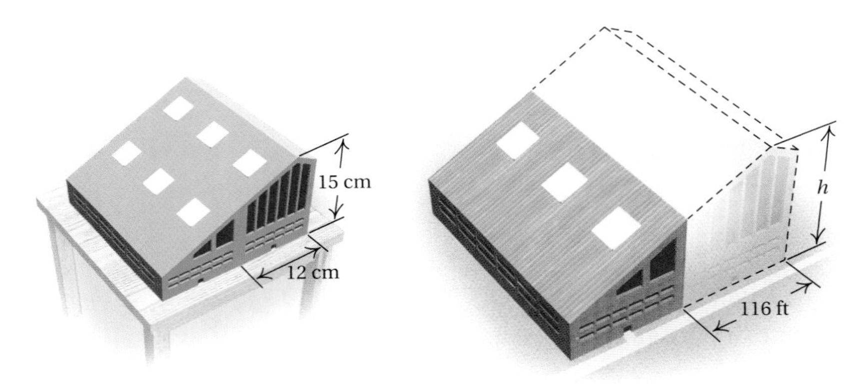

We let h = the height of the addition. Then we translate to a proportion.

$$\text{Width in model} \rightarrow \frac{12}{116} = \frac{15}{h} \leftarrow \text{Height in model}$$
$$\text{Actual width} \rightarrow \qquad\qquad \leftarrow \text{Actual height}$$

Solve: $12 \cdot h = 116 \cdot 15$ Equating cross products

$$\frac{12 \cdot h}{12} = \frac{116 \cdot 15}{12}$$ Dividing by 12 on both sides

$$h = \frac{116 \cdot 15}{12} = 145$$

Thus the height of the addition will be 145 ft.

Do Exercise 5.

Study Tips

TIME MANAGEMENT (PART 3)

Here are some additional tips to help you with time management. (See also the Study Tips on time management in Sections 1.5 and 5.1.)

- **Are you a morning or an evening person?** If you are an evening person, it might be best to avoid scheduling early-morning classes. If you are a morning person, do the opposite, but go to bed earlier to compensate. Nothing can drain your study time and effectiveness like fatigue.
- **Keep on schedule.** Your course syllabus provides a plan for the semester's schedule. Use a write-on calendar, daily planner, laptop computer, or personal digital assistant to outline your time for the semester. Be sure to note deadlines involving term papers and exams so you can begin a task early, breaking it down into smaller segments that can be accomplished more easily.
- **Balance your class schedule.** You may be someone who prefers large blocks of time for study on the off days. In that case, it might be advantageous for you to take courses that meet only three days a week. Keep in mind, however, that this might be a problem when tests in more than one course are scheduled for the same day.

5.5 EXERCISE SET

For Extra Help

MathXL | MyMathLab | InterAct Math | Math Tutor Center | Digital Video Tutor CD 3 Videotape 5 | Student's Solutions Manual

a The triangles in each exercise are similar. Find the missing lengths.

1.

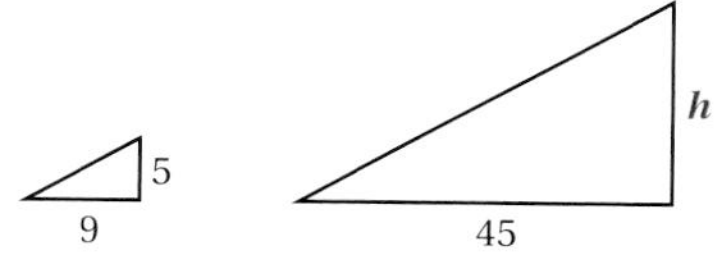

2.

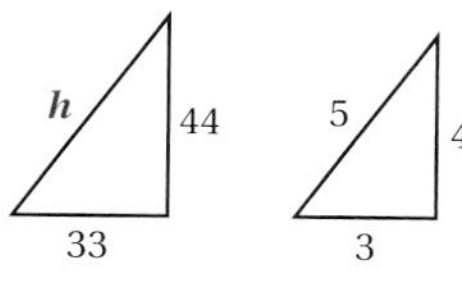

3.

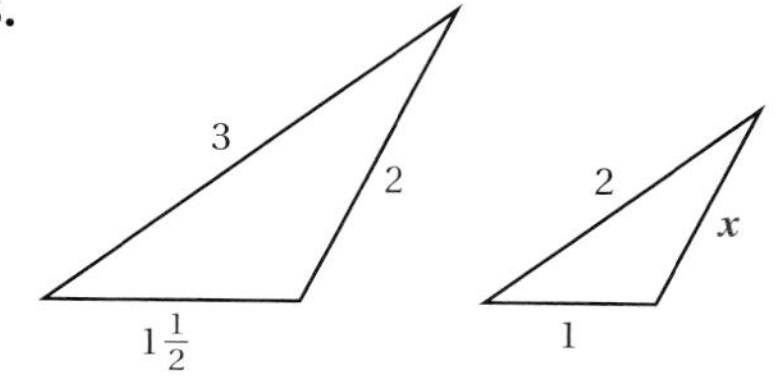

4.

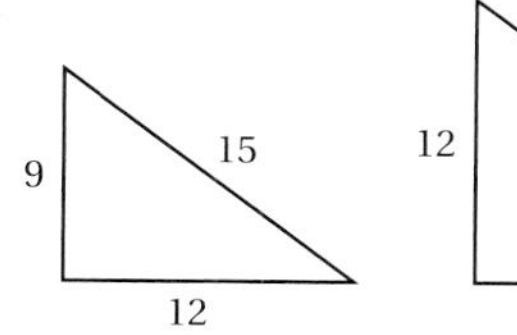

5.

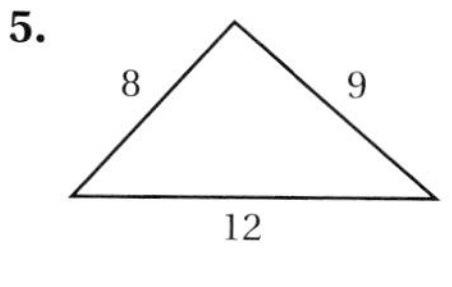

6.

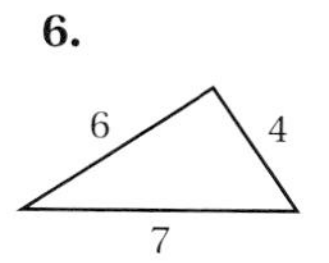

7.

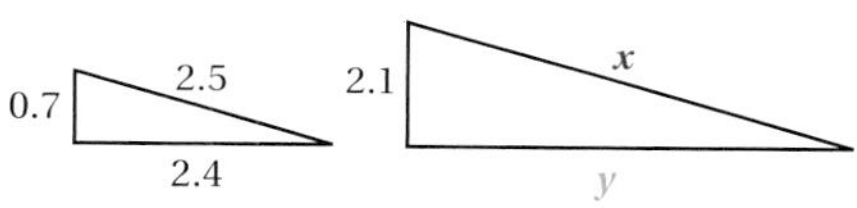

8.

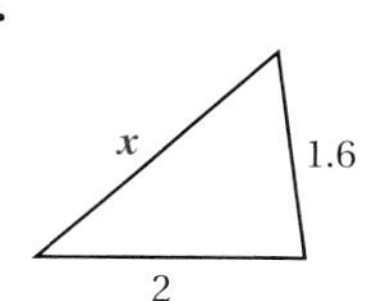

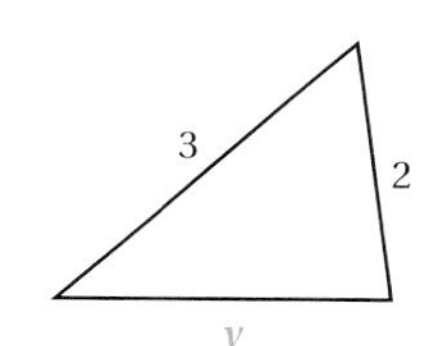

9. When a tree 8 m high casts a shadow 5 m long, how long a shadow is cast by a person 2 m tall?

10. How high is a flagpole that casts a 42-ft shadow at the same time that a $5\frac{1}{2}$-ft woman casts a 7-ft shadow?

11. How high is a tree that casts a 27-ft shadow at the same time that a 4-ft fence post casts a 3-ft shadow?

12. How high is a tree that casts a 32-ft shadow at the same time that an 8-ft light pole casts a 9-ft shadow?

13. Find the height h of the wall.

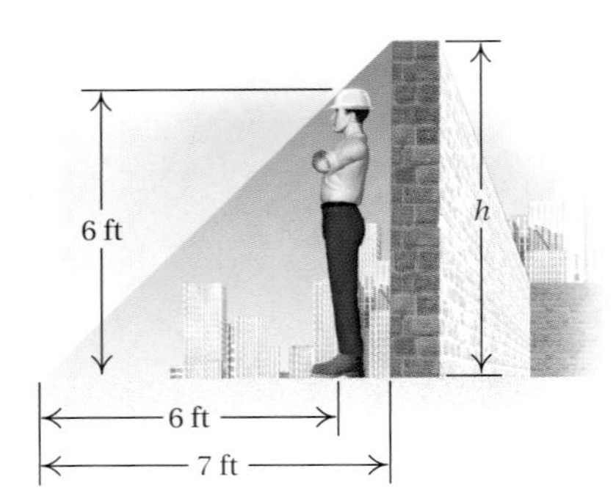

14. Find the length L of the lake. Assume that the ratio of L to 120 yd is the same as the ratio of 720 yd to 30 yd.

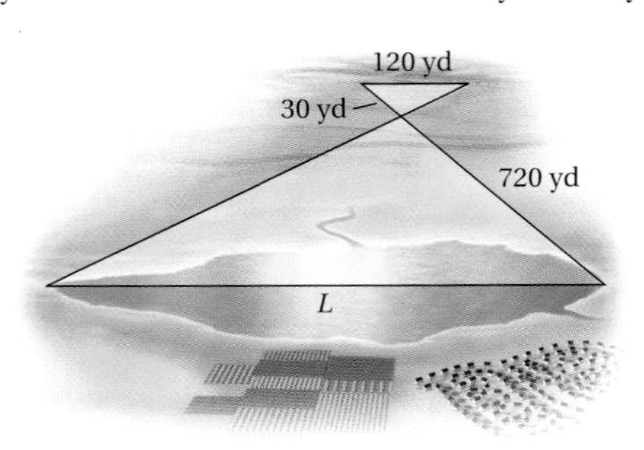

15. Find the distance across the river. Assume that the ratio of d to 25 ft is the same as the ratio of 40 ft to 10 ft.

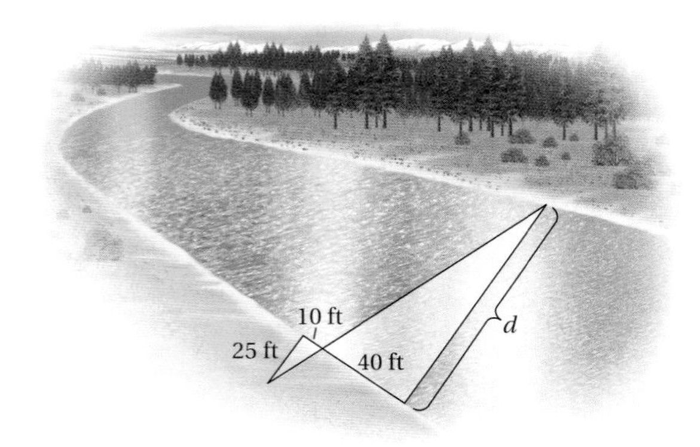

16. To measure the height of a hill, a string is drawn tight from level ground to the top of the hill. A 3-ft stick is placed under the string, touching it at point P, a distance of 5 ft from point G, where the string touches the ground. The string is then detached and found to be 120 ft long. How high is the hill?

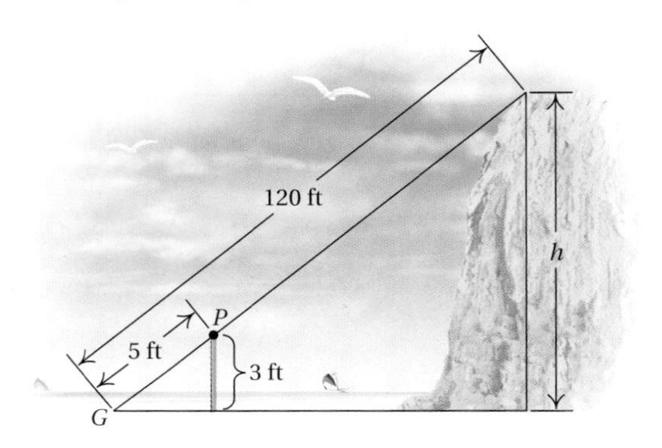

b In each of Exercises 17–26, the sides in each pair of figures are proportional. Find the missing lengths.

17.

18.

19.

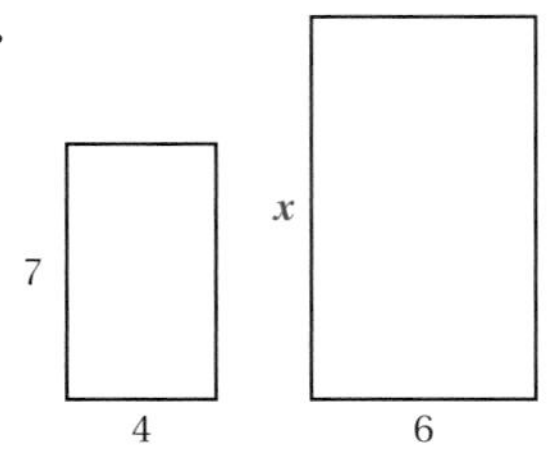

20.

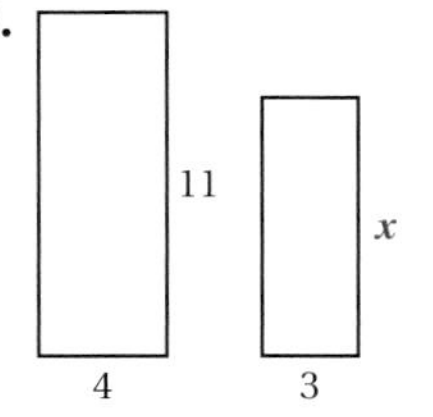

21.

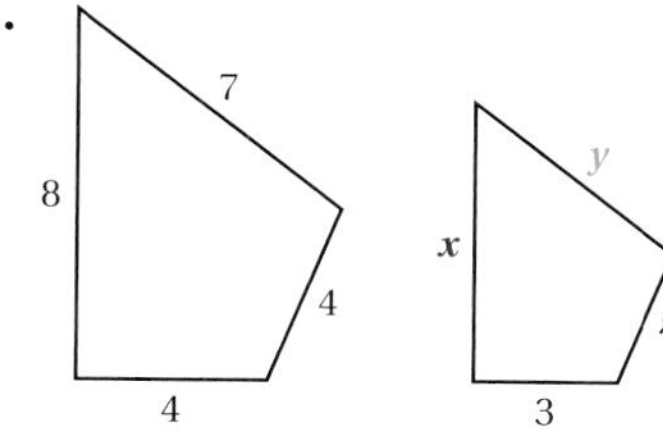

22.

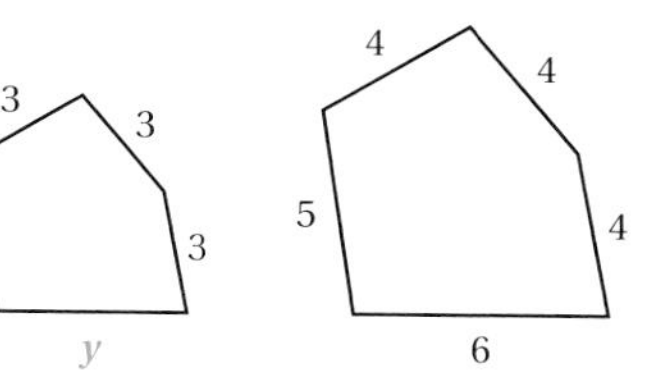

23.

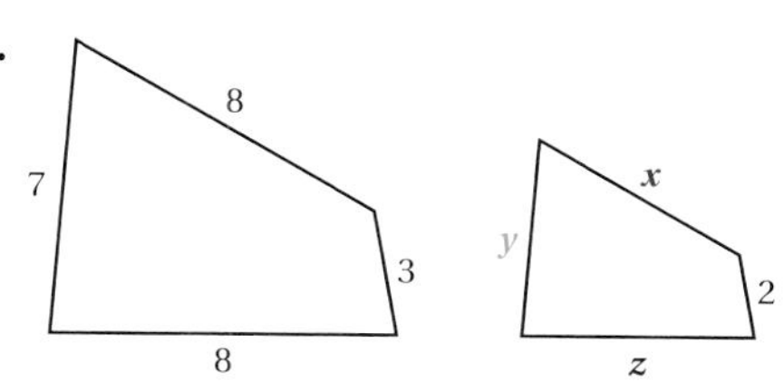

24.

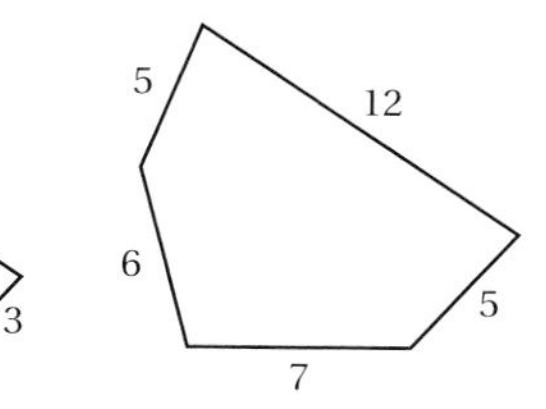

25.

26.

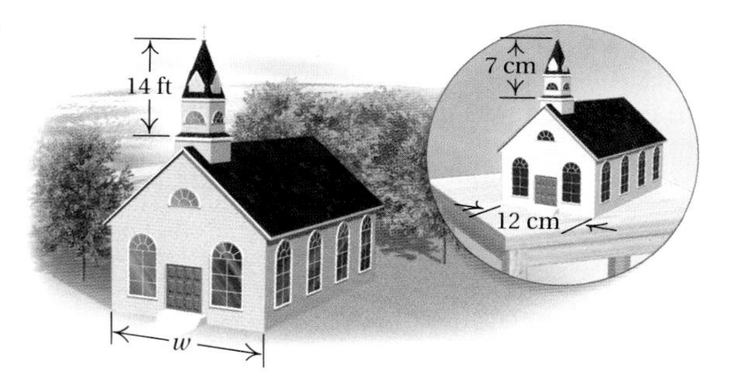

27. A scale model of an addition to an athletic facility is 15 cm wide at the base and rises to a height of 19 cm. If the actual base is to be 120 ft, what will be the height of the addition?

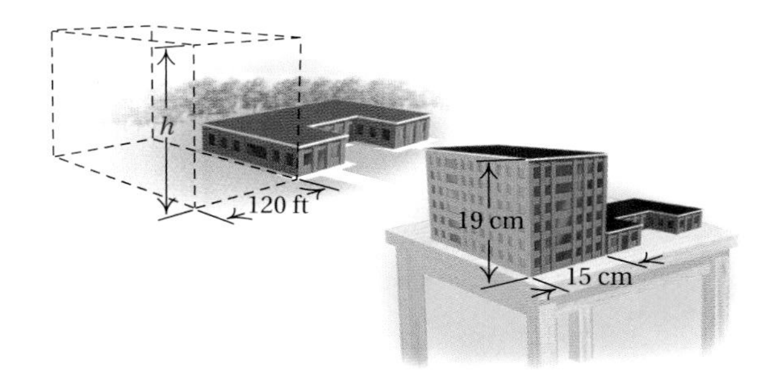

28. Refer to the figures in Exercise 27. If a model skylight is 3 cm wide, how wide will the actual skylight be?

29. $\mathbf{D_W}$ Is it possible for two triangles to have two pairs of sides that are proportional without the triangles being similar? Why or why not?

30. $\mathbf{D_W}$ Design for a classmate a problem involving similar triangles for which

$$\frac{18}{128.95} = \frac{x}{789.89}.$$

SKILL MAINTENANCE

31. *Expense Needs.* A student has $34.97 to spend for a book at $49.95, a CD at $14.88, and a sweatshirt at $29.95. How much more money does the student need to make these purchases? [4.7a]

32. Divide: $80.892 \div 8.4$. [4.4a]

Multiply. [4.3a]

33. 8.4×80.892

34. 0.01×274.568

35. 100×274.568

36. 0.002×274.568

Find decimal notation and round to the nearest thousandth, if appropriate. [4.5a, b]

37. $\frac{17}{20}$

38. $\frac{73}{40}$

39. $\frac{10}{11}$

40. $\frac{43}{51}$

SYNTHESIS

Hockey Goals. An official hockey goal is 6 ft wide. To make scoring more difficult, goalies often locate themselves far in front of the goal to "cut down the angle." In Exercises 41 and 42, suppose that a slapshot from point A is attempted and that the goalie is 2.7 ft wide. Determine how far from the goal the goalie should be located if point A is the given distance from the goal. (*Hint*: First find how far the goalie should be from point A.)

41. 🖩 25 ft

42. 🖩 35 ft

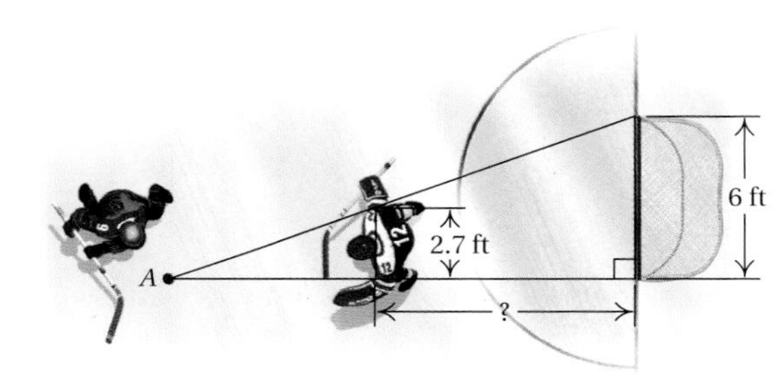

43. A miniature basketball hoop is built for the model referred to in Exercise 27. An actual hoop is 10 ft high. How high should the model hoop be?

🖩 Solve. Round the answer to the nearest thousandth.

44. $\frac{8664.3}{10{,}344.8} = \frac{x}{9776.2}$

45. $\frac{12.0078}{56.0115} = \frac{789.23}{y}$

🖩 The triangles in each exercise are similar triangles. Find the lengths not given.

46.

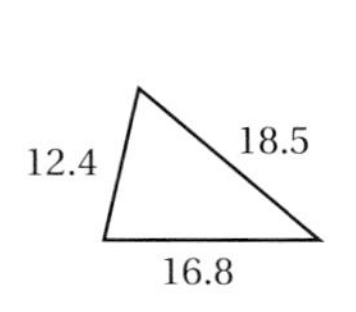

x
y
423.36

47.

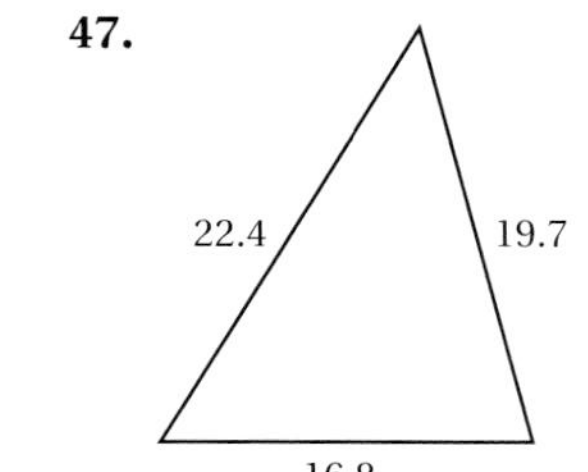

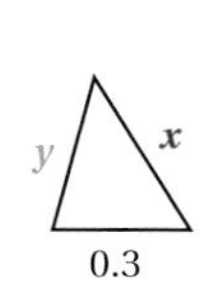

5 Summary and Review

The review that follows is meant to prepare you for a chapter exam. It consists of two parts. The first part, Concept Reinforcement, is designed to increase understanding of the concepts through true/false exercises. The second part is the Review Exercises. These provide practice exercises for the exam, together with references to section objectives so you can go back and review. Before beginning, stop and look back over the skills you have obtained. What skills in mathematics do you have now that you did not have before studying this chapter?

CONCEPT REINFORCEMENT

Determine whether the statement is true or false. Answers are given at the back of the book.

_______ **1.** The proportion $\frac{a}{b} = \frac{c}{d}$ can also be written as $\frac{c}{a} = \frac{d}{b}$.

_______ **2.** Lengths of corresponding sides of similar triangles have the same ratio.

_______ **3.** The larger size of a product always has the lower unit price.

_______ **4.** If $\frac{x}{t} = \frac{y}{s}$, then $xy = ts$.

_______ **5.** A ratio is a quotient of two quantities.

Review Exercises

Write fraction notation for the ratio. Do not simplify. [5.1a]

1. 47 to 84

2. 46 to 1.27

3. 83 to 100

4. 0.72 to 197

5. *Kona Jack's Restaurants.* Kona Jack's is a seafood restaurant chain in Indianapolis. Each year they sell 12,480 lb of tuna and 16,640 lb of salmon. [5.1a]

a) Write fraction notation for the ratio of tuna sold to salmon sold.

b) Write fraction notation for the ratio of salmon sold to the total number of pounds of both kinds of fish.

Source: Kona Jack's Restaurants

Find the ratio of the first number to the second number and simplify. [5.1b]

6. 9 to 12

7. 3.6 to 6.4

8. *Gas Mileage.* The Chrysler PT Cruiser will go 377 mi on 14.5 gal of gasoline in highway driving. What is the rate in miles per gallon? [5.2a]

Source: DaimlerChrysler Corporation

9. *CD-ROM Spin Rate.* A 12x CD-ROM on a computer will spin 472,500 revolutions if left running for 75 min. What is the rate of its spin in revolutions per minute (rpm)? [5.2a]

Source: *Electronic Engineering Times*, June 1997

10. A lawn requires 319 gal of water for every 500 ft^2. What is the rate in gallons per square foot? [5.2a]

11. *Turkey Servings.* A 25-lb turkey serves 18 people. Find the rate in servings per pound. [5.2a]

12. *Calcium Supplement.* The price for a particular calcium supplement is $12.99 for 300 tablets. Find the unit price in cents per tablet. [5.2b]

13. *Pillsbury Orange Breakfast Rolls.* The price for these breakfast rolls is $1.97 for 13.9 oz. Find the unit price in cents per ounce. [5.2b]

In each of Exercises 14 and 15, find the unit prices. Then determine in each case which has the lowest unit price. [5.2b]

14. *Paper Towels.*

PACKAGE	PRICE	UNIT PRICE PER SHEET
8 rolls, 60 (2 ply) sheets per roll	$6.38	
15 rolls, 60 (2 ply) sheets per roll	$13.99	
6 big rolls, 165 (2 ply) sheets per roll	$10.99	

15. *Crisco Oil.*

PACKAGE	PRICE	UNIT PRICE
32 oz	$2.19	
48 oz	$2.49	
64 oz	$3.59	
128 oz	$7.09	

Determine whether the two pairs of numbers are proportional. [5.3a]

16. 9, 15 and 36, 59

17. 24, 37 and 40, 46.25

Solve. [5.3b]

18. $\frac{8}{9} = \frac{x}{36}$

19. $\frac{6}{x} = \frac{48}{56}$

20. $\frac{120}{\frac{3}{7}} = \frac{7}{x}$

21. $\frac{4.5}{120} = \frac{0.9}{x}$

Solve. [5.4a]

22. If 3 dozen eggs cost $2.67, how much will 5 dozen eggs cost?

23. *Quality Control.* A factory manufacturing computer circuits found 39 defective circuits in a lot of 65 circuits. At this rate, how many defective circuits can be expected in a lot of 585 circuits?

24. *Exchanging Money.* On 5 August 2005, 1 U.S. dollar was worth about 0.808 European Monetary Unit (Euro).

a) How much would 250 U.S. dollars be worth in Euros?

b) Jamal was traveling in France and saw a sweatshirt that cost 50 Euros. How much would it cost in U.S. dollars?

25. A train travels 448 mi in 7 hr. At this rate, how far will it travel in 13 hr?

26. Fifteen acres are required to produce 54 bushels of tomatoes. At this rate, how many acres are required to produce 97.2 bushels of tomatoes?

27. *Garbage Production.* It is known that 5 people produce 13 kg of garbage in one day. San Diego, California, has 1,266,753 people. How many kilograms of garbage are produced in San Diego in one day?

28. *Snow to Water.* Under typical conditions, $1\frac{1}{2}$ ft of snow will melt to 2 in. of water. To how many inches of water will $4\frac{1}{2}$ ft of snow melt?

29. *Lawyers in Michigan.* In Michigan, there are 2.3 lawyers for every 1000 people. The population of Detroit is 911,402. How many lawyers would you expect there to be in Detroit?

Source: U.S. Bureau of the Census

Each pair of triangles in Exercises 30 and 31 is similar. Find the missing length(s). [5.5a]

30.

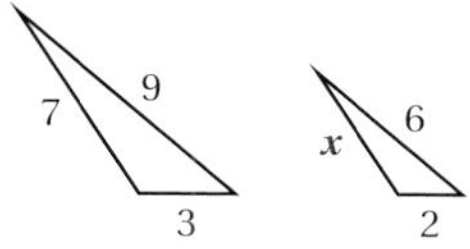

31.

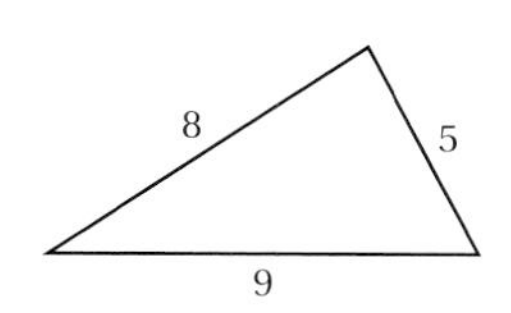

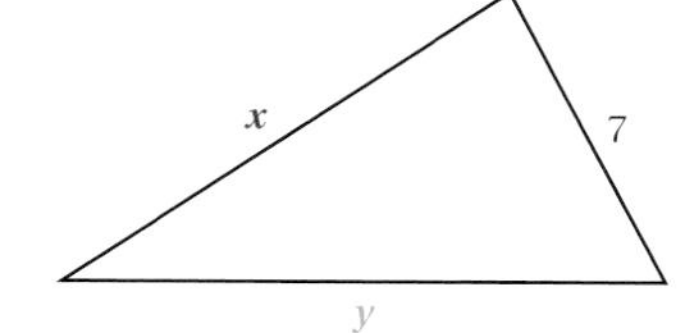

32. How high is a billboard that casts a 25-ft shadow at the same time that an 8-ft sapling casts a 5-ft shadow? [5.5a]

33. The lengths in the figures below are proportional. Find the missing lengths. [5.5b]

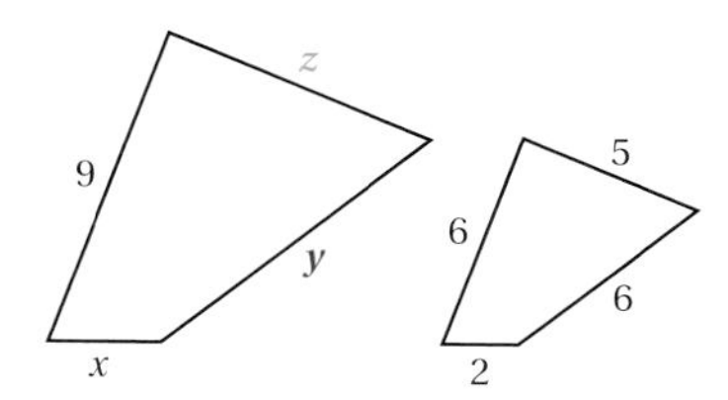

34. D_W If you were a college president, which would you prefer: a low or high faculty-to-student ratio? Why? [5.1a]

35. D_W Write a proportion problem for a classmate to solve. Design the problem so that the solution is "Leslie would need 16 gal of gasoline in order to travel 368 mi." [5.4a]

SYNTHESIS

36. It takes Yancy Martinez 10 min to type two-thirds of a page of his term paper. At this rate, how long will it take him to type a 7-page term paper? [5.4a]

37. The following triangles are similar. Find the missing lengths. [5.5a]

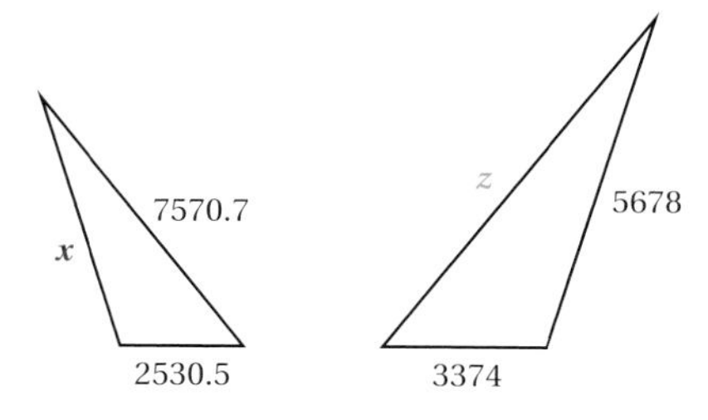

38. Shine-and-Glo Painters uses 2 gal of finishing paint for every 3 gal of primer. Each gallon of finishing paint covers 450 ft^2. If a surface of 4950 ft^2 needs both primer and finishing paint, how many gallons of each should be purchased? [5.4a]

5 Chapter Test

For Extra Help

Work It Out! Chapter Test Video on CD

Write fraction notation for the ratio. Do not simplify.

1. 85 to 97

2. 0.34 to 124

Find the ratio of the first number to the second number and simplify.

3. 18 to 20

4. 0.75 to 0.96

5. What is the rate in feet per second?

10 feet, 16 seconds

6. *Ham Servings.* A 12-lb shankless ham contains 16 servings. What is the rate in servings per pound?

7. *Gas Mileage.* The 2005 Chevrolet Malibu Maxx will go 319 mi on 14.5 gal of gasoline in city driving. What is the rate in miles per gallon?

Source: General Motors Corporation

8. *Laundry Detergent.* A box of Cheer laundry detergent powder sells at \$6.29 for 81 oz. Find the unit price in cents per ounce.

9. The following table lists prices for various packages of Tide laundry detergent powder. Find the unit price of each package. Then determine which has the lowest unit price.

PACKAGE	PRICE	UNIT PRICE
33 oz	\$3.69	
87 oz	\$6.22	
131 oz	\$10.99	
263 oz	\$17.99	

Determine whether the two pairs of numbers are proportional.

10. 7, 8 and 63, 72

11. 1.3, 3.4 and 5.6, 15.2

Solve.

12. $\frac{9}{4} = \frac{27}{x}$

13. $\frac{150}{2.5} = \frac{x}{6}$

14. $\frac{x}{100} = \frac{27}{64}$

15. $\frac{68}{y} = \frac{17}{25}$

Solve.

16. *Distance Traveled.* An ocean liner traveled 432 km in 12 hr. At this rate, how far would the boat travel in 42 hr?

17. *Time Loss.* A watch loses 2 min in 10 hr. At this rate, how much will it lose in 24 hr?

18. *Map Scaling.* On a map, 3 in. represents 225 mi. If two cities are 7 in. apart on the map, how far are they apart in reality?

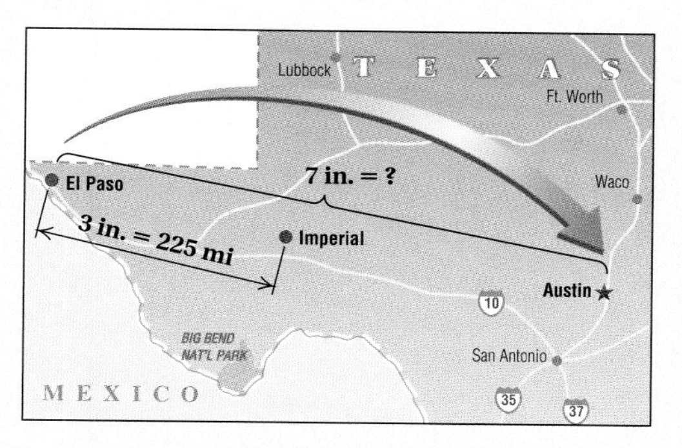

19. *Tower Height.* A birdhouse built on a pole that is 3 m high casts a shadow 5 m long. At the same time, the shadow of a tower is 110 m long. How high is the tower?

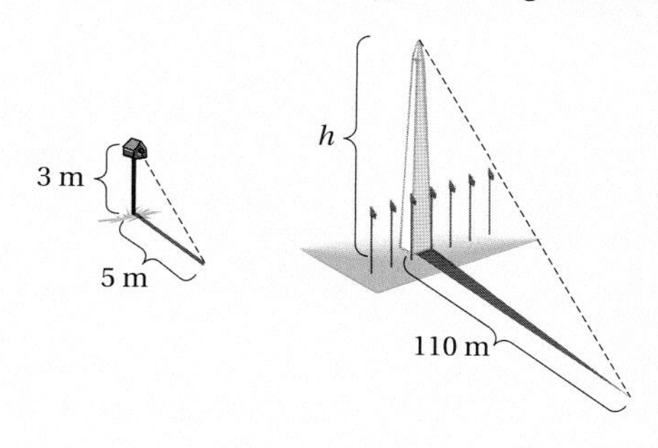

20. *Exchanging Money.* On 5 August 2005, 1 U.S. dollar was worth about 7.775 Hong Kong dollars.

a) How much would 450 U.S. dollars be worth in Hong Kong dollars?

b) Mitchell was traveling in Hong Kong and saw a DVD player that cost 795 Hong Kong dollars. How much would it cost in U.S. dollars?

21. *Automobile Violations.* In a recent year, the Indianapolis Police Department employed 1088 officers and made 37,493 arrests. At this rate, how many arrests could be made if the number of officers were increased to 2500?

Source: *Indianapolis Star,* 12-31-00

The lengths in each pair of figures are proportional. Find the missing lengths.

22.

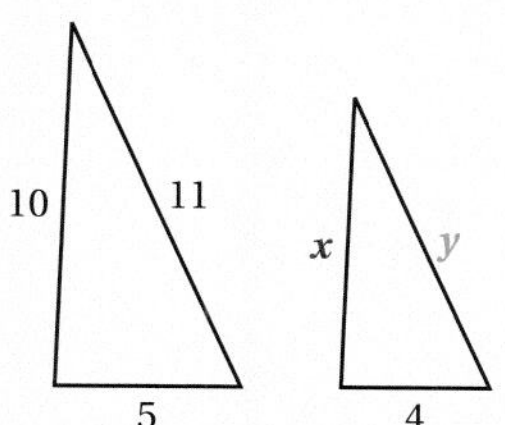

23.

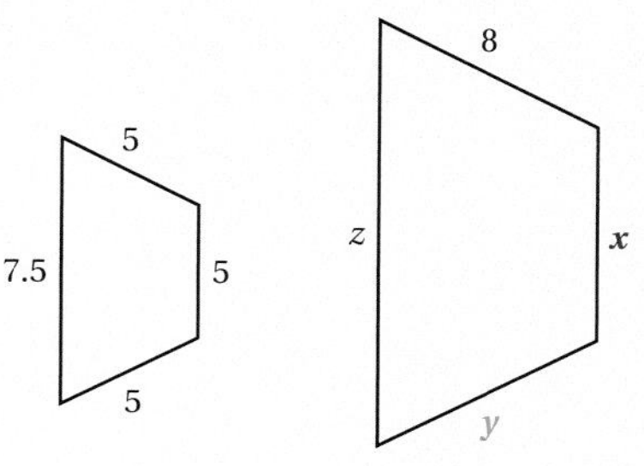

SYNTHESIS

24. Nancy Morano-Smith wants to win a season football ticket from the local bookstore. Her goal is to guess the number of marbles in an 8-gal jar. She knows that there are 128 oz in a gallon. She goes home and fills an 8-oz jar with 46 marbles. How many marbles should she guess are in the jar?

CHAPTERS 1–5

Cumulative Review

1. *Baseball Salaries.* Alex Rodriguez now plays with the New York Yankees under a 10-yr contract for $252 million that he originally signed with the Texas Rangers in 2000.

a) Find standard notation for the dollar amount of this contract.

b) How many billion dollars was this contract?

c) What is the average amount of money he makes each year?

d) In 2000, Rodriguez had 554 at-bats. How much money did he average for each at-bat?

Source: Major League Baseball

2. *Gas Mileage.* The 2005 Volkswagen Jetta 2.5 L will go 319 mi on 14.5 gal of gasoline in city driving. What is the rate in miles per gallon?

Source: Volkswagen of America, Inc.

Add and simplify.

3.
$$\begin{array}{r} 27.68 \\ 3.019 \\ +\ 483.297 \\ \hline \end{array}$$

4.
$$\begin{array}{r} 2\frac{1}{3} \\ +4\frac{5}{12} \\ \hline \end{array}$$

5. $\frac{6}{35} + \frac{5}{28}$

Subtract and simplify.

6.
$$\begin{array}{r} 40.2 \\ -\ 9.709 \\ \hline \end{array}$$

7. $73.82 - 0.908$

8. $\frac{4}{15} - \frac{3}{20}$.

Multiply and simplify.

9.
$$\begin{array}{r} 37.64 \\ \times\ 5.9 \\ \hline \end{array}$$

10. 5.678×100

11. $2\frac{1}{3} \cdot 1\frac{2}{7}$

Divide and simplify.

12. $2.3\overline{)98.9}$

13. $54\overline{)48{,}546}$

14. $\frac{7}{11} \div \frac{14}{33}$

15. Write expanded notation: 30,074.

16. Write a word name for 120.07.

Which number is larger?

17. 0.7, 0.698

18. 0.799, 0.8

19. Find the prime factorization of 144.

20. Find the LCM of 28 and 35.

21. What part is shaded?

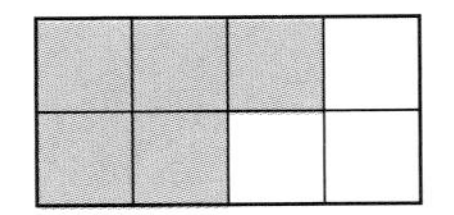

22. Simplify: $\frac{90}{144}$.

Calculate.

23. $\frac{3}{5} \times 9.53$

24. $\frac{1}{3} \times 0.645 - \frac{3}{4} \times 0.048$

25. Write fraction notation for the ratio 0.3 to 15.

26. Determine whether the pairs 3, 9 and 25, 75 are proportional.

27. What is the rate in meters per second?

660 meters, 12 seconds

28. The following table lists prices for various brands of liquid dish soap. Find the unit price of each brand. Then determine which has the lowest unit price.

BRAND	PACKAGE	PRICE	UNIT PRICE
Palmolive	13 oz	\$1.53	
Dawn	42.7 oz	\$3.99	
Dawn	14.7 oz	\$1.43	
Joy	28 oz	\$1.78	
Joy	42.7 oz	\$2.99	

Solve.

29. $\frac{14}{25} = \frac{x}{54}$

30. $423 = 16 \cdot t$

31. $\frac{2}{3} \cdot y = \frac{16}{27}$

32. $\frac{7}{16} = \frac{56}{x}$

33. $34.56 + n = 67.9$

34. $t + \frac{7}{25} = \frac{5}{7}$

Solve.

35. A particular kind of fettuccini Alfredo has 520 calories in 1 cup. How many calories are there in $\frac{3}{4}$ cup?

36. *Exchanging Money.* On 26 April 2005, 1 U.S. dollar was worth about 6.085 South African Rand.

a) How much would 220 U.S. dollars be worth in Rand?
b) Monica was traveling in South Africa and saw a camera that cost 2050 Rand. How much would it cost in U.S. dollars?

37. *Gas Mileage.* A Greyhound tour bus traveled 347.6 mi, 249.8 mi, and 379.5 mi on three separate trips. What was the total mileage of the bus?

38. A machine can stamp out 925 washers in 5 min. The company owning the machine needs 1295 washers by the end of the morning. How long will it take to stamp them out?

39. A 46-oz juice can contains $5\frac{3}{4}$ cups of juice. A recipe calls for $3\frac{1}{2}$ cups of juice. How many cups are left over?

40. It takes a carpenter $\frac{2}{3}$ hr to hang a door. How many doors can the carpenter hang in 8 hr?

41. *The Leaning Tower of Pisa.* At the time of this writing, the Leaning Tower of Pisa was still standing. It is 184.5 ft tall but leans about 17 ft out from its base. Each year, it leans about an additional $\frac{1}{20}$ in., or $\frac{1}{240}$ ft.

a) After how many years will it lean the same length as it is tall?
b) At most how many years do you think the Tower will stand?

42. *Airplane Tire Costs.* A Boeing 747-400 jumbo jet has 2 nose tires and 16 rear tires. Each tire costs about $20,000.

a) What is the total cost of a new set of tires for such a plane?
b) Suppose an airline has a fleet of 400 such planes. What is the total cost of a new set of tires for all the planes?
c) Suppose the airline has to change tires every month. What would be the total cost for tires for the airline for an entire year?

Source: *World-Traveler*, October 2000

43. *Car Travel.* A car travels 337.62 mi in 8 hr. How far does it travel in 1 hr?

44. *Shuttle Orbits.* A recent space shuttle made 16 orbits a day during an 8.25-day mission. How many orbits were made during the entire mission?

For each of Exercises 45–47, choose the correct answer from the selections given.

45. How many even prime numbers are there?

a) 5 **b)** 3 **c)** 2
d) 1 **e)** None

46. The gas mileage of a car is 28.16 miles per gallon. How many gallons per mile is this?

a) $\frac{704}{25}$ **b)** $\frac{25}{704}$ **c)** $\frac{2816}{100}$
d) $\frac{250}{704}$ **e)** None

47. By what number do you multiply the side s of a square to find its perimeter?

a) s itself **b)** 4 **c)** 2
d) 8 **e)** None

SYNTHESIS

48. A soccer goalie wishing to block an opponent's shot moves toward the shooter to reduce the shooter's view of the goal. If the goalie can only defend a region 10 ft wide, how far in front of the goal should the goalie be? (See the figure at right.)

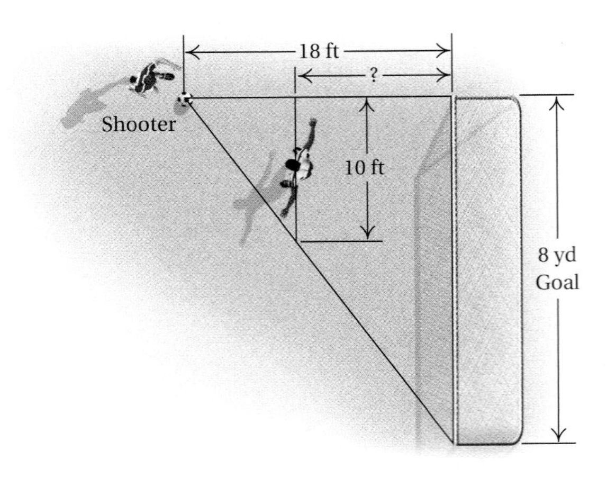

6

Percent Notation

Real-World Application

There are 1,168,195 people in the United States in active military service. The numbers in the four armed services are as follows: Air Force, 314,477; Army, 391,126; Marine Corps, 135,324; Navy, 327,268. What percent of the total does each branch represent? Round the answers to the nearest tenth of a percent.

Source: U.S. Department of Defense

This problem appears as Exercise 2 in Section 6.5.

6.1 PERCENT NOTATION

Objectives

a Write three kinds of notation for a percent.

b Convert between percent notation and decimal notation.

a Understanding Percent Notation

Of all the surface area of the earth, 70% of it is covered by water. What does this mean? It means that of every 100 square miles of the earth's surface area, 70 square miles are covered by water. Thus, 70% is a ratio of 70 to 100, or $\frac{70}{100}$.

Source: *The Handy Geography Answer Book*

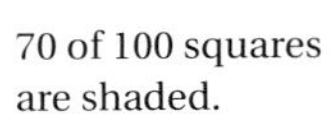

70% or $\frac{70}{100}$ or 0.70 of the large square is shaded.

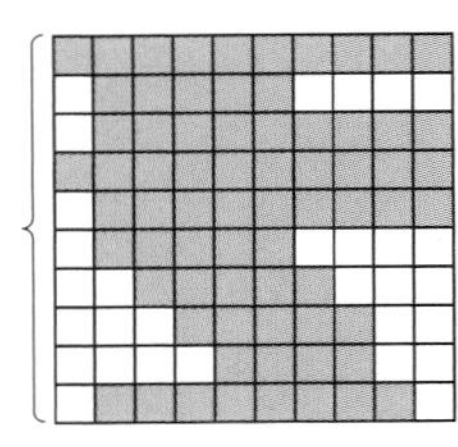

Percent notation is used extensively in our everyday lives. Here are some examples:

44% of all Americans use at least one prescribed medicine.

35.5% of Massachusetts residents age 25 and over have a bachelor's degree.

18% of household personal vehicles are pickup trucks.

50.3% of all paper used in the United States is recycled.

54% of all kids prefer white bread.

0.08% blood alcohol level is a standard used by some states as the legal limit for drunk driving.

Percent notation is often represented in pie charts to show how the parts of a quantity are related. For example, the circle graph below illustrates the percentage of vehicles manufactured in the most popular car colors during 2003 in North America.

Most Popular Car Colors, 2003 (SUV / Truck / Van)

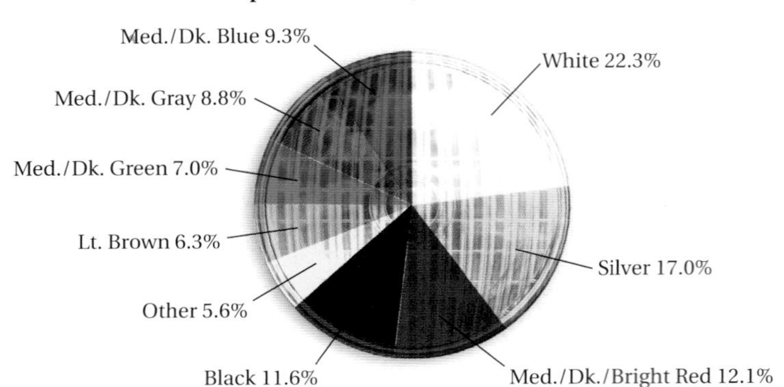

Source: DuPont Herberts Automotive Systems, Troy, Michigan, 2003, DuPont Automotive Color Survey Results

Write three kinds of notation as in Examples 1 and 2.

1. 70%

2. 23.4%

3. 100%

It is thought that the Roman emperor Augustus began percent notation by taxing goods sold at a rate of $\frac{1}{100}$. In time, the symbol "%" evolved by interchanging the parts of the symbol "100" to "0/0" and then to "%."

Answers on page A-13

PERCENT NOTATION

The notation ***n*%** means "n per hundred."

This definition leads us to the following equivalent ways of defining percent notation.

NOTATION FOR $n\%$

Percent notation, $n\%$, can be expressed using:

ratio $\rightarrow$ $n\% =$ the ratio of n to 100 $= \frac{n}{100}$,

fraction notation $\rightarrow$ $n\% = n \times \frac{1}{100}$, or

decimal notation $\rightarrow$ $n\% = n \times 0.01$.

From 1998 to 2008, the number of jobs for professional chefs will increase by 13.4%.

Source: *Handbook of U.S. Labor Statistics*

EXAMPLE 1 Write three kinds of notation for 35%.

Using ratio: $35\% = \frac{35}{100}$ A ratio of 35 to 100

Using fraction notation: $35\% = 35 \times \frac{1}{100}$ Replacing % with $\times \frac{1}{100}$

Using decimal notation: $35\% = 35 \times 0.01$ Replacing % with $\times$ 0.01

EXAMPLE 2 Write three kinds of notation for 67.8%.

Using ratio: $67.8\% = \frac{67.8}{100}$ A ratio of 67.8 to 100

Using fraction notation: $67.8\% = 67.8 \times \frac{1}{100}$ Replacing % with $\times \frac{1}{100}$

Using decimal notation: $67.8\% = 67.8 \times 0.01$ Replacing % with $\times$ 0.01

Do Exercises 1–3 on the preceding page.

b Converting Between Percent Notation and Decimal Notation

Consider 78%. To convert to decimal notation, we can think of percent notation as a ratio and write

$78\% = \frac{78}{100}$ Using the definition of percent as a ratio

$= 0.78.$ Dividing

Similarly,

$4.9\% = \frac{4.9}{100}$ Using the definition of percent as a ratio

$= 0.049.$ Dividing

We could also convert 78% to decimal notation by replacing "%" with "$\times$ 0.01" and write

$78\% = 78 \times 0.01$ Replacing % with $\times$ 0.01

$= 0.78.$ Multiplying

CALCULATOR CORNER

Converting from Percent Notation to Decimal Notation Many calculators have a [%] key that can be used to convert from percent notation to decimal notation. This is often the second operation associated with a particular key and is accessed by first pressing a [2nd] or [SHIFT] key. To convert 57.6% to decimal notation, for example, you might press [5] [7] [.] [6] [2nd] [%] or [5] [7] [.] [6] [SHIFT] [%]. The display would read [0.576], so 57.6% = 0.576. Read the user's manual to determine whether your calculator can do this conversion.

Exercises: Use a calculator to find decimal notation.

1. 14%
2. 0.069%
3. 43.8%
4. 125%

Find decimal notation.

4. 34%

5. 78.9%

6. $6\frac{5}{8}\%$

Find decimal notation for the percent notation in the sentence.

7. Pickup Trucks. Of all household personal vehicles in the United States, 18% are pickup trucks.

Source: Department of Transportation, 2001 National Household Travel Survey

8. Blood Alcohol Level. A blood alcohol level of 0.08% is a standard used by some states as the legal limit for drunk driving.

Answers on page A-13

Similarly,

$$4.9\% = 4.9 \times 0.01 \quad \text{Replacing \% with } \times 0.01$$
$$= 0.049. \quad \text{Multiplying}$$

Dividing by 100 amounts to moving the decimal point two places to the left, which is the same as multiplying by 0.01. This leads us to a quick way to convert from percent notation to decimal notation—we drop the percent symbol and move the decimal point two places to the left.

To convert from percent notation to decimal notation, — 36.5%

a) replace the percent symbol % with × 0.01, and — 36.5 × 0.01

b) multiply by 0.01, which means move the decimal point two places to the left. — 0.36.5 Move 2 places to the left.

36.5% = 0.365

EXAMPLE 3 Find decimal notation for 99.44%.

a) Replace the percent symbol with × 0.01. — 99.44 × 0.01

b) Move the decimal point two places to the left. — 0.99.44

Thus, 99.44% = 0.9944.

EXAMPLE 4 The interest rate on a $2\frac{1}{2}$-year certificate of deposit is $6\frac{3}{8}\%$. Find decimal notation for $6\frac{3}{8}\%$.

a) Convert $6\frac{3}{8}$ to decimal notation and replace the percent symbol with × 0.01. — $6\frac{3}{8}\%$, 6.375 × 0.01

b) Move the decimal point two places to the left. — 0.06.375

Thus, $6\frac{3}{8}\% = 0.06375$.

Do Exercises 4–8.

To convert 0.38 to percent notation, we can first write fraction notation, as follows:

$$0.38 = \frac{38}{100} \quad \text{Converting to fraction notation}$$
$$= 38\%. \quad \text{Using the definition of percent as a ratio}$$

Note that 100% = 100 × 0.01 = 1. Thus to convert 0.38 to percent notation, we can multiply by 1, using 100% as a symbol for 1.

$$0.38 = 0.38 \times 1$$
$$= 0.38 \times 100\%$$
$$= 0.38 \times 100 \times 0.01 \quad \text{Replacing 100\% with } 100 \times 0.01$$
$$= (0.38 \times 100) \times 0.01 \quad \text{Using the associative law of multiplication}$$

Then

$38 \times 0.01 = 38\%$. Replacing "$\times$ 0.01" with the % symbol

Even more quickly, since $0.38 = 0.38 \times 100\%$, we can simply multiply 0.38 by 100 and write the % symbol.

To convert from decimal notation to percent notation, we multiply by 100%—that is, we move the decimal point two places to the right and write a percent symbol.

To convert from decimal notation to percent notation, multiply by 100%. That is,

$0.675 = 0.675 \times 100\%$

a) move the decimal point two places to the right, and — 0.67.5 Move 2 places to the right.

b) write a % symbol. — 67.5%

$0.675 = 67.5\%$

EXAMPLE 5 Find percent notation for 1.27.

a) Move the decimal point two places to the right. — 1.27.

b) Write a % symbol. — 127%

Thus, $1.27 = 127\%$.

EXAMPLE 6 Of the time that people declare as sick leave, 0.21 is actually used for family issues. Find percent notation for 0.21.

Source: CCH Inc.

a) Move the decimal point two places to the right. — 0.21.

b) Write a % symbol. — 21%

Thus, $0.21 = 21\%$.

EXAMPLE 7 Find percent notation for 5.6.

a) Move the decimal point two places to the right, adding an extra zero. — 5.60.

b) Write a % symbol. — 560%

Thus, $5.6 = 560\%$.

EXAMPLE 8 Of those who play golf, 0.149 play 8–24 rounds per year. Find percent notation for 0.149.

Source: U.S. Golf Association

a) Move the decimal point two places to the right. — 0.14.9

b) Write a % symbol. — 14.9%

Thus, $0.149 = 14.9\%$.

Do Exercises 9–13.

Find percent notation.

9. 0.24

10. 3.47

11. 1

Find percent notation for the decimal notation in the sentence.

12. High School Graduate. The highest level of education for 0.321 of persons 25 and over in the United States is high school graduation.

Source: U.S. Department of Commerce, Bureau of the Census, *Current Population Survey*

13. Golf. Of those who play golf, 0.253 play 25–49 rounds per year.

Source: U.S. Golf Association

Answers on page A-13

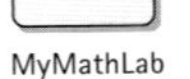

MathXL MyMathLab InterAct Math Math Tutor Center Digital Video Tutor CD 3 Videotape 6 Student's Solutions Manual

6.1 EXERCISE SET

For Extra Help

a Write three kinds of notation as in Examples 1 and 2 on p. 343.

1. 90%

2. 58.7%

3. 12.5%

4. 130%

b Find decimal notation.

5. 67%

6. 17%

7. 45.6%

8. 76.3%

9. 59.01%

10. 30.02%

11. 10%

12. 80%

13. 1%

14. 100%

15. 200%

16. 300%

17. 0.1%

18. 0.4%

19. 0.09%

20. 0.12%

21. 0.18%

22. 5.5%

23. 23.19%

24. 87.99%

25. $14\frac{7}{8}\%$

26. $93\frac{1}{8}\%$

27. $56\frac{1}{2}\%$

28. $61\frac{3}{4}\%$

Find decimal notation for the percent notation(s) in the sentence.

29. *Pediatricians.* By 2020, the population of children in the United States is expected to increase by about 9% while the number of pediatricians will leap 58%.

Source: Study by Dr. Scott Shipman, an Oregon Health and Science University pediatrician

30. *Cancer Survival.* In 2005, 78.9% of children with cancer survive at least 5 years compared to about 60% in 1975.

Source: National Cancer Institute

31. *Eating Out.* On a given day, 44% of all adults eat in a restaurant.

Source: *AARP Bulletin*, November 2004

32. Of those who play golf, 18.6% play 100 or more rounds per year.

Source: U.S. Golf Association

33. *Knitting and Crocheting.* Of all women ages 25 to 34, 36% know how to knit or crochet.

Source: Research Inc. for Craft Yarn Council of America

34. According to a recent survey, 95.1% of those asked to name what sports they participate in chose swimming.

Source: Sporting Goods Manufacturers

Find percent notation.

35. 0.47

36. 0.87

37. 0.03

38. 0.01

39. 8.7

40. 4

41. 0.334

42. 0.889

43. 0.75

44. 0.99

45. 0.4

46. 0.5

47. 0.006

48. 0.008

49. 0.017

50. 0.024

51. 0.2718

52. 0.8911

53. 0.0239

54. 0.00073

Find percent notation for the decimal notation(s) in the sentence.

55. *Hours of Sleep.* In 2005, only 0.26 of people get eight or more hours of sleep a night on weekdays. This rate has declined from 0.38 in 2001.

Source: National Sleep Foundation's 2005 Sleep in America Poll

56. According to a recent survey, 0.526 of those asked to name what sports they participate in chose bowling.

Source: Sporting Goods Manufacturers

57. *Bachelor's Degrees.* For 0.177 of the United States population 25 and over, the bachelor's degree is the highest level of educational attainment.

Source: U.S. Department of Commerce, Bureau of the Census, *Current Population Survey;* National Center for Educational Statistics, *Digest of Education Statistics,* 2003

58. About 0.69 of all newspapers are recycled.

Sources: American Forest and Paper Association; Newspaper Association of America

59. *65 and Over.* In Clearwater, Florida, 0.215 of the residents are 65 or over.

Source: U.S. Census Bureau

60. Of those people living in North Carolina, 0.1134 will die of heart disease.

Source: American Association of Retired Persons

61. **DW** *Winning Percentage.* During the 2004 regular baseball season, the Boston Red Sox won 98 of 162 games and went on to win the World Series. Find the ratio of number of wins to total number of games played in the regular season and convert it to decimal notation. Such a rate is often called a "winning percentage." Explain why.

62. **DW** Athletes sometimes speak of "giving 110%" effort. Does this make sense? Explain.

SKILL MAINTENANCE

Convert to a mixed numeral. [3.4a]

63. $\frac{100}{3}$

64. $\frac{75}{2}$

65. $\frac{75}{8}$

66. $\frac{297}{16}$

67. $\frac{567}{98}$

68. $\frac{2345}{21}$

Convert to decimal notation. [4.5a]

69. $\frac{2}{3}$

70. $\frac{1}{3}$

71. $\frac{5}{6}$

72. $\frac{17}{12}$

73. $\frac{8}{3}$

74. $\frac{15}{16}$

6.2 PERCENT AND FRACTION NOTATION

Objectives

a Convert from fraction notation to percent notation.

b Convert from percent notation to fraction notation.

a Converting from Fraction Notation to Percent Notation

Consider the fraction notation $\frac{7}{8}$. To convert to percent notation, we use two skills we already have. We first find decimal notation by dividing:

$$\frac{7}{8} = 0.875 \qquad \begin{array}{r} 0.875 \\ 8\overline{)7.000} \\ \underline{6\,4} \\ 60 \\ \underline{56} \\ 40 \\ \underline{40} \\ 0 \end{array}$$

Then we convert the decimal notation to percent notation. We move the decimal point two places to the right

$0.87.5$

and write a % symbol:

$$\frac{7}{8} = 87.5\%, \text{ or } 87\frac{1}{2}\%.$$

To convert from fraction notation to percent notation,

$\frac{3}{5}$ Fraction notation

a) find decimal notation by division, and

$$\begin{array}{r} 0.6 \\ 5\overline{)3.0} \\ \underline{3\,0} \\ 0 \end{array}$$

b) convert the decimal notation to percent notation.

$0.6 = 0.60 = 60\%$ Percent notation

$\frac{3}{5} = 60\%$

EXAMPLE 1 Find percent notation for $\frac{9}{16}$.

a) We first find decimal notation by division.

$$\begin{array}{r} 0.5625 \\ 16\overline{)9.0000} \\ \underline{8\,0} \\ 100 \\ \underline{96} \\ 40 \\ \underline{32} \\ 80 \\ \underline{80} \\ 0 \end{array} \qquad \frac{9}{16} = 0.5625$$

Study Tips

BEING A TUTOR

Try being a tutor for a fellow student. Understanding and retention of concepts can be maximized for yourself if you explain the material to someone else.

Find percent notation.

1. $\frac{1}{4}$

2. $\frac{5}{8}$

b) Next, we convert the decimal notation to percent notation. We move the decimal point two places to the right and write a % symbol.

0.56.25

$\frac{9}{16} = 56.25\%$, or $56\frac{1}{4}\%$

Don't forget the % symbol.

Do Exercises 1 and 2.

CALCULATOR CORNER

Converting from Fraction Notation to Percent Notation A calculator can be used to convert from fraction notation to percent notation. We simply perform the division on the calculator and then use the percent key. To convert $\frac{17}{40}$ to percent notation, for example, we press [1] [7] [÷] [4] [0] [2nd] [%], or [1] [7] [÷] [4] [0] [SHIFT] [%]. The display reads [42.5], so $\frac{17}{40} = 42.5\%$. Read the user's manual to determine whether your calculator can do this conversion.

Exercises: Use a calculator to find percent notation. Round to the nearest hundredth of a percent.

1. $\frac{13}{25}$
2. $\frac{5}{13}$
3. $\frac{43}{39}$
4. $\frac{12}{7}$
5. $\frac{217}{364}$
6. $\frac{2378}{8401}$

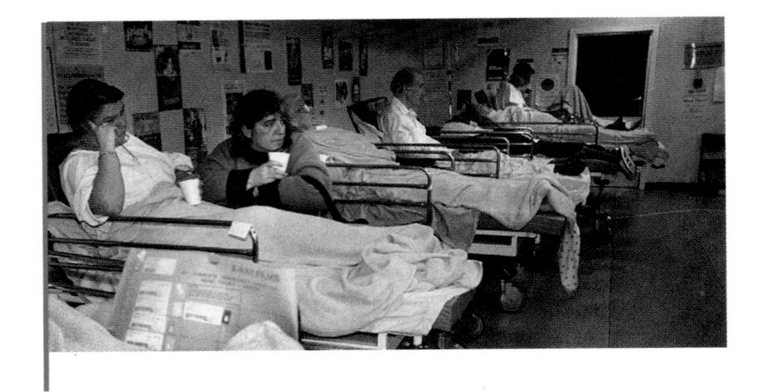

EXAMPLE 2 *Without Health Insurance.* Approximately $\frac{1}{6}$ of all people in the United States are without health insurance. Find percent notation for $\frac{1}{6}$.

Source: U.S. Bureau of the Census, *Current Population Survey*, March 2003

a) Find decimal notation by division.

$$\begin{array}{r} 0.1\,6\,6 \\ 6\,\overline{)\,1.0\,0\,0} \\ \underline{6} \\ 4\,0 \\ \underline{3\,6} \\ 4\,0 \\ \underline{3\,6} \\ 4 \end{array}$$

We get a repeating decimal: $0.16\overline{6}$.

b) Convert the answer to percent notation.

$0.16.\overline{6}$

$\frac{1}{6} = 16.\overline{6}\%$, or $16\frac{2}{3}\%$

Answers on page A-13

Do Exercises 3 and 4.

In some cases, division is not the fastest way to convert. The following are some optional ways in which conversion might be done.

EXAMPLE 3 Find percent notation for $\frac{69}{100}$.

We use the definition of percent as a ratio.

$$\frac{69}{100} = 69\%$$

EXAMPLE 4 Find percent notation for $\frac{17}{20}$.

We multiply by 1 to get 100 in the denominator. We think of what we have to multiply 20 by in order to get 100. That number is 5, so we multiply by 1 using $\frac{5}{5}$.

$$\frac{17}{20} \cdot \frac{5}{5} = \frac{85}{100} = 85\%$$

Note that this shortcut works only when the denominator is a factor of 100.

Do Exercises 5 and 6.

b Converting from Percent Notation to Fraction Notation

To convert from percent notation to fraction notation,

a) use the definition of percent as a ratio, and

b) simplify, if possible.

30%	Percent notation
$\frac{30}{100}$	
$\frac{3}{10}$	Fraction notation

EXAMPLE 5 Find fraction notation for 75%.

$$75\% = \frac{75}{100} \quad \text{Using the definition of percent}$$

$$\left.\begin{aligned} &= \frac{3 \cdot 25}{4 \cdot 25} = \frac{3}{4} \cdot \frac{25}{25} \\ &= \frac{3}{4} \end{aligned}\right\} \quad \text{Simplifying}$$

3. Water is the single most abundant chemical in the body. The human body is about $\frac{2}{3}$ water. Find percent notation for $\frac{2}{3}$.

4. Find percent notation: $\frac{5}{6}$.

Find percent notation.

5. $\frac{57}{100}$

6. $\frac{19}{25}$

Answers on page A-13

Find fraction notation.

7. 60%

8. 3.25%

9. $66\frac{2}{3}\%$

10. Complete this table.

FRACTION NOTATION	$\frac{1}{5}$		
DECIMAL NOTATION		$0.83\overline{3}$	
PERCENT NOTATION			$37\frac{1}{2}\%$

Answers on page A-13

Study Tips

MEMORIZING

Memorizing is a very helpful tool in the study of mathematics. Don't underestimate its power as you memorize the table of decimal, fraction, and percent notation on the inside back cover. We will discuss memorizing more later.

EXAMPLE 6 Find fraction notation for 62.5%.

$$62.5\% = \frac{62.5}{100} \quad \text{Using the definition of percent}$$
$$= \frac{62.5}{100} \times \frac{10}{10} \quad \text{Multiplying by 1 to eliminate the decimal point in the numerator}$$
$$= \frac{625}{1000}$$
$$= \frac{5 \cdot 125}{8 \cdot 125} = \frac{5}{8} \cdot \frac{125}{125}$$
$$= \frac{5}{8} \quad \text{Simplifying}$$

EXAMPLE 7 Find fraction notation for $16\frac{2}{3}\%$.

$$16\frac{2}{3}\% = \frac{50}{3}\% \quad \text{Converting from the mixed numeral to fraction notation}$$
$$= \frac{50}{3} \times \frac{1}{100} \quad \text{Using the definition of percent}$$
$$= \frac{50 \cdot 1}{3 \cdot 50 \cdot 2} = \frac{1}{6} \cdot \frac{50}{50}$$
$$= \frac{1}{6} \quad \text{Simplifying}$$

The table on the inside back cover lists decimal, fraction, and percent equivalents used so often that it would speed up your work if you memorized them. For example, $\frac{1}{3} = 0.\overline{3}$, so we say that the **decimal equivalent** of $\frac{1}{3}$ is $0.\overline{3}$, or that $0.\overline{3}$ has the **fraction equivalent** $\frac{1}{3}$.

EXAMPLE 8 Find fraction notation for $16.\overline{6}\%$.

We can use the table on the inside back cover or recall that $16.\overline{6}\% = 16\frac{2}{3}\% = \frac{1}{6}$. We can also recall from our work with repeating decimals in Chapter 4 that $0.\overline{6} = \frac{2}{3}$. Then we have $16.\overline{6}\% = 16\frac{2}{3}\%$ and can proceed as in Example 7.

Do Exercises 7–10.

CALCULATOR CORNER

Applications of Ratio and Percent: The Price–Earnings Ratio and Stock Yields

The Price–Earnings Ratio If the total earnings of a company one year were $5,000,000 and 100,000 shares of stock were issued, the earnings per share were $50. At one time, the price per share of Coca-Cola was $58.125 and the earnings per share were $0.76. The **price–earnings ratio,** *P/E*, is the price of the stock divided by the earnings per share. For the Coca-Cola stock, the price–earnings ratio, *P/E*, is given by

$$\frac{P}{E} = \frac{58.125}{0.76} \approx 76.48. \quad \text{Dividing, using a calculator, and rounding to the nearest hundredth}$$

Stock Yields At one time, the price per share of Coca-Cola stock was $58.125 and the company was paying a yearly dividend of $0.68 per share. It is helpful to those interested in stocks to know what percent the dividend is of the price of the stock. The percent is called the **yield.** For the Coca-Cola stock, the yield is given by

$$\text{Yield} = \frac{\text{Dividend}}{\text{Price per share}} = \frac{0.68}{58.125} \approx 0.0117 \quad \text{Dividing and rounding to the nearest ten-thousandth}$$

$$= 1.17\% \quad \text{Converting to percent notation}$$

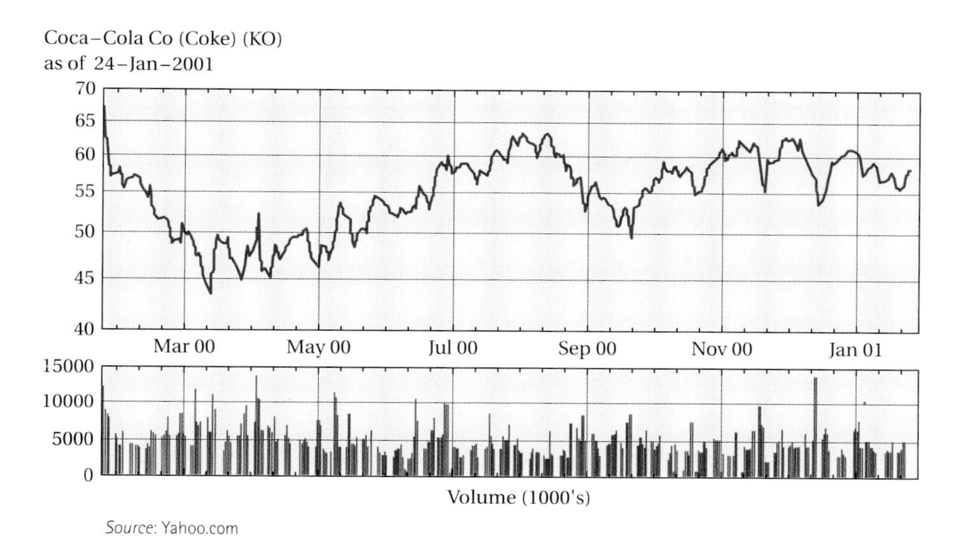

Source: Yahoo.com

Exercises: Compute the price–earnings ratio and the yield for each stock listed below.

	STOCK	PRICE PER SHARE	EARNINGS	DIVIDEND	P/E	YIELD
1.	Pepsi (PEP)	$42.75	$1.40	$0.56		
2.	Pearson (PSO)	$25.00	$0.78	$0.30		
3.	Quaker Oats (OAT)	$92.375	$2.68	$1.10		
4.	Texas Insts (TEX)	$42.875	$1.62	$0.43		
5.	Ford Motor Co (F)	$27.5625	$2.30	$1.19		
6.	Wendy's Intl (WEN)	$25.75	$1.47	$0.23		

6.2 EXERCISE SET

For Extra Help

MathXL | MyMathLab | InterAct Math | Math Tutor Center | Digital Video Tutor CD 3 Videotape 6 | Student's Solutions Manual

a Find percent notation.

1. $\frac{41}{100}$

2. $\frac{36}{100}$

3. $\frac{5}{100}$

4. $\frac{1}{100}$

5. $\frac{2}{10}$

6. $\frac{7}{10}$

7. $\frac{3}{10}$

8. $\frac{9}{10}$

9. $\frac{1}{2}$

10. $\frac{3}{4}$

11. $\frac{7}{8}$

12. $\frac{1}{8}$

13. $\frac{4}{5}$

14. $\frac{2}{5}$

15. $\frac{2}{3}$

16. $\frac{1}{3}$

17. $\frac{1}{6}$

18. $\frac{5}{6}$

19. $\frac{3}{16}$

20. $\frac{11}{16}$

21. $\frac{13}{16}$

22. $\frac{7}{16}$

23. $\frac{4}{25}$

24. $\frac{17}{25}$

25. $\frac{1}{20}$

26. $\frac{31}{50}$

27. $\frac{17}{50}$

28. $\frac{3}{20}$

Find percent notation for the fraction notation in the sentence.

29. *Driving While Drowsy.* Almost half of U.S. drivers say they have driven while drowsy. Of this group, $\frac{2}{5}$ fight off sleep by opening a window, and $\frac{9}{50}$ drink a caffeinated beverage.

Source: Harris Interactive for Tylenol PM

30. *Paved Roads.* About $\frac{13}{20}$ of the roads and streets in the United States are paved.

Source: U.S. Department of Transportation, *Highway Statistics*

In Exercises 31–36, write percent notation for the fractions in this pie chart.

Time Workers Spend Sorting Unsolicited e-mail and Spam

Source: Data from InsightExpress

31. $\frac{11}{50}$

32. $\frac{3}{100}$

33. $\frac{1}{20}$

34. $\frac{59}{100}$

35. $\frac{9}{100}$

36. $\frac{1}{50}$

b Find fraction notation. Simplify.

37. 85%

38. 55%

39. 62.5%

40. 12.5%

41. $33\frac{1}{3}\%$

42. $83\frac{1}{3}\%$

43. $16.\overline{6}\%$

44. $66.\overline{6}\%$

45. 7.25%

46. 4.85%

47. 0.8%

48. 0.2%

49. $25\frac{3}{8}\%$

50. $48\frac{7}{8}\%$

51. $78\frac{2}{9}\%$

52. $16\frac{5}{9}\%$

53. $64\frac{7}{11}\%$

54. $73\frac{3}{11}\%$

55. 150%

56. 110%

57. 0.0325%

58. 0.419%

59. $33.\overline{3}\%$

60. $83.\overline{3}\%$

Find fraction notation for the percent notation in the following bar graph.

Organ Transplants in the United States, 2003

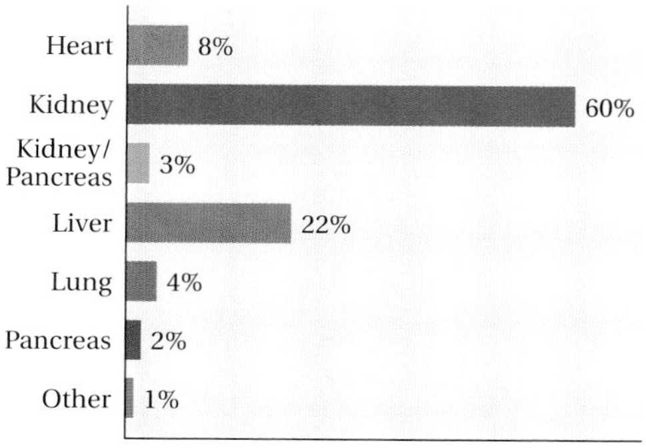

Heart	8%
Kidney	60%
Kidney/Pancreas	3%
Liver	22%
Lung	4%
Pancreas	2%
Other	1%

Source: National Organ Procurement and Transplantation Network

61. 8%

62. 22%

63. 60%

64. 4%

65. 2%

66. 3%

Find fraction notation for the percent notation in the sentence.

67. A 1.9-oz serving of Raisin Bran Crunch® cereal with $\frac{1}{2}$ cup of skim milk satisfies 35% of the minimum daily requirements for Vitamin B_{12}.

Source: Kellogg's USA, Inc.

68. A 1-cup serving of Wheaties® cereal with $\frac{1}{2}$ cup of skim milk satisfies 15% of the minimum daily requirements for calcium.

Source: General Mills Sales, Inc.

69. Of all those who are 85 or older, 47% have Alzheimer's disease.

Source: Alzheimer's Association

70. In 2003, 24.4% of Americans 18 and older smoked cigarettes.

Source: U.S. Centers for Disease Control and Prevention

Complete the table.

71.

FRACTION NOTATION	DECIMAL NOTATION	PERCENT NOTATION
$\frac{1}{8}$		12.5%, or $12\frac{1}{2}\%$
$\frac{1}{6}$		
		20%
	0.25	
		$33.\overline{3}\%$, or $33\frac{1}{3}\%$
		37.5% or $37\frac{1}{2}\%$
		40%
$\frac{1}{2}$		

72.

FRACTION NOTATION	DECIMAL NOTATION	PERCENT NOTATION
$\frac{3}{5}$		
	0.625	
$\frac{2}{3}$		
	0.75	75%
$\frac{4}{5}$		
$\frac{5}{6}$		$83.\overline{3}\%$, or $83\frac{1}{3}\%$
$\frac{7}{8}$		87.5%, or $87\frac{1}{2}\%$
		100%

73.

FRACTION NOTATION	DECIMAL NOTATION	PERCENT NOTATION
	0.5	
$\frac{1}{3}$		
		25%
		$16.\overline{6}\%$, or $16\frac{2}{3}\%$
	0.125	
$\frac{3}{4}$		
	$0.8\overline{3}$	
$\frac{3}{8}$		

74.

FRACTION NOTATION	DECIMAL NOTATION	PERCENT NOTATION
		40%
		62.5%, or $62\frac{1}{2}\%$
	0.875	
$\frac{1}{1}$		
	0.6	
	$0.\overline{6}$	
$\frac{1}{5}$		

75. $\mathbf{D_W}$ What do the following have in common? Explain.

$\frac{23}{16}$, $1\frac{875}{2000}$, 1.4375, $\frac{207}{144}$, $1\frac{7}{16}$, 143.75%, $1\frac{4375}{10{,}000}$

76. $\mathbf{D_W}$ Is it always best to convert from fraction notation to percent notation by first finding decimal notation? Why or why not?

SKILL MAINTENANCE

Solve.

77. $13 \cdot x = 910$ [1.7b]

78. $15 \cdot y = 75$ [1.7b]

79. $0.05 \times b = 20$ [4.4b]

80. $3 = 0.16 \times b$ [4.4b]

81. $\frac{24}{37} = \frac{15}{x}$ [5.3b]

82. $\frac{17}{18} = \frac{x}{27}$ [5.3b]

83. $\frac{9}{10} = \frac{x}{5}$ [5.3b]

84. $\frac{7}{x} = \frac{4}{5}$ [5.3b]

Convert to a mixed numeral. [3.4a]

85. $\frac{100}{3}$

86. $\frac{75}{2}$

87. $\frac{250}{3}$

88. $\frac{123}{6}$

89. $\frac{345}{8}$

90. $\frac{373}{6}$

91. $\frac{75}{4}$

92. $\frac{67}{9}$

Convert from a mixed numeral to fraction notation. [3.4a]

93. $1\frac{1}{17}$

94. $20\frac{9}{10}$

95. $101\frac{1}{2}$

96. $32\frac{3}{8}$

SYNTHESIS

Write percent notation.

97. $\frac{41}{369}$

98. $\frac{54}{999}$

99. $2.5\overline{74631}$

100. $3.2\overline{93847}$

Write decimal notation.

101. $\frac{14}{9}\%$

102. $\frac{19}{12}\%$

103. $\frac{729}{7}\%$

104. $\frac{637}{6}\%$

6.3 SOLVING PERCENT PROBLEMS USING PERCENT EQUATIONS

Objectives

a Translate percent problems to percent equations.

b Solve basic percent problems.

a Translating to Equations

To solve a problem involving percents, it is helpful to translate first to an equation. To distinguish the method in Section 6.3 from that of Section 6.4, we will call these *percent equations.*

EXAMPLE 1 Translate:

23%	of	5	is	what?
↓	↓	↓	↓	↓
23%	·	5	=	a

This is a *percent equation.*

KEY WORDS IN PERCENT TRANSLATIONS

"**Of**" translates to "·", or "×".

"**Is**" translates to "=".

"**What**" translates to any letter.

"**%**" translates to "$\times \frac{1}{100}$" or "× 0.01".

EXAMPLE 2 Translate:

What	is	11%	of	49?
↓	↓	↓	↓	↓
a	=	11%	·	49

Any letter can be used.

Do Exercises 1 and 2.

EXAMPLE 3 Translate:

3	is	10%	of	what?
↓	↓	↓	↓	↓
3	=	10%	·	b

EXAMPLE 4 Translate:

45%	of	what	is	23?
↓	↓	↓	↓	↓
45%	×	b	=	23

Do Exercises 3 and 4.

EXAMPLE 5 Translate:

10	is	what percent	of	20?
↓	↓	↓	↓	↓
10	=	p	×	20

Translate to an equation. Do not solve.

1. 12% of 50 is what?

2. What is 40% of 60?

Translate to an equation. Do not solve.

3. 45 is 20% of what?

4. 120% of what is 60?

Answers on page A-14

EXAMPLE 6 Translate:

$$\underbrace{\text{What percent}}_{p} \;\; \underset{\cdot}{\text{of}} \;\; \underset{50}{50} \;\; \underset{=}{\text{is}} \;\; \underset{7}{7?}$$

Do Exercises 5 and 6.

Translate to an equation. Do not solve.

5. 16 is what percent of 40?

6. What percent of 84 is 10.5?

b Solving Percent Problems

In solving percent problems, we use the *Translate* and *Solve* steps in the problem-solving strategy used throughout this text.

Percent problems are actually of three different types. Although the method we present does *not* require that you be able to identify which type you are solving, it is helpful to know them.

We know that

15 is 25% of 60, or $15 = 25\% \times 60$.

We can think of this as:

Amount = Percent number × Base.

Each of the three types of percent problems depends on which of the three pieces of information is missing.

1. **Finding the *amount* (the result of taking the percent)**

   Example: **What** is 25% of 60?

   Translation: $a = 25\% \cdot 60$

2. **Finding the *base* (the number you are taking the percent of)**

   Example: 15 is 25% of **what number?**

   Translation: $15 = 25\% \cdot b$

3. **Finding the *percent number* (the percent itself)**

   Example: 15 is **what percent** of 60?

   Translation: $15 = p \cdot 60$

7. Solve:

What is 12% of 50?

Answers on page A-14

FINDING THE AMOUNT

EXAMPLE 7 What is 4.6% of 105,000,000?

Translate: $a = 4.6\% \times 105{,}000{,}000$.

Solve: The letter is by itself. To solve the equation, we just convert 4.6% to decimal notation and multiply:

$$a = 4.6\% \times 105{,}000{,}000$$
$$= 0.046 \times 105{,}000{,}000 = 4{,}830{,}000.$$

Thus, 4,830,000 is 4.6% of 105,000,000. The answer is 4,830,000.

There are 105,000,000 households in the United States and approximately 4.6% own a pet bird. How many households own at least one pet bird?

Do Exercise 7.

8. Solve:

64% of $55 is what?

EXAMPLE 8 120% of $42 is what?

Translate: $120\% \times 42 = a.$

Solve: The letter is by itself. To solve the equation, we carry out the calculation:

$$a = 120\% \times 42$$
$$= 1.2 \times 42 \qquad 120\% = 1.2$$
$$= 50.4.$$

Thus, 120% of $42 is $50.40. The answer is $50.40.

Do Exercise 8.

In a survey of a group of people, it was found that 5%, or 20 people, chose strawberry as their favorite ice cream. How many people were surveyed?

Source: International Ice Cream Association

FINDING THE BASE

EXAMPLE 9 5% of what is 20?

Translate: $5\% \times b = 20.$

Solve: This time the letter is *not* by itself. To solve the equation, we divide by 5% on both sides:

$$\frac{5\% \times b}{5\%} = \frac{20}{5\%} \qquad \text{Dividing by 5\% on both sides}$$
$$b = \frac{20}{0.05} \qquad 5\% = 0.05$$
$$b = 400.$$

Thus, 5% of 400 is 20. The answer is 400.

Solve.

9. 20% of what is 45?

10. $60 is 120% of what?

EXAMPLE 10 $3 is 16% of what?

Translate: $3 is 16% of what?

$$3 = 16\% \times b.$$

Solve: Again, the letter is *not* by itself. To solve the equation, we divide by 16% on both sides:

$$\frac{3}{16\%} = \frac{16\% \times b}{16\%} \qquad \text{Dividing by 16\% on both sides}$$
$$\frac{3}{0.16} = b \qquad 16\% = 0.16$$
$$18.75 = b.$$

Thus, $3 is 16% of $18.75. The answer is $18.75.

Do Exercises 9 and 10.

Answers on page A-14

FINDING THE PERCENT NUMBER

In solving these problems, you *must* remember to convert to percent notation after you have solved the equation.

EXAMPLE 11 13 is what percent of 260?

Translate: 13 is $\underbrace{\text{what percent}}$ of 260?

$$13 = p \times 260.$$

Solve: To solve the equation, we divide by 260 on both sides and convert the result to percent notation:

$p \cdot 260 = 13$

$\frac{p \cdot 260}{260} = \frac{13}{260}$ Dividing by 260 on both sides

$p = 0.05$ Converting to decimal notation

$p = 5\%.$ Converting to percent notation

Thus, 13 is 5% of 260. The answer is 5%.

Do Exercise 11.

Workers in the United States on the average get 13 vacation days per year. What percent of the 260 work days per year are vacation days?

11. Solve:

16 is what percent of 40?

EXAMPLE 12 What percent of \$50 is \$16?

Translate: $\underbrace{\text{What percent}}$ of \$50 is \$16?

$$p \times 50 = 16.$$

Solve: To solve the equation, we divide by 50 on both sides and convert the answer to percent notation:

$\frac{p \times 50}{50} = \frac{16}{50}$ Dividing by 50 on both sides

$p = \frac{16}{50}$

$p = 0.32$

$p = 32\%.$ Converting to percent notation

Thus, 32% of \$50 is \$16. The answer is 32%.

Do Exercise 12.

12. Solve:

What percent of \$84 is \$10.50?

Caution!

When a question asks "what percent?", be sure to give the answer in percent notation.

Answers on page A-14

CALCULATOR CORNER

Using Percents in Computations Many calculators have a [%] key that can be used in computations. (See the Calculator Corner on page 343.) For example, to find 11% of 49, we press [1] [1] [2nd] [%] [×] [4] [9] [=] or [4] [9] [×] [1] [1] [SHIFT] [%]. The display reads [5.39], so 11% of 49 is 5.39.

In Example 9, we perform the computation 20/5%. To use the [%] key in this computation, we press [2] [0] [÷] [5] [2nd] [%] [=], or [2] [0] [÷] [5] [SHIFT] [%]. The result is 400.

We can also use the [%] key to find the percent number in a problem. In Example 11, for instance, we answer the question "13 is what percent of 260?" On a calculator, we press [1] [3] [÷] [2] [6] [0] [2nd] [%] [=], or [1] [3] [÷] [2] [6] [0] [SHIFT] [%]. The result is 5, so 13 is 5% of 260.

Exercises: Use a calculator to find each of the following.

1. What is 5% of 24?
2. What is 12.6% of $40?
3. What is 19% of 256?
4. 140% of $16 is what?
5. 0.04% of 28 is what?
6. 33% of $90 is what?
7. 8% of what is 36?
8. $45 is 4.5% of what?
9. 23 is what percent of 920?
10. What percent of $442 is $53.04?

6.3 EXERCISE SET

For Extra Help

MathXL

MyMathLab

InterAct Math

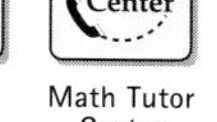
Math Tutor Center

Digital Video Tutor CD 3 Videotape 6

Student's Solutions Manual

a Translate to an equation. Do not solve.

1. What is 32% of 78?

2. 98% of 57 is what?

3. 89 is what percent of 99?

4. What percent of 25 is 8?

5. 13 is 25% of what?

6. 21.4% of what is 20?

b Translate to an equation and solve.

7. What is 85% of 276?

8. What is 74% of 53?

9. 150% of 30 is what?

10. 100% of 13 is what?

11. What is 6% of \$300?

12. What is 4% of \$45?

13. 3.8% of 50 is what?

14. $33\frac{1}{3}$% of 480 is what?
(*Hint:* $33\frac{1}{3}\% = \frac{1}{3}$.)

15. \$39 is what percent of \$50?

16. \$16 is what percent of \$90?

17. 20 is what percent of 10?

18. 60 is what percent of 20?

19. What percent of \$300 is \$150?

20. What percent of \$50 is \$40?

21. What percent of 80 is 100?

22. What percent of 60 is 15?

23. 20 is 50% of what?

24. 57 is 20% of what?

25. 40% of what is \$16?

26. 100% of what is \$74?

27. 56.32 is 64% of what?

28. 71.04 is 96% of what?

29. 70% of what is 14?

30. 70% of what is 35?

31. What is $62\frac{1}{2}\%$ of 10?

32. What is $35\frac{1}{4}\%$ of 1200?

33. What is 8.3% of \$10,200?

34. What is 9.2% of \$5600?

35. DW Write a question that could be translated to the equation

$$25 = 4\% \times b.$$

36. DW Suppose we know that 40% of 92 is 36.8. What is a quick way to find 4% of 92? 400% of 92? Explain.

SKILL MAINTENANCE

Write fraction notation. [4.1b]

37. 0.09

38. 1.79

39. 0.875

40. 0.125

41. 0.9375

42. 0.6875

Write decimal notation. [4.1b]

43. $\frac{89}{100}$

44. $\frac{7}{100}$

45. $\frac{3}{10}$

46. $\frac{17}{1000}$

SYNTHESIS

Solve.

47. What is 7.75% of \$10,880?
Estimate ____________
Calculate ____________

48. 50,951.775 is what percent of 78,995?
Estimate ____________
Calculate ____________

49. \$2496 is 24% of what amount?
Estimate ____________
Calculate ____________

50. What is 38.2% of \$52,345.79?
Estimate ____________
Calculate ____________

51. 40% of $18\frac{3}{4}\%$ of \$25,000 is what?

6.4 SOLVING PERCENT PROBLEMS USING PROPORTIONS*

Objectives

a Translate percent problems to proportions.

b Solve basic percent problems.

a Translating to Proportions

A percent is a ratio of some number to 100. For example, 47% is the ratio $\frac{47}{100}$. The numbers 68,859,700 and 146,510,000 have the same ratio as 47 and 100.

$$\frac{47}{100} = \frac{68{,}859{,}700}{146{,}510{,}000}$$

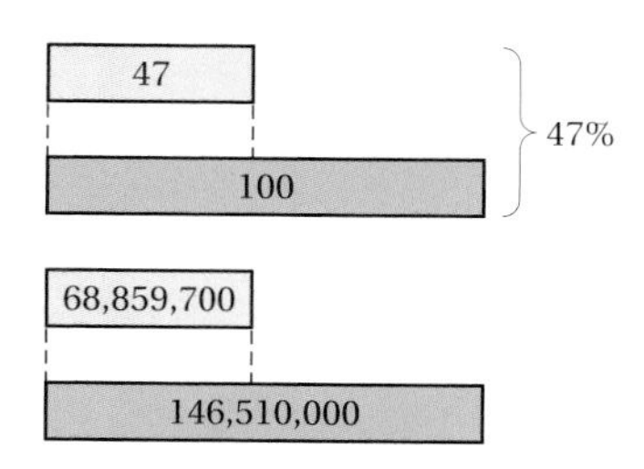

In the United States, 47% of the labor force are women. In 2003, there were 146,510,000 people in the labor force. This means that 68,859,700 were women.

Sources: U.S. Department of Labor, Bureau of Labor Statistics

To solve a percent problem using a proportion, we translate as follows:

$$\text{Number} \rightarrow \frac{N}{100} = \frac{a}{b} \leftarrow \begin{matrix}\text{Amount}\\ \text{Base}\end{matrix}$$

(100 → 100)

You might find it helpful to read this as "part is to whole as part is to whole."

For example, 60% of 25 is 15 translates to

$$\frac{60}{100} = \frac{15}{25}. \quad \begin{matrix}\leftarrow \text{Amount}\\ \leftarrow \text{Base}\end{matrix}$$

A clue in translating is that the base, b, corresponds to 100 and usually follows the wording "percent of." Also, $N\%$ always translates to $N/100$. Another aid in translating is to make a comparison drawing. To do this, we start with the percent side and list 0% at the top and 100% near the bottom. Then we estimate where the specified percent—in this case, 60%—is located. The corresponding quantities are then filled in. The base—in this case, 25—always corresponds to 100% and the amount—in this case, 15—corresponds to the specified percent.

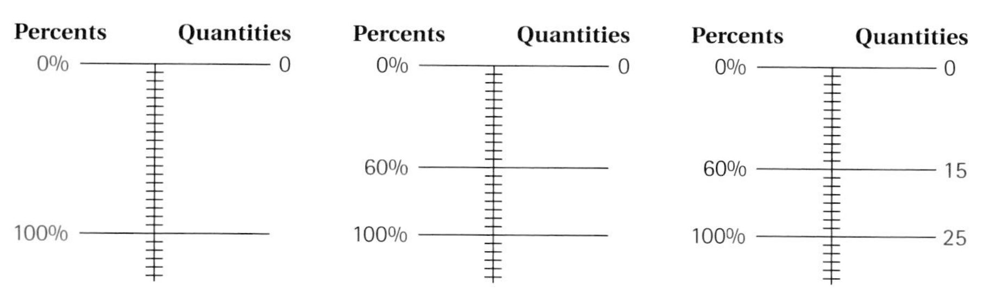

The proportion can then be read easily from the drawing: $\frac{60}{100} = \frac{15}{25}$.

*Note: This section presents an alternative method for solving basic percent problems. You can use either equations or proportions to solve percent problems, but you might prefer one method over the other, or your instructor may direct you to use one method over the other.

Translate to a proportion. Do not solve.

1. 12% of 50 is what?

2. What is 40% of 60?

3. 130% of 72 is what?

Translate to a proportion. Do not solve.

4. 45 is 20% of what?

5. 120% of what is 60?

Answers on page A-14

EXAMPLE 1 Translate to a proportion.

23% of 5 is what?

$$\frac{23}{100} = \frac{a}{5}$$

Percents Quantities
0% 0
23% a
100% 5

EXAMPLE 2 Translate to a proportion.

What is 124% of 49?

$$\frac{124}{100} = \frac{a}{49}$$

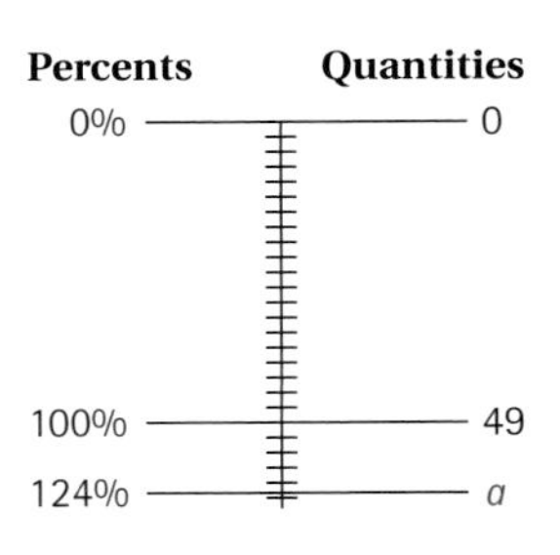

Do Exercises 1–3.

EXAMPLE 3 Translate to a proportion.

3 is 10% of what?

$$\frac{10}{100} = \frac{3}{b}$$

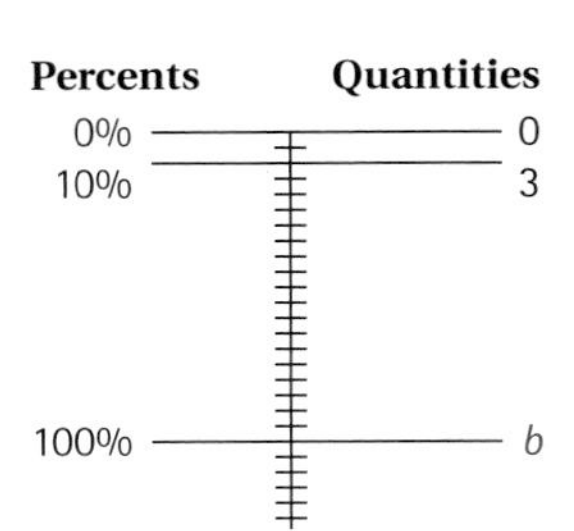

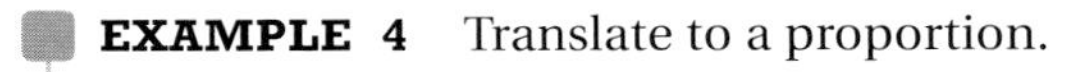

EXAMPLE 4 Translate to a proportion.

45% of what is 23?

$$\frac{45}{100} = \frac{23}{b}$$

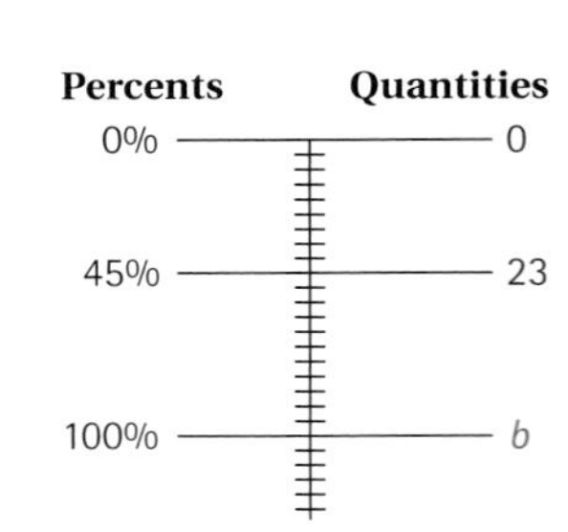

Do Exercises 4 and 5.

EXAMPLE 5 Translate to a proportion.

10 is what percent of 20?

$$\frac{N}{100} = \frac{10}{20}$$

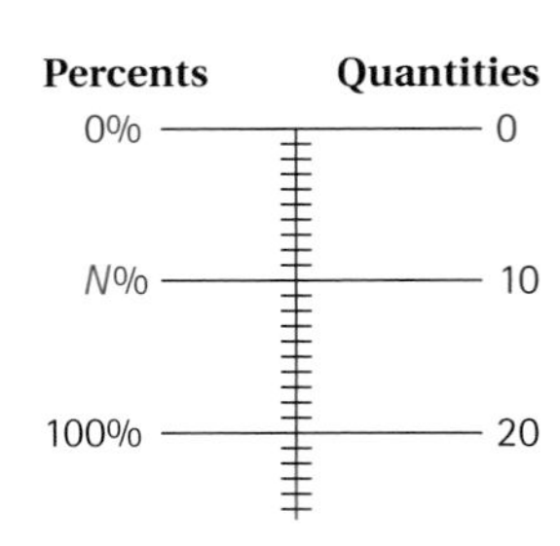

EXAMPLE 6 Translate to a proportion.

What percent of 50 is 7?

$$\frac{N}{100} = \frac{7}{50}$$

Percents | Quantities
0% — 0
N% — 7
100% — 50

Do Exercises 6 and 7.

Translate to a proportion. Do not solve.

6. 16 is what percent of 40?

7. What percent of 84 is 10.5?

b Solving Percent Problems

After a percent problem has been translated to a proportion, we solve as in Section 5.3.

EXAMPLE 7 5% of what is $20?

Translate: $\frac{5}{100} = \frac{20}{b}$

Solve: $5 \cdot b = 100 \cdot 20$ Equating cross products

$\frac{5 \cdot b}{5} = \frac{100 \cdot 20}{5}$ Dividing by 5

$b = \frac{2000}{5}$

$b = 400$ Simplifying

Thus, 5% of $400 is $20. The answer is $400.

Percents | Quantities
0% — 0
5% — 20
100% — b

Do Exercise 8.

8. Solve:

20% of what is $45?

EXAMPLE 8 120% of 42 is what?

Translate: $\frac{120}{100} = \frac{a}{42}$

Solve: $120 \cdot 42 = 100 \cdot a$ Equating cross products

$\frac{120 \cdot 42}{100} = \frac{100 \cdot a}{100}$ Dividing by 100

$\frac{5040}{100} = a$

$50.4 = a$ Simplifying

Thus, 120% of 42 is 50.4. The answer is 50.4.

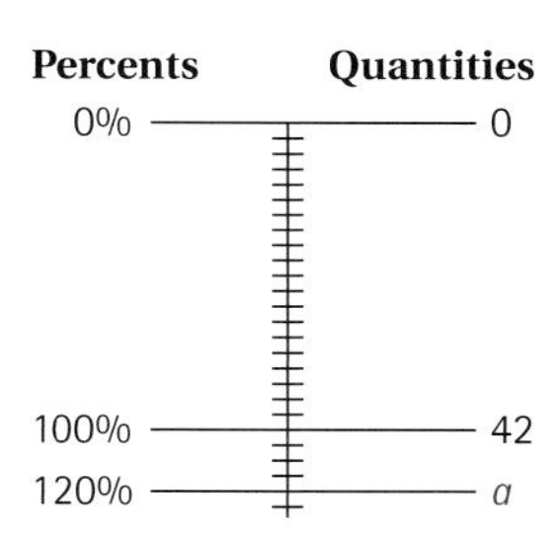

Do Exercises 9 and 10.

Solve.

9. 64% of 55 is what?

10. What is 12% of 50?

Answers on page A-14

11. Solve:

60 is 120% of what?

12. Solve:

$12 is what percent of $40?

13. Solve:

What percent of 84 is 10.5?

Answers on page A-14

EXAMPLE 9 3 is 16% of what?

Translate: $\frac{3}{b} = \frac{16}{100}$

Solve:

$$3 \cdot 100 = b \cdot 16 \qquad \text{Equating cross products}$$

$$\frac{3 \cdot 100}{16} = \frac{b \cdot 16}{16} \qquad \text{Dividing by 16}$$

$$\frac{300}{16} = b \qquad \text{Multiplying and simplifying}$$

$$18.75 = b \qquad \text{Dividing}$$

Thus, 3 is 16% of 18.75. The answer is 18.75.

Percents	Quantities
0%	0
16%	3
100%	b

Do Exercise 11.

EXAMPLE 10 $10 is what percent of $20?

Translate: $\frac{10}{20} = \frac{N}{100}$

Solve:

$$10 \cdot 100 = 20 \cdot N \qquad \text{Equating cross products}$$

$$\frac{10 \cdot 100}{20} = \frac{20 \cdot N}{20} \qquad \text{Dividing by 20}$$

$$\frac{1000}{20} = N \qquad \text{Multiplying and simplifying}$$

$$50 = N \qquad \text{Dividing}$$

Thus, $10 is 50% of $20. The answer is 50%.

Percents	Quantities
0%	0
N%	$10
100%	$20

Do Exercise 12.

EXAMPLE 11 What percent of 50 is 16?

Translate: $\frac{N}{100} = \frac{16}{50}$

Solve:

$$50 \cdot N = 100 \cdot 16 \qquad \text{Equating cross products}$$

$$\frac{50 \cdot N}{50} = \frac{100 \cdot 16}{50} \qquad \text{Dividing by 50}$$

$$N = \frac{1600}{50} \qquad \text{Multiplying and simplifying}$$

$$N = 32 \qquad \text{Dividing}$$

Thus, 32% of 50 is 16. The answer is 32%.

Percents	Quantities
0%	0
N%	16
100%	50

Do Exercise 13.

EXERCISE SET

For Extra Help

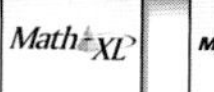

MathXL MyMathLab InterAct Math Math Tutor Center Digital Video Tutor CD 3 Videotape 6 Student's Solutions Manual

a Translate to a proportion. Do not solve.

1. What is 37% of 74?

2. 66% of 74 is what?

3. 4.3 is what percent of 5.9?

4. What percent of 6.8 is 5.3?

5. 14 is 25% of what?

6. 133% of what is 40?

b Translate to a proportion and solve.

7. What is 76% of 90?

8. What is 32% of 70?

9. 70% of 660 is what?

10. 80% of 920 is what?

11. What is 4% of 1000?

12. What is 6% of 2000?

13. 4.8% of 60 is what?

14. 63.1% of 80 is what?

15. $24 is what percent of $96?

16. $14 is what percent of $70?

17. 102 is what percent of 100?

18. 103 is what percent of 100?

19. What percent of $480 is $120?

20. What percent of $80 is $60?

21. What percent of 160 is 150?

22. What percent of 33 is 11?

23. $18 is 25% of what?

24. $75 is 20% of what?

25. 60% of what is 54?

26. 80% of what is 96?

27. 65.12 is 74% of what?

28. 63.7 is 65% of what?

29. 80% of what is 16?

30. 80% of what is 10?

31. What is $62\frac{1}{2}\%$ of 40?

32. What is $43\frac{1}{4}\%$ of 2600?

33. What is 9.4% of $8300?

34. What is 8.7% of $76,000?

35. **D_W** In your own words, list steps that a classmate could use to solve any percent problem in this section.

36. **D_W** In solving Example 10, a student simplifies $\frac{10}{20}$ before solving. Is this a good idea? Why or why not?

SKILL MAINTENANCE

Solve. [5.3b]

37. $\frac{x}{188} = \frac{2}{47}$

38. $\frac{15}{x} = \frac{3}{800}$

39. $\frac{4}{7} = \frac{x}{14}$

40. $\frac{612}{t} = \frac{72}{244}$

41. $\frac{5000}{t} = \frac{3000}{60}$

42. $\frac{75}{100} = \frac{n}{20}$

43. $\frac{x}{1.2} = \frac{36.2}{5.4}$

44. $\frac{y}{1\frac{1}{2}} = \frac{2\frac{3}{4}}{22}$

Solve.

45. A recipe for muffins calls for $\frac{1}{2}$ qt of buttermilk, $\frac{1}{3}$ qt of skim milk, and $\frac{1}{16}$ qt of oil. How many quarts of liquid ingredients does the recipe call for? [3.2b]

46. The Ferristown School District purchased $\frac{3}{4}$ ton (T) of clay. If the clay is to be shared equally among the district's 6 art departments, how much will each art department receive? [2.7d]

SYNTHESIS

Solve.

47. What is 8.85% of $12,640?
Estimate ____________
Calculate ____________

48. 78.8% of what is 9809.024?
Estimate ____________
Calculate ____________

6.5 APPLICATIONS OF PERCENT

Objectives

a Solve applied problems involving percent.

b Solve applied problems involving percent of increase or decrease.

a Applied Problems Involving Percent

Applied problems involving percent are not always stated in a manner easily translated to an equation. In such cases, it is helpful to rephrase the problem before translating. Sometimes it also helps to make a drawing.

EXAMPLE 1 *Presidential Deaths in Office.* George W. Bush was inaugurated as the 43rd President of the United States in 2001. Since Grover Cleveland was both the 22nd and the 24th presidents, there have been only 42 different presidents. Of the 42 presidents, 8 have died in office: William Henry Harrison, Zachary Taylor, Abraham Lincoln, James A. Garfield, William McKinley, Warren G. Harding, Franklin D. Roosevelt, and John F. Kennedy. What percent have died in office?

1. **Familiarize.** The question asks for a percent of the presidents who have died in office. We note that 42 is approximately 40 and 8 is $\frac{1}{5}$, or 20%, of 40, so our answer is close to 20%. We let $p =$ the percent who have died in office.
2. **Translate.** There are two ways in which we can translate this problem.

Percent equation (see Section 6.3):

$$\underbrace{8}_{8}\ \underbrace{\text{is}}_{=}\ \underbrace{\text{what percent}}_{p}\ \underbrace{\text{of}}_{\cdot}\ \underbrace{42?}_{42}$$

$$8 = p \cdot 42$$

Proportion (see Section 6.4):

$$\frac{N}{100} = \frac{8}{42}$$

For proportions, $N\% = p$.

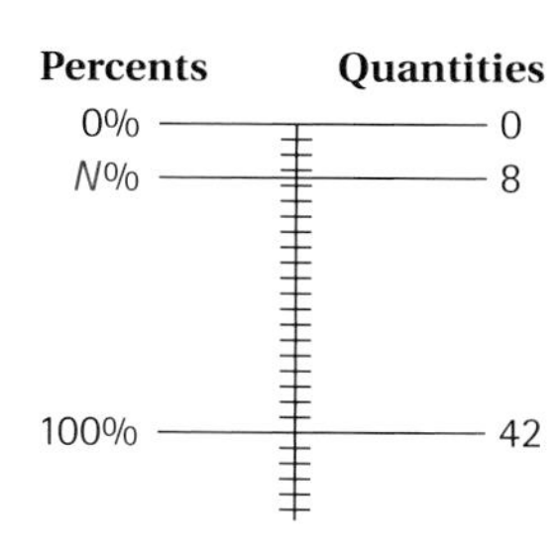

Study Tips

MAKING APPLICATIONS REAL

Newspapers and magazines are full of mathematical applications. Some of the easiest ones to find in the area of Basic Mathematics are about percent. Find such an application and share it with your class. As you obtain more skills in mathematics, you will find yourself observing the world from a different perspective, seeing mathematics everywhere. Math courses become more interesting when we connect the concepts to the real world.

1. **Presidential Assassinations in Office.** Of the 42 U.S. presidents, 4 have been assassinated in office. These were Garfield, McKinley, Lincoln, and Kennedy. What percent have been assassinated in office?

3. **Solve.** We now have two ways in which to solve this problem.

Percent equation (see Section 6.3):

$$8 = p \cdot 42$$

$$\frac{8}{42} = \frac{p \cdot 42}{42} \quad \text{Dividing by 42 on both sides}$$

$$\frac{8}{42} = p$$

$$0.190 \approx p \quad \text{Finding decimal notation and rounding to the nearest thousandth}$$

$$19.0\% \approx p \quad \text{Remember to find percent notation.}$$

Note here that the solution, p, includes the % symbol.

Proportion (see Section 6.4):

$$\frac{N}{100} = \frac{8}{42}$$

$$N \cdot 42 = 100 \cdot 8 \quad \text{Equating cross products}$$

$$\frac{N \cdot 42}{42} = \frac{800}{42} \quad \text{Dividing by 42 on both sides}$$

$$N = \frac{800}{42}$$

$$N \approx 19.0 \quad \text{Dividing and rounding to the nearest tenth}$$

We use the solution of the proportion to express the answer to the problem as 19.0%. Note that in the proportion method, $N\% = p$.

4. **Check.** To check, we note that the answer 19.0% is close to 20%, as estimated in the *Familiarize* step.

5. **State.** About 19.0% of the U.S. presidents have died in office.

Do Exercise 1.

EXAMPLE 2 *Transportation to Work.* In the 15 largest cities in the United States, there are about 130,000,000 workers 16 years and over. Approximately 76.3% drive to work alone. How many workers in these 15 cities drive to work alone?

Transportation to Work (in the 15 Largest Cities in the United States)

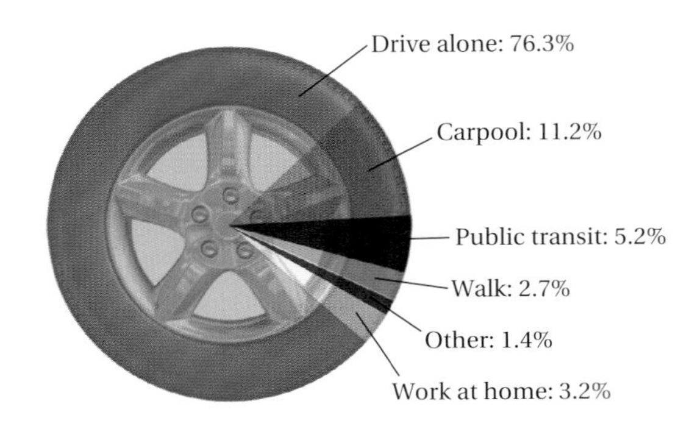

Source: U.S. Bureau of the Census

Answer on page A-14

1. **Familiarize.** We can make a drawing of a pie chart to help familiarize ourselves with the problem. We let b = the total number of workers who drive to work alone.

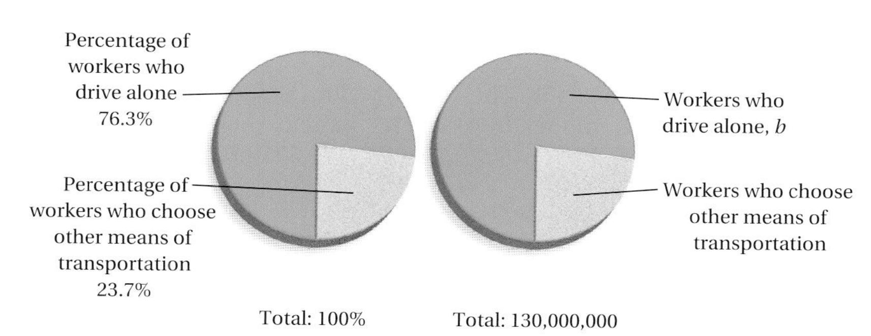

2. **Translate.** There are two ways in which we can translate this problem.

Percent equation:

What number is 76.3% of 130,000,000?

$$b = 76.3\% \cdot 130{,}000{,}000$$

Proportion:

$$\frac{76.3}{100} = \frac{b}{130{,}000{,}000}$$

Percents | **Quantities**
0% — 0
76.3% — b
100% — 130,000,000

3. **Solve.** We now have two ways in which to solve this problem.

Percent equation:

$$b = 76.3\% \cdot 130{,}000{,}000$$

We convert 76.3% to decimal notation and multiply:

$$b = 0.763 \times 130{,}000{,}000 = 99{,}190{,}000.$$

Proportion:

$$\frac{76.3}{100} = \frac{b}{130{,}000{,}000}$$

$$76.3 \times 130{,}000{,}000 = 100 \cdot b \qquad \text{Equating cross products}$$

$$\frac{76.3 \cdot 130{,}000{,}000}{100} = \frac{100 \cdot b}{100} \qquad \text{Dividing by 100}$$

$$\frac{9{,}919{,}000{,}000}{100} = b$$

$$99{,}190{,}000 = b \qquad \text{Simplifying}$$

4. **Check.** To check, we can repeat the calculations. We can also do a partial check by estimating. Since 76.3% is about 80%, or $\frac{4}{5}$, and $\frac{4}{5}$ of 130,000,000 is 104,000,000 and 104,000,000 is close to 99,190,000, our answer is reasonable.

5. **State.** The number of workers in the 15 largest cities who drive to work alone is 99,190,000.

Do Exercise 2.

2. **Transportation to Work.** There are about 130,000,000 workers 16 years or over in the 15 largest cities in the United States. Approximately 11.2% carpool to work. How many workers in these 15 cities carpool to work?
Source: U.S. Bureau of the Census

Answer on page A-14

3. **Percent of Increase.** The value of a car is \$36,875. The price is increased by 4%.
   a) How much is the increase?
   b) What is the new price?

4. **Percent of Decrease.** The value of a car is \$36,875. The car depreciates in value by 25% after one year.
   a) How much is the decrease?
   b) What is the depreciated value of the car?

Answers on page A-14

b Percent of Increase or Decrease

Percent is often used to state increase or decrease. Let's consider an example of each, using the price of a car as the original number.

PERCENT OF INCREASE

One year a car sold for \$20,455. The manufacturer decides to raise the price of the following year's model by 6%. The increase is $0.06 \times \$20,455$, or \$1227.30. The new price is $\$20,455 + \$1227.30$, or \$21,682.30. The *percent of increase* is 6%.

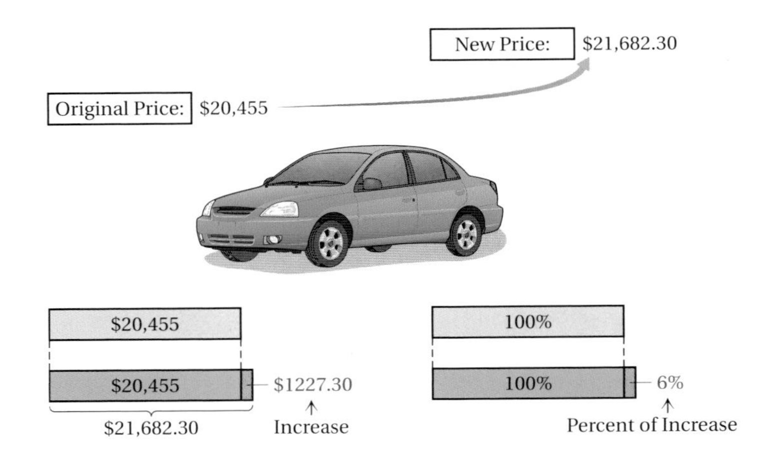

PERCENT OF DECREASE

Lisa buys the car listed above for \$20,455. After one year, the car depreciates in value by 25%. This is $0.25 \times \$20,455$, or \$5113.75. This lowers the value of the car to

$$\$20,455 - \$5113.75, \quad \text{or} \quad \$15,341.25.$$

Note that the new price is thus 75% of the original price. If Lisa decides to sell the car after a year, \$15,341.25 might be the most she could expect to get for it. The *percent of decrease* is 25%, and the decrease is \$5113.75.

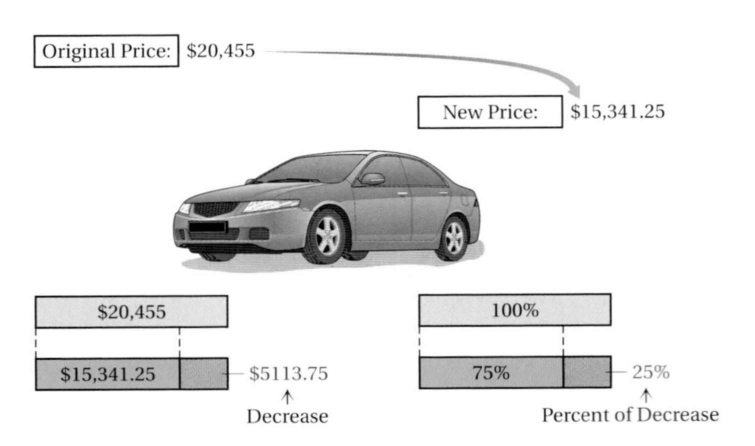

Do Exercises 3 and 4.

When a quantity is decreased by a certain percent, we say we have **percent of decrease.**

EXAMPLE 3 *Chain Link Fence.* For one week only, Sam's Farm Supply had 4 ft × 50 ft rolls of galvanized chain link fence on sale for \$39.99. The regular price was \$49.99 per roll. What was the percent of decrease?

1. Familiarize. We find the amount of decrease and then make a drawing.

4 9.9 9	Retail price
− 3 9.9 9	Sale price
1 0.0 0	Decrease

\$49.99 — 100%

\$39.99 — \$10.00 — ?%

2. Translate. There are two ways in which we can translate this problem.

Percent equation:

10.00 is what percent of 49.99?

$$10.00 = p \times 49.99$$

Proportion:

$$\frac{N}{100} = \frac{10.00}{49.99}$$

For proportions, $N\% = p$.

Percents	Quantities
0%	0
N%	10.00
100%	49.99

3. Solve. We have two ways in which to solve this problem.

Percent equation:

$$10.00 = p \times 49.99$$

$$\frac{10.00}{49.99} = \frac{p \times 49.99}{49.99} \quad \text{Dividing by 49.99 on both sides}$$

$$\frac{10.00}{49.99} = p$$

$$0.20 \approx p$$

$$20\% \approx p \quad \text{Converting to percent notation}$$

Proportion:

$$\frac{N}{100} = \frac{10.00}{49.99}$$

$$49.99 \times N = 100 \times 10 \quad \text{Equating cross products}$$

$$\frac{49.99 \times N}{49.99} = \frac{100 \times 10}{49.99} \quad \text{Dividing by 49.99 on both sides}$$

$$N = \frac{1000}{49.99}$$

$$N \approx 20$$

We use the solution of the proportion to express the answer to the problem as 20%.

5. Volume of Mail. The volume of U.S. mail decreased from about 208 billion pieces of mail in 2000 to 202 billion pieces in 2003. What was the percent of decrease?

Source: U.S. Postal Service

Answer on page A-14

4. **Check.** To check, we note that, with a 20% decrease, the reduced (or sale) price should be 80% of the retail (or original) price. Since

$$80\% \times 49.99 = 0.80 \times 49.99 = 39.992 \approx 39.99,$$

our answer checks.

5. **State.** The percent of decrease in the price of the roll of fence was 20%.

Do Exercise 5 on the preceding page.

When a quantity is increased by a certain percent, we say we have **percent of increase.**

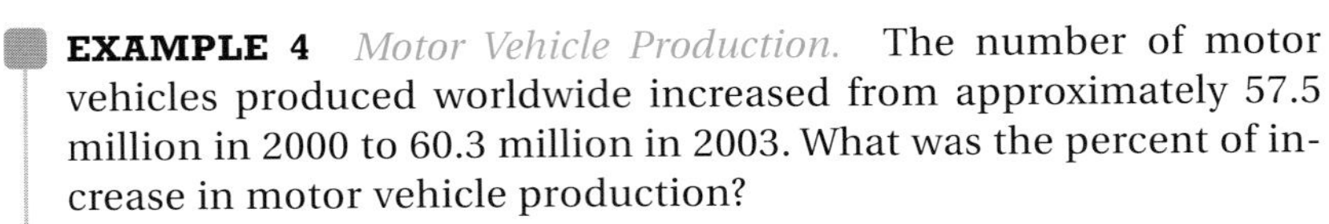

EXAMPLE 4 *Motor Vehicle Production.* The number of motor vehicles produced worldwide increased from approximately 57.5 million in 2000 to 60.3 million in 2003. What was the percent of increase in motor vehicle production?

Source: Ward's Communications, *Ward's Motor Vehicle Facts & Figures,* 2004

1. **Familiarize.** We first note that the increase in the number of vehicles produced was 60.3 − 57.5 million, or 2.8 million. A drawing can help us visualize the situation.

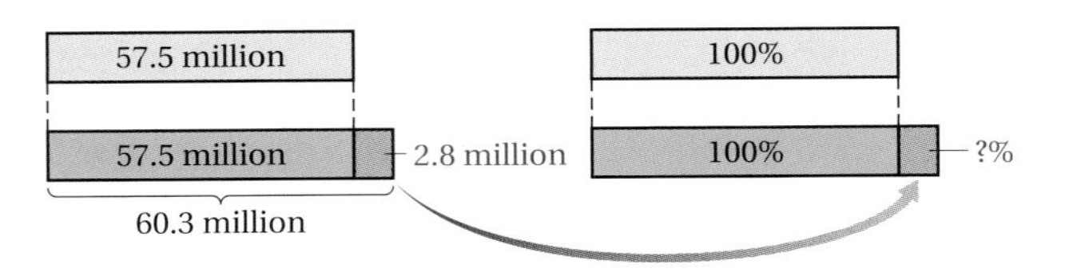

We are asking this question: The increase is what percent of the *original* amount? We let $p =$ the percent of increase.

2. **Translate.** There are two ways in which we can translate this problem.

Percent equation:

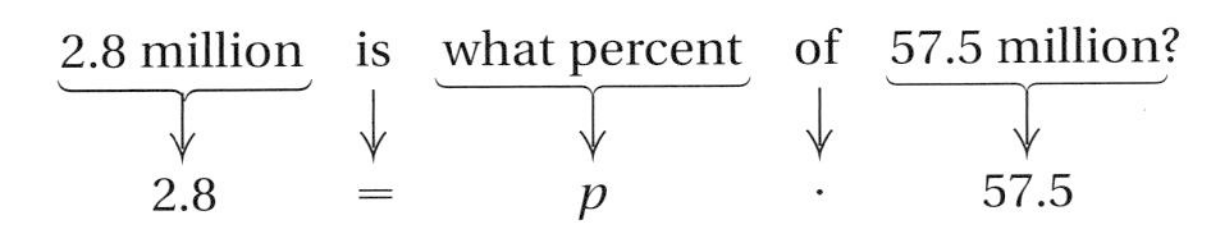

Proportion:

$$\frac{N}{100} = \frac{2.8}{57.5}$$

For proportions, $N\% = p$.

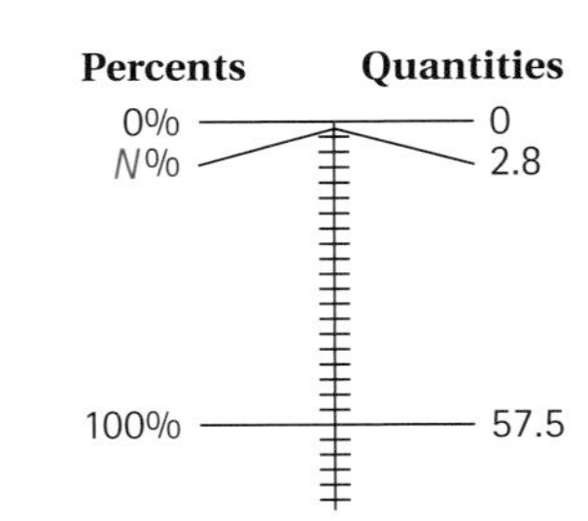

3. Solve. We have two ways in which to solve this problem.

Percent equation:

$$2.8 = p \cdot 57.5$$

$$\frac{2.8}{57.5} = \frac{p \times 57.5}{57.5} \quad \text{Dividing by 57.5 on both sides}$$

$$\frac{2.8}{57.5} = p$$

$$0.049 \approx p$$

$$4.9\% \approx p \quad \text{Converting to percent notation}$$

Proportion:

$$\frac{N}{100} = \frac{2.8}{57.5}$$

$$57.5 \times N = 100 \times 2.8 \quad \text{Equating cross products}$$

$$\frac{57.5 \times N}{57.5} = \frac{100 \times 2.8}{57.5} \quad \text{Dividing by 57.5 on both sides}$$

$$N = \frac{280}{57.5}$$

$$N \approx 4.9$$

We use the solution of the proportion to express the answer to the problem as 4.9%.

4. Check. To check, we take 4.9% of 57.5:

$$4.9\% \times 57.5 = 0.049 \times 57.5 = 2.8175.$$

Since we rounded the percent, this approximation is close enough to 2.8 to be a good check.

5. State. The percent of increase in the number of motor vehicles produced is 4.9%.

Do Exercise 6.

6. Patents Issued. The number of patents issued per year by the U.S. government increased from 107,332 in 1993 to 189,597 in 2003. What was the percent of increase over the ten-year period?

Source: U.S. Patent and Trademark Office

Answer on page A-14

Translating for Success

1. *Distance Walked.* After knee replacement, Alex walked $\frac{1}{8}$ mi each morning and $\frac{1}{5}$ mi each afternoon. How much farther did he walk in the afternoon?

2. *Stock Prices.* A stock sold for \$5 per share on Monday and only \$$2\frac{1}{8}$ per share on Friday. What was the percent of decrease from Monday to Friday?

3. *SAT Score.* After attending a class titled *Improving Your SAT Scores,* Jacob raised his total score from 884 to 1040. What was the percent of increase?

4. *Change in Population.* The population of a small farming community decreased from 1040 to 884. What was the percent of decrease?

5. *Lawn Mowing.* During the summer, brothers Steve and Rob earned money for college by mowing lawns. The largest lawn they mowed was $2\frac{1}{8}$ acres. Steve can mow $\frac{1}{5}$ acre per hour, and Rob can mow only $\frac{1}{8}$ acre per hour. Working together, how many acres did they mow per hour?

The goal of these matching questions is to practice step (2), *Translate,* of the five-step problem-solving process. Translate each word problem to an equation and select a correct translation from equations A–O.

A. $x + \frac{1}{5} = \frac{1}{8}$

B. $250 = x \cdot 1040$

C. $884 = x \cdot 1040$

D. $\frac{250}{16.25} = \frac{1000}{x}$

E. $156 = x \cdot 1040$

F. $16.25 = 250 \cdot x$

G. $\frac{1}{5} + \frac{1}{8} = x$

H. $2\frac{1}{8} = x \cdot 5$

I. $5 = 2\frac{7}{8} \cdot x$

J. $\frac{1}{8} + x = \frac{1}{5}$

K. $1040 = x \cdot 884$

L. $\frac{250}{16.25} = \frac{x}{1000}$

M. $2\frac{7}{8} = x \cdot 5$

N. $x \cdot 884 = 156$

O. $x = 16.25 \cdot 250$

Answers on page A-14

6. *Land Sale.* Cole sold $2\frac{1}{8}$ acre from the 5 acres he inherited from his uncle. What percent did he sell?

7. *Travel Expenses.* A magazine photographer is reimbursed 16.25¢ per mile for business travel up to 1000 miles per week. In a recent week, he traveled only 250 miles. What was the total reimbursement for travel?

8. *Trip Expenses.* The total expenses for Claire's recent business trip were \$1040. She put \$884 on her charge card a paid the balance in cash. Wh percent did she place on her charge card?

9. *Cost of Copies.* During the f summer session at a commu college, the campus copy cen advertised 250 copies for \$16 At this rate, what is the cost c 1000 copies?

10. *Cost of Insurance.* Followin raise in the cost of health insurance, 250 of a company 1040 employees dropped the health coverage. What perce the employees canceled thei insurance?

EXERCISE SET

For Extra Help

MathXL MyMathLab InterAct Math Math Tutor Center Digital Video Tutor CD 3 Videotape 6 Student's Solutions Manual

a Solve.

1. *Wild Horses.* There are 27,369 wild horses on land managed by the Federal Bureau of Land Management. It is estimated that 48.4% of this total is in Nevada. How many wild horses are in Nevada?

Source: Bureau of Land Management, 2005

2. *U.S. Armed Forces* There are 1,168,195 people in the United States in active military service. The numbers in the four armed services are listed in the table below. What percent of the total does each branch represent? Round the answers to the nearest tenth of a percent.

U.S. ARMED FORCES WORLDWIDE, 2004

Total	1,168,195
Air Force	314,477
Army	391,126
Marine Corps	135,324
Navy	327,268

Source: U.S. Department of Defense

3. *Car Value.* The base price of a Nissan 350Z is \$34,000. This vehicle is expected to retain 62% of its value at the end of three years and 52% at the end of five years. What is the value of a Nissan 350Z after three years? after five years?

Source: November/December *Kelley Blue Book Residual Values Guide*

4. *Panda Survival.* Breeding the much-loved panda bear in captivity has been quite difficult for zookeepers.

a) From 1964 to 1997, of 133 panda cubs born in captivity, only 90 lived to be one month old. What percent lived to be one month old?

b) In 1999, Mark Edwards of the San Diego Zoo developed a nutritional formula on which 18 of 20 newborns lived to be one month old. What percent lived to be one month old?

5. *Overweight and Obese.* Of the 294 million people in the United States, 60% are considered overweight and 25% are considered obese. How many are overweight? How many are obese?

Source: U.S. Centers for Disease Control

6. *Smoking and Diabetes.* Of the 294 million people in the United States, 26% are smokers. How many are smokers?

Source: SAMHSA, Office of Applied Studies, National Survey on Drug Use and Health

7. A lab technician has 680 mL of a solution of water and acid; 3% is acid. How many milliliters are acid? water?

8. A lab technician has 540 mL of a solution of alcohol and water; 8% is alcohol. How many milliliters are alcohol? water?

9. *Mississippi River.* The Mississippi River, which extends from Minneapolis, Minnesota, to the Gulf of Mexico, is 2348 miles long. Approximately 77% of the river is navigable. How many miles of the river are navigable?

Source: National Oceanic and Atmospheric Administration

10. *Immigrants.* In 2003, 705,827 immigrants entered the United States. Of this total, 16.4% were from Mexico and 7.1% were from India. How many immigrants came from Mexico? from India?

Source: U.S. Department of Justice, *2003 Yearbook of Immigration Statistics*

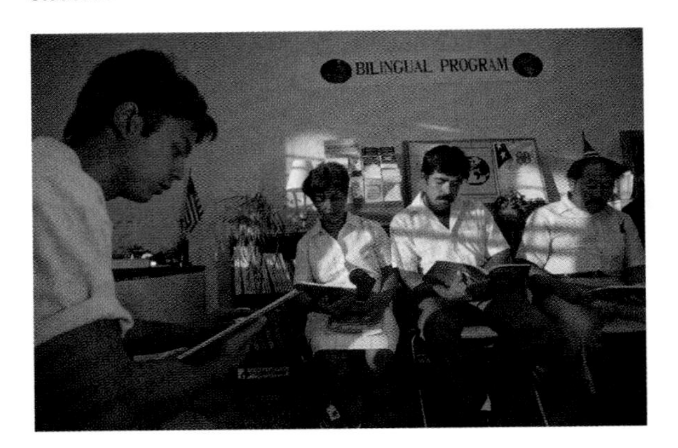

11. *Hispanic Population.* The Hispanic population is growing rapidly in the United States. In 2003, the population of the United States was about 291,000,000 and 13.7% of this total was Hispanic. How many Hispanic people lived in the United States in 2003?

Source: U.S. Bureau of the Census

12. *Age 65 and Over.* By 2010, it is predicted that 13.2% of the U.S. population will be 65 and over. If the population of the United States in 2010 is 307,000,000, how many people will be 65 and over?

Source: U.S. Bureau of the Census

13. *Test Results.* On a test of 40 items, Christina got 91% correct. (There was partial credit on some items.) How many items did she get correct? incorrect?

14. *Test Results.* On a test of 80 items, Pedro got 93% correct. (There was partial credit on some items.) How many items did he get correct? incorrect?

15. *Test Results.* On a test, Maj Ling got 86%, or 81.7, of the items correct. (There was partial credit on some items.) How many items were on the test?

16. *Test Results.* On a test, Juan got 85%, or 119, of the items correct. How many items were on the test?

17. *TV Usage.* Of the 8760 hr in a year, most television sets are on for 2190 hr. What percent is this?

18. *Colds from Kissing.* In a medical study, it was determined that if 800 people kiss someone who has a cold, only 56 will actually catch a cold. What percent is this?

Source: U.S. Centers for Disease Control

19. *Maximum Heart Rate.* Treadmill tests are often administered to diagnose heart ailments. A guideline in such a test is to try to get you to reach your *maximum heart rate,* in beats per minute. The maximum heart rate is found by subtracting your age from 220 and then multiplying by 85%. What is the maximum heart rate of someone whose age is 25? 36? 48? 55? 76? Round to the nearest one.

20. It costs an oil company $40,000 a day to operate two refineries. Refinery A accounts for 37.5% of the cost, and refinery B for the rest of the cost.

a) What percent of the cost does it take to run refinery B?

b) What is the cost of operating refinery A? refinery B?

b

Solve.

21. *Savings Increase.* The amount in a savings account increased from $200 to $216. What was the percent of increase?

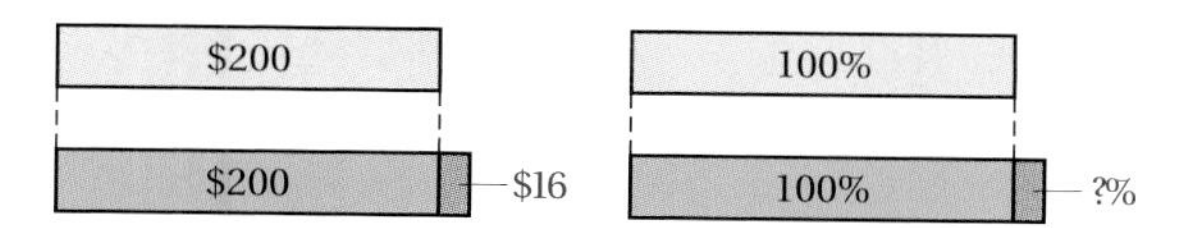

22. *Population Increase.* The population of a small mountain town increased from 840 to 882. What was the percent of increase?

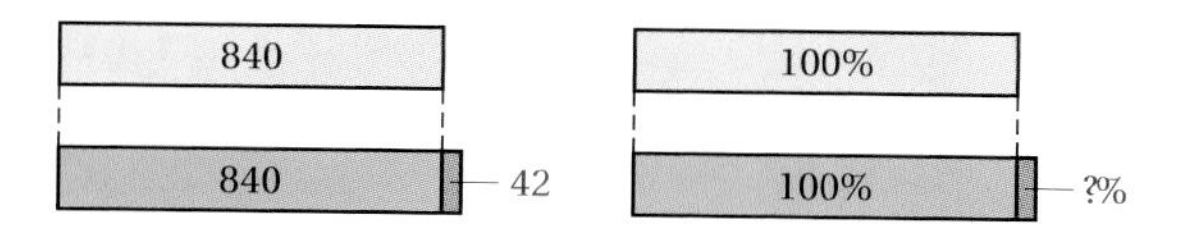

23. During a sale, a dress decreased in price from $90 to $72. What was the percent of decrease?

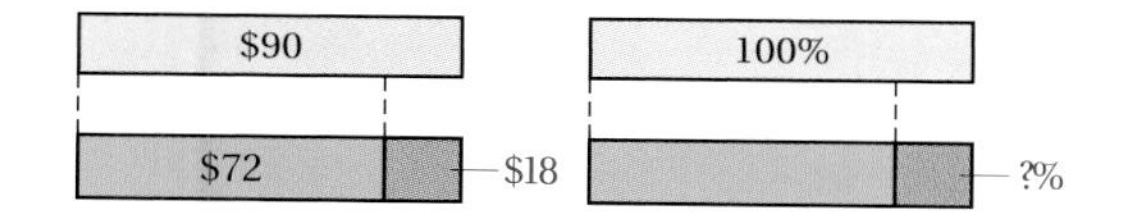

24. A person on a diet goes from a weight of 125 lb to a weight of 110 lb. What is the percent of decrease?

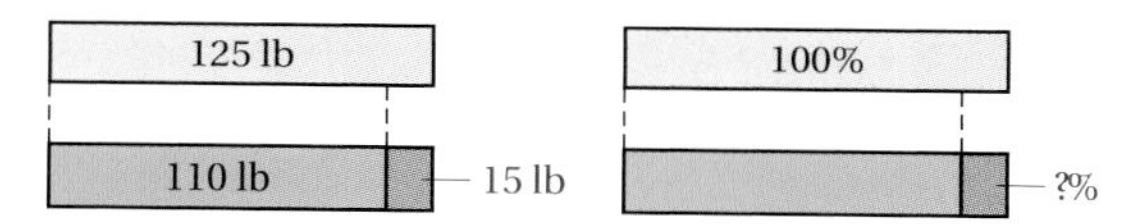

25. *Population Increase.* The population of the state of Nevada increased from 1,201,833 in 1990 to 2,241,154 in 2003. What is the percent of increase?

Source: U.S. Bureau of the Census

26. *Population Increase.* The population of the state of Utah increased from 1,722,850 in 1990 to 2,351,467 in 2003. What is the percent of increase?

Source: U.S. Bureau of the Census

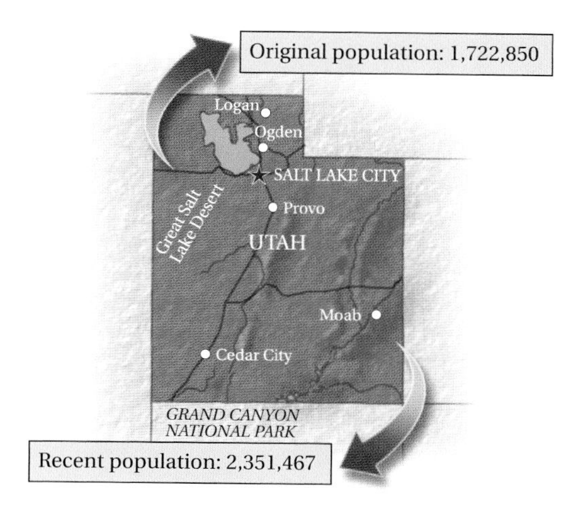

27. A person earns \$28,600 one year and receives a 5% raise in salary. What is the new salary?

28. A person earns \$43,200 one year and receives an 8% raise in salary. What is the new salary?

29. *Car Depreciation.* Irwin buys a car for \$21,566. It depreciates 25% each year that he owns it. What is the depreciated value of the car after 1 yr? after 2 yr?

30. *Car Depreciation.* Janice buys a car for \$22,688. It depreciates 25% each year that she owns it. What is the depreciated value of the car after 1 yr? after 2 yr?

31. *Doormat.* A bordered coir doormat 42 in. × 24 in. × $1\frac{1}{2}$ in. has a retail price of \$89.95. Over the holidays it is on sale for \$65.49. What is the percent of decrease?

32. *Business Tote.* A leather business tote retails for \$239.99. An insurance company bought 30 totes for its sales staff at a reduced price of \$184.95. What was the percent of decrease in the price?

33. *Two-by-Four.* A cross-section of a standard or nominal "two-by-four" board actually measures $1\frac{1}{2}$ in. by $3\frac{1}{2}$ in. The rough board is 2 in. by 4 in. but is planed and dried to the finished size. What percent of the wood is removed in planing and drying?

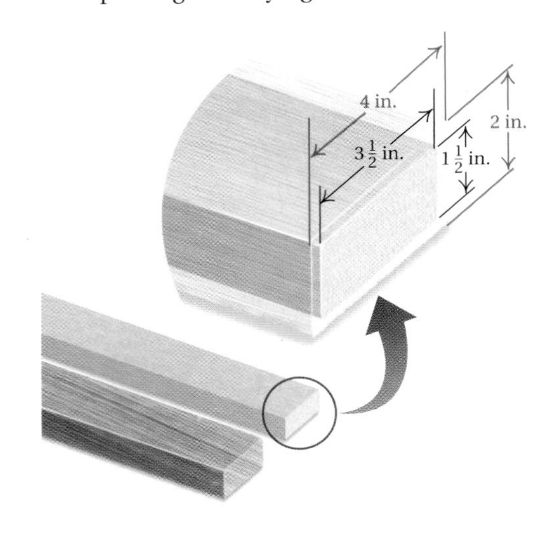

34. *Tipping.* Diners frequently add a 15% tip when charging a meal to a credit card. What is the total amount charged if the cost of the meal, without tip, is \$18? \$34? \$49?

35. *Population Decrease.* Between 1990 and 2000, the population of Detroit, Michigan, decreased from 1,028,000 to 951,000.

a) What is the percent of decrease?

b) If this percent of decrease over a 10-yr period repeated itself in the following decade, what would the population be in 2010?

Source: U.S. Bureau of the Census

36. *World Population.* World population is increasing by 1.14% each year. In 2004, it was 6.39 billion. How much will it be in 2006? 2010? 2015?

Source: *The World Factbook*, 2004

Life Insurance Rates for Smokers and Nonsmokers. The following table provides data showing how yearly rates (premiums) for a $500,000 term life insurance policy are increased for smokers. Complete the missing numbers in the table.

TYPICAL INSURANCE PREMIUMS (DOLLARS)

	AGE	RATE FOR NONSMOKER	RATE FOR SMOKER	PERCENT INCREASE FOR SMOKER
	35	$ 345	$ 630	83%
37.	40	$ 430	$ 735	
38.	45	$ 565		84%
39.	50	$ 780		100%
40.	55	$ 985		117%
41.	60	$1645	$2955	
42.	65	$2943	$5445	

Source: Pacific Life PL Protector Term Life Portfolio, OYT Rates

Population Increase. The following table provides data showing how the populations of various states increased from 1990 to 2003. Complete the missing numbers in the table.

	AGE	POPULATION IN 1990	POPULATION IN 2003	CHANGE	PERCENT CHANGE
43.	Alaska	550,043	648,818		
44.	Connecticut	3,287,116		196,256	
45.	Montana		917,621	118,556	
46.	Texas		22,118,509	5,131,999	
47.	Colorado	3,294,394		1,256,294	
48.	Pennsylvania	11,881,643	12,365,455		

Source: U.S. Bureau of the Census

49. *Car Depreciation.* A car generally depreciates 25% of its original value in the first year. A car is worth \$27,300 after the first year. What was its original cost?

50. *Car Depreciation.* Given normal use, an American-made car will depreciate 25% of its original cost the first year and 14% of its remaining value in the second year. What is the value of a car at the end of the second year if its original cost was \$36,400? \$28,400? \$26,800?

51. *Strike Zone.* In baseball, the *strike zone* is normally a 17-in. by 30-in. rectangle. Some batters give the pitcher an advantage by swinging at pitches thrown out of the strike zone. By what percent is the area of the strike zone increased if a 2-in. border is added to the outside?

Source: Major League Baseball

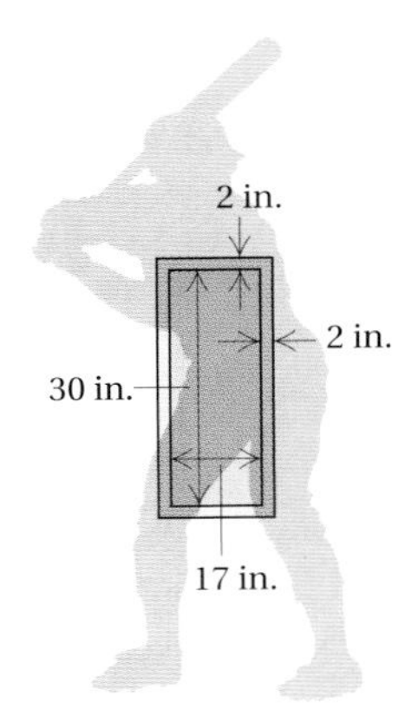

52. Tony has planted grass seed on a 24-ft by 36-ft area in his back yard. He has also installed a 6-ft by 8-ft garden. By what percent does the garden reduce the area he will have to mow?

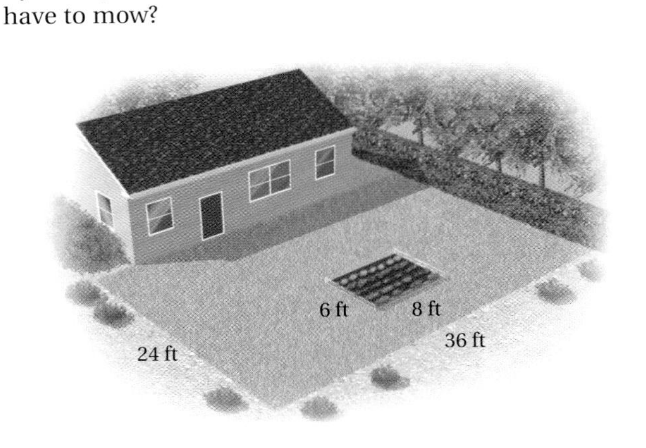

53. D_W Which is better for a wage earner, and why: a 10% raise followed by a 5% raise a year later, or a 5% raise followed by a 10% raise a year later?

54. D_W A worker receives raises of 3%, 6%, and then 9%. By what percent has the original salary increased? Explain.

SKILL MAINTENANCE

Convert to decimal notation. [4.1b], [4.5a]

55. $\frac{25}{11}$ **56.** $\frac{11}{25}$ **57.** $\frac{27}{8}$ **58.** $\frac{43}{9}$ **59.** $\frac{23}{25}$

60. $\frac{20}{24}$ **61.** $\frac{14}{32}$ **62.** $\frac{2317}{1000}$ **63.** $\frac{34{,}809}{10{,}000}$ **64.** $\frac{27}{40}$

SYNTHESIS

65. *Adult Height.* It has been determined that at the age of 10, a girl has reached 84.4% of her final adult growth. Cynthia is 4 ft, 8 in. at the age of 10. What will be her final adult height?

Source: *Dunlop Illustrated Encyclopedia of Facts.* New York: Sterling Publishing, 1970.

66. *Adult Height.* It has been determined that at the age of 15, a boy has reached 96.1% of his final adult height. Claude is 6 ft, 4 in. at the age of 15. What will be his final adult height?

Source: *Dunlop Illustrated Encyclopedia of Facts.* New York: Sterling Publishing, 1970.

67. If p is 120% of q, then q is what percent of p?

68. A coupon allows a couple to have dinner and then have \$10 subtracted from the bill. Before subtracting \$10, however, the restaurant adds a tip of 15%. If the couple is presented with a bill for \$44.05, how much would the dinner (without tip) have cost without the coupon?

6.6 SALES TAX, COMMISSION, AND DISCOUNT

Objectives

a Solve applied problems involving sales tax and percent.

b Solve applied problems involving commission and percent.

c Solve applied problems involving discount and percent.

a Sales Tax

Sales tax computations represent a special type of percent of increase problem. The sales tax rate in Maryland is 5%. This means that the tax is 5% of the purchase price. Suppose the purchase price on a coat is \$124.95. The sales tax is then 5% of \$124.95, or 0.05×124.95, or 6.2475, or about \$6.25.

BILL:		
Purchase price	=	\$124.95
Sales tax (5% of \$124.95)	=	+ 6.25
Total price		\$131.20

The total that you pay is the price plus the sales tax:

\$124.95 + \$6.25, or \$131.20.

SALES TAX

Sales tax = Sales tax rate × Purchase price

Total price = Purchase price + Sales tax

EXAMPLE 1 *Florida Sales Tax.* The sales tax rate in Florida is 6%. How much tax is charged on the purchase of 4 inflatable rafts at \$89.99 each? What is the total price?

\$89.99 plus 6% each

a) We first find the cost of the rafts. It is

$4 \times \$89.99 = \$359.96.$

b) The sales tax on items costing \$359.96 is

Sales tax rate × Purchase price
↓ ↓
6% × \$359.96

or 0.06×359.96, or 21.5976. Thus the tax is \$21.60 (rounded to the nearest cent).

c) The total price is given by the purchase price plus the sales tax:

\$359.96 + \$21.60, or \$381.56.

To check, note that the total price is the purchase price plus 6% of the purchase price. Thus the total price is 106% of the purchase price. Since $1.06 \times 359.96 \approx 381.56$, we have a check. The sales tax is \$21.60 and the total price is \$381.56.

Do Exercises 1 and 2.

1. California Sales Tax. The sales tax rate in California is 7.25%. How much tax is charged on the purchase of a refrigerator that sells for \$668.95? What is the total price?

2. Louisiana Sales Tax. Sam buys 5 hardcover copies of Dean Koontz's novel *From the Corner of His Eye* for \$26.95 each. The sales tax rate in Louisiana is 4%. How much sales tax will be charged? What is the total price?

Answers on page A-14

3. The sales tax is $50.94 on the purchase of a night table that costs $849. What is the sales tax rate?

4. The sales tax on a portable navigation system is $59.94 and the sales tax rate is 6%. Find the purchase price (the price before taxes are added).

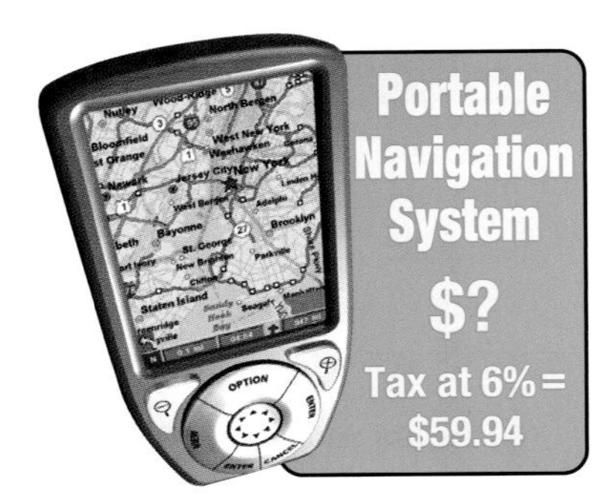

EXAMPLE 2 The sales tax is $43.96 on the purchase of this wooden play center, which costs $1099. What is the sales tax rate?

Rephrase: Sales tax is what percent of purchase price?

Translate: $\$43.96 = r \times \$1099$

To solve the equation, we divide by 1099 on both sides:

$$\frac{43.96}{1099} = \frac{r \times 1099}{1099}$$
$$\frac{43.96}{1099} = r$$
$$0.04 = r$$
$$4\% = r.$$

The sales tax rate is 4%.

Do Exercise 3.

EXAMPLE 3 The sales tax on a Tiffany torchiere lamp is $8.93 and the sales tax rate is 5%. Find the purchase price (the price before taxes are added).

Rephrase: Sales tax is 5% of what?

Translate: $\$8.93 = 5\% \times b,$ or $8.93 = 0.05 \times b.$

To solve, we divide by 0.05 on both sides:

$$\frac{8.93}{0.05} = \frac{0.05 \times b}{0.05}$$
$$\frac{8.93}{0.05} = b$$
$$\$178.60 = b.$$

The purchase price is $178.60.

Do Exercise 4.

Answers on page A-14

b Commission

When you work for a **salary,** you receive the same amount of money each week or month. When you work for a **commission,** you are paid a percentage of the total sales for which you are responsible.

COMMISSION
Commission = Commission rate × Sales

EXAMPLE 4 *Exercise Equipment Sales.* A salesperson's commission rate is 16%. What is the commission from the sale of \$9700 worth of exercise equipment?

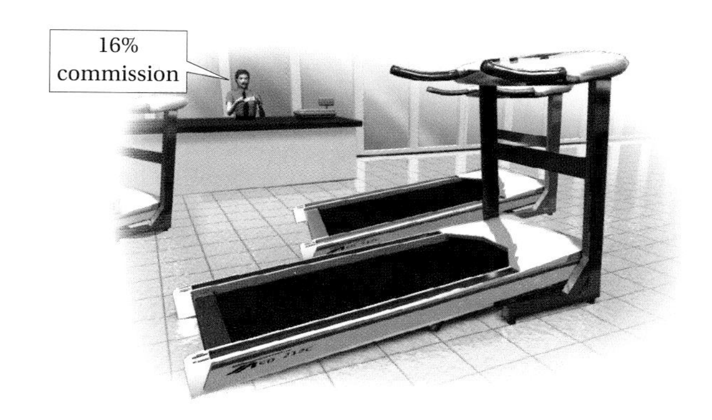

Commission	=	*Commission rate*	×	*Sales*
C	=	16%	×	9700
C	=	0.16	×	9700
C	=	1552		

The commission is \$1552.

Do Exercise 5.

5. Raul's commission rate is 30%. What is the commission from the sale of \$18,760 worth of air conditioners?

EXAMPLE 5 *Farm Machinery Sales.* Dawn earns a commission of \$30,000 selling \$600,000 worth of farm machinery. What is the commission rate?

Commission	=	*Commission rate*	×	*Sales*
30,000	=	r	×	600,000

Answer on page A-14

6. Liz earns a commission of $3000 selling $24,000 worth of *NSYNC concert tickets. What is the commission rate?

7. Ben's commission rate is 16%. He receives a commission of $268 from sales of clothing. How many dollars worth of clothing were sold?

Answers on page A-14

To solve this equation, we divide by 600,000 on both sides:

$$\frac{30{,}000}{600{,}000} = \frac{r \times 600{,}000}{600{,}000}$$

$$\frac{1}{20} = r$$

$$0.05 = r$$

$$5\% = r.$$

The commission rate is 5%.

Do Exercise 6.

EXAMPLE 6 *Motorcycle Sales.* Joyce's commission rate is 12%. She receives a commission of $936 on the sale of a motorcycle. How much did the motorcycle cost?

Commission	=	*Commission rate*	×	*Sales*	
936	=	12%	×	S,	or $936 = 0.12 \times S$

To solve this equation, we divide by 0.12 on both sides:

$$\frac{936}{0.12} = \frac{0.12 \times S}{0.12}$$

$$\frac{936}{0.12} = S$$

$$7800 = S.$$

The motorcycle cost $7800.

Do Exercise 7.

C Discount

Suppose that the regular price of a rug is $60, and the rug is on sale at 25% off. Since 25% of $60 is $15, the sale price is $60 − $15, or $45. We call $60 the **original,** or **marked price,** 25% the **rate of discount,** $15 the **discount,** and $45 the **sale price.** Note that discount problems are a type of percent of decrease problem.

DISCOUNT AND SALE PRICE

Discount = Rate of discount × Original price

Sale price = Original price − Discount

EXAMPLE 7 A masonite door marked \$1389 is on sale at $33\frac{1}{3}\%$ off. What is the discount? the sale price?

DOOR
$1389 orig.
Save
33 1/3 %
Sale price = ?

a) *Discount* = *Rate of discount* × *Original price*

$$D = 33\frac{1}{3}\% \times 1389$$

$$D = \frac{1}{3} \times 1389$$

$$D = \frac{1389}{3} = 463$$

b) *Sale price* = *Original price* − *Discount*

$$S = 1389 - 463$$

$$S = 926$$

The discount is \$463 and the sale price is \$926.

Do Exercise 8.

EXAMPLE 8 *Antique Pricing.* An antique table is marked down from \$620 to \$527. What is the rate of discount?

We first find the discount by subtracting the sale price from the original price:

$$620 - 527 = 93.$$

The discount is \$93.

Next, we use the equation for discount:

Discount = *Rate of discount* × *Original price*

$$93 = r \times 620.$$

To solve, we divide by 620 on both sides:

$$\frac{93}{620} = \frac{r \times 620}{620}$$

$$\frac{93}{620} = r$$

$$0.15 = r$$

$$15\% = r.$$

The discount rate is 15%.

To check, note that a 15% discount rate means that 85% of the original price is paid:

$$0.85 \times 620 = 527.$$

Do Exercise 9.

8. A suit marked \$540 is on sale at $33\frac{1}{3}\%$ off. What is the discount? the sale price?

9. A pair of hiking boots is reduced from \$75 to \$60. Find the rate of discount.

Answers on page A-14

EXERCISE SET

For Extra Help

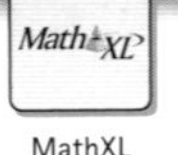

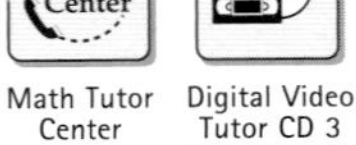

MathXL MyMathLab InterAct Math Math Tutor Center Digital Video Tutor CD 3 Videotape 6 Student's Solutions Manual

a Solve.

1. *Tennessee Sales Tax.* The sales tax rate in Tennessee is 7%. How much sales tax would be charged on a lawn mower that costs $279?

2. *Arizona Sales Tax.* The sales tax rate in Arizona is 5.6%. How much sales tax would be charged on a lawn mower that costs $279?

3. *Kansas Sales Tax.* The sales tax rate in Kansas is 5.3%. How much sales tax would be charged on a video game, Quest of the Planets, which sells for $49.99?

4. *New Jersey Sales Tax.* The sales tax rate in New Jersey is 6%. How much sales tax would be charged on a copy of John Grisham's novel *A Painted House*, which sells for $27.95?

Source: Borders Bookstore; Andrea Sutcliffe, *Numbers*

5. *Utah Sales Tax.* The sales tax rate in Utah is 4.75%. How much tax is charged on a purchase of 5 telephones at $69 apiece? What is the total price?

6. *New York Sales Tax.* The sales tax rate in New York is 4.25%. How much tax is charged on a purchase of 5 teapots at $37.99 apiece? What is the total price?

7. The sales tax is $48 on the purchase of a dining room set that sells for $960. What is the sales tax rate?

8. The sales tax is $15 on the purchase of a diamond ring that sells for $500. What is the sales tax rate?

9. The sales tax is $35.80 on the purchase of a refrigerator–freezer that sells for $895. What is the sales tax rate?

10. The sales tax is $9.12 on the purchase of a patio set that sells for $456. What is the sales tax rate?

11. The sales tax on a used car is $100 and the sales tax rate is 5%. Find the purchase price (the price before taxes are added).

12. The sales tax on the purchase of a new boat is $112 and the sales tax rate is 2%. Find the purchase price.

13. The sales tax on a dining room set is $28 and the sales tax rate is 3.5%. Find the purchase price.

14. The sales tax on a portable DVD player is $24.75 and the sales tax rate is 5.5%. Find the purchase price.

15. The sales tax rate in Austin is 2% for the city and county and 6.25% for the state. Find the total amount paid for 2 shower units at $332.50 apiece.

16. The sales tax rate in Omaha is 1.5% for the city and 5% for the state. Find the total amount paid for 3 air conditioners at $260 apiece.

17. The sales tax is $1030.40 on an automobile purchase of $18,400. What is the sales tax rate?

18. The sales tax is $979.60 on an automobile purchase of $15,800. What is the sales tax rate?

b Solve.

19. Katrina's commission rate is 6%. What is the commission from the sale of $45,000 worth of furnaces?

20. Jose's commission rate is 32%. What is the commission from the sale of $12,500 worth of sailboards?

21. Mitchell earns $120 selling $2400 worth of television sets in a consignment shop. What is the commission rate?

22. Donna earns $408 selling $3400 worth of shoes. What is the commission rate?

23. An art gallery's commission rate is 40%. They receive a commission of $392. How many dollars worth of artwork were sold?

24. A real estate agent's commission rate is 7%. She receives a commission of $5600 on the sale of a home. How much did the home sell for?

25. A real estate commission is 6%. What is the commission on the sale of a $98,000 home?

26. A real estate commission is 8%. What is the commission on the sale of a piece of land for $68,000?

27. Bonnie earns $280.80 selling $2340 worth of tee shirts. What is the commission rate?

28. Chuck earns $1147.50 selling $7650 worth of ski passes. What is the commission rate?

29. Miguel's commission is increased according to how much he sells. He receives a commission of 5% for the first $2000 and 8% on the amount over $2000. What is the total commission on sales of $6000?

30. Lucinda earns a salary of $500 a month, plus a 2% commission on sales. One month, she sold $990 worth of encyclopedias. What were her wages that month?

C Find what is missing.

	MARKED PRICE	RATE OF DISCOUNT	DISCOUNT	SALE PRICE
31.	$300	10%		
32.	$2000	40%		
33.	$17	15%		
34.	$20	25%		
35.		10%	$12.50	
36.		15%	$65.70	
37.	$600		$240	
38.	$12,800		$1920	

39. Find the discount and the rate of discount for the car seat in this ad.

Sale
Car Seat
$149.99
Was $179.99

40. Find the discount and the rate of discount for the wicker chair in this ad.

41. Find the marked price and the rate of discount for the stacked tool storage units in this ad.

42. Find the marked price and the rate of discount for the basketball system in this ad.

43. $\mathbf{D}_\mathbf{W}$ Is the following ad mathematically correct? Why or why not?

44. $\mathbf{D}_\mathbf{W}$ An item that is no longer on sale at "25% off" receives a price tag that is $33\frac{1}{3}\%$ more than the sale price. Has the item price been restored to its original price? Why or why not?

45. $\mathbf{D}_\mathbf{W}$ Which is better, a discount of 40% or a discount of 20% followed by another of 20%? Explain.

46. $\mathbf{D}_\mathbf{W}$ You take 40% of 50% of a number. What percent of the number could you take to obtain the same result making only one multiplication? Explain your answer.

SKILL MAINTENANCE

Solve. [5.3b]

47. $\frac{x}{12} = \frac{24}{16}$

48. $\frac{7}{2} = \frac{11}{x}$

Solve. [4.4b]

49. $0.64 \cdot x = 170$

50. $28.5 = 25.6 \times y$

Find decimal notation. [4.5a]

51. $\frac{5}{9}$

52. $\frac{23}{11}$

53. $\frac{11}{12}$

54. $\frac{13}{7}$

55. $\frac{15}{7}$

56. $\frac{19}{12}$

Convert to standard notation. [4.3b]

57. \$4.03 trillion

58. 5.8 million

59. 42.7 million

60. 6.09 trillion

SYNTHESIS

61. 🖩 *Magazine Subscriptions.* In a recent subscription drive, *People* offered a subscription of 52 weekly issues for a price of \$1.89 per issue. They advertised that this was a savings of 29.7% off the newsstand price. What was the newsstand price?

Source: *People Magazine*

62. 🖩 Gordon receives a 10% commission on the first \$5000 in sales and 15% on all sales beyond \$5000. If Gordon receives a commission of \$2405, how much did he sell? Use a calculator and trial and error if you wish.

63. Herb collects baseball memorabilia. He bought two autographed plaques, but became short of funds and had to sell them quickly for \$200 each. On one, he made a 20% profit and on the other, he lost 20%. Did he make or lose money on the sale?

64. Tee shirts are being sold at the mall for \$5 each, or 3 for \$10. If you buy three tee shirts, what is the rate of discount?

6.7 SIMPLE AND COMPOUND INTEREST

Objectives

a Solve applied problems involving simple interest.

b Solve applied problems involving compound interest.

1. What is the simple interest on \$4300 invested at an interest rate of 7% for 1 year?

2. What is the simple interest on a principal of \$4300 invested at an interest rate of 7% for $\frac{3}{4}$ year?

Answers on page A-15

a Simple Interest

Suppose you put \$1000 into an investment for 1 year. The \$1000 is called the **principal.** If the **interest rate** is 8%, in addition to the principal, you get back 8% of the principal, which is

8% of \$1000, or 0.08×1000, or \$80.00.

The \$80.00 is called the **simple interest.** It is, in effect, the price that a financial institution pays for the use of the money over time.

SIMPLE INTEREST FORMULA

The **simple interest *I*** on principal P, invested for t years at interest rate r, is given by

$$I = P \cdot r \cdot t.$$

EXAMPLE 1 What is the simple interest on \$2500 invested at an interest rate of 6% for 1 year?

We use the formula $I = P \cdot r \cdot t$:

$$\begin{aligned} I = P \cdot r \cdot t &= \$2500 \times 6\% \times 1 \\ &= \$2500 \times 0.06 \\ &= \$150. \end{aligned}$$

The simple interest for 1 year is \$150.

Do Exercise 1.

EXAMPLE 2 What is the simple interest on a principal of \$2500 invested at an interest rate of 6% for $\frac{1}{4}$ year?

We use the formula $I = P \cdot r \cdot t$:

$$\begin{aligned} I = P \cdot r \cdot t &= \$2500 \times 6\% \times \frac{1}{4} \\ &= \frac{\$2500 \times 0.06}{4} \\ &= \$37.50. \end{aligned}$$

We could instead have found $\frac{1}{4}$ of 6% and then multiplied by 2500.

The simple interest for $\frac{1}{4}$ year is \$37.50.

Do Exercise 2.

When time is given in days, we generally divide it by 365 to express the time as a fractional part of a year.

EXAMPLE 3 To pay for a shipment of tee shirts, New Wave Designs borrows \$8000 at $9\frac{3}{4}\%$ for 60 days. Find (a) the amount of simple interest that is due and (b) the total amount that must be paid after 60 days.

a) We express 60 days as a fractional part of a year:

$$I = P \cdot r \cdot t = \$8000 \times 9\frac{3}{4}\% \times \frac{60}{365}$$
$$= \$8000 \times 0.0975 \times \frac{60}{365}$$
$$\approx \$128.22.$$

The interest due for 60 days is \$128.22.

b) The total amount to be paid after 60 days is the principal plus the interest:

$\$8000 + \$128.22 = \$8128.22.$

The total amount due is \$8128.22.

Do Exercise 3.

3. The Glass Nook borrows \$4800 at $8\frac{1}{2}\%$ for 30 days. Find (a) the amount of simple interest due and (b) the total amount that must be paid after 30 days.

b Compound Interest

When interest is paid *on interest,* we call it **compound interest.** This is the type of interest usually paid on investments. Suppose you have \$5000 in a savings account at 6%. In 1 year, the account will contain the original \$5000 plus 6% of \$5000. Thus the total in the account after 1 year will be

106% of \$5000, or $1.06 \times \$5000$, or \$5300.

Now suppose that the total of \$5300 remains in the account for another year. At the end of this second year, the account will contain the \$5300 plus 6% of \$5300. The total in the account would thus be

106% of \$5300, or $1.06 \times \$5300$, or \$5618.

Note that in the second year, interest is earned on the first year's interest. When this happens, we say that interest is **compounded annually.**

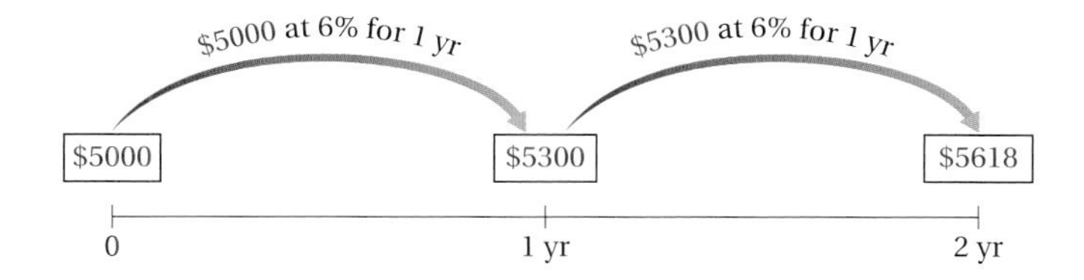

Answer on page A-15

4. Find the amount in an account if \$2000 is invested at 9%, compounded annually, for 2 years.

Answer on page A-15

EXAMPLE 4 Find the amount in an account if \$2000 is invested at 8%, compounded annually, for 2 years.

a) After 1 year, the account will contain 108% of \$2000:

$$1.08 \times \$2000 = \$2160.$$

b) At the end of the second year, the account will contain 108% of \$2160:

$$1.08 \times \$2160 = \$2332.80.$$

The amount in the account after 2 years is \$2332.80.

Do Exercise 4.

Suppose that the interest in Example 4 were **compounded semi-annually**—that is, every half year. Interest would then be calculated twice a year at a rate of 8% ÷ 2, or 4% each time. The approach used in Example 4 can then be adapted, as follows.

After the first $\frac{1}{2}$ year, the account will contain 104% of \$2000:

$$1.04 \times \$2000 = \$2080.$$

After a second $\frac{1}{2}$ year (1 full year), the account will contain 104% of \$2080:

$$1.04 \times \$2080 = \$2163.20.$$

After a third $\frac{1}{2}$ year $\left(1\frac{1}{2}\text{ full years}\right)$, the account will contain 104% of \$2163.20:

$$1.04 \times \$2163.20 = \$2249.728$$
$$\approx \$2249.73. \qquad \text{Rounding to the nearest cent}$$

Finally, after a fourth $\frac{1}{2}$ year (2 full years), the account will contain 104% of \$2249.73:

$$1.04 \times \$2249.73 = \$2339.7192$$
$$\approx \$2339.72. \qquad \text{Rounding to the nearest cent}$$

Let's summarize our results and look at them another way:

End of 1st $\frac{1}{2}$ year $\rightarrow 1.04 \times 2000 = 2000 \times (1.04)^1$;
End of 2nd $\frac{1}{2}$ year $\rightarrow 1.04 \times (1.04 \times 2000) = 2000 \times (1.04)^2$;
End of 3rd $\frac{1}{2}$ year $\rightarrow 1.04 \times (1.04 \times 1.04 \times 2000) = 2000 \times (1.04)^3$;
End of 4th $\frac{1}{2}$ year $\rightarrow 1.04 \times (1.04 \times 1.04 \times 1.04 \times 2000) = 2000 \times (1.04)^4$.

Note that each multiplication was by 1.04 and that

$$\$2000 \times 1.04^4 \approx \$2339.72. \qquad \text{Using a calculator and rounding to the nearest cent}$$

We have illustrated the following result.

COMPOUND INTEREST FORMULA

If a principal P has been invested at interest rate r, compounded n times a year, in t years it will grow to an amount A given by

$$A = P \cdot \left(1 + \frac{r}{n}\right)^{n \cdot t}.$$

Let's apply this formula to confirm our preceding discussion, where the amount invested is $P = \$2000$, the number of years is $t = 2$, and the number of compounding periods each year is $n = 2$. Substituting into the compound interest formula, we have

$$A = P \cdot \left(1 + \frac{r}{n}\right)^{n \cdot t} = 2000 \cdot \left(1 + \frac{8\%}{2}\right)^{2 \cdot 2}$$
$$= 2000 \cdot \left(1 + \frac{0.08}{2}\right)^{4} = 2000(1.04)^4$$
$$= 2000 \times 1.16985856 \approx \$2339.72.$$

If you are using a calculator, you could perform this computation in one step.

EXAMPLE 5 The Ibsens invest \$4000 in an account paying $5\frac{5}{8}\%$, compounded quarterly. Find the amount in the account after $2\frac{1}{2}$ years.

The compounding is quarterly, so n is 4. We substitute \$4000 for P, $5\frac{5}{8}\%$, or 0.05625, for r, 4 for n, and $2\frac{1}{2}$, or $\frac{5}{2}$, for t and compute A:

$$A = P \cdot \left(1 + \frac{r}{n}\right)^{n \cdot t} = \$4000 \cdot \left(1 + \frac{5\frac{5}{8}\%}{4}\right)^{4 \cdot 5/2}$$
$$= \$4000 \cdot \left(1 + \frac{0.05625}{4}\right)^{10}$$
$$= \$4000(1.0140625)^{10}$$
$$\approx \$4599.46.$$

The amount in the account after $2\frac{1}{2}$ years is \$4599.46.

Do Exercise 5.

5. A couple invests \$7000 in an account paying $6\frac{3}{8}\%$, compounded semiannually. Find the amount in the account after $1\frac{1}{2}$ years.

Answer on page A-15

CALCULATOR CORNER

Compound Interest A calculator is useful in computing compound interest. Not only does it do computations quickly but it also eliminates the need to round until the computation is completed. This minimizes "round-off errors" that occur when rounding is done at each stage of the computation. We must keep order of operations in mind when computing compound interest.

To find the amount due on a \$20,000 loan made for 25 days at 11% interest, compounded daily, we would compute $20{,}000\left(1 + \frac{0.11}{365}\right)^{25}$. To do this on a calculator, we press [2] [0] [0] [0] [0] [×] [(] [1] [+] [.] [1] [1] [÷] [3] [6] [5] [)] [y^x] (or [x^y]) [2] [5] [=]. Without parentheses, we would first find $1 + \frac{0.11}{365}$, raise this result to the 25th power, and then multiply by 20,000. To do this, we press [1] [+] [.] [1] [1] [÷] [3] [6] [5] [=] [y^x] (or [x^y]) [2] [5] [=] [×] [2] [0] [0] [0] [0]. In either case, the result is 20,151.23, rounded to the nearest cent.

Some calculators have business keys that allow such computations to be done more quickly.

Exercises:

1. Find the amount due on a \$16,000 loan made for 62 days at 13% interest, compounded daily.
2. An investment of \$12,500 is made for 90 days at 8.5% interest, compounded daily. How much is the investment worth after 90 days?

EXERCISE SET

For Extra Help

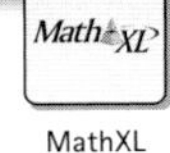

MathXL | MyMathLab | InterAct Math | Math Tutor Center | Digital Video Tutor CD 3 Videotape 6 | Student's Solutions Manual

a Find the simple interest.

	PRINCIPAL	RATE OF INTEREST	TIME	SIMPLE INTEREST
1.	\$200	4%	1 year	
2.	\$450	2%	1 year	
3.	\$2000	8.4%	$\frac{1}{2}$ year	
4.	\$200	7.7%	$\frac{1}{2}$ year	
5.	\$4300	10.56%	$\frac{1}{4}$ year	
6.	\$8000	9.42%	$\frac{1}{6}$ year	
7.	\$20,000	$4\frac{5}{8}\%$	1 year	
8.	\$100,000	$3\frac{7}{8}\%$	1 year	
9.	\$50,000	$5\frac{3}{8}\%$	$\frac{1}{4}$ year	
10.	\$80,000	$6\frac{3}{4}\%$	$\frac{1}{12}$ year	

Solve. Assume that simple interest is being calculated in each case.

11. CopiPix, Inc. borrows \$10,000 at 9% for 60 days. Find (a) the amount of interest due and (b) the total amount that must be paid after 60 days.

12. Sal's Laundry borrows \$8000 at 10% for 90 days. Find (a) the amount of interest due and (b) the total amount that must be paid after 90 days.

13. Animal Instinct, a pet supply shop, borrows \$6500 at 5% for 90 days. Find (a) the amount of interest due and (b) the total amount that must be paid after 90 days.

14. Andante's Cafe borrows \$4500 at 12% for 60 days. Find (a) the amount of interest due and (b) the total amount that must be paid after 60 days.

15. Jean's Garage borrows \$5600 at 10% for 30 days. Find (a) the amount of interest due and (b) the total amount that must be paid after 30 days.

16. Shear Delights, a hair salon, borrows \$3600 at 4% for 30 days. Find (a) the amount of interest due and (b) the total amount that must be paid after 30 days.

b Interest is compounded annually. Find the amount in the account after the given length of time. Round to the nearest cent.

	PRINCIPAL	RATE OF INTEREST	TIME	AMOUNT IN THE ACCOUNT
17.	\$400	5%	2 years	
18.	\$450	4%	2 years	
19.	\$2000	8.8%	4 years	
20.	\$4000	7.7%	4 years	
21.	\$4300	10.56%	6 years	
22.	\$8000	9.42%	6 years	
23.	\$20,000	$6\frac{5}{8}\%$	25 years	
24.	\$100,000	$5\frac{7}{8}\%$	30 years	

Interest is compounded semiannually. Find the amount in the account after the given length of time. Round to the nearest cent.

	PRINCIPAL	RATE OF INTEREST	TIME	AMOUNT IN THE ACCOUNT
25.	\$4000	6%	1 year	
26.	\$1000	5%	1 year	
27.	\$20,000	8.8%	4 years	
28.	\$40,000	7.7%	4 years	
29.	\$5000	10.56%	6 years	
30.	\$8000	9.42%	8 years	
31.	\$20,000	$7\frac{5}{8}\%$	25 years	
32.	\$100,000	$4\frac{7}{8}\%$	30 years	

Solve.

33. A family invests \$4000 in an account paying 6%, compounded monthly. How much is in the account after 5 months?

34. A couple invests \$2500 in an account paying 3%, compounded monthly. How much is in the account after 6 months?

35. A couple invests $1200 in an account paying 10%, compounded quarterly. How much is in the account after 1 year?

36. The O'Hares invest $6000 in an account paying 8%, compounded quarterly. How much is in the account after 18 months?

37. DW Which is a better investment and why: $1000 invested at $7\frac{3}{4}\%$ simple interest for 1 year, or $1000 invested at 7% compounded monthly for 1 year?

38. DW A firm must choose between borrowing $5000 at 10% for 30 days and borrowing $10,000 at 8% for 60 days. Give arguments in favor of and against each option.

SKILL MAINTENANCE

VOCABULARY REINFORCEMENT

In each of Exercises 39–46, fill in the blank with the correct term from the given list. Some of the choices may not be used.

39. If the product of two numbers is 1, they are ______________ of each other. [2.7a]

40. A number is ______________ if its ones digit is even and the sum of its digits is divisible by 3. [2.2a]

41. The number 0 is the ______________ identity. [1.2a]

42. A(n) ______________ is the ratio of price to the number of units. [5.2b]

43. The distance around an object is its ______________. [1.2b]

44. A number is ______________ if the sum of its digits is divisible by 3. [2.2a]

45. A natural number that has exactly two different factors, only itself and 1, is called a ______________ number. [2.1c]

46. When two pairs of numbers have the same ratio, they are ______________. [5.3a]

- divisible by 3
- divisible by 4
- divisible by 6
- divisible by 9
- perimeter
- area
- unit rate
- reciprocals
- proportional
- composite
- prime
- additive
- multiplicative

SYNTHESIS

Effective Yield. The *effective yield* is the yearly rate of simple interest that corresponds to a rate for which interest is compounded two or more times a year. For example, if P is invested at 12%, compounded quarterly, we would multiply P by $(1 + 0.12/4)^4$, or 1.03^4. Since $1.03^4 \approx 1.126$, the 12% compounded quarterly corresponds to an effective yield of approximately 12.6%. In Exercises 47 and 48, find the effective yield for the indicated account.

47. The account pays 9% compounded monthly.

48. The account pays 10% compounded daily.

6.8 INTEREST RATES ON CREDIT CARDS AND LOANS

Objective

a Solve applied problems involving interest rates on credit cards and loans.

a Credit Cards and Loans

Look at the following graphs. They offer good reason for a study of the real-world applications of percent, interest, loans, and credit cards.

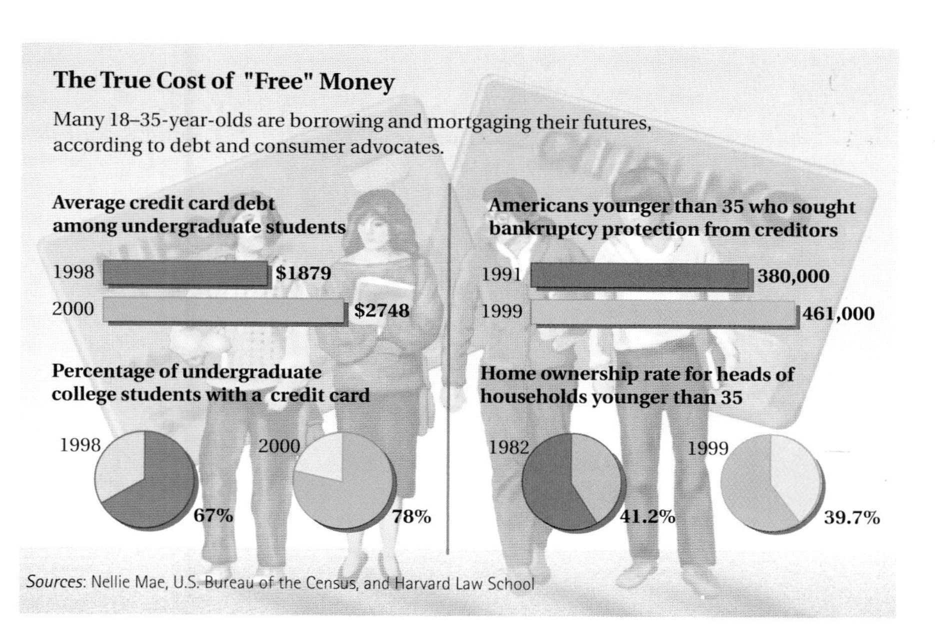

Comparing interest rates is essential if one is to become financially responsible. A small change in an interest rate can make a *large* difference in the cost of a loan. When you make a payment on a loan, do you know how much of that payment is interest and how much is applied to reducing the principal?

We begin with an example involving credit cards. A balance carried on a credit card is a type of loan. In a recent year in the United States, 100,000 young adults declared bankruptcy because of excessive credit card debt. The money you obtain through the use of a credit card is not "free" money. There is a price (interest) to be paid for the privilege.

EXAMPLE 1 *Credit Cards.* After the holidays, Sarah has a balance of $3216.28 on a credit card with an annual percentage rate (APR) of 19.7%. She decides not to make additional purchases with this card until she has paid off the balance.

a) Many credit cards require a minimum monthly payment of 2% of the balance. What is Sarah's minimum payment on a balance of $3216.28? Round the answer to the nearest dollar.

b) Find the amount of interest and the amount applied to reduce the principal in the minimum payment found in part (a).

c) If Sarah had transferred her balance to a card with an APR of 12.5%, how much of her first payment would be interest and how much would be applied to reduce the principal?

d) Compare the amounts for 12.5% from part (c) with the amounts for 19.7% from part (b).

1. Credit Cards. After the holidays, Jamal has a balance of \$4867.59 on a credit card with an annual percentage rate (APR) of 21.3%. He decides not to make additional purchases with this card until he has paid off the balance.

a) Many credit cards require a minimum monthly payment of 2% of the balance. What is Jamal's minimum payment on a balance of \$4867.59? Round the answer to the nearest dollar.

b) Find the amount of interest and the amount applied to reduce the principal in the minimum payment found in part (a).

c) If Jamal had transferred his balance to a card with an APR of 13.6%, how much of his first payment would be interest and how much would be applied to reduce the principal?

d) Compare the amounts for 13.6% from part (c) with the amounts for 21.3% from part (b).

Answers on page A-15

We solve as follows.

a) We multiply the balance of \$3216.28 by 2%:

$0.02 \times \$3216.28 = \$64.3256.$ Sarah's minimum payment, rounded to the nearest dollar, is \$64.

b) The amount of interest on \$3216.28 at 19.7% for one month* is given by

$$I = P \cdot r \cdot t = \$3216.28 \times 0.197 \times \frac{1}{12} \approx \$52.80.$$

We subtract to find the amount applied to reduce the principal in the first payment:

$$\begin{aligned}\text{Amount applied to reduce the principal} &= \text{Minimum payment} - \text{Interest for the month}\\ &= \$64 - \$52.80\\ &= \$11.20.\end{aligned}$$

Thus the principal of \$3216.28 is decreased by only \$11.20 with the first payment. (Sarah still owes \$3205.08.)

c) The amount of interest on \$3216.28 at 12.5% for one month is

$$I = P \cdot r \cdot t = \$3216.28 \times 0.125 \times \frac{1}{12} \approx \$33.50.$$

We subtract to find the amount applied to reduce the principal in the first payment:

$$\begin{aligned}\text{Amount applied to reduce the principal} &= \text{Minimum payment} - \text{Interest for the month}\\ &= \$64 - \$33.50\\ &= \$30.50.\end{aligned}$$

Thus the principal of \$3216.28 is decreased by \$30.50 with the first payment. (Sarah still owes \$3185.78.)

d) Let's organize the information for both rates in the following table.

BALANCE BEFORE FIRST PAYMENT	FIRST MONTH'S PAYMENT	% APR	AMOUNT OF INTEREST	AMOUNT APPLIED TO PRINCIPAL	BALANCE AFTER FIRST PAYMENT
\$3216.28	\$64	19.7%	\$52.80	\$11.20	\$3205.08
3216.28	64	12.5	33.50	30.50	3185.78

Difference in balance after first payment → \$19.30

At 19.7%, the interest is \$52.80 and the principal is decreased by \$11.20. At 12.5%, the interest is \$33.50 and the principal is decreased by \$30.50. Thus the principal is decreased by \$30.50 − \$11.20, or \$19.30 more with the 12.5% rate than with the 19.7% rate. Thus the interest at 19.7% is \$19.30 greater than the interest at 12.5%.

*Actually, the interest on a credit card is computed daily with a rate called a daily percentage rate (DPR). The DPR for Example 1 would be 19.7%/365 ≈ 0.054%. When no payments or additional purchases are made during the month, the difference in total interest for the month is minimal and we will not deal with it here.

Do Exercise 1 on the preceding page.

Even though the mathematics of the information in the chart below is beyond the scope of this text, it is interesting to compare how long it takes to pay off the balance of Example 1 if Sarah continues to pay $64 for each payment with how long it takes if she pays double that amount, $128, for each payment. Financial consultants frequently tell clients that if they want to take control of their debt, they should pay double the minimum payment.

RATE	PAYMENT PER MONTH	NUMBER OF PAYMENTS TO PAY OFF DEBT	TOTAL PAID BACK	ADDITIONAL COST OF PURCHASES
19.7%	$64	107, or 8 yr 11 mo	$6848	$3631.72
19.7	128	33, or 2 yr 9 mo	4224	1007.72
12.5	64	72, or 6 yr	4608	1391.72
12.5	128	29, or 2 yr 5 mo	3712	495.72

As with most loans, if you pay an extra amount toward the principal with each payment, the length of the loan can be greatly reduced. Note that at the rate of 19.7%, it will take Sarah almost 9 yr to pay off her debt if she pays only $64 per month and does not make additional purchases. If she transfers her balance to a card with a 12.5% rate and pays $128 per month, she could eliminate her debt in approximately $2\frac{1}{2}$ yr. You can see how debt can get out of control if you continue to make purchases and pay only the minimum payment. The debt will never be eliminated.

The Federal Stafford Loan program provides educational loans to students at interest rates that are much lower than those on credit cards. Payments on a loan do not begin until 6 months after graduation. At that time, the student has 10 years, or 120 monthly payments, to pay off the loan.

EXAMPLE 2 *Federal Stafford Loans.* After graduation, the balance on Taylor's Stafford loan is $28,650. If the rate on his loan is 3.37%, he will make 120 payments of approximately $282 each to pay off the loan.

a) Find the amount of interest and the amount of principal in the first payment.

b) If the interest rate were 5.25%, he would make 120 monthly payments of approximately $307 each. How much more of the first payment is interest if the loan is 5.25% rather than 3.37%?

c) Compare the total amount of interest on the loan at 3.37% with the amount on the loan at 5.25%. How much more would Taylor pay in interest on the 5.25% loan than on the 3.37% loan?

We solve as follows.

a) We use the formula $I = P \cdot r \cdot t$, substituting $28,650 for P, 0.0337 for r, and 1/12 for t:

$$I = \$28{,}650 \times 0.0337 \times \frac{1}{12}$$
$$\approx \$80.46.$$

The amount of interest in the first payment is $80.46. The payment is $282. We subtract to determine the amount applied to the principal:

$$\$282 - \$80.46 = \$201.54.$$

With the first payment, the principal will be reduced by $201.54.

2. Federal Stafford Loans. After graduation, the balance on Maggie's Stafford loan is $32,680. To pay off the loan at 3.37%, she will make 120 payments of approximately $321 each.

a) Find the amount of interest and the amount of principal in the first payment.

b) If the interest rate were 5.5%, she would make 120 payments of approximately $355 each. How much more of the first payment is interest if the loan is 5.5% rather than 3.37%?

c) Compare the total amount of interest on the loan at 3.37% with the amount of interest on the loan at 5.5%. How much more would Maggie pay in interest on the 5.5% loan than on the 3.37% loan?

Answers on page A-15

b) The interest at 5.25% would be

$$I = \$28{,}650 \times 0.0525 \times \frac{1}{12}$$
$$\approx \$125.34.$$

At the rate of 5.25%, the additional interest in the first payment is

$$\$125.34 - \$80.46 = \$44.88.$$

The higher interest rate results in an additional $44.88 in interest in the first payment.

c) For the 3.37% loan, there will be 120 payments of $282 each:

$$120 \times \$282 = \$33{,}840.$$

The total amount of interest at this rate is

$$\$33{,}840 - \$28{,}650 = \$5190.$$

For the 5.25% loan, there will be 120 payments of $307 each:

$$120 \times \$307 = \$36{,}840.$$

The total amount of interest at this rate is

$$\$36{,}840 - \$28{,}650 = \$8190.$$

At the rate of 5.25%, Taylor would pay

$$\$8190 - \$5190 = \$3000$$

more in interest than at the rate of 3.37%.

Do Exercise 2.

EXAMPLE 3 *Home Loans.* The Sawyers recently purchased their first home. They borrowed $153,000 at $6\frac{5}{8}\%$ for 30 years (360 payments). Their monthly payment (excluding insurance and taxes) is $979.68.

a) How much of the first payment is interest and how much is applied to reduce the principal?

b) If the Sawyers pay the entire 360 payments, how much interest will be paid on the loan?

We solve as follows.

a) To find the amount of interest paid in the first payment, we use the formula $I = P \cdot r \cdot t$:

$$I = P \cdot r \cdot t = \$153{,}000 \times 0.06625 \times \frac{1}{12} \approx \$844.69.$$

The amount applied to the principal is

$$\$979.68 - \$844.69, \text{ or } \$134.99.$$

b) Over the 30-year period, the total paid will be

$$360 \times \$979.68, \text{ or } \$352{,}684.80.$$

The total amount of interest paid over the lifetime of the loan is

$$\$352{,}684.80 - \$153{,}000, \text{ or } \$199{,}684.80.$$

Do Exercises 3 and 4 on the following page.

AMORTIZATION TABLES

If we make 360 calculations as in Example 3(a) and continue with a decreased principal as in Margin Exercise 3, we can create an *amortization table*, part of which is shown below. Such tables are also found in reference books. The beginning, middle, and last part of the loan described are shown. Look over the table and note how small a portion of the payment reduces the principal at the beginning of the loan and how that portion increases throughout the lifetime of the loan. Do you see again why a loan is not "free"?

MORTGAGE AMORTIZATION PROGRAM

MORTGAGE AMOUNT: $153,000
INTEREST RATE: 6.625%
NUMBER OF YEARS: 30
MONTHLY PAYMENTS ARE: $979.68

PAYMENT	PRINCIPAL	INTEREST	BALANCE	
1	$134.99	$844.69	$152,865.01	← Example 3
2	$135.73	$843.94	$152,729.28	← Margin Exercise 3
3	$136.48	$843.19	$152,592.80	
4	$137.24	$842.44	$152,455.56	
5	$137.99	$841.68	$152,317.56	
6	$138.76	$840.92	$152,178.81	
7	$139.52	$840.15	$152,039.29	
8	$140.29	$839.38	$151,898.99	
9	$141.07	$838.61	$151,757.93	
10	$141.85	$837.83	$151,616.08	
11	$142.63	$837.05	$151,473.45	
12	$143.42	$836.26	$151,330.04	
⋮	⋮	⋮ Interest for 12 periods =	$10,086.14	
174	$349.91	$629.77	$113,721.56	
175	$351.84	$627.84	$113,369.72	
176	$353.78	$625.90	$113,015.94	
177	$355.73	$623.94	$112,660.20	
178	$357.70	$621.98	$112,302.51	
179	$359.67	$620.00	$111,942.83	
180	$361.66	$618.02	$111,581.18	
181	$363.65	$616.02	$111,217.52	
182	$365.66	$614.01	$110,851.86	
183	$367.68	$611.99	$110,484.18	
184	$369.71	$609.96	$110,114.47	
185	$371.75	$607.92	$109,742.71	
186	$373.80	$605.87	$109,368.91	
⋮	⋮	⋮ Interest for 12 periods =	$8033.22	
349	$917.04	$62.63	$10,427.84	
350	$922.11	$57.57	$9,505.74	
351	$927.20	$52.48	$8,578.54	
352	$932.32	$47.36	$7,646.23	
353	$937.46	$42.21	$6,708.76	
354	$942.64	$37.04	$5,766.13	
355	$947.84	$31.83	$4,818.28	
356	$953.07	$26.60	$3,865.21	
357	$958.34	$21.34	$2,906.87	
358	$963.63	$16.05	$1,943.24	
359	$968.95	$10.73	$974.30	
360	$974.30	$0.00	$0.00	
		Interest for 12 periods =	$405.84	
		Total interest for 360 periods =	$199,684.80	

Refer to Example 3 for Margin Exercises 3 and 4.

3. **Home Loans.** Since the principal has been reduced by the first payment, at the time of the second payment of the Sawyers' 30-year loan, the new principal is the decreased principal

$$\$153,000 - \$134.99,$$

or

$$\$152,865.01.$$

Use $152,865.01 as the principal, and determine how much of the second payment is interest and how much is applied to reduce the principal. (In effect, repeat Example 3(a) using the new principal.)

4. **Home Loans.** The Sawyers decide to change the period of their home loan from 30 years to 15 years. Their monthly payment increases to $1343.33.

a) How much of the first payment is interest and how much is applied to reduce the principal?

b) If the Sawyers pay the entire 180 payments, how much interest will be paid on this loan?

c) Compare the amount of interest to pay off the 15-yr loan with the amount of interest to pay off the 30-yr loan.

Answers on page A-15

5. Refinancing a Home Loan. Consider Example 4 for a 15-yr loan. The new monthly payment is \$1250.14.

a) How much of the first payment is interest and how much is applied to reduce the principal?

b) If the Sawyers pay the entire 180 payments, how much interest will be paid on this loan?

c) Compare the amount of interest to pay off the 15-yr loan at $5\frac{1}{2}\%$ with the amount of interest to pay off the 15-yr loan at $6\frac{5}{8}\%$ in Margin Exercise 4.

Answers on page A-15

EXAMPLE 4 *Refinancing a Home Loan.* Refer to Example 3. Ten months after the Sawyers buy their home financed at a rate of $6\frac{5}{8}\%$, the rates drop to $5\frac{1}{2}\%$. After much consideration, they decide to refinance even though the new loan will cost them \$1200 in refinance charges. They have reduced the principal a small amount in the 10 payments they have made, but they decide to again borrow \$153,000 for 30 years at the new rate. Their new monthly payment is \$868.72.

a) How much of the first payment is interest and how much is applied to the principal?

b) Compare the amounts at $5\frac{1}{2}\%$ found in part (a) with the amounts at $6\frac{5}{8}\%$ found in Example 3(a).

c) With the lower house payment, how long will it take the Sawyers to recoup the refinance charge of \$1200?

d) If the Sawyers pay the entire 360 payments, how much interest will be paid on this loan? How much less is the total interest at $5\frac{1}{2}\%$ than at $6\frac{5}{8}\%$?

We solve as follows.

a) To find the interest paid in the first payment, we use the formula $I = P \cdot r \cdot t$:

$$I = P \cdot r \cdot t = \$153{,}000 \times 0.055 \times \frac{1}{12} = \$701.25.$$

The amount applied to the principal is

\$868.72 − \$701.25, or \$167.47.

b) We compare the amounts found in part (a) with the amounts found in Example 3(a):

Rate	*Monthly payment*	*Interest in first payment*	*Amount applied to principal*
$6\frac{5}{8}\%$	\$979.68	\$844.69	\$134.99
$5\frac{1}{2}\%$	\$868.72	\$701.25	\$167.47

At $5\frac{1}{2}\%$, the amount of interest in the first payment is \$844.69 − \$701.25, or \$143.44, less than at $6\frac{5}{8}\%$. The amount applied to the principal is \$167.47 − \$134.99, or \$32.48, more.

c) The monthly payment at $5\frac{1}{2}\%$ is \$979.68 − \$868.72, or \$110.96, less than the payment at $6\frac{5}{8}\%$. The total savings each month is approximately \$111. We can divide the cost of the refinancing by this monthly savings to determine the number of months it will take to recoup the \$1200 refinancing charge: $\$1200 \div \$111 \approx 11$. It will take the Sawyers approximately 11 months to break even.

d) Over the 30-year period, the total paid will be

360 × \$868.72, or \$312,739.20.

The total amount of interest paid over the lifetime of the loan is

\$312,739.20 − \$153,000, or \$159,739.20.

The total interest paid at $5\frac{1}{2}\%$ is

\$199,684.80 (see Example 3) − \$159,739.20, or \$39,945.60,

less than the total interest paid at $6\frac{5}{8}\%$. Thus the $5\frac{1}{2}\%$ loan saves the Sawyers approximately \$40,000 in interest charges over the 30 years.

Do Exercise 5.

6.8 EXERCISE SET

For Extra Help

MathXL | MyMathLab | InterAct Math | Math Tutor Center | Digital Video Tutor CD 3 Videotape 6 | Student's Solutions Manual

a Solve.

1. *Credit Cards.* At the end of his freshman year of college, Antonio has a balance of \$4876.54 on a credit card with an annual percentage rate (APR) of 21.3%. He decides not to make additional purchases with his card until he has paid off the balance.

a) Many credit cards require a minimum monthly payment of 2% of the balance. What is Antonio's minimum payment on a balance of \$4876.54? Round the answer to the nearest dollar.
b) Find the amount of interest and the amount applied to reduce the principal in the minimum payment found in part (a).
c) If Antonio had transferred his balance to a card with an APR of 12.6%, how much of his first payment would be interest and how much would be applied to reduce the principal?
d) Compare the amounts for 12.6% from part (c) with the amounts for 21.3% from part (b).

2. *Credit Cards.* At the end of her junior year of college, Becky had a balance of \$5328.88 on a credit card with an annual percentage rate (APR) of 18.7%. She decides not to make additional purchases with this card until she has paid off the balance.

a) Many credit cards require a minimum monthly payment of 2% of the balance. What is Becky's minimum payment on a balance of \$5328.88? Round the answer to the nearest dollar.
b) Find the amount of interest and the amount applied to reduce the principal in the minimum payment found in part (a).
c) If Becky had transferred her balance to a card with an APR of 13.2%, how much of her first payment would be interest and how much would be applied to reduce the principal?
d) Compare the amounts for 13.2% from part (c) with the amounts for 18.7% from part (b).

3. *Federal Stafford Loans.* After graduation, the balance on Grace's Stafford loan is \$44,560. To pay off the loan at 3.37%, she will make 120 payments of approximately \$437.93 each.

a) Find the amount of interest and the amount applied to reduce the principal in the first payment.
b) If the interest rate were 4.75%, she would make 120 monthly payments of approximately \$467.20 each. How much more of the first payment is interest if the loan is 4.75% rather than 3.37%?
c) Compare the total amount of interest on a loan at 3.37% with the amount on the loan at 4.75%. How much more would Grace pay on the 4.75% loan than on the 3.37% loan?

4. *Federal Stafford Loans.* After graduation, the balance on Ricky's Stafford loan is \$38,970. To pay off the loan at 3.37%, he will make 120 payments of approximately \$382.99 each.

a) Find the amount of interest and the amount applied to reduce the principal in the first payment.
b) If the interest rate were 5.4%, he would make 120 monthly payments of approximately \$421 each. How much more of the first payment is interest if the loan is 5.4% rather than 3.37%?
c) Compare the total amount of interest on the loan at 3.37% with the amount on the loan at 5.4%. How much more would Ricky pay on the 5.4% loan than on the 3.37% loan?

5. *Home Loan.* The Martinez family recently purchased a home. They borrowed \$164,000 at $6\frac{1}{4}\%$ for 30 years (360 payments). Their monthly payment (excluding insurance and taxes) is \$1009.78.

a) How much of the first payment is interest and how much is applied to reduce the principal?
b) If this family pays the entire 360 payments, how much interest will be paid on the loan?
c) Determine the new principal after the first payment. Use that new principal to determine how much of the second payment is interest and how much is applied to reduce the principal.

6. *Home Loan.* The Kaufmans recently purchased a home. They borrowed \$136,000 at 5.75% for 30 years (360 payments). Their monthly payment (excluding insurance and taxes) is \$793.66.

a) How much of the first payment is interest and how much is applied to reduce the principal?
b) If the Kaufmans pay the entire 360 payments, how much interest will be paid on the loan?
c) Determine the new principal after the first payment. Use that new principal to determine how much of the second payment is interest and how much is applied to reduce the principal.

7. *Refinancing a Home Loan.* Refer to Exercise 5. The Martinez family decides to change the period of their home loan to 15 years. Their monthly payment increases to $1406.17.

a) How much of the first payment is interest and how much is applied to reduce the principal?
b) If the Martinez family pays the entire 180 payments, how much interest will be paid on the loan?
c) Compare the amount of interest to pay off the 15-yr loan with the amount of interest to pay off the 30-yr loan.

8. *Refinancing a Home Loan.* Refer to Exercise 6. The Kaufmans decide to change the period of their home loan to 15 years. Their monthly payment increased to $1129.36.

a) How much of the first payment is interest and how much is applied to reduce the principal?
b) If the Kaufmans pay the entire 180 payments, how much interest will be paid on the loan?
c) Compare the amount of interest to pay off the 15-yr loan with the amount of interest to pay off the 30-yr loan.

Complete the following table, assuming monthly payments as given.

	INTEREST RATE	HOME MORTGAGE	TIME OF LOAN	MONTHLY PAYMENT	PRINCIPAL AFTER FIRST PAYMENT	PRINCIPAL AFTER SECOND PAYMENT
9.	6.98%	$100,000	360 mos	$663.96		
10.	6.98%	$100,000	180 mos	$897.71		
11.	8.04%	$100,000	180 mos	$957.96		
12.	8.04%	$100,000	360 mos	$736.55		
13.	7.24%	$150,000	360 mos	$1022.25		
14.	7.24%	$75,000	180 mos	$684.22		
15.	7.24%	$200,000	180 mos	$1824.60		
16.	7.24%	$180,000	360 mos	$1226.70		

17. *Dealership's Car Loan Offer.* For a trip to Colorado, Michael and Rebecca buy a 2005 Toyota Sienna van whose selling price is $23,950. For financing, they accept the promotion from the manufacturer that offers a 48-month loan at 2.9% with 10% down. Their monthly payment is $454.06.

a) What is the down payment? the amount borrowed?
b) How much of the first payment is interest and how much is applied to reduce the principal?
c) What is the total interest cost of the loan if they pay all of the 48 payments?

18. *New-Car Loan.* After working at her first job for 2 years, Janice buys a new Saturn for $16,385. She makes a down payment of $1385 and finances $15,000 for 4 years at a new-car loan rate of 8.99%. Her monthly payment is $373.20.

a) How much of her first payment is interest and how much is applied to reduce the principal?
b) Find the principal balance at the beginning of the second month and determine how much less interest she will pay in the second payment than in the first.
c) What is the total interest cost of the loan if she pays all of the 48 payments?

19. *Used-Car Loan.* Twin brothers, Jerry and Terry, each take a job at the college cafeteria in order to have the money to make payments on the purchase of a 2002 Chrysler PT Cruiser for $11,900. They make a down payment of 5% and finance the remainder at 9.3% for 3 years. (Used-car loan rates are generally higher than new-car loan rates.) Their monthly payment is $361.08.

a) What is the down payment? the amount borrowed?

b) How much of the first payment is interest and how much is applied to reduce the principal?

c) If they pay all 36 payments, how much interest will they pay for the loan?

20. *Used-Car Loan.* For his construction job, Clint buys a 2003 Dodge Ram 1500 truck for $13,800. He makes a down payment of $1380 and finances the remainder for 4 years at 8.8%. The monthly payment is $307.89.

a) How much is financed?

b) How much of the first payment is interest and how much is applied to reduce the principal?

c) If he pays all 48 payments, how much interest will he pay for the loan?

21. **D$_W$** Based on the skills of mathematics you have obtained in this section, discuss the significant new ideas you now have about interest rates and credit cards that you didn't have before.

22. **D$_W$** Examine the information in the graphs at the beginning of the section. Discuss how a knowledge of this section might have been of help to some of these students.

23. **D$_W$** Compare the following two purchases and describe a situation in which each purchase is the best choice.

Purchase A: A new car for $15,145. The loan is for $14,500 at 6.9% for 4 years. The monthly payment is $346.55.

Purchase B: A used car for $10,600. The loan is for $9300 at 12.5% for 3 years. The monthly payment is $311.12.

24. **D$_W$** Look over the examples and exercises in this section. What seems to happen to the monthly payment on a loan if the time of payment changes from 30 years to 15 years, assuming the interest rate stays the same? Discuss the pros and cons of both time periods.

SKILL MAINTENANCE

Solve. Round the answer to the nearest hundredth where appropriate. [5.3b]

25. $\frac{5}{8} = \frac{x}{28}$

26. $\frac{5}{8} = \frac{17.5}{y}$

27. $\frac{13}{16} = \frac{81.25}{N}$

28. $\frac{9}{16} = \frac{p}{100}$

29. $\frac{1284}{t} = \frac{3456}{5000}$

30. $\frac{12.8}{32.5} = \frac{x}{2000}$

31. $\frac{56.3}{78.4} = \frac{t}{100}$

32. $\frac{28}{x} = \frac{8}{5}$

33. $\frac{16}{9} = \frac{100}{p}$

34. $\frac{t}{1284} = \frac{5000}{3456}$

6 Summary and Review

The review that follows is meant to prepare you for a chapter exam. It consists of three parts. The first part, Concept Reinforcement, is designed to increase understanding through true/false exercises. The second part is a list of important properties and formulas. The third part is the Review Exercises. These provide practice exercises for the exam, together with references to section objectives so you can go back and review. Before beginning, stop and look back over the skills you have obtained. What skills in mathematics do you have now that you did not have before studying this chapter?

CONCEPT REINFORCEMENT

Determine whether the statement is true or false. Answers are given at the back of the book.

______ **1.** A fixed principal invested for four years will earn more interest when interest is compounded quarterly than when interest is compounded semi-annually.

______ **2.** Of the numbers 0.5%, $\frac{5}{1000}\%$, $\frac{1}{2}\%$, $\frac{1}{5}$, and $0.\overline{1}$, the largest number is $0.\overline{1}$.

______ **3.** If principal A equals principal B and principal A is invested for 2 years at 4% compounded quarterly while principal B is invested for 4 years at 2% compounded semi-annually, the interest earned from each investment is the same.

______ **4.** The symbol % is equivalent to $\times \frac{1}{10}$.

IMPORTANT PROPERTIES AND FORMULAS

Commission = Commission rate × Sales

Discount = Rate of discount × Original price

Sale price = Original price − Discount

Simple Interest: $I = P \cdot r \cdot t$

Compound Interest: $A = P \cdot \left(1 + \frac{r}{n}\right)^{n \cdot t}$

Review Exercises

Find percent notation for the decimal notation in the sentence in Exercises 1 and 2. [6.1b]

1. Of all the snacks eaten on Super Bowl Sunday, 0.56 of them are chips and salsa.

Source: Korbel Research and Pace Foods

2. Of all the vehicles in Mexico City, 0.017 of them are taxis.

Source: *The Handy Geography Answer Book*

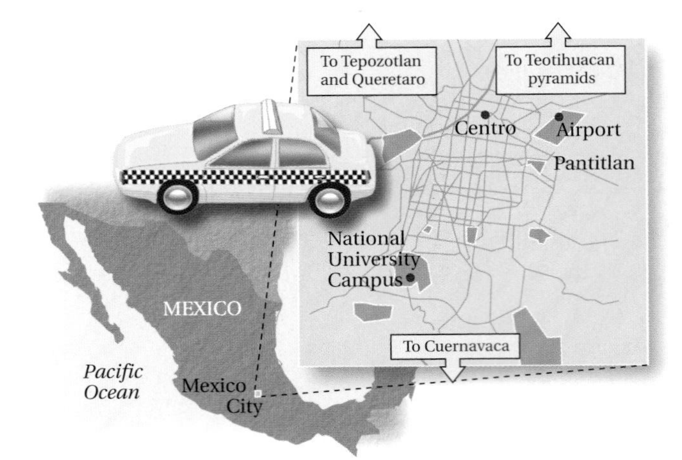

Find percent notation. [6.2a]

3. $\frac{3}{8}$

4. $\frac{1}{3}$

Find decimal notation. [6.1b]

5. 73.5%

6. $6\frac{1}{2}\%$

Find fraction notation. [6.2b]

7. 24%

8. 6.3%

Translate to a percent equation. Then solve. [6.3a, b]

9. 30.6 is what percent of 90?

10. 63 is 84 percent of what?

11. What is $38\frac{1}{2}\%$ of 168?

Translate to a proportion. Then solve. [6.4a, b]

12. 24 percent of what is 16.8?

13. 42 is what percent of 30?

14. What is 10.5% of 84?

Solve. [6.5a, b]

15. *Favorite Ice Creams.* According to a recent survey, 8.9% of those interviewed chose chocolate as their favorite ice cream flavor and 4.2% chose butter pecan. Of the 2500 students in a freshman class, how many would choose chocolate as their favorite ice cream? butter pecan?

Source: International Ice Cream Association

16. *Prescriptions.* Of the 295 million people in the United States, 123.64 million take at least one kind of prescription drug per day. What percent take at least one kind of prescription drug per day?

Source: American Society of Health-System Pharmacies

17. *Water Output.* The average person expels 200 mL of water per day by sweating. This is 8% of the total output of water from the body. How much is the total output of water?

Source: Elaine N. Marieb, *Essentials of Human Anatomy and Physiology*, 6th ed. Boston: Addison Wesley Longman, Inc., 2000

18. *Test Scores.* Jason got a 75 on a math test. He was allowed to go to the math lab and take a retest. He increased his score to 84. What was the percent of increase?

19. *Test Scores.* Jenny got an 81 on a math test. By taking a retest in the math lab, she increased her score by 15%. What was her new score?

Solve. [6.6a, b, c]

20. A state charges a meals tax of $4\frac{1}{2}\%$. What is the meals tax charged on a dinner party costing $320?

21. In a certain state, a sales tax of $378 is collected on the purchase of a used car for $7560. What is the sales tax rate?

22. Kim earns $753.50 selling $6850 worth of televisions. What is the commission rate?

23. An air conditioner has a marked price of $350. It is placed on sale at 12% off. What are the discount and the sale price?

24. A fax machine priced at $305 is discounted at the rate of 14%. What are the discount and the sale price?

25. An insurance salesperson receives a 7% commission. If $42,000 worth of life insurance is sold, what is the commission?

26. Find the rate of discount.

Our lowest price of the season!
18-ft metal frame
$399.69
Reg. 489.99
Family-size metal-frame pool

Solve. [6.7a, b]

27. What is the simple interest on $1800 at 6% for $\frac{1}{3}$ year?

28. The Dress Shack borrows $24,000 at 10% simple interest for 60 days. Find (a) the amount of interest due and (b) the total amount that must be paid after 60 days.

29. What is the simple interest on $2200 principal at the interest rate of 5.5% for 1 year?

30. The Kleins invest $7500 in an investment account paying an annual interest rate of 12%, compounded monthly. How much is in the account after 3 months?

31. Find the amount in an investment account if $8000 is invested at 9%, compounded annually, for 2 years.

Solve. [6.8a]

32. *Credit Cards.* At the end of her junior year of college, Judy has a balance of $6428.74 on a credit card with an annual percentage rate (APR) of 18.7%. She decides not to make additional purchases with this card until she has paid off the balance.

a) Many credit cards require a minimum payment of 2% of the balance. What is Judy's minimum payment on a balance of $6428.74? Round the answer to the nearest dollar.
b) Find the amount of interest and the amount applied to reduce the principal in the minimum payment found in part (a).
c) If Judy had transferred her balance to a card with an APR of 13.2%, how much of her first payment would be interest and how much would be applied to reduce the principal?
d) Compare the amounts for 13.2% from part (c) with the amounts for 18.7% from part (b).

33. **DW** Ollie buys a microwave oven during a 10%-off sale. The sale price that Ollie paid was $162. To find the original price, Ollie calculates 10% of $162 and adds that to $162. Is this correct? Why or why not? [6.6c]

34. **DW** Which is the better deal for a consumer and why: a discount of 40% or a discount of 20% followed by another of 22%? [6.6c]

SYNTHESIS

35. *Land Area of the United States.* After Hawaii and Alaska became states, the total land area of the United States increased from 2,963,681 mi^2 to 3,540,939 mi^2. What was the percent of increase? [6.5b]

36. Rhonda's Dress Shop reduces the price of a dress by 40% during a sale. By what percent must the store increase the sale price, after the sale, to get back to the original price? [6.6c]

37. A $200 coat is marked up 20%. After 30 days, it is marked down 30% and sold. What was the final selling price of the coat? [6.6c]

6 Chapter Test

For Extra Help

Work It Out! Chapter Test Video on CD

1. *Bookmobiles.* Since 1991, the number of bookmobiles has decreased by approximately 6.4%. Find decimal notation for 6.4%.

   Source: American Library Association

2. *Gravity.* The gravity of Mars is 0.38 as strong as Earth's. Find percent notation for 0.38.

   Source: www.marsinstitute.info/epo/mermarsfacts.html

3. Find percent notation for $\frac{11}{8}$.

4. Find fraction notation for 65%.

5. Translate to a percent equation. Then solve.

   What is 40% of 55?

6. Translate to a proportion. Then solve.

   What percent of 80 is 65?

Solve.

7. *Cruise Ship Passengers.* Of the passengers on a typical cruise ship, 16% are in the 25–34 age group and 23% are in the 35–44 age group. A cruise ship has 2500 passengers. How many are in the 25–34 age group? the 35–44 age group?

   Source: Polk

8. *Batting Averages.* Luis Castillo, second baseman for the Florida Marlins, got 180 hits during the 2000 baseball season. This was about 33.4% of his at-bats. How many at-bats did he have?

   Source: Major League Baseball

9. *Airline Profits.* Profits of the entire U. S. Airline industry decreased from \$5.5 billion in 1999 to \$2.7 billion in 2000. Find the percent of decrease.

   Source: Air Transport Association

10. There are 6.6 billion people living in the world today. It is estimated that the total number who have ever lived is about 120 billion. What percent of people who have ever lived are alive today?

    Source: *The Handy Geography Answer Book*

11. *Maine Sales Tax.* The sales tax rate in Maine is 5%. How much tax is charged on a purchase of \$324? What is the total price?

12. Gwen's commission rate is 15%. What is the commission from the sale of \$4200 worth of merchandise?

13. The marked price of a DVD player is \$200 and the item is on sale at 20% off. What are the discount and the sale price?

14. What is the simple interest on a principal of \$120 at the interest rate of 7.1% for 1 year?

15. The Burnham Parents–Teachers Association invests \$5200 at 6% simple interest. How much is in the account after $\frac{1}{2}$ year?

16. Find the amount in an account if \$1000 is invested at $5\frac{3}{8}\%$, compounded annually, for 2 years.

17. The Suarez family invests $10,000 at an annual interest rate of 4.9%, compounded monthly. How much is in the account after 3 years?

18. *Job Opportunities.* The table below lists job opportunities, in millions, in 2002 and projected increases to 2012. Find the missing numbers.

OCCUPATION	NUMBER OF JOBS IN 2002 (in millions)	NUMBER OF JOBS IN 2012 (in millions)	CHANGE	PERCENT OF INCREASE
Retail salespersons	4.1	4.8	0.7	17.1%
Registered nurses	2.3		0.6	
Post-secondary teachers	1.6	2.2		
Food preparation and service workers		2.4	0.4	
Restaurant servers		2.5		19.0%

Source: Department of Labor

19. Find the discount and the discount rate of the washer-dryer duet in this ad.

SHORT TIME ONLY
$1675
Was $1950
No interest for 18 months

20. *Home Loan.* Complete the following table, assuming the monthly payment as given.

Interest Rate	7.4%
Mortgage	$120,000
Time of Loan	360 mos
Monthly Payment	$830.86
Principal after First Payment	
Principal after Second Payment	

SYNTHESIS

21. By selling a home without using a realtor, Juan and Marie can avoid paying a 7.5% commission. They receive an offer of $180,000 from a potential buyer. In order to give a comparable offer, for what price would a realtor need to sell the house? Round to the nearest hundred.

22. Karen's commission rate is 16%. She invests her commission from the sale of $15,000 worth of merchandise at the interest rate of 12%, compounded quarterly. How much is Karen's investment worth after 6 months?

CHAPTERS 1–6

Cumulative Review/ Final Examination

This cumulative review also serves as a final examination for the entire book. A question that may occur at this point is what notation to use for a particular problem or exercise. Although there is no particular rule, especially as you use mathematics outside the classroom, here is the guideline that we follow: Use the notation given in the problem. That is, if the problem is given using mixed numerals, give the answer in mixed numerals. If the problem is given in decimal notation, give the answer in decimal notation.

1. It is expected that by 2010, 53% of all food expenses will occur away from home. Find decimal notation for 53%.

Source: National Restaurant Association

2. During the 2004 baseball season, Johan Santana of the Minnesota Twins had a 0.375 batting average. Find percent notation for 0.375.

Source: Major League Baseball

3. Find percent notation: $\frac{9}{8}$.

4. Find decimal notation: $\frac{13}{6}$.

5. Write fraction notation for the ratio 5 to 0.5.

6. Find the rate in kilometers per hour: 350 km, 15 hr.

Use $<$, $>$, or $=$ for $\square$ to write a true sentence.

7. $\frac{5}{7} \square \frac{6}{8}$

8. $\frac{6}{14} \square \frac{15}{25}$

Estimate the sum or difference by rounding to the nearest hundred.

9. $263{,}961 + 32{,}090 + 127.89$

10. $73{,}510 - 23{,}450$

Calculate.

11. $46 - [4(6 + 4 \div 2) + 2 \times 3 - 5]$

12. $[0.8(1.5 - 9.8 \div 49) + (1 + 0.1)^2] \div 1.5$

Compute and simplify.

13. $\frac{6}{5} + 1\frac{5}{6}$

14. $46.9 + 2.84$

15. $\begin{array}{r} 487{,}094 \\ 6{,}936 \\ +\ 21{,}120 \\ \hline \end{array}$

16. $35 - 34.98$

17. $3\frac{1}{3} - 2\frac{2}{3}$

18. $\frac{8}{9} - \frac{6}{7}$

19. $\frac{7}{9} \cdot \frac{3}{14}$

20. $\begin{array}{r} 236{,}984 \\ \times\ 3{,}600 \\ \hline \end{array}$

21. $\begin{array}{r} 46.012 \\ \times\ 0.03 \\ \hline \end{array}$

22. $6\frac{3}{5} \div 4\frac{2}{5}$

23. $431.2 \div 35.2$

24. $15 \overline{)1850}$

Solve.

25. $36 \cdot x = 3420$

26. $y + 142.87 = 151$

27. $\frac{2}{15} \cdot t = \frac{6}{5}$

28. $\frac{3}{4} + x = \frac{5}{6}$

29. $\frac{y}{25} = \frac{24}{15}$

30. $\frac{16}{n} = \frac{21}{11}$

Solve.

31. *Box-Office Revenue.* The table below shows the weekend gross (box-office revenue), in millions, of the movie *The Family Man.*

WEEKEND	GROSS (In Millions)
1	\$15.1
2	16.4
3	9.1
4	6.6
5	3.3

a) Convert each amount to standard notation.
b) Find the total amount earned for the five weekends.
c) Find the average amount earned.
d) Find the percent of increase from the first weekend to the second.
e) Find the percent of decrease from the second weekend to the third.

Source: Exhibitor Relations

32. *Consumer Credit.* The outstanding consumer credit in the United States in 2003 was \$2025.5 billion. This was an increase of \$263.2 billion over the outstanding consumer credit in 2003.

a) How much was the outstanding consumer credit in 2002?
b) What was the percent of increase?

Source: Federal Reserve Board

33. *e-mails.* The typical Internet user receives 17,000 e-mail messages per year. How many messages are received by the typical user each day? Use 365 days in one year.

Source: Dave Barry, *The Miami Herald*

34. At one point in the 2000–2001 NBA season, the Utah Jazz had won 26 out of 40 games. At this rate, how many games would they win in the entire season of 82 games?

Source: National Basketball Association

35. *Neckties.* A total of \$212.50 was paid for 5 neckties at an upscale men's store. How much did each necktie cost?

36. *Unit Price.* A 200-oz bottle of Gain liquid laundry detergent costs \$9.99. What is the unit price?

37. Patty walked $\frac{7}{10}$ mi to school and then $\frac{8}{10}$ mi to the library. How far did she walk?

38. On a map, 1 in. represents 80 mi. How much does $\frac{3}{4}$ in. represent?

39. *Compound Interest.* The Bakers invest \$8500 in an investment account paying 8%, compounded monthly. How much is in the account after 5 years?

40. *Ribbons.* How many pieces of ribbon $1\frac{4}{5}$ yd long can be cut from a length of ribbon 9 yd long?

41. *Job Opportunities.* The table below shows job opportunities, in thousands, in 1998 and projected to 2008. Find the missing numbers.

OCCUPATION	NUMBER OF JOBS IN 1998 (in Thousands)	NUMBER OF JOBS IN 2008 (in Thousands)	CHANGE	% INCREASE
Court clerk	100	112	12	12%
Office manager	1611		313	
Office clerk	3021	3484		
Teacher assistant		1567	375	
Host/hostess		351		18.2%

Source: Handbook of U.S. Labor Statistics

For each of Exercises 42–45, choose the correct answer from the selections given.

42. The population of the state of Kentucky increased from 3,685,296 in 1990 to 4,041,769 in 2000. What was the percent increase?

Source: U.S. Bureau of the Census

a) 9.04% **b)** 9.7% **c)** 7.9%
d) 8.8% **e)** None

43. Find decimal notation: $\frac{8}{13}$.

a) 0.615 **b)** 0.615384 **c)** 0.615385
d) 0.6153846154 **e)** None

44. Antonio bought a sweater vest for $59.95 and a pair of blue jeans for $39.50 and paid for them with a $100 bill. How much change did he receive?

a) $95 **b)** $55 **c)** $99.45
d) $0.55 **e)** None

45. Subtract and simplify: $\frac{14}{25} - \frac{3}{20}$.

a) $\frac{11}{500}$ **b)** $\frac{11}{5}$ **c)** $\frac{41}{100}$
d) $\frac{205}{500}$ **e)** None

SYNTHESIS

46. *Nutrition Facts.* Food companies are required by law to provide nutrition facts on packaging. But when choosing a product, one must be careful that a proper comparison is made. Consider, for example, the following nutrition information on a box of Wheaties compared to a box of Kellogg's Complete. Note that Wheaties defines 1 serving as 1 cup and Complete defines 1 serving as $\frac{3}{4}$ cup.

Sources: General Mills; Kellogg's

a) Use proportions to rewrite the nutrition facts for a box of Complete so that it is based on a serving size being 1 cup. Then compare the nutrition facts between the cereals.

b) Which cereal has the most calories per serving?

c) Which cereal has the most fat per serving?

d) Which cereal has the most sodium per serving?

e) Which cereal has the most potassium per serving?

Wheaties

Nutrition Facts
Serving Size 1 cup (30g)
Servings Per Container About 17

Amount Per Serving	Wheaties	with 1/2 cup skim milk
Calories	110	150
Calories from Fat	10	10
	% Daily Value **	
Total Fat 1g*	1%	2%
Saturated Fat 0g	0%	0%
Polyunsaturated Fat 0g		
Monounsaturated Fat 0g		
Cholesterol 0mg	0%	1%
Sodium 220mg	9%	12%
Potassium 110mg	3%	9%
Total Carbohydrate 24g	8%	10%
Dietary Fiber 3g	12%	12%
Sugars 4g		
Other Carbohydrate 17g		
Protein 3g		

Not the same

Complete

Nutrition Facts
Serving Size 3/4 cup (29g/1.1 oz.)
Servings Per Container About 17

Amount Per Serving	Cereal	Cereal with 1/2 Cup Vitamins A & D Fat Free Milk
Calories	90	130
Calories from Fat	5	5
	% Daily Value **	
Total Fat 0.5g*	1%	1%
Saturated Fat 0g	0%	0%
Polyunsaturated Fat 0g		
Monounsaturated Fat 0g		
Cholesterol 0mg	0%	0%
Sodium 210mg	9%	11%
Potassium 170mg	5%	11%
Total Carbohydrate 23g	8%	10%
Dietary Fiber 5g	20%	20%
Soluble Fiber 1g		
Insoluble Fiber 4g		
Sugars 5g		
Other Carbohydrate 13g		
Protein 3g		

Answers

CHAPTER 1

Margin Exercises, Section 1.1, pp. 2–5

1. 2 ten thousands **2.** 2 hundred thousands
3. 2 millions **4.** 2 ten millions
5. 2 tens **6.** 2 hundreds
7. 2 hundred thousands; 8 ten thousands; 0 thousands; 2 hundreds; 1 ten; 9 ones
8. 1 thousand + 8 hundreds + 9 tens + 5 ones
9. 2 ten thousands + 3 thousands + 4 hundreds + 1 ten + 6 ones
10. 3 thousands + 0 hundreds + 3 tens + 1 one, or 3 thousands + 3 tens + 1 one
11. 4 thousands + 1 hundred + 8 tens + 0 ones, or 4 thousands + 1 hundred + 8 tens
12. 1 hundred thousand + 5 ten thousands + 4 thousands + 6 hundreds + 1 ten + 6 ones
13. Forty-nine **14.** Sixteen **15.** Thirty-eight
16. Two hundred four
17. Forty-four thousand, one hundred fifty-five
18. One million, eight hundred seventy-nine thousand, two hundred four **19.** Six billion, four hundred forty-nine million **20.** 213,105,329

Exercise Set 1.1, p. 6

1. 5 thousands **3.** 5 hundreds **5.** 2 **7.** 1
9. 5 thousands + 7 hundreds + 0 tens + 2 ones, or 5 thousands + 7 hundreds + 2 ones
11. 9 ten thousands + 3 thousands + 9 hundreds + 8 tens + 6 ones
13. 2 thousands + 0 hundreds + 5 tens + 8 ones, or 2 thousands + 5 tens + 8 ones
15. 1 thousand + 2 hundreds + 6 tens + 8 ones
17. 4 hundred thousands + 0 ten thousands + 5 thousands + 6 hundreds + 9 tens + 8 ones, or 4 hundred thousands + 5 thousands + 6 hundreds + 9 tens + 8 ones
19. 2 hundred thousands + 7 ten thousands + 2 thousands + 1 hundred + 6 tens + 1 one
21. 1 million + 1 hundred thousand + 8 ten thousands + 0 thousands + 2 hundreds + 1 ten + 2 ones, or 1 million + 1 hundred thousand + 8 ten thousands + 2 hundreds + 1 ten + 2 ones **23.** Eighty-five
25. Eighty-eight thousand **27.** One hundred twenty-three thousand, seven hundred sixty-five **29.** Seven billion, seven hundred fifty-four million, two hundred eleven thousand, five hundred seventy-seven
31. 2,233,812 **33.** 8,000,000,000 **35.** Five hundred sixty-six thousand, two hundred eighty **37.** seventy-six million, eighty-six thousand, seven hundred ninety-two
39. 9,460,000,000,000 **41.** 64,186,000 **43.** ^{D}W
45. 138

Margin Exercises, Section 1.2, pp. 10–11

1. 13,465 **2.** 9745 **3.** 16,182 **4.** 27,474 **5.** 29 in.
6. 62 ft **7.** 16 in.; 26 in.

Calculator Corner, p. 12

1. 55 **2.** 121 **3.** 1602 **4.** 734 **5.** 1932 **6.** 864

Exercise Set 1.2, p. 13

1. 387 **3.** 5198 **5.** 164 **7.** 100 **9.** 8503 **11.** 5266
13. 4466 **15.** 6608 **17.** 34,432 **19.** 101,310
21. 230 **23.** 18,424 **25.** 31,685 **27.** 114 mi
29. 570 ft **31.** ^{D}W **33.** 8 ten thousands
34. Five billion, two hundred ninety-four million, two hundred forty-seven thousand
35. $1 + 99 = 100, 2 + 98 = 100, \ldots, 49 + 51 = 100$. Then $49 \cdot 100 = 4900$ and $4900 + 50 + 100 = 5050$.

Margin Exercises, Section 1.3, pp. 16–19

1. $7 = 2 + 5$, or $7 = 5 + 2$ **2.** $17 = 9 + 8$, or $17 = 8 + 9$
3. $5 = 13 - 8$; $8 = 13 - 5$ **4.** $11 = 14 - 3$; $3 = 14 - 11$
5. 3801 **6.** 6328 **7.** 4747 **8.** 56 **9.** 205
10. 658 **11.** 2851 **12.** 1546

Calculator Corner, p. 18

1. 28 **2.** 47 **3.** 67 **4.** 119 **5.** 2128 **6.** 2593

Exercise Set 1.3, p. 20

1. 7 = 3 + 4, or 7 = 4 + 3 **3.** 13 = 5 + 8, or 13 = 8 + 5 **5.** 23 = 14 + 9, or 23 = 9 + 14 **7.** 43 = 27 + 16, or 43 = 16 + 27 **9.** 6 = 15 − 9; 9 = 15 − 6 **11.** 8 = 15 − 7; 7 = 15 − 8 **13.** 17 = 23 − 6; 6 = 23 − 17 **15.** 23 = 32 − 9; 9 = 32 − 23 **17.** 44 **19.** 533 **21.** 39 **23.** 234 **25.** 5382 **27.** 3831 **29.** 7748 **31.** 43,028 **33.** 56 **35.** 454 **37.** 3749 **39.** 2191 **41.** 95,974 **43.** 9989 **45.** 4206 **47.** 10,305 **49.** DW **51.** 1024 **52.** 12,732 **53.** 90,283 **54.** 29,364 **55.** 1345 **56.** 924 **57.** 22,692 **58.** 10,920 **59.** Six million, three hundred seventy-five thousand, six hundred two **60.** 7 ten thousands **61.** 3; 4

Margin Exercises, Section 1.4, pp. 22–26

1. 40 **2.** 50 **3.** 70 **4.** 100 **5.** 40 **6.** 80 **7.** 90 **8.** 140 **9.** 470 **10.** 240 **11.** 290 **12.** 600 **13.** 800 **14.** 800 **15.** 9300 **16.** 8000 **17.** 8000 **18.** 19,000 **19.** 69,000 **20.** 48,970; 49,000; 49,000 **21.** 269,580; 269,600; 270,000 **22.** Eliminate the power sunroof and the power package. Answers may vary. **23.** **(a)** $18,300; **(b)** yes **24.** 70 + 20 + 40 + 70 = 200 **25.** 700 + 700 + 200 + 200 = 1800 **26.** 9300 − 6700 = 2600 **27.** 23,000 − 12,000 = 11,000 **28.** $<$ **29.** $>$ **30.** $>$ **31.** $<$ **32.** $<$ **33.** $>$

Exercise Set 1.4, p. 27

1. 50 **3.** 470 **5.** 730 **7.** 900 **9.** 100 **11.** 1000 **13.** 9100 **15.** 32,900 **17.** 6000 **19.** 8000 **21.** 45,000 **23.** 373,000 **25.** 180 **27.** 5720 **29.** 220; incorrect **31.** 890; incorrect **33.** 16,500 **35.** 5200 **37.** $11,200 **39.** $18,900; no **41.** Answers will vary depending on the options chosen. **43.** 1600 **45.** 1500 **47.** 31,000 **49.** 69,000 **51.** $<$ **53.** $>$ **55.** $<$ **57.** $>$ **59.** $>$ **61.** $>$ **63.** $1,800,607 < 2,136,068$, or $2,136,068 > 1,800,607$ **65.** $6482 > 4827$, or $4827 < 6482$ **67.** DW **69.** 7 thousands + 9 hundreds + 9 tens + 2 ones **70.** 2 ten millions + 3 millions **71.** Two hundred forty-six billion, six hundred five million, four thousand, thirty-two **72.** One million, five thousand, one hundred **73.** 86,754 **74.** 13,589 **75.** 48,824 **76.** 4415 **77.** Left to the student **79.** Left to the student

Margin Exercises, Section 1.5, pp. 32–36

1. 116 **2.** 148 **3.** 4938 **4.** 6740 **5.** 1035 **6.** 3024 **7.** 46,252 **8.** 205,065 **9.** 144,432 **10.** 287,232 **11.** 14,075,720 **12.** 391,760 **13.** 17,345,600 **14.** 56,200 **15.** 562,000 **16.** **(a)** 1081; **(b)** 1081; **(c)** same **17.** 40 **18.** 15 **19.** $15,300 **20.** 840 × 250 = 210,000; 800 × 200 = 160,000 **21.** 45 sq ft

Calculator Corner, p. 35

1. 448 **2.** 21,970 **3.** 6380 **4.** 39,564 **5.** 180,480 **6.** 2,363,754

Exercise Set 1.5, p. 37

1. 870 **3.** 2,340,000 **5.** 520 **7.** 564 **9.** 1527 **11.** 64,603 **13.** 4770 **15.** 3995 **17.** 46,080 **19.** 14,652 **21.** 207,672 **23.** 798,408 **25.** 20,723,872 **27.** 362,128 **29.** 20,064,048 **31.** 25,236,000 **33.** 302,220 **35.** 49,101,136 **37.** 50 · 70 = 3500 **39.** 30 · 30 = 900 **41.** 900 · 300 = 270,000 **43.** 400 · 200 = 80,000 **45.** **(a)** $2,840,000; **(b)** $2,850,000 **47.** 529,984 sq mi **49.** 8100 sq ft **51.** DW **53.** 12,685 **54.** 10,834 **55.** 427,477 **56.** 111,110 **57.** 1241 **58.** 8889 **59.** 254,119 **60.** 66,444 **61.** 6,376,000 **62.** 6,375,600 **63.** 247,464 sq ft

Margin Exercises, Section 1.6, pp. 40–46

1. $\frac{54}{6} = 9$; $\begin{array}{r} 9 \\ 6\overline{)54} \end{array}$ **2.** 15 = 5 · 3, or 15 = 3 · 5 **3.** 72 = 9 · 8, or 72 = 8 · 9 **4.** 6 = 12 ÷ 2; 2 = 12 ÷ 6 **5.** 6 = 42 ÷ 7; 7 = 42 ÷ 6 **6.** 6; 6 · 9 = 54 **7.** 6 R 7; 6 · 9 = 54, 54 + 7 = 61 **8.** 4 R 5; 4 · 12 = 48, 48 + 5 = 53 **9.** 6 R 13; 6 · 24 = 144, 144 + 13 = 157 **10.** 59 R 3 **11.** 1475 R 5 **12.** 1015 **13.** 134 **14.** 63 R 12 **15.** 807 R 4 **16.** 1088 **17.** 360 R 4 **18.** 800 R 47

Calculator Corner, p. 47

1. 3 R 11 **2.** 28 **3.** 124 R 2 **4.** 131 R 18 **5.** 283 R 57 **6.** 843 R 187

Exercise Set 1.6, p. 48

1. 18 = 3 · 6, or 18 = 6 · 3 **3.** 22 = 22 · 1, or 22 = 1 · 22 **5.** 54 = 6 · 9, or 54 = 9 · 6 **7.** 37 = 1 · 37, or 37 = 37 · 1 **9.** 9 = 45 ÷ 5; 5 = 45 ÷ 9 **11.** 37 = 37 ÷ 1; 1 = 37 ÷ 37 **13.** 8 = 64 ÷ 8 **15.** 11 = 66 ÷ 6; 6 = 66 ÷ 11 **17.** 12 **19.** 1 **21.** 22 **23.** Not defined **25.** 55 R 2 **27.** 108 **29.** 307 **31.** 753 R 3 **33.** 74 R 1 **35.** 92 R 2 **37.** 1703 **39.** 987 R 5 **41.** 12,700 **43.** 127 **45.** 52 R 52 **47.** 29 R 5 **49.** 40 R 12 **51.** 90 R 22 **53.** 29 **55.** 105 R 3 **57.** 1609 R 2 **59.** 1007 R 1 **61.** 23 **63.** 107 R 1 **65.** 370 **67.** 609 R 15 **69.** 304 **71.** 3508 R 219 **73.** 8070 **75.** DW **77.** perimeter **78.** equation, inequality **79.** digits, periods **80.** additive **81.** dividend **82.** factors, product **83.** minuend **84.** associative **85.** 54, 122; 33, 2772; 4, 8 **87.** 30 buses

Margin Exercises, Section 1.7, pp. 52–55

1. 7 **2.** 5 **3.** No **4.** Yes **5.** 5 **6.** 10 **7.** 5 **8.** 22 **9.** 22,490 **10.** 9022 **11.** 570 **12.** 3661 **13.** 8 **14.** 45 **15.** 77 **16.** 3311 **17.** 6114 **18.** 8 **19.** 16 **20.** 644 **21.** 96 **22.** 94

Exercise Set 1.7, p. 56

1. 14 **3.** 0 **5.** 29 **7.** 0 **9.** 8 **11.** 14 **13.** 1035 **15.** 25 **17.** 450 **19.** 90,900 **21.** 32 **23.** 143 **25.** 79 **27.** 45 **29.** 324 **31.** 743 **33.** 37 **35.** 66

37. 15 **39.** 48 **41.** 175 **43.** 335 **45.** 104 **47.** 45 **49.** 4056 **51.** 17,603 **53.** 18,252 **55.** 205 **57.** DW **59.** $7 = 15 - 8$; $8 = 15 - 7$ **60.** $6 = 48 \div 8$; $8 = 48 \div 6$ **61.** $<$ **62.** $>$ **63.** $>$ **64.** $<$ **65.** 142 R 5 **66.** 142 **67.** 334 **68.** 334 R 11 **69.** 347

Margin Exercises, Section 1.8, pp. 59–66

1. 11,277 adoptions **2.** 8441 adoptions **3.** 19,089 adoptions **4.** $1874 **5.** $171 **6.** $5572 **7.** 9180 sq in. **8.** 378 cartons with 1 can left over **9.** 79 gal **10.** 181 seats

Translating for Success, p. 67

1. E **2.** M **3.** D **4.** G **5.** A **6.** O **7.** F **8.** K **9.** J **10.** H

Exercise Set 1.8, p. 68

1. 33,042 performances **3.** 704 performances **5.** 2054 miles **7.** 18 rows **9.** 792,316 degrees; 1,291,900 degrees **11.** 192,268 degrees **13.** $69,277 **15.** 240 mi **17.** 168 hr **19.** 225 squares **21.** $24,456 **23.** 35 weeks; 2 episodes left over **25.** 236 gal **27.** 3616 mi **29.** 12,804 gal **31.** $14,445 **33.** $247 **35.** **(a)** 4200 sq ft; **(b)** 268 ft **37.** $29,105,000,000 **39.** 151,500 **41.** 563 packages; 7 bars left over **43.** 384 mi; 27 in. **45.** 21 columns **47.** 56 full cartons; 11 books left over. If 1355 books are shipped, it will take 57 cartons. **49.** 32 $10 bills **51.** $400 **53.** 280 min; or 4 hr 40 min **55.** 525 min, or 8 hr 45 min **57.** 106 bones **59.** 3000 sq in. **61.** DW **63.** 234,600 **64.** 234,560 **65.** 235,000 **66.** 22,000 **67.** 16,000 **68.** 8000 **69.** 4000 **70.** 320,000 **71.** 720,000 **72.** 46,800,000 **73.** 792,000 mi; 1,386,000 mi

Margin Exercises, Section 1.9, pp. 75–80

1. 5^4 **2.** 5^5 **3.** 10^2 **4.** 10^4 **5.** 10,000 **6.** 100 **7.** 512 **8.** 32 **9.** 51 **10.** 30 **11.** 584 **12.** 84 **13.** 4; 1 **14.** 52; 52 **15.** 29 **16.** 1880 **17.** 253 **18.** 93 **19.** 1880 **20.** 305 **21.** 75 **22.** 4 **23.** 1496 ft **24.** 46 **25.** 4

Calculator Corner, p. 76

1. 243 **2.** 15,625 **3.** 20,736 **4.** 2048

Calculator Corner, p. 78

1. 49 **2.** 85 **3.** 36 **4.** 0 **5.** 73 **6.** 49

Exercise Set 1.9, p. 81

1. 3^4 **3.** 5^2 **5.** 7^5 **7.** 10^3 **9.** 49 **11.** 729 **13.** 20,736 **15.** 121 **17.** 22 **19.** 20 **21.** 100 **23.** 1 **25.** 49 **27.** 5 **29.** 434 **31.** 41 **33.** 88 **35.** 4 **37.** 303 **39.** 20 **41.** 70 **43.** 295 **45.** 32 **47.** 906 **49.** 62 **51.** 102 **53.** 32 **55.** $94 **57.** 401 **59.** 110 **61.** 7 **63.** 544 **65.** 708 **67.** 27 **69.** DW **71.** 452 **72.** 835 **73.** 13 **74.** 37 **75.** 2342 **76.** 4898 **77.** 25 **78.** 100 **79.** 104,286 mi^2 **80.** 98 gal **81.** 24; $1 + 5 \cdot (4 + 3) = 36$ **83.** 7; $12 \div (4 + 2) \cdot 3 - 2 = 4$

Concept Reinforcement, p. 84

1. False **2.** True **3.** True **4.** True **5.** False **6.** False

Summary and Review: Chapter 1, p. 84

1. 8 thousands **2.** 3 **3.** 2 thousands + 7 hundreds + 9 tens + 3 ones **4.** 5 ten thousands + 6 thousands + 0 hundreds + 7 tens + 8 ones, or 5 ten thousands + 6 thousands + 7 tens + 8 ones **5.** 4 millions + 0 hundred thousands + 0 ten thousands + 7 thousands + 1 hundred + 0 tens + 1 one, or 4 millions + 7 thousands + 1 hundred + 1 one **6.** Sixty-seven thousand, eight hundred nineteen **7.** Two million, seven hundred eighty-one thousand, four hundred twenty-seven **8.** 1 billion, sixty-five million, seventy thousand, six hundred seven **9.** 476,588 **10.** 2,000,400,000 **11.** 14,272 **12.** 66,024 **13.** 21,788 **14.** 98,921 **15.** $10 = 6 + 4$, or $10 = 4 + 6$ **16.** $8 = 11 - 3$; $3 = 11 - 8$ **17.** 5148 **18.** 1689 **19.** 2274 **20.** 17,757 **21.** 345,800 **22.** 345,760 **23.** 346,000 **24.** 300,000 **25.** $41{,}300 + 19{,}700 = 61{,}000$ **26.** $38{,}700 - 24{,}500 = 14{,}200$ **27.** $400 \cdot 700 = 280{,}000$ **28.** $>$ **29.** $<$ **30.** 5,100,000 **31.** 6,276,800 **32.** 506,748 **33.** 27,589 **34.** 5,331,810 **35.** $56 = 8 \cdot 7$, or $56 = 7 \cdot 8$ **36.** $4 = 52 \div 13$; $13 = 52 \div 4$ **37.** 12 R 3 **38.** 5 **39.** 913 R 3 **40.** 384 R 1 **41.** 4 R 46 **42.** 54 **43.** 452 **44.** 5008 **45.** 4389 **46.** 8 **47.** 45 **48.** 58 **49.** 0 **50.** 4^3 **51.** 10,000 **52.** 36 **53.** 65 **54.** 233 **55.** 56 **56.** 32 **57.** 260 **58.** 165 **59.** $502 **60.** $484 **61.** 1982 **62.** 19 cartons **63.** 14 beehives **64.** $13,585 **65.** $27,598 **66.** 137 beakers filled; 13 mL left over **67.** 98 ft^2; 42 ft **68.** DW A vat contains 1152 oz of hot sauce. If 144 bottles are to be filled equally, how much will each bottle contain? Answers may vary. **69.** DW No; if subtraction were associative, then $a - (b - c) = (a - b) - c$ for any a, b, and c. But, for example,

$$12 - (8 - 4) = 12 - 4 = 8,$$

whereas

$$(12 - 8) - 4 = 4 - 4 = 0.$$

Since $8 \neq 0$, this example shows that subtraction is not associative. **70.** $d = 8$ **71.** $a = 8, b = 4$ **72.** 6 days

Test: Chapter 1, p. 87

1. [1.1a] 5 **2.** [1.1b] 8 thousands + 8 hundreds + 4 tens + 3 ones **3.** [1.1c] Thirty-eight million, four hundred three thousand, two hundred seventy-seven

4. [1.2a] 9989 **5.** [1.2a] 63,791 **6.** [1.2a] 34
7. [1.2a] 10,515 **8.** [1.3b] 3630 **9.** [1.3b] 1039
10. [1.3b] 6848 **11.** [1.3b] 5175 **12.** [1.5a] 41,112
13. [1.5a] 5,325,600 **14.** [1.5a] 2405
15. [1.5a] 534,264 **16.** [1.6b] 3 R 3 **17.** [1.6b] 70
18. [1.6b] 97 **19.** [1.6b] 805 R 8
20. [1.8a] 1852 12-packs; 7 cakes left over
21. [1.8a] 1,256,615 mi^2 **22.** **(a)** [1.2b], [1.5c] 300 in., 5000 in^2; 264 in., 3872 in^2; 228 in., 2888 in^2;
(b) [1.8a] 2112 in^2 **23.** [1.8a] 206,330 voters
24. [1.8a] 1808 lb **25.** [1.8a] 20 staplers **26.** [1.7b] 46
27. [1.7b] 13 **28.** [1.7b] 14 **29.** [1.7b] 381
30. [1.4a] 35,000 **31.** [1.4a] 34,580 **32.** [1.4a] 34,600
33. [1.4b] $23{,}600 + 54{,}700 = 78{,}300$
34. [1.4b] $54{,}800 - 23{,}600 = 31{,}200$
35. [1.5b] $800 \cdot 500 = 400{,}000$ **36.** [1.4c] $>$
37. [1.4c] $<$ **38.** [1.9a] 12^4 **39.** [1.9b] 343
40. [1.9b] 100,000 **41.** [1.9b] 625 **42.** [1.9c] 31
43. [1.9c] 98 **44.** [1.9c] 2 **45.** [1.9d] 216
46. [1.9c] 18 **47.** [1.9c] 92 **48.** [1.5c], [1.8a] 336 in^2
49. [1.9c] 9 **50.** [1.8a] 80 payments

CHAPTER 2

Margin Exercises, Section 2.1, pp. 91–95

1. Yes **2.** No **3.** 1, 2, 5, 10 **4.** 1, 3, 5, 9, 15, 45
5. 1, 2, 31, 62 **6.** 1, 2, 3, 4, 6, 8, 12, 24 **7.** $5 = 1 \cdot 5$; $45 = 9 \cdot 5$; $100 = 20 \cdot 5$ **8.** $10 = 1 \cdot 10$; $60 = 6 \cdot 10$; $110 = 11 \cdot 10$ **9.** 5, 10, 15, 20, 25, 30, 35, 40, 45, 50
10. Yes **11.** Yes **12.** No **13.** 2, 13, 19, 41, 73 are prime; 6, 12, 65, 99 are composite; 1 is neither **14.** $2 \cdot 3$
15. $2 \cdot 2 \cdot 3$ **16.** $3 \cdot 3 \cdot 5$ **17.** $2 \cdot 7 \cdot 7$
18. $2 \cdot 3 \cdot 3 \cdot 7$ **19.** $2 \cdot 2 \cdot 2 \cdot 2 \cdot 3 \cdot 3$
20. $2 \cdot 2 \cdot 2 \cdot 5 \cdot 7 \cdot 7$ **21.** $5 \cdot 5 \cdot 7 \cdot 11$

Calculator Corner, p. 92

1. Yes **2.** No **3.** No **4.** Yes **5.** No **6.** Yes
7. Yes **8.** No **9.** Yes **10.** No

Exercise Set 2.1, p. 96

1. No **3.** Yes **5.** 1, 2, 3, 6, 9, 18
7. 1, 2, 3, 6, 9, 18, 27, 54 **9.** 1, 2, 4 **11.** 1
13. 1, 2, 7, 14, 49, 98 **15.** 1, 3, 5, 15, 17, 51, 85, 255
17. 4, 8, 12, 16, 20, 24, 28, 32, 36, 40
19. 20, 40, 60, 80, 100, 120, 140, 160, 180, 200
21. 3, 6, 9, 12, 15, 18, 21, 24, 27, 30
23. 12, 24, 36, 48, 60, 72, 84, 96, 108, 120
25. 10, 20, 30, 40, 50, 60, 70, 80, 90, 100
27. 9, 18, 27, 36, 45, 54, 63, 72, 81, 90 **29.** No **31.** Yes
33. Yes **35.** No **37.** No **39.** Neither
41. Composite **43.** Prime **45.** Prime **47.** $2 \cdot 2 \cdot 2$
49. $2 \cdot 7$ **51.** $2 \cdot 3 \cdot 7$ **53.** $5 \cdot 5$ **55.** $2 \cdot 5 \cdot 5$
57. $13 \cdot 13$ **59.** $2 \cdot 2 \cdot 5 \cdot 5$ **61.** $5 \cdot 7$
63. $2 \cdot 2 \cdot 2 \cdot 3 \cdot 3$ **65.** $7 \cdot 11$ **67.** $2 \cdot 2 \cdot 7 \cdot 103$
69. $3 \cdot 17$ **71.** $2 \cdot 2 \cdot 2 \cdot 2 \cdot 3 \cdot 5 \cdot 5$ **73.** $3 \cdot 7 \cdot 13$
75. $2 \cdot 3 \cdot 11 \cdot 17$ **77.** DW **79.** 26 **80.** 256
81. 425 **82.** 4200 **83.** 0 **84.** 22 **85.** 1 **86.** 3
87. \$612 **88.** 201 min, or 3 hr 21 min **89.** Row 1: 48, 90, 432, 63; row 2: 7, 2, 2, 10, 8, 6, 21, 10; row 3: 9, 18, 36, 14, 12, 11, 21; row 4: 29, 19, 42

Margin Exercises, Section 2.2, pp. 98–101

1. Yes **2.** No **3.** Yes **4.** No **5.** Yes **6.** No
7. Yes **8.** No **9.** Yes **10.** No **11.** No **12.** Yes
13. No **14.** Yes **15.** No **16.** Yes **17.** No **18.** Yes
19. No **20.** Yes **21.** Yes **22.** No **23.** No **24.** Yes
25. Yes **26.** No **27.** No **28.** Yes **29.** No **30.** Yes
31. Yes **32.** No

Exercise Set 2.2, p. 102

1. 46, 224, 300, 36, 45,270, 4444, 256, 8064, 21,568
3. 224, 300, 36, 4444, 256, 8064, 21,568
5. 300, 36, 45,270, 8064 **7.** 36, 45,270, 711, 8064
9. 324, 42, 501, 3009, 75, 2001, 402, 111,111, 1005
11. 55,555, 200, 75, 2345, 35, 1005 **13.** 324 **15.** 200
17. 313,332, 7624, 111,126, 876, 1110, 5128, 64,000, 9990
19. 313,332, 111,126, 876, 1110, 9990
21. 9990 **23.** 1110, 64,000, 9990 **25.** DW **27.** 138
28. 139 **29.** 874 **30.** 56 **31.** 26 **32.** 13 **33.** 234
34. 4003 **35.** 45 gal **36.** 4320 min
37. $2 \cdot 2 \cdot 2 \cdot 3 \cdot 5 \cdot 5 \cdot 13$ **39.** $2 \cdot 2 \cdot 3 \cdot 3 \cdot 7 \cdot 11$
41. 95,238

Margin Exercises, Section 2.3, pp. 104–109

1. Numerator: 83; denominator: 100
2. Numerator: 27; denominator: 50
3. Numerator: 11; denominator: 25
4. Numerator: 21; denominator: 1000
5. $\frac{1}{2}$ **6.** $\frac{1}{3}$ **7.** $\frac{1}{3}$ **8.** $\frac{2}{3}$ **9.** $\frac{3}{4}$ **10.** $\frac{15}{16}$ **11.** $\frac{15}{15}$ **12.** $\frac{5}{4}$
13. $\frac{7}{4}$ **14.** Clocks: $\frac{3}{5}$; thermometers: $\frac{2}{5}$ **15.** $\frac{98}{64}$; $\frac{98}{162}$; $\frac{64}{162}$
16. 1 **17.** 1 **18.** 1 **19.** 1 **20.** 1 **21.** 1 **22.** 0
23. 0 **24.** 0 **25.** 0 **26.** Not defined
27. Not defined **28.** 8 **29.** 10 **30.** 346 **31.** 23

Exercise Set 2.3, p. 110

1. Numerator: 3; denominator: 4
3. Numerator: 11; denominator: 2
5. Numerator: 0; denominator: 7
7. $\frac{2}{4}$ **9.** $\frac{1}{8}$ **11.** $\frac{4}{3}$ **13.** $\frac{12}{16}$ **15.** $\frac{38}{16}$ **17.** $\frac{3}{4}$ **19.** $\frac{12}{12}$
21. $\frac{4}{8}$ **23.** $\frac{6}{12}$ **25.** $\frac{5}{8}$ **27.** $\frac{4}{7}$ **29.** **(a)** $\frac{2}{8}$; **(b)** $\frac{6}{8}$
31. **(a)** $\frac{3}{8}$; **(b)** $\frac{5}{8}$ **33.** **(a)** $\frac{3}{7}$; **(b)** $\frac{3}{4}$; **(c)** $\frac{4}{7}$; **(d)** $\frac{4}{3}$
35. **(a)** $\frac{35}{10{,}000}$; **(b)** $\frac{50}{10{,}000}$; **(c)** $\frac{44}{10{,}000}$; **(d)** $\frac{63}{10{,}000}$; **(e)** $\frac{43}{10{,}000}$; **(f)** $\frac{21}{10{,}000}$
37. **(a)** $\frac{390}{13}$; **(b)** $\frac{13}{390}$ **39.** $\frac{850}{1000}$ **41.** 0 **43.** 7 **45.** 1
47. 1 **49.** 0 **51.** 1 **53.** 1 **55.** 1 **57.** 1 **59.** 18
61. Not defined **63.** Not defined **65.** DW
67. 34,560 **68.** 34,600 **69.** 35,000 **70.** 30,000
71. \$15,691 **72.** 96 gal **73.** 2203

74. 848 **75.** 37,239 **76.** 11,851 **77.** $\frac{1}{6}$ **79.** $\frac{2}{16}$, or $\frac{1}{8}$
81. **83.**

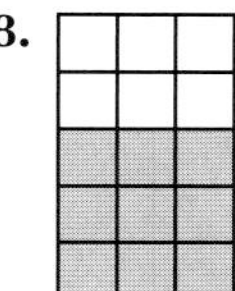

Margin Exercises, Section 2.4, pp. 115–119

1. $\frac{2}{3}$ **2.** $\frac{5}{8}$ **3.** $\frac{10}{3}$ **4.** $\frac{33}{8}$ **5.** $\frac{46}{5}$
6.

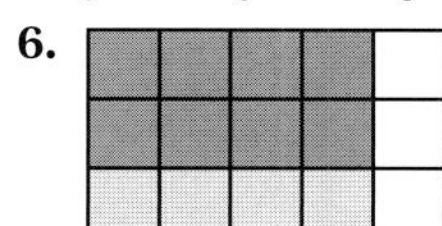

7. $\frac{15}{56}$ **8.** $\frac{32}{15}$ **9.** $\frac{3}{100}$ **10.** $\frac{14}{3}$
11. $\frac{3}{8}$ **12.** $\frac{8}{81}\text{ ft}^2$ **13.** $\frac{3}{40}$

Exercise Set 2.4, p. 120

1. $\frac{3}{5}$ **3.** $\frac{5}{8}$ **5.** $\frac{8}{11}$ **7.** $\frac{70}{9}$ **9.** $\frac{2}{5}$ **11.** $\frac{6}{5}$ **13.** $\frac{21}{4}$
15. $\frac{85}{6}$ **17.** $\frac{1}{6}$ **19.** $\frac{1}{40}$ **21.** $\frac{2}{15}$ **23.** $\frac{4}{15}$ **25.** $\frac{9}{16}$ **27.** $\frac{14}{39}$
29. $\frac{7}{100}$ **31.** $\frac{49}{64}$ **33.** $\frac{1}{1000}$ **35.** $\frac{182}{285}$ **37.** $\frac{3}{8}$ cup **39.** $\frac{21}{32}$
41. $\frac{12}{25}\text{ m}^2$ **43.** $\frac{7}{16}$ L **45.** DW **47.** 204 **48.** 700
49. 3001 **50.** 204 R 8 **51.** 8 thousands
52. 8 millions **53.** 8 ones **54.** 8 hundreds **55.** 3
56. 81 **57.** 50 **58.** 6399 **59.** $\frac{71,269}{180,433}$ **61.** $\frac{56}{1125}$

Margin Exercises, Section 2.5, pp. 122–126

1. $\frac{8}{16}$ **2.** $\frac{30}{50}$ **3.** $\frac{52}{100}$ **4.** $\frac{200}{75}$ **5.** $\frac{12}{9}$ **6.** $\frac{18}{24}$ **7.** $\frac{90}{100}$
8. $\frac{9}{45}$ **9.** $\frac{56}{49}$ **10.** $\frac{1}{4}$ **11.** $\frac{5}{6}$ **12.** 5 **13.** $\frac{4}{3}$ **14.** $\frac{7}{8}$
15. $\frac{89}{78}$ **16.** $\frac{8}{7}$ **17.** $\frac{1}{4}$ **18.** $\frac{7}{24}$ **19.** $\frac{37}{80}$
20. $\frac{20}{100} = \frac{1}{5}$; $\frac{42}{100} = \frac{21}{50}$; $\frac{8}{100} = \frac{2}{25}$; $\frac{16}{100} = \frac{4}{25}$; $\frac{14}{100} = \frac{7}{50}$
21. $=$ **22.** $\neq$

Calculator Corner, p. 125

1. $\frac{14}{15}$ **2.** $\frac{7}{8}$ **3.** $\frac{138}{167}$ **4.** $\frac{7}{25}$

Exercise Set 2.5, p. 127

1. $\frac{5}{10}$ **3.** $\frac{20}{32}$ **5.** $\frac{27}{30}$ **7.** $\frac{28}{32}$ **9.** $\frac{20}{48}$ **11.** $\frac{51}{54}$ **13.** $\frac{75}{45}$
15. $\frac{42}{132}$ **17.** $\frac{1}{2}$ **19.** $\frac{3}{4}$ **21.** $\frac{1}{5}$ **23.** 3 **25.** $\frac{3}{4}$ **27.** $\frac{7}{8}$
29. $\frac{6}{5}$ **31.** $\frac{1}{3}$ **33.** 6 **35.** $\frac{1}{3}$ **37.** $\frac{4}{75}$ **39.** $\frac{45}{112}$ **41.** $=$
43. $\neq$ **45.** $=$ **47.** $\neq$ **49.** $=$ **51.** $\neq$ **53.** $=$
55. $\neq$ **57.** DW **59.** 4992 ft^2; 284 ft **60.** $928
61. 11 **62.** 32 **63.** 186 **64.** 2737 **65.** 5 **66.** 89
67. 3520 **68.** 9001 **69.** $\frac{137}{149}$ **71.** $\frac{2}{5}$; $\frac{3}{5}$
73. No. $\frac{262}{704} \neq \frac{135}{373}$ because $262 \cdot 373 \neq 704 \cdot 135$.

Margin Exercises, Section 2.6, p. 130

1. $\frac{7}{12}$ **2.** $\frac{1}{3}$ **3.** 6 **4.** $\frac{5}{2}$ **5.** 14 lb

Exercise Set 2.6, p. 131

1. $\frac{1}{3}$ **3.** $\frac{1}{8}$ **5.** $\frac{1}{10}$ **7.** $\frac{1}{6}$ **9.** $\frac{27}{10}$ **11.** $\frac{14}{9}$ **13.** 1
15. 1 **17.** 1 **19.** 1 **21.** 2 **23.** 4 **25.** 9 **27.** 9
29. $\frac{26}{5}$ **31.** $\frac{98}{5}$ **33.** 60 **35.** 30 **37.** $\frac{1}{5}$ **39.** $\frac{9}{25}$
41. $\frac{11}{40}$ **43.** $\frac{5}{14}$ **45.** $\frac{5}{8}$ in. **47.** 30 mph
49. 625 addresses **51.** $\frac{1}{3}$ cup **53.** $115,500
55. 160 mi **57.** Food: $9000; housing: $7200; clothing: $3600; savings: $4000; taxes: $9000; other expenses: $3200
59. DW **61.** 35 **62.** 85 **63.** 125 **64.** 120
65. 4989 **66.** 8546 **67.** 6498 **68.** 6407 **69.** 4673
70. 5338 **71.** $\frac{129}{485}$ **73.** $\frac{1}{12}$ **75.** $\frac{1}{168}$

Margin Exercises, Section 2.7, pp. 135–139

1. $\frac{5}{2}$ **2.** $\frac{7}{10}$ **3.** $\frac{1}{9}$ **4.** 5 **5.** $\frac{8}{7}$ **6.** $\frac{8}{3}$ **7.** $\frac{1}{10}$ **8.** 100
9. 1 **10.** $\frac{14}{15}$ **11.** $\frac{4}{5}$ **12.** 32 **13.** 320 loops
14. 252 mi

Translating for Success, p. 140

1. C **2.** H **3.** A **4.** N **5.** O **6.** F **7.** I **8.** L
9. D **10.** M

Exercise Set 2.7, p. 141

1. $\frac{6}{5}$ **3.** $\frac{1}{6}$ **5.** 6 **7.** $\frac{3}{10}$ **9.** $\frac{4}{5}$ **11.** $\frac{4}{15}$ **13.** 4 **15.** 2
17. $\frac{1}{8}$ **19.** $\frac{3}{7}$ **21.** 8 **23.** 35 **25.** 1 **27.** $\frac{2}{3}$ **29.** $\frac{9}{4}$
31. 144 **33.** 75 **35.** 2 **37.** $\frac{3}{5}$ **39.** 315
41. 75 times **43.** 32 pairs **45.** 24 bowls **47.** 16 L
49. 288 km; 108 km **51.** $\frac{1}{16}$ in. **53.** DW
55. associative **56.** factors **57.** prime
58. denominator **59.** additive **60.** reciprocals
61. whole **62.** equation **63.** $\frac{9}{19}$ **65.** 36 **67.** $\frac{3}{8}$

Concept Reinforcement, p. 144

1. True **2.** False **3.** True **4.** True **5.** False
6. True **7.** True **8.** False

Summary and Review: Chapter 2, p. 144

1. 1, 2, 3, 4, 5, 6, 10, 12, 15, 20, 30, 60
2. 1, 2, 4, 8, 11, 16, 22, 44, 88, 176
3. 8, 16, 24, 32, 40, 48, 56, 64, 72, 80
4. Yes **5.** No **6.** Prime **7.** Neither **8.** Composite
9. $2 \cdot 5 \cdot 7$ **10.** $2 \cdot 3 \cdot 5$ **11.** $3 \cdot 3 \cdot 5$ **12.** $2 \cdot 3 \cdot 5 \cdot 5$
13. $2 \cdot 2 \cdot 2 \cdot 3 \cdot 3 \cdot 3 \cdot 3$ **14.** $2 \cdot 3 \cdot 5 \cdot 5 \cdot 5 \cdot 7$
15. 4344, 600, 93, 330, 255,555, 780, 2802, 711
16. 140, 182, 716, 2432, 4344, 600, 330, 780, 2802
17. 140, 716, 2432, 4344, 600, 780 **18.** 2432, 4344, 600
19. 140, 95, 475, 600, 330, 255,555, 780
20. 4344, 600, 330, 780, 2802 **21.** 255,555, 711
22. 140, 600, 330, 780 **23.** Numerator: 2; denominator: 7
24. $\frac{3}{5}$ **25.** $\frac{7}{6}$ **26.** $\frac{2}{7}$ **27.** (a) $\frac{3}{5}$; (b) $\frac{5}{3}$; (c) $\frac{3}{8}$ **28.** 0
29. 1 **30.** 48 **31.** 6 **32.** $\frac{2}{3}$ **33.** $\frac{1}{4}$ **34.** 1 **35.** 0
36. $\frac{2}{5}$ **37.** 18 **38.** 4 **39.** $\frac{1}{3}$ **40.** Not defined

41. Not defined **42.** $\frac{11}{23}$ **43.** $\frac{2}{7}$ **44.** $\frac{39}{40}$ **45.** $\frac{32}{225}$
46. $\frac{3}{100}$; $\frac{8}{100} = \frac{2}{25}$; $\frac{10}{100} = \frac{1}{10}$; $\frac{15}{100} = \frac{3}{20}$; $\frac{21}{100}$; $\frac{43}{100}$ **47.** $\neq$ **48.** $=$
49. $\neq$ **50.** $=$ **51.** $\frac{3}{2}$ **52.** 56 **53.** $\frac{5}{2}$ **54.** 24 **55.** $\frac{2}{3}$
56. $\frac{1}{14}$ **57.** $\frac{2}{3}$ **58.** $\frac{1}{22}$ **59.** $\frac{3}{20}$ **60.** $\frac{10}{7}$ **61.** $\frac{5}{4}$ **62.** $\frac{1}{3}$
63. 9 **64.** $\frac{36}{47}$ **65.** $\frac{9}{2}$ **66.** 2 **67.** $\frac{11}{6}$ **68.** $\frac{1}{4}$ **69.** $\frac{9}{4}$
70. 300 **71.** 1 **72.** $\frac{4}{9}$ **73.** $\frac{3}{10}$ **74.** 240 **75.** 9 days
76. 1000 km **77.** $\frac{1}{3}$ cup; 2 cups **78.** \$15 **79.** 60 bags
80. 256,000,000 metric tons
81. D_W To simplify fraction notation, first factor the numerator and the denominator into prime numbers. Examine the factorizations for factors common to both the numerator and the denominator. Factor the fraction, with each pair of like factors forming a factor of 1. Remove the factors of 1, and multiply the remaining factors in the numerator and in the denominator, if necessary.
82. D_W Taking $\frac{1}{2}$ of a number is equivalent to multiplying the number by $\frac{1}{2}$. Dividing by $\frac{1}{2}$ is equivalent to multiplying by the reciprocal of $\frac{1}{2}$, or 2. Thus taking $\frac{1}{2}$ of a number is not the same as dividing by $\frac{1}{2}$.
83. D_W $9432 = 9 \cdot 1000 + 4 \cdot 100 + 3 \cdot 10 + 2 \cdot 1 = 9(999 + 1) + 4(99 + 1) + 3(9 + 1) + 2 \cdot 1 = 9 \cdot 999 + 9 \cdot 1 + 4 \cdot 99 + 4 \cdot 1 + 3 \cdot 9 + 3 \cdot 1 + 2 \cdot 1$. Since 999, 99, and 9 are each a multiple of 9, $9 \cdot 999$, $4 \cdot 99$, and $3 \cdot 9$ are multiples of 9. This leaves $9 \cdot 1 + 4 \cdot 1 + 3 \cdot 1 + 2 \cdot 1$, or $9 + 4 + 3 + 2$. If $9 + 4 + 3 + 2$, the sum of the digits, is divisible by 9, then 9432 is divisible by 9.
84. $a = 11{,}176$; $b = 9887$ **85.** 13, 11, 101, 37

Test: Chapter 2, p. 147

1. [2.1a] 1, 2, 3, 4, 5, 6, 10, 12, 15, 20, 25, 30, 50, 60, 75, 100, 150, 300 **2.** [2.1c] Prime **3.** [2.1c] Composite
4. [2.1d] $2 \cdot 3 \cdot 3$ **5.** [2.1d] $2 \cdot 2 \cdot 3 \cdot 5$ **6.** [2.2a] Yes
7. [2.2a] No **8.** [2.2a] No **9.** [2.2a] Yes
10. [2.3a] Numerator: 4; denominator: 5 **11.** [2.3a] $\frac{3}{4}$
12. [2.3a] $\frac{3}{7}$ **13.** [2.3a] **(a)** $\frac{336}{497}$; **(b)** $\frac{161}{497}$ **14.** [2.3b] 26
15. [2.3b] 1 **16.** [2.3b] 0 **17.** [2.5b] $\frac{1}{2}$ **18.** [2.5b] 6
19. [2.5b] $\frac{1}{14}$ **20.** [2.3b] Not defined
21. [2.3b] Not defined **22.** [2.5b] $\frac{1}{4}$ **23.** [2.5b] $\frac{2}{3}$
24. [2.5c] $=$ **25.** [2.5c] $\neq$ **26.** [2.6a] 32 **27.** [2.6a] $\frac{3}{2}$
28. [2.6a] $\frac{5}{2}$ **29.** [2.6a] $\frac{1}{10}$ **30.** [2.6a] $\frac{2}{9}$ **31.** [2.7a] $\frac{8}{5}$
32. [2.7a] 4 **33.** [2.7a] $\frac{1}{18}$ **34.** [2.7b] $\frac{3}{10}$ **35.** [2.7b] $\frac{8}{5}$
36. [2.7b] 18 **37.** [2.7b] $\frac{18}{7}$ **38.** [2.7c] 64 **39.** [2.7c] $\frac{7}{4}$
40. [2.6b] 4375 students **41.** [2.7d] $\frac{3}{40}$ m **42.** [2.7d] 5 qt
43. [2.6b] $\frac{3}{4}$ in. **44.** [2.6b] $\frac{15}{8}$ tsp **45.** [2.6b] $\frac{7}{48}$ acre
46. [2.6a], [2.7b] $\frac{7}{960}$ **47.** [2.7c] $\frac{7}{5}$

CHAPTER 3

Margin Exercises, Section 3.1, pp. 150–155

1. 45 **2.** 40 **3.** 30 **4.** 24 **5.** 10 **6.** 80 **7.** 40
8. 360 **9.** 864 **10.** 2520 **11.** $2^3 \cdot 3^2 \cdot 5 \cdot 7$, or 2520
12. $2^3 \cdot 5$, or 40; $2^3 \cdot 3^2 \cdot 5$, or 360; $2^5 \cdot 3^3$, or 864 **13.** 18
14. 24 **15.** 36 **16.** 210 **17.** 600 **18.** 3780
19. 600

Exercise Set 3.1, p. 156

1. 4 **3.** 50 **5.** 40 **7.** 54 **9.** 150 **11.** 120
13. 72 **15.** 420 **17.** 144 **19.** 288 **21.** 30
23. 105 **25.** 72 **27.** 60 **29.** 36 **31.** 900 **33.** 48
35. 50 **37.** 143 **39.** 420 **41.** 378 **43.** 810
45. 2160 **47.** 9828 **49.** Every 60 yr **51.** Every 420 yr
53. D_W **55.** 90 days **56.** 704 performances
57. 33,135 **58.** 6939 **59.** $\frac{2}{3}$ **60.** $\frac{8}{7}$ **61.** 18,900
63. 5 in. by 24 in.

Margin Exercises, Section 3.2, pp. 158–161

1. $\frac{4}{5}$ **2.** 1 **3.** $\frac{1}{2}$ **4.** $\frac{3}{4}$ **5.** $\frac{5}{6}$ **6.** $\frac{29}{24}$ **7.** $\frac{5}{9}$ **8.** $\frac{413}{1000}$
9. $\frac{197}{210}$ **10.** $\frac{65}{72}$ **11.** $\frac{9}{10}$ mi

Exercise Set 3.2, p. 162

1. 1 **3.** $\frac{3}{4}$ **5.** $\frac{3}{2}$ **7.** $\frac{7}{24}$ **9.** $\frac{3}{2}$ **11.** $\frac{19}{24}$ **13.** $\frac{9}{10}$
15. $\frac{29}{18}$ **17.** $\frac{31}{100}$ **19.** $\frac{41}{60}$ **21.** $\frac{189}{100}$ **23.** $\frac{7}{8}$ **25.** $\frac{13}{24}$
27. $\frac{17}{24}$ **29.** $\frac{3}{4}$ **31.** $\frac{437}{500}$ **33.** $\frac{53}{40}$ **35.** $\frac{391}{144}$ **37.** $\frac{5}{6}$ lb
39. $\frac{23}{12}$ mi **41.** 690 kg; $\frac{14}{23}$ cement; $\frac{5}{23}$ stone; $\frac{4}{23}$ sand; 1
43. $\frac{5}{8}$" **45.** $\frac{13}{12}$ lb **47.** D_W **49.** 210,528
50. 4,194,000 **51.** 3,387,807 **52.** 352,350
53. 537,179 **54.** 5 **55.** 84 **56.** 510,314
57. 21,468,755 **58.** 33,112,603 **59.** 12 tickets
60. $\frac{3}{64}$ acre **61.** $\frac{4}{15}$; \$320

Margin Exercises, Section 3.3, pp. 165–168

1. $\frac{1}{2}$ **2.** $\frac{3}{8}$ **3.** $\frac{1}{2}$ **4.** $\frac{1}{12}$ **5.** $\frac{13}{18}$ **6.** $\frac{1}{2}$ **7.** $\frac{9}{112}$ **8.** $<$
9. $>$ **10.** $>$ **11.** $>$ **12.** $<$ **13.** $\frac{1}{6}$ **14.** $\frac{11}{40}$
15. $\frac{5}{24}$ mi

Translating for Success, p. 169

1. J **2.** E **3.** D **4.** B **5.** I **6.** N **7.** A **8.** C
9. L **10.** F

Exercise Set 3.3, p. 170

1. $\frac{2}{3}$ **3.** $\frac{3}{4}$ **5.** $\frac{5}{8}$ **7.** $\frac{1}{24}$ **9.** $\frac{1}{2}$ **11.** $\frac{9}{14}$ **13.** $\frac{3}{5}$ **15.** $\frac{7}{10}$
17. $\frac{17}{60}$ **19.** $\frac{53}{100}$ **21.** $\frac{26}{75}$ **23.** $\frac{9}{100}$ **25.** $\frac{13}{24}$ **27.** $\frac{1}{10}$
29. $\frac{1}{24}$ **31.** $\frac{13}{16}$ **33.** $\frac{31}{75}$ **35.** $\frac{13}{75}$ **37.** $<$ **39.** $>$
41. $<$ **43.** $<$ **45.** $>$ **47.** $>$ **49.** $<$ **51.** $\frac{1}{15}$
53. $\frac{2}{15}$ **55.** $\frac{1}{2}$ **57.** $\frac{5}{12}$ hr **59.** $\frac{19}{24}$ cup
61. $\frac{1}{4}$ of the business **63.** $\frac{11}{20}$ lb **65.** D_W **67.** 1
68. Not defined **69.** Not defined **70.** 4 **71.** $\frac{4}{21}$
72. $\frac{3}{2}$ **73.** 21 **74.** $\frac{1}{32}$ **75.** 17 days **76.** 9 cups
77. $\frac{14}{3553}$ **79.** $\frac{21}{40}$ km **81.** $\frac{19}{24}$ **83.** $\frac{145}{144}$ **85.** $>$
87. *Day 1*: Cut off $\frac{1}{7}$ of bar and pay him. *Day 2*: Cut off $\frac{2}{7}$ of the bar. Trade him for the $\frac{1}{7}$. *Day 3*: Give him back the $\frac{1}{7}$.

Day 4: Trade him the $\frac{4}{7}$ for his $\frac{3}{7}$. *Day 5*: Give him the $\frac{1}{7}$ again. *Day 6*: Trade him the $\frac{2}{7}$ for the $\frac{1}{7}$. *Day 7*: Give him the $\frac{1}{7}$ again. This assumes that he does not spend parts of the gold bar immediately.

Margin Exercises, Section 3.4, pp. 173–176

1. $1\frac{2}{3}$ **2.** $2\frac{3}{4}$ **3.** $8\frac{3}{4}$ **4.** $12\frac{2}{3}$ **5.** $\frac{22}{5}$ **6.** $\frac{61}{10}$ **7.** $\frac{29}{6}$ **8.** $\frac{37}{4}$ **9.** $\frac{62}{3}$ **10.** $2\frac{1}{3}$ **11.** $1\frac{1}{10}$ **12.** $18\frac{1}{3}$ **13.** $807\frac{2}{3}$ **14.** $134\frac{23}{45}$

Calculator Corner, p. 176

1. $1476\frac{1}{6}$ **2.** $676\frac{4}{9}$ **3.** $800\frac{51}{56}$ **4.** $13{,}031\frac{1}{2}$ **5.** $51{,}626\frac{9}{11}$ **6.** $7330\frac{7}{32}$ **7.** $134\frac{1}{15}$ **8.** $2666\frac{130}{213}$ **9.** $3571\frac{51}{112}$ **10.** $12\frac{169}{454}$

Exercise Set 3.4, p. 177

1. $\frac{57}{4}, \frac{27}{4}, \frac{9}{4}$ **3.** $3\frac{5}{8}, 2\frac{3}{4}$ **5.** $\frac{17}{3}$ **7.** $\frac{13}{4}$ **9.** $\frac{81}{8}$ **11.** $\frac{51}{10}$ **13.** $\frac{103}{5}$ **15.** $\frac{59}{6}$ **17.** $\frac{73}{10}$ **19.** $\frac{13}{8}$ **21.** $\frac{51}{4}$ **23.** $\frac{43}{10}$ **25.** $\frac{203}{100}$ **27.** $\frac{200}{3}$ **29.** $\frac{279}{50}$ **31.** $3\frac{3}{5}$ **33.** $4\frac{2}{3}$ **35.** $4\frac{1}{2}$ **37.** $5\frac{7}{10}$ **39.** $7\frac{4}{7}$ **41.** $7\frac{1}{2}$ **43.** $11\frac{1}{2}$ **45.** $1\frac{1}{2}$ **47.** $7\frac{57}{100}$ **49.** $43\frac{1}{8}$ **51.** $108\frac{5}{8}$ **53.** $618\frac{1}{5}$ **55.** $40\frac{4}{7}$ **57.** $55\frac{1}{51}$ **59.** DW **61.** 45,800 **62.** 45,770 **63.** $\frac{8}{15}$ **64.** $\frac{21}{25}$ **65.** $\frac{16}{27}$ **66.** $\frac{583}{669}$ **67.** 18 **68.** $\frac{5}{2}$ **69.** $\frac{1}{4}$ **70.** $\frac{2}{5}$ **71.** 24 **72.** 49 **73.** $\frac{2560}{3}$ **74.** $\frac{4}{3}$ **75.** $237\frac{19}{541}$ **77.** $8\frac{2}{3}$ **79.** $52\frac{2}{7}$

Margin Exercises, Section 3.5, pp. 180–184

1. $7\frac{2}{5}$ **2.** $12\frac{1}{10}$ **3.** $13\frac{7}{12}$ **4.** $1\frac{1}{2}$ **5.** $3\frac{1}{6}$ **6.** $3\frac{5}{18}$ **7.** $3\frac{2}{3}$ **8.** $232\frac{3}{20}$ mi **9.** $4\frac{5}{6}$ ft **10.** $354\frac{23}{24}$ gal

Exercise Set 3.5, p. 185

1. $28\frac{3}{4}$ **3.** $185\frac{7}{8}$ **5.** $6\frac{1}{2}$ **7.** $2\frac{11}{12}$ **9.** $14\frac{7}{12}$ **11.** $12\frac{1}{10}$ **13.** $16\frac{5}{24}$ **15.** $21\frac{1}{2}$ **17.** $27\frac{7}{8}$ **19.** $27\frac{13}{24}$ **21.** $1\frac{3}{5}$ **23.** $4\frac{1}{10}$ **25.** $21\frac{17}{24}$ **27.** $12\frac{1}{4}$ **29.** $15\frac{3}{8}$ **31.** $7\frac{5}{12}$ **33.** $13\frac{3}{8}$ **35.** $11\frac{5}{18}$ **37.** $5\frac{3}{8}$ yd **39.** $7\frac{5}{12}$ lb **41.** $6\frac{5}{12}$ in. **43.** $2\frac{7}{8}$ in., $4\frac{7}{8}$ in. **45.** $19\frac{1}{16}$ in. **47.** $95\frac{1}{5}$ mi **49.** $36\frac{1}{2}$ in. **51.** $20\frac{1}{8}$ in. **53.** $78\frac{1}{12}$ in. **55.** $3\frac{4}{5}$ hr **57.** $28\frac{3}{4}$ yd **59.** $7\frac{3}{8}$ ft **61.** $1\frac{9}{16}$ in. **63.** DW **65.** 16 packages **66.** 286 cartons; 2 oz left over **67.** Yes **68.** No **69.** No **70.** Yes **71.** No **72.** Yes **73.** Yes **74.** Yes **75.** $\frac{10}{13}$ **76.** $\frac{1}{10}$ **77.** $8568\frac{786}{1189}$ **79.** $5\frac{3}{4}$ ft

Margin Exercises, Section 3.6, pp. 190–193

1. 20 **2.** $1\frac{7}{8}$ **3.** $12\frac{4}{5}$ **4.** $8\frac{1}{3}$ **5.** 16 **6.** $7\frac{3}{7}$ **7.** $1\frac{7}{8}$ **8.** $\frac{7}{10}$ **9.** $227\frac{1}{2}$ mi **10.** 20 mpg **11.** $240\frac{3}{4}$ ft^2

Calculator Corner, p. 194

1. $\frac{7}{12}$ **2.** $\frac{11}{10}$ **3.** $\frac{35}{16}$ **4.** $\frac{3}{10}$ **5.** $10\frac{2}{15}$ **6.** $1\frac{1}{28}$ **7.** $10\frac{11}{15}$ **8.** $2\frac{91}{115}$

Translating for Success, p. 195

1. O **2.** K **3.** F **4.** D **5.** H **6.** G **7.** L **8.** E **9.** M **10.** J

Exercise Set 3.6, p. 196

1. $22\frac{2}{3}$ **3.** $2\frac{5}{12}$ **5.** $8\frac{1}{6}$ **7.** $9\frac{31}{40}$ **9.** $24\frac{91}{100}$ **11.** $975\frac{4}{5}$ **13.** $6\frac{1}{4}$ **15.** $1\frac{1}{5}$ **17.** $3\frac{9}{16}$ **19.** $1\frac{1}{8}$ **21.** $1\frac{8}{43}$ **23.** $\frac{9}{40}$ **25.** 45,000 beagles **27.** About 4,500,000 **29.** $62\frac{1}{2}$ ft^2 **31.** $13\frac{1}{3}$ tsp **33.** $343\frac{3}{4}$ lb **35.** 68°F **37.** About 1,800,750 **39.** *Cake*: $1\frac{1}{8}$ cups all-purpose flour, $\frac{3}{8}$ cup sugar, $\frac{3}{8}$ cup cold butter, $\frac{3}{8}$ cup sour cream, $\frac{1}{4}$ teaspoon baking powder, $\frac{1}{4}$ teaspoon baking soda, $\frac{1}{2}$ egg, $\frac{1}{2}$ teaspoon almond extract; *filling*: $\frac{1}{2}$ package (4 ounces) cream cheese, $\frac{1}{8}$ cup sugar, $\frac{1}{2}$ egg, $\frac{3}{8}$ cup peach preserves, $\frac{1}{4}$ cup sliced almonds; *cake*: 9 cups all-purpose flour, 3 cups sugar, 3 cups cold butter, 3 cups sour cream, 2 teaspoons baking powder, 2 teaspoons baking soda, 4 eggs, 4 teaspoons almond extract; *filling*: 4 packages (8 ounces each) cream cheese, 1 cup sugar 4 eggs, 3 cups peach preserves, 2 cups sliced almonds **41.** 15 mpg **43.** 4 cu ft **45.** $16\frac{1}{2}$ servings **47.** $35\frac{115}{256}$ sq in. **49.** $59{,}538\frac{1}{8}$ sq ft **51.** DW **53.** divisor, quotient, dividend **54.** common **55.** composite **56.** divisible, divisible **57.** multiplications, divisions, additions, subtractions **58.** addends **59.** numerator **60.** reciprocal **61.** $360\frac{60}{473}$ **63.** $35\frac{57}{64}$ **65.** $\frac{4}{9}$ **67.** $1\frac{4}{5}$

Margin Exercises, Section 3.7, pp. 201–204

1. $\frac{1}{2}$ **2.** $\frac{3}{10}$ **3.** $20\frac{2}{3}$, or $\frac{62}{3}$ **4.** $9\frac{7}{8}$ in. **5.** $\frac{5}{9}$ **6.** $\frac{31}{40}$ **7.** $\frac{27}{56}$ **8.** 0 **9.** 1 **10.** $\frac{1}{2}$ **11.** 1 **12.** 12; answers may vary **13.** 32; answers may vary **14.** 27; answers may vary **15.** 15; answers may vary **16.** 1; answers may vary **17.** 1,000,000; answers may vary **18.** $22\frac{1}{2}$ **19.** 132 **20.** 37

Exercise Set 3.7, p. 205

1. $\frac{1}{24}$ **3.** $\frac{2}{5}$ **5.** $\frac{4}{7}$ **7.** $\frac{59}{30}$, or $1\frac{29}{30}$ **9.** $\frac{3}{20}$ **11.** $\frac{211}{8}$, or $26\frac{3}{8}$ **13.** $\frac{7}{16}$ **15.** $\frac{1}{36}$ **17.** $\frac{3}{8}$ **19.** $\frac{37}{48}$ **21.** $\frac{25}{72}$ **23.** $\frac{103}{16}$, or $6\frac{7}{16}$ **25.** $2\frac{41}{128}$ lb **27.** $7\frac{23}{50}$ sec **29.** $\frac{17}{6}$, or $2\frac{5}{6}$ **31.** $\frac{8395}{84}$, or $99\frac{79}{84}$ **33.** 0 **35.** 0 **37.** $\frac{1}{2}$ **39.** $\frac{1}{2}$ **41.** 0 **43.** 1 **45.** 6 **47.** 12 **49.** 19 **51.** 6 **53.** 12 **55.** 16 **57.** 3 **59.** 13 **61.** 2 **63.** $1\frac{1}{2}$ **65.** $\frac{1}{2}$ **67.** $271\frac{1}{2}$ **69.** 3 **71.** 100 **73.** $29\frac{1}{2}$ **75.** DW **77.** 3402 **78.** 1,038,180 **79.** 59 R 77 **80.** 348 **81.** 783 **82.** $\frac{8}{3}$ **83.** $\frac{3}{8}$ **84.** Prime: 5, 7, 23, 43; composite: 9, 14; neither: 1 **85.** 16 people **86.** 43 mg

87. (a) $13 \cdot 9\frac{1}{4} + 8\frac{1}{4} \cdot 7\frac{1}{4}$; (b) $\frac{2881}{16}$, or $180\frac{1}{16}$ in^2; (c) Multiply before adding. **89.** $a = 2, b = 8$
91. The largest is $\frac{4}{3} + \frac{5}{2} = \frac{23}{6}$.

Concept Reinforcement, p. 209

1. True **2.** True **3.** False **4.** True **5.** True **6.** False

Summary and Review: Chapter 3, p. 209

1. 36 **2.** 90 **3.** 30 **4.** 1404 **5.** $\frac{63}{40}$ **6.** $\frac{19}{48}$ **7.** $\frac{25}{12}$ **8.** $\frac{891}{1000}$ **9.** $\frac{1}{3}$ **10.** $\frac{1}{8}$ **11.** $\frac{5}{27}$ **12.** $\frac{11}{18}$ **13.** $>$ **14.** $>$ **15.** $\frac{19}{40}$ **16.** $\frac{2}{5}$ **17.** $\frac{15}{2}$ **18.** $\frac{67}{8}$ **19.** $\frac{13}{3}$ **20.** $\frac{75}{7}$ **21.** $2\frac{1}{3}$ **22.** $6\frac{3}{4}$ **23.** $12\frac{3}{5}$ **24.** $3\frac{1}{2}$ **25.** $877\frac{1}{3}$ **26.** $456\frac{5}{23}$ **27.** $10\frac{2}{5}$ **28.** $11\frac{11}{15}$ **29.** $10\frac{2}{3}$ **30.** $8\frac{1}{4}$ **31.** $7\frac{7}{9}$ **32.** $4\frac{11}{15}$ **33.** $4\frac{3}{20}$ **34.** $13\frac{3}{8}$ **35.** 16 **36.** $3\frac{1}{2}$ **37.** $2\frac{21}{50}$ **38.** 6 **39.** 12 **40.** $1\frac{7}{17}$ **41.** $\frac{1}{8}$ **42.** $\frac{9}{10}$ **43.** $4\frac{1}{4}$ yd **44.** $177\frac{3}{4}$ in^2 **45.** $50\frac{1}{4}$ in^2
46. *Serving 2*: $\frac{1}{8}$ cup extra-virgin olive oil, $\frac{3}{4}$ pound fresh red snapper fillets, $\frac{1}{6}$ cup kalamata olives, $1\frac{1}{4}$ tablespoons capers, $\frac{1}{2}$ cup canned tomatoes, $1\frac{1}{2}$ tablespoons chopped shallots, $\frac{1}{4}$ tablespoon fresh rosemary leaves, $\frac{1}{4}$ tablespoon minced garlic, $\frac{1}{6}$ cup white wine; *serving 12*: $\frac{3}{4}$ cup extra-virgin olive oil, $4\frac{1}{2}$ pounds fresh red snapper fillets, 1 cup kalamata olives, $7\frac{1}{2}$ tablespoons capers, 3 cups canned tomatoes, 9 tablespoons chopped shallots, $1\frac{1}{2}$ tablespoons fresh rosemary leaves, $1\frac{1}{2}$ tablespoons minced garlic, 1 cup white wine
47. $1\frac{73}{100}$ in. **48.** 24 lb **49.** About $69\frac{3}{8}$ kg **50.** $8\frac{3}{8}$ cups **51.** $63\frac{2}{3}$ pies; $19\frac{1}{3}$ pies **52.** 1 **53.** $\frac{7}{40}$ **54.** 3 **55.** $\frac{77}{240}$ **56.** $\frac{1}{2}$ **57.** 0 **58.** 1 **59.** 7 **60.** 10 **61.** 2 **62.** $28\frac{1}{2}$
63. **DW** It might be necessary to find the least common denominator before adding or subtracting. The least common denominator is the least common multiple of the denominators.
64. **DW** Suppose that a room has dimensions $15\frac{3}{4}$ ft by $28\frac{5}{8}$ ft. The equation $2 \cdot 15\frac{3}{4} + 2 \cdot 28\frac{5}{8} = 88\frac{3}{4}$ gives the perimeter of the room, in feet. Answers may vary.
65. 12 min **66.** $\frac{6}{3} + \frac{5}{4} = 3\frac{1}{4}$

Test: Chapter 3, p. 212

1. [3.1a] 48 **2.** [3.1a] 600 **3.** [3.2a] 3 **4.** [3.2a] $\frac{37}{24}$ **5.** [3.2a] $\frac{921}{1000}$ **6.** [3.3a] $\frac{1}{3}$ **7.** [3.3a] $\frac{1}{12}$ **8.** [3.3a] $\frac{77}{120}$ **9.** [3.3c] $\frac{15}{4}$ **10.** [3.3c] $\frac{1}{4}$ **11.** [3.3b] $>$ **12.** [3.4a] $\frac{7}{2}$ **13.** [3.4a] $\frac{79}{8}$ **14.** [3.4a] $4\frac{1}{2}$ **15.** [3.4a] $8\frac{2}{9}$ **16.** [3.4b] $162\frac{7}{11}$ **17.** [3.5a] $14\frac{1}{5}$ **18.** [3.5a] $14\frac{5}{12}$ **19.** [3.5b] $4\frac{7}{24}$ **20.** [3.5b] $6\frac{1}{6}$ **21.** [3.6a] 39 **22.** [3.6a] $4\frac{1}{2}$ **23.** [3.6b] 2 **24.** [3.6b] $\frac{1}{36}$ **25.** [3.6c] About 105 kg **26.** [3.6c] 80 books **27.** [3.5c] (a) 3 in.; (b) $4\frac{1}{2}$ in. **28.** [3.3d] $\frac{1}{16}$ in. **29.** [3.7a] $6\frac{11}{36}$ ft **30.** [3.7a] $3\frac{1}{2}$ **31.** [3.7a] $\frac{3}{4}$ **32.** [3.7b] 0 **33.** [3.7b] 1 **34.** [3.7b] 4 **35.** [3.7b] $18\frac{1}{2}$ **36.** [3.7b] 16 **37.** [3.7b] $1214\frac{1}{2}$
38. [3.1a] (a) 24, 48, 72; (b) 24
39. [3.3b], [3.5c] Rebecca walks $\frac{17}{56}$ mi farther.

Cumulative Review: Chapters 1–3, p. 214

1. [3.3d] (a) $\frac{1}{48}$ in.; (b) $\frac{1}{12}$ in. **2.** [2.7d] 61
3. (a) [3.5c] $14\frac{13}{24}$ mi; (b) [3.6c] $4\frac{61}{72}$ mi
4. [3.6c], [3.5c] (a) $142\frac{1}{4}$ ft^2; (b) 54 ft **5.** [1.8a] 31 people
6. [1.8a] \$108 **7.** [2.6b] $\frac{2}{5}$ tsp; 4 tsp **8.** [3.6c] 39 lb
9. [3.6c] 16 pieces **10.** [3.2b] $\frac{33}{20}$ mi **11.** [1.1a] 5
12. [1.1b] 6 thousands + 7 tens + 5 ones
13. [1.1c] Twenty-nine thousand, five hundred
14. [2.3a] $\frac{5}{16}$ **15.** [1.2a] 899 **16.** [1.2a] 8982
17. [3.2a] $\frac{5}{12}$ **18.** [3.5a] $8\frac{1}{4}$ **19.** [1.3b] 5124
20. [1.3b] 4518 **21.** [3.3a] $\frac{5}{12}$ **22.** [3.5b] $1\frac{1}{6}$
23. [1.5a] 5004 **24.** [1.5a] 293,232 **25.** [2.6a] $\frac{3}{2}$
26. [2.6a] 15 **27.** [3.6a] $7\frac{1}{3}$ **28.** [1.6b] 715
29. [1.6b] 56 R 11 **30.** [3.4b] $56\frac{11}{45}$
31. [3.7a] $\frac{1377}{100}$, or $13\frac{77}{100}$ **32.** [2.7b] $\frac{4}{7}$ **33.** [3.6b] $7\frac{1}{3}$
34. [1.4a] 38,500 **35.** [3.1a] 72 **36.** [2.2a] No
37. [2.1a] 1, 2, 4, 8, 16 **38.** [3.3b] $>$ **39.** [2.5c] $=$
40. [3.3b] $<$ **41.** [2.5b] $\frac{4}{5}$ **42.** [2.3b] 0 **43.** [2.5b] 32
44. [3.4a] $\frac{37}{8}$ **45.** [3.4a] $5\frac{2}{3}$ **46.** [1.7b] 93 **47.** [3.3c] $\frac{5}{9}$
48. [2.7c] $\frac{12}{7}$ **49.** [1.7b] 905 **50.** [3.7b] 1 **51.** [3.7b] $\frac{1}{2}$
52. [3.7b] 0 **53.** [3.7b] 30 **54.** [3.7b] 1 **55.** [3.7b] 42
56. [2.1a, b, c, d], [2.2a]
Factors of 68: 1, 2, 4, 17, 34, 68
Factorization of 68: $2 \cdot 34$, or $2 \cdot 2 \cdot 17$
Prime factorization of 68: $2 \cdot 2 \cdot 17$
Numbers divisible by 6: 12, 54, 72, 300
Numbers divisible by 8: 8, 16, 24, 32, 40, 48, 64, 864
Numbers divisible by 5: 70, 95, 215
Prime numbers: 2, 3, 17, 19, 23, 31, 47, 101
57. [2.6b] (d) **58.** [2.7d] (a) **59.** [3.6c] (a)
60. [3.6c] (a) **61.** [3.2a] (a) $\frac{1}{2}, \frac{2}{3}, \frac{3}{4}, \frac{4}{5}$; (b) $\frac{9}{10}$
62. [2.1c] 2003

CHAPTER 4

Margin Exercises, Section 4.1, pp. 219–224

1. Eighty and thirty-one hundredths; seventy-seven and forty-three hundredths **2.** Two and six thousand seven hundred sixty-seven hundred-thousandths
3. Two hundred forty-five and eighty-nine hundredths
4. Thirty-four and sixty-four ten-thousandths
5. Thirty-one thousand, seventy-nine and seven hundred sixty-four thousandths
6. $\frac{896}{1000}$ **7.** $\frac{2378}{100}$ **8.** $\frac{56{,}789}{10{,}000}$ **9.** $\frac{19}{10}$ **10.** 7.43 **11.** 0.406 **12.** 6.7089 **13.** 0.9 **14.** 0.057 **15.** 0.083 **16.** 4.3 **17.** 283.71 **18.** 456.013 **19.** 2.04 **20.** 0.06 **21.** 0.58 **22.** 1 **23.** 0.8989 **24.** 21.05

25. 2.8 **26.** 13.9 **27.** 234.4 **28.** 7.0 **29.** 0.64 **30.** 7.83 **31.** 34.68 **32.** 0.03 **33.** 0.943 **34.** 8.004 **35.** 43.112 **36.** 37.401 **37.** 7459.355 **38.** 7459.35 **39.** 7459.4 **40.** 7459 **41.** 7460 **42.** 7500 **43.** 7000

Exercise Set 4.1, p. 225

1. Sixty-three and five hundredths **3.** Twenty-six and fifty-nine hundredths **5.** Eight and thirty-five hundredths **7.** Eighty-six and eighty-nine hundredths **9.** Thirty-four and eight hundred ninety-one thousandths **11.** $\frac{83}{10}$ **13.** $\frac{356}{100}$ **15.** $\frac{4603}{100}$ **17.** $\frac{13}{100{,}000}$ **19.** $\frac{10{,}008}{10{,}000}$ **21.** $\frac{20{,}003}{1000}$ **23.** 0.8 **25.** 8.89 **27.** 3.798 **29.** 0.0078 **31.** 0.00019 **33.** 0.376193 **35.** 99.44 **37.** 3.798 **39.** 2.1739 **41.** 8.953073 **43.** 0.58 **45.** 0.91 **47.** 0.001 **49.** 235.07 **51.** $\frac{4}{100}$ **53.** 0.4325 **55.** 0.1 **57.** 0.5 **59.** 2.7 **61.** 123.7 **63.** 0.89 **65.** 0.67 **67.** 1.00 **69.** 0.09 **71.** 0.325 **73.** 17.002 **75.** 10.101 **77.** 9.999 **79.** 800 **81.** 809.473 **83.** 809 **85.** 34.5439 **87.** 34.54 **89.** 35 **91.** D_W **93.** 6170 **94.** 6200 **95.** 6000 **96.** $2 \cdot 2 \cdot 2 \cdot 2 \cdot 5 \cdot 5 \cdot 5$, or $2^4 \cdot 5^3$ **97.** $2 \cdot 3 \cdot 3 \cdot 5 \cdot 17$, or $2 \cdot 3^2 \cdot 5 \cdot 17$ **98.** $2 \cdot 7 \cdot 11 \cdot 13$ **99.** $2 \cdot 2 \cdot 2 \cdot 7 \cdot 7 \cdot 11$, or $2^3 \cdot 7^2 \cdot 11$ **101.** 2.000001, 2.0119, 2.018, 2.0302, 2.1, 2.108, 2.109 **103.** 6.78346 **105.** 0.03030

Margin Exercises, Section 4.2, pp. 228–232

1. 10.917 **2.** 34.2079 **3.** 4.969 **4.** 3.5617 **5.** 9.40544 **6.** 912.67 **7.** 2514.773 **8.** 10.754 **9.** 0.339 **10.** 0.5345 **11.** 0.5172 **12.** 7.36992 **13.** 1194.22 **14.** 4.9911 **15.** 38.534 **16.** 14.164 **17.** 2133.5
18. The "balance forward" column should read:
$3078.92
2738.23
2659.67
2890.47
2877.33
2829.33
2868.91
2766.04
2697.45
2597.45

Calculator Corner, p. 230

1. 317.645 **2.** 506.553 **3.** 17.15 **4.** 49.08 **5.** 4.4 **6.** 33.83 **7.** 454.74 **8.** 0.99

Exercise Set 4.2, p. 233

1. 334.37 **3.** 1576.215 **5.** 132.560 **7.** 50.0248 **9.** 40.007 **11.** 977.955 **13.** 771.967 **15.** 8754.8221 **17.** 49.02 **19.** 85.921 **21.** 2.4975 **23.** 3.397 **25.** 8.85 **27.** 3.37 **29.** 1.045 **31.** 3.703 **33.** 0.9902 **35.** 99.66 **37.** 4.88 **39.** 0.994 **41.** 17.802 **43.** 51.13 **45.** 32.7386 **47.** 4.0622 **49.** 11.65 **51.** 384.68 **53.** 582.97 **55.** 15,335.3
57. The balance forward should read:
$ 9704.56
9677.12
10,677.12
10,553.17
10,429.15
10,416.72
12,916.72
12,778.94
12,797.82
9997.82

59. D_W **61.** 35,000 **62.** 34,000 **63.** $\frac{1}{6}$ **64.** $\frac{34}{45}$ **65.** 6166 **66.** 5366 **67.** $16\frac{1}{2}$ servings **68.** $60\frac{1}{5}$ mi **69.** 345.8

Margin Exercises, Section 4.3, pp. 237–240

1. 529.48 **2.** 5.0594 **3.** 34.2906 **4.** 0.348 **5.** 0.0348 **6.** 0.00348 **7.** 0.000348 **8.** 34.8 **9.** 348 **10.** 3480 **11.** 34,800 **12.** 3,700,000 **13.** 6,100,000,000 **14.** 2,700,000,000 **15.** 1569¢ **16.** 17¢ **17.** $0.35 **18.** $5.77

Calculator Corner, p. 239

1. 48.6 **2.** 6930.5 **3.** 142.803 **4.** 0.5076 **5.** 7916.4 **6.** 20.4153

Exercise Set 4.3, p. 241

1. 60.2 **3.** 6.72 **5.** 0.252 **7.** 0.522 **9.** 237.6 **11.** 583,686.852 **13.** 780 **15.** 8.923 **17.** 0.09768 **19.** 0.782 **21.** 521.6 **23.** 3.2472 **25.** 897.6 **27.** 322.07 **29.** 55.68 **31.** 3487.5 **33.** 50.0004 **35.** 114.42902 **37.** 13.284 **39.** 90.72 **41.** 0.0028728 **43.** 0.72523 **45.** 1.872115 **47.** 45,678 **49.** 2888¢ **51.** 66¢ **53.** $0.34 **55.** $34.45 **57.** 258,700,000,000 **59.** 748,900,000 **61.** D_W **63.** $11\frac{1}{5}$ **64.** $\frac{35}{72}$ **65.** $2\frac{7}{15}$ **66.** $7\frac{2}{15}$ **67.** 342 **68.** 87 **69.** 4566 **70.** 1257 **71.** 87 **72.** 1176 R 14 **73.** 10^{21} = 1 sextillion **75.** 10^{24} = 1 septillion

Margin Exercises, Section 4.4, pp. 244–250

1. 0.6 **2.** 1.5 **3.** 0.47 **4.** 0.32 **5.** 3.75 **6.** 0.25 **7.** **(a)** 375; **(b)** 15 **8.** 4.9 **9.** 12.8 **10.** 15.625 **11.** 12.78 **12.** 0.001278 **13.** 0.09847 **14.** 67.832 **15.** 0.78314 **16.** 1105.6 **17.** 0.04 **18.** 0.2426 **19.** 593.44 **20.** 5967.5 m

Calculator Corner, p. 247

1. 14.3 **2.** 2.56 **3.** 200 **4.** 0.75 **5.** 20 **6.** 0.064 **7.** 15.7 **8.** 75.8

Exercise Set 4.4, p. 251

1. 2.99 **3.** 23.78 **5.** 7.48 **7.** 7.2 **9.** 1.143 **11.** 4.041 **13.** 0.07 **15.** 70 **17.** 20 **19.** 0.4 **21.** 0.41 **23.** 8.5 **25.** 9.3 **27.** 0.625 **29.** 0.26 **31.** 15.625 **33.** 2.34 **35.** 0.47 **37.** 0.2134567 **39.** 21.34567 **41.** 1023.7 **43.** 9.3 **45.** 0.0090678 **47.** 45.6 **49.** 2107 **51.** 303.003 **53.** 446.208 **55.** 24.14 **57.** 13.0072 **59.** 19.3204 **61.** 473.188278 **63.** 10.49 **65.** 911.13 **67.** 205 **69.** $1288.36 **71.** $206.34 billion **73.** DW **75.** $\frac{6}{7}$ **76.** $\frac{7}{8}$ **77.** $\frac{19}{73}$ **78.** $\frac{19}{73}$ **79.** $2 \cdot 2 \cdot 3 \cdot 3 \cdot 19$, or $2^2 \cdot 3^2 \cdot 19$ **80.** $2 \cdot 3 \cdot 3 \cdot 3 \cdot 3$, or $2 \cdot 3^4$ **81.** $3 \cdot 3 \cdot 223$, or $3^2 \cdot 223$ **82.** $5 \cdot 401$ **83.** $15\frac{1}{8}$ **84.** $5\frac{7}{8}$ **85.** 6.254194585 **87.** 1000 **89.** 100

Margin Exercises, Section 4.5, pp. 255–259

1. 0.8 **2.** 0.45 **3.** 0.275 **4.** 1.32 **5.** 0.4 **6.** 0.375 **7.** $0.1\overline{6}$ **8.** $0.\overline{6}$ **9.** $0.\overline{45}$ **10.** $1.\overline{09}$ **11.** $0.\overline{428571}$ **12.** 0.7; 0.67; 0.667 **13.** 0.8; 0.81; 0.808 **14.** 6.2; 6.25; 6.245 **15.** 0.510 **16.** 24.2 mpg **17.** 12.1 million digital cameras **18.** 0.72 **19.** 0.552 **20.** 9.6575

Exercise Set 4.5, p. 260

1. 0.23 **3.** 0.6 **5.** 0.325 **7.** 0.2 **9.** 0.85 **11.** 0.375 **13.** 0.975 **15.** 0.52 **17.** 20.016 **19.** 0.25 **21.** 1.16 **23.** 1.1875 **25.** $0.2\overline{6}$ **27.** $0.\overline{3}$ **29.** $1.\overline{3}$ **31.** $1.1\overline{6}$ **33.** $0.\overline{571428}$ **35.** $0.91\overline{6}$ **37.** 0.3; 0.27; 0.267 **39.** 0.3; 0.33; 0.333 **41.** 1.3; 1.33; 1.333 **43.** 1.2; 1.17; 1.167 **45.** 0.6; 0.57; 0.571 **47.** 0.9; 0.92; 0.917 **49.** 0.2; 0.18; 0.182 **51.** 0.3; 0.28; 0.278 **53.** **(a)** 0.429; **(b)** 0.75; **(c)** 0.571; **(d)** 1.333 **55.** 15.8 mpg **57.** 17.8 mpg **59.** 15.2 mph **61.** $24.5625; $24.56 **63.** $3.734375; $3.73 **65.** $59.875; $59.88 **67.** 11.06 **69.** 8.4 **71.** $417.51\overline{6}$ **73.** 0 **75.** 2.8125 **77.** 0.20425 **79.** 317.14 **81.** 0.1825 **83.** 18 **85.** 2.736 **87.** DW **89.** 21 **90.** $238\frac{7}{8}$ **91.** 10 **92.** $\frac{43}{52}$ **93.** $50\frac{5}{24}$ **94.** $30\frac{7}{10}$ **95.** $1\frac{1}{2}$ **96.** $14\frac{13}{24}$ **97.** $1\frac{1}{24}$ cups **98.** $1\frac{33}{100}$ in. **99.** $0.\overline{142857}$ **101.** $0.\overline{428571}$ **103.** $0.\overline{714285}$ **105.** $0.\overline{1}$ **107.** $0.\overline{001}$

Margin Exercises, Section 4.6, pp. 264–266

1. (b) **2.** (a) **3.** (d) **4.** (b) **5.** (a) **6.** (d) **7.** (b) **8.** (c) **9.** (b) **10.** (b) **11.** (c) **12.** (a) **13.** (c) **14.** (c)

Exercise Set 4.6, p. 267

1. (d) **3.** (c) **5.** (a) **7.** (c) **9.** 1.6 **11.** 6 **13.** 60 **15.** 2.3 **17.** 180 **19.** (a) **21.** (c) **23.** (b) **25.** (b) **27.** $1800 \div 9 = 200$ posts; answers may vary **29.** DW **31.** repeating **32.** multiple **33.** distributive **34.** solution **35.** multiplicative **36.** commutative **37.** denominator; multiple **38.** divisible; divisible **39.** Yes **41.** No **43.** **(a)** +, ×; **(b)** +, ×, −

Margin Exercises, Section 4.7, pp. 270–276

1. $2.5 billion **2.** 199.1 lb **3.** $51.26 **4.** $368.75 **5.** 96.52 cm^2 **6.** $1.33 **7.** 28.6 mpg **8.** $716,667 **9.** $158,760

Translating for Success, p. 277

1. I **2.** C **3.** N **4.** A **5.** G **6.** B **7.** D **8.** O **9.** F **10.** M

Exercise Set 4.7, p. 278

1. $17.8 billion **3.** $19.15 **5.** 102.8°F **7.** $64,333,333.33 **9.** Area: 8.125 cm^2; perimeter: 11.5 cm **11.** 22,691.5 mi **13.** 20.2 mpg **15.** $30 **17.** 11.9752 ft^3 **19.** 78.1 cm **21.** 28.5 cm **23.** 2.31 cm **25.** 876 calories **27.** $1171.74 **29.** 227.75 ft^2 **31.** 0.362 **33.** $53.04 **35.** 2152.56 yd^2 **37.** 10.8¢ **39.** $906.50 **41.** 5.665 billion **43.** 1.4°F **45.** $266,791 **47.** $165,565 **49.** $1,131,429 **51.** DW **53.** 6335 **54.** $\frac{31}{24}$ **55.** $6\frac{5}{6}$ **56.** $\frac{23}{15}$ **57.** $\frac{1}{24}$ **58.** 2803 **59.** $\frac{2}{15}$ **60.** $1\frac{5}{6}$ **61.** $\frac{129}{251}$ **62.** $\frac{5}{16}$ **63.** $\frac{13}{25}$ **64.** $\frac{25}{19}$ **65.** 28 min **66.** $7\frac{1}{5}$ min **67.** 186 calories **68.** 30 calories **69.** $17.28

Concept Reinforcement, p. 285

1. False **2.** True **3.** True **4.** False **5.** True

Summary and Review: Chapter 4, p. 285

1. 6,590,000 **2.** 6,900,000 **3.** Three and forty-seven hundredths **4.** Thirty-one thousandths **5.** Twenty-seven and eleven hundred-thousandths **6.** Seven millionths **7.** $\frac{9}{100}$ **8.** $\frac{4561}{1000}$ **9.** $\frac{89}{1000}$ **10.** $\frac{30,227}{10,000}$ **11.** 0.034 **12.** 4.2603 **13.** 27.91 **14.** 867.006 **15.** 0.034 **16.** 0.91 **17.** 0.741 **18.** 1.041 **19.** 17.4 **20.** 17.43 **21.** 17.429 **22.** 17 **23.** 574.519 **24.** 0.6838 **25.** 229.1 **26.** 45.551 **27.** 29.2092 **28.** 790.29 **29.** 29.148 **30.** 70.7891 **31.** 12.96 **32.** 0.14442 **33.** 4.3 **34.** 0.02468 **35.** 7.5 **36.** 0.45 **37.** 45.2 **38.** 1.022 **39.** 0.2763 **40.** 1389.2 **41.** 496.2795 **42.** 6.95 **43.** 42.54 **44.** 4.9911 **45.** $15.52 **46.** 1.9 lb **47.** $912.68 **48.** $307.49 **49.** 14.5 mpg **50.** **(a)** 102.6 lb; **(b)** 14.7 lb **51.** 272 **52.** 216 **53.** 4 **54.** $125 **55.** 2.6 **56.** 1.28 **57.** 2.75 **58.** 3.25 **59.** $1.1\overline{6}$ **60.** $1.\overline{54}$ **61.** 1.5 **62.** 1.55 **63.** 1.545 **64.** $82.73 **65.** $4.87 **66.** 2493¢ **67.** 986¢ **68.** 1.8045 **69.** 57.1449 **70.** 15.6375 **71.** $41.537\overline{3}$ **72.** DW Multiply by 1 to get a denominator that is a power of 10:

$$\frac{44}{125} = \frac{44}{125} \cdot \frac{8}{8} = \frac{352}{1000} = 0.352.$$

We can also divide to find that $\frac{44}{125} = 0.352$. **73.** DW Each decimal place in the decimal notation corresponds to one zero in the power of ten in the fraction

notation. When the fractions are multiplied, the number of zeros in the denominator of the product is the sum of the number of zeros in the denominators of the factors. So the number of decimal places in the product is the sum of the number of decimal places in the factors.
74. **(a)** $2.56 \times 6.4 \div 51.2 - 17.4 + 89.7 = 72.62$;
(b) $(11.12 - 0.29) \times 3^4 = 877.23$
75. $\frac{1}{3} + \frac{2}{3} = 0.33333333\ldots + 0.66666666\ldots$
$= 0.99999999\ldots.$
Therefore, $1 = 0.99999999\ldots$ because $\frac{1}{3} + \frac{2}{3} = 1$.
76. $2 = 1.\overline{9}$ **77.** $a = 5, b = 9$

Test: Chapter 4, p. 288

1. [4.3b] 8,900,000,000 **2.** [4.3b] 3,756,000
3. [4.1a] Two and thirty-four hundredths
4. [4.1a] One hundred five and five ten-thousandths
5. [4.1b] $\frac{91}{100}$ **6.** [4.1b] $\frac{2769}{1000}$ **7.** [4.1b] 0.074
8. [4.1b] 3.7047 **9.** [4.1b] 756.09 **10.** [4.1b] 91.703
11. [4.1c] 0.162 **12.** [4.1c] 0.078 **13.** [4.1c] 0.9
14. [4.1d] 6 **15.** [4.1d] 5.68 **16.** [4.1d] 5.678
17. [4.1d] 5.7 **18.** [4.2a] 0.7902 **19.** [4.2a] 186.5
20. [4.2a] 1033.23 **21.** [4.2b] 48.357 **22.** [4.2b] 19.0901
23. [4.2b] 152.8934 **24.** [4.3a] 0.03 **25.** [4.3a] 0.21345
26. [4.3a] 73,962 **27.** [4.4a] 4.75 **28.** [4.4a] 30.4
29. [4.4a] 0.19 **30.** [4.4a] 0.34689 **31.** [4.4a] 34,689
32. [4.4b] 84.26 **33.** [4.2c] 8.982 **34.** [4.7a] $314.99
35. [4.7a] 28.3 mpg **36.** [4.7a] $6572.45
37. [4.7a] $181.93 **38.** [4.7a] 58.24 million passengers
39. [4.6a] 198 **40.** [4.6a] 4 **41.** [4.5a] 1.6
42. [4.5a] 0.88 **43.** [4.5a] 5.25 **44.** [4.5a] 0.75
45. [4.5a] $1.\overline{2}$ **46.** [4.5a] $2.\overline{142857}$ **47.** [4.5b] 2.1
48. [4.5b] 2.14 **49.** [4.5b] 2.143 **50.** [4.3b] $9.49
51. [4.4c] 40.0065 **52.** [4.4c] 384.8464 **53.** [4.5c] 302.4
54. [4.5c] $52.339\overline{4}$ **55.** [4.7a] $35
56. [4.1b, c] $\frac{2}{3}, \frac{5}{7}, \frac{15}{19}, \frac{11}{13}, \frac{17}{20}, \frac{13}{15}$

CHAPTER 5

Margin Exercises, Section 5.1, pp. 292–295

1. $\frac{5}{11}$, or 5:11 **2.** $\frac{57.3}{86.1}$, or 57.3:86.1 **3.** $\frac{6\frac{3}{4}}{7\frac{2}{5}}$, or $6\frac{3}{4}:7\frac{2}{5}$
4. $\frac{739}{12}$ **5.** $\frac{12}{14}$ **6.** $\frac{71}{214.1}; \frac{214.1}{71}$ **7.** $\frac{38.2}{56.1}$ **8.** $\frac{205}{278}; \frac{278}{205}; \frac{278}{483}$
9. 18 is to 27 as 2 is to 3 **10.** 3.6 is to 12 as 3 is to 10
11. 1.2 is to 1.5 as 4 is to 5 **12.** $\frac{9}{16}$

Exercise Set 5.1, p. 296

1. $\frac{4}{5}$ **3.** $\frac{178}{572}$ **5.** $\frac{0.4}{12}$ **7.** $\frac{3.8}{7.4}$ **9.** $\frac{56.78}{98.35}$ **11.** $\frac{8\frac{3}{4}}{9\frac{5}{6}}$ **13.** $\frac{4}{1}$
15. $\frac{356}{100{,}000}; \frac{173}{100{,}000}$ **17.** $\frac{93.2}{1000}$ **19.** $\frac{190}{547}; \frac{547}{190}$ **21.** $\frac{60}{100}; \frac{100}{60}$
23. $\frac{2}{3}$ **25.** $\frac{3}{4}$ **27.** $\frac{12}{25}$ **29.** $\frac{7}{9}$ **31.** $\frac{2}{3}$ **33.** $\frac{14}{25}$ **35.** $\frac{1}{2}$
37. $\frac{3}{4}$ **39.** $\frac{478}{213}, \frac{213}{478}$ **41.** **DW** **43.** = **44.** ≠ **45.** ≠
46. = **47.** 50 **48.** 9.5 **49.** 14.5 **50.** 152
51. $6\frac{7}{20}$ cm **52.** $17\frac{11}{20}$ cm **53.** $\frac{30}{47}$ **55.** 1:2:3

Margin Exercises, Section 5.2, pp. 300–301

1. 5 mi/hr, or 5 mph **2.** 12 mi/hr, or 12 mph
3. $\frac{89}{13}$ km/h, or 6.85 km/h **4.** 1100 ft/sec **5.** 4 ft/sec
6. $\frac{121}{8}$ ft/sec, or 15.125 ft/sec **7.** $\frac{34 \text{ home runs}}{92 \text{ strikeouts}} \approx$ 0.370 home run per strikeout **8.** $\frac{714 \text{ home runs}}{1330 \text{ strikeouts}} \approx$ 0.537 home run per strikeout; Babe Ruth's rate is approximately 0.167 higher **9.** 7.45¢/oz
10. 24.143¢/oz; 25.9¢/oz; 24.17¢/oz; the 7-oz package has the lowest unit price

Exercise Set 5.2, p. 302

1. 40 km/h **3.** 7.48 mi/sec **5.** 24 mpg **7.** 23 mpg
9. $\frac{32{,}270 \text{ people}}{0.75 \text{ sq mi}}$; about 43,027 people/sq mi
11. 25 mph; 0.04 hr/mi **13.** About 18.3 points/game
15. 0.623 gal/ft^2 **17.** 186,000 mi/sec **19.** 124 km/h
21. 25 beats/min **23.** 19.185¢/oz; 15.709¢/oz; 25.4 oz
25. 11.5¢/oz; 13.833¢/oz; 16 oz
27. 18.174¢/oz; 15.275¢/oz; 34.5 oz
29. 10.5¢/oz; 11.607¢/oz; 12.475¢/oz; 12.484¢/oz; 18 oz
31. 8.58¢/oz; 5.29¢/oz; 5.245¢/oz; 5.263¢/oz; 200 fl oz
33. **DW** **35.** 1.7 million **36.** $37\frac{1}{2}$ servings
37. 30 tests **38.** 8.9 billion **39.** 109.608 **40.** 67,819
41. 5833.56 **42.** 466,190.4
43. 6-oz: 10.83¢/oz; 5.5-oz: 10.91¢/oz

Margin Exercises, Section 5.3, pp. 307–310

1. Yes **2.** No **3.** No **4.** Yes **5.** No **6.** Yes
7. 14 **8.** $11\frac{1}{4}$ **9.** 10.5 **10.** 2.64 **11.** 10.8
12. $\frac{125}{42}$, or $2\frac{41}{42}$

Calculator Corner, p. 310

1. Left to the student **2.** Left to the student **3.** 27.5625
4. 25.6 **5.** 15.140625 **6.** 40.03952941
7. 39.74857143 **8.** 119

Exercise Set 5.3, p. 311

1. No **3.** Yes **5.** Yes **7.** No **9.** 0.61; 0.66; 0.69; 0.66; the completion rates (rounded to the nearest hundredth) are the same for Brees and Roethlisberger
11. 45 **13.** 12 **15.** 10 **17.** 20 **19.** 5 **21.** 18
23. 22 **25.** 28 **27.** $9\frac{1}{3}$ **29.** $2\frac{8}{9}$ **31.** 0.06 **33.** 5
35. 1 **37.** 1 **39.** 14 **41.** $2\frac{3}{16}$ **43.** $\frac{51}{16}$, or $3\frac{3}{16}$
45. 12.5725 **47.** $\frac{1748}{249}$, or $7\frac{5}{249}$ **49.** **DW** **51.** quotient
52. sum **53.** average **54.** dollars, cents
55. subtrahend **56.** terminating **57.** commutative
58. cross products **59.** Approximately 2731.4
61. **(a)** Ruth: 1.863 strikeouts per home run; Schmidt: 3.436 strikeouts per home run; **(b)** Schmidt

Margin Exercises, Section 5.4, pp. 314–318

1. 445 calories **2.** 15 gal **3.** 8 shirts **4.** 38 in. or less **5.** 9.5 in. **6.** 2074 deer

Translating for Success, p. 319

1. N **2.** I **3.** A **4.** K **5.** J **6.** F **7.** M **8.** B **9.** G **10.** E

Exercise Set 5.4, p. 320

1. 11.04 hr **3.** 880 calories **5.** 177 million, or 177,000,000 **7.** 9.75 gal **9.** 175 bulbs **11.** 2975 ft^2 **13.** 450 pages **15.** **(a)** 23.63445 British pounds; **(b)** $16,450.56 **17.** **(a)** 21,206 Japanese yen; **(b)** $29.99 **19.** **(a)** About 112 gal; **(b)** 3360 mi **21.** 60 students **23.** 13,500 mi **25.** 120 lb **27.** 64 gal **29.** 100 oz **31.** 954 deer **33.** 58.1 mi **35.** **(a)** 56 games; **(b)** about 2197 points **37.** **DW** **39.** Neither **40.** Composite **41.** Prime **42.** Composite **43.** Prime **44.** $2 \cdot 2 \cdot 2 \cdot 101$, or $2^3 \cdot 101$ **45.** $2 \cdot 2 \cdot 7$, or $2^2 \cdot 7$ **46.** $2 \cdot 433$ **47.** $3 \cdot 31$ **48.** $2 \cdot 2 \cdot 5 \cdot 101$, or $2^2 \cdot 5 \cdot 101$ **49.** 17 positions **51.** 2150 earned runs **53.** CD player: $150; receiver: $450; speakers: $300

Margin Exercises, Section 5.5, pp. 325–328

1. 15 **2.** 24.75 ft **3.** 7.5 ft **4.** 21 cm **5.** 29 ft

Exercise Set 5.5, p. 329

1. 25 **3.** $\frac{4}{3}$, or $1\frac{1}{3}$ **5.** $x = \frac{27}{4}$, or $6\frac{3}{4}$; $y = 9$ **7.** $x = 7.5$; $y = 7.2$ **9.** 1.25 m **11.** 36 ft **13.** 7 ft **15.** 100 ft **17.** 4 **19.** $10\frac{1}{2}$ **21.** $x = 6$; $y = 5.25$; $z = 3$ **23.** $x = 5\frac{1}{3}$, or $5.\overline{3}$; $y = 4\frac{2}{3}$, or $4.\overline{6}$; $z = 5\frac{1}{3}$, or $5.\overline{3}$ **25.** 20 ft **27.** 152 ft **29.** **DW** **31.** $59.81 **32.** 9.63 **33.** 679.4928 **34.** 2.74568 **35.** 27,456.8 **36.** 0.549136 **37.** 0.85 **38.** 1.825 **39.** 0.909 **40.** 0.843 **41.** 13.75 ft **43.** 1.25 cm **45.** 3681.437 **47.** $x = 0.4$; $y \approx 0.35$

Concept Reinforcement, p. 333

1. True **2.** True **3.** False **4.** False **5.** True

Summary and Review: Chapter 5, p. 333

1. $\frac{47}{84}$ **2.** $\frac{46}{1.27}$ **3.** $\frac{83}{100}$ **4.** $\frac{0.72}{197}$ **5.** **(a)** $\frac{12,480}{16,640}$, or $\frac{3}{4}$; **(b)** $\frac{16,640}{29,120}$, or $\frac{4}{7}$ **6.** $\frac{3}{4}$ **7.** $\frac{9}{16}$ **8.** 26 mpg **9.** 6300 rpm **10.** 0.638 gal/ft^2 **11.** 0.72 serving/lb **12.** 4.33¢/tablet **13.** 14.173¢/oz **14.** 1.329¢/sheet; 1.554¢/sheet; 1.110¢/sheet; 6 big rolls **15.** 6.844¢/oz; 5.188¢/oz; 5.609¢/oz; 5.539¢/oz; 48 oz **16.** No **17.** No **18.** 32 **19.** 7 **20.** $\frac{1}{40}$ **21.** 24 **22.** $4.45 **23.** 351 circuits **24.** **(a)** 202 Euros; **(b)** $61.88 **25.** 832 mi **26.** 27 acres **27.** Approximately 3,293,558 kg **28.** 6 in. **29.** Approximately 2096 lawyers **30.** $x = \frac{14}{3}$, or $4\frac{2}{3}$ **31.** $x = \frac{56}{5}$, or $11\frac{1}{5}$; $y = \frac{63}{5}$, or $12\frac{3}{5}$ **32.** 40 ft **33.** $x = 3$; $y = 9$; $z = \frac{15}{2}$, or $7\frac{1}{2}$ **34.** **DW** In terms of cost, a low faculty-to-student ratio is less expensive than a high faculty-to-student ratio. In terms of quality of education and student satisfaction, a high faculty-to-student ratio is more desirable. A college president must balance the cost and quality issues. **35.** **DW** Leslie used 4 gal of gasoline to drive 92 mi. At the same rate, how many gallons would be needed to travel 368 mi? **36.** 105 min, or 1 hr 45 min **37.** $x = 4258.5$; $z \approx 10,094.3$ **38.** Finishing paint: 11 gal; primer: 16.5 gal

Test: Chapter 5, p. 336

1. [5.1a] $\frac{85}{97}$ **2.** [5.1a] $\frac{0.34}{124}$ **3.** [5.1b] $\frac{9}{10}$ **4.** [5.1b] $\frac{25}{32}$ **5.** [5.2a] 0.625 ft/sec **6.** [5.2a] $1\frac{1}{3}$ servings/lb **7.** [5.2a] 22 mpg **8.** [5.2b] About 7.765¢/oz **9.** [5.2b] 11.182¢/oz; 7.149¢/oz; 8.389¢/oz; 6.840¢/oz; 263 oz **10.** [5.3a] Yes **11.** [5.3a] No **12.** [5.3b] 12 **13.** [5.3b] 360 **14.** [5.3b] 42.1875 **15.** [5.3b] 100 **16.** [5.4a] 1512 km **17.** [5.4a] 4.8 min **18.** [5.4a] 525 mi **19.** [5.5a] 66 m **20.** [5.4a] **(a)** 3498.75 Hong Kong dollars; **(b)** $102.25 **21.** [5.4a] About 86,151 arrests **22.** [5.5a] $x = 8$; $y = 8.8$ **23.** [5.5b] $x = 8$; $y = 8$; $z = 12$ **24.** [5.4a] 5888

Cumulative Review: Chapters 1–5, p. 338

1. [1.6b], [4.3b], [4.4a] **(a)** $252,000,000; **(b)** $0.252 billion; **(c)** $25,200,000; **(d)** $45,487.36 **2.** [5.2a] 22 mpg **3.** [4.2a] 513.996 **4.** [3.5a] $6\frac{3}{4}$ **5.** [3.2a] $\frac{7}{20}$ **6.** [4.2b] 30.491 **7.** [4.2b] 72.912 **8.** [3.3a] $\frac{7}{60}$ **9.** [4.3a] 222.076 **10.** [4.3a] 567.8 **11.** [3.6a] 3 **12.** [4.4a] 43 **13.** [1.6b] 899 **14.** [2.7b] $\frac{3}{2}$ **15.** [1.1b] 3 ten thousands + 7 tens + 4 ones **16.** [4.1a] One hundred twenty and seven hundredths **17.** [4.1c] 0.7 **18.** [4.1c] 0.8 **19.** [2.1d] $2 \cdot 2 \cdot 2 \cdot 2 \cdot 3 \cdot 3$, or $2^4 \cdot 3^2$ **20.** [3.1a] 140 **21.** [2.3a] $\frac{5}{8}$ **22.** [2.5b] $\frac{5}{8}$ **23.** [4.5c] 5.718 **24.** [4.5c] 0.179 **25.** [5.1a] $\frac{0.3}{15}$ **26.** [5.3a] Yes **27.** [5.2a] 55 m/sec **28.** [5.2b] 11.769¢/oz; 9.344¢/oz; 9.728¢/oz; 6.357¢/oz; 7.002¢/oz; 28-oz Joy **29.** [5.3b] 30.24 **30.** [4.4b] 26.4375 **31.** [2.7c] $\frac{8}{9}$ **32.** [5.3b] 128 **33.** [4.2c] 33.34 **34.** [3.3c] $\frac{76}{175}$ **35.** [2.6b] 390 cal **36.** [5.4a] **(a)** 1338.7 Rand; **(b)** $336.89 **37.** [4.7a] 976.9 mi **38.** [5.4a] 7 min **39.** [3.5c] $2\frac{1}{4}$ cups **40.** [2.7d] 12 doors **41.** [4.2b], [2.7d] **(a)** 40,200 yr; **(b)** answers may vary **42.** [1.8a] **(a)** $360,000; **(b)** $144,000,000; **(c)** $1,728,000,000 **43.** [4.7a], [5.2a] 42.2025 mi **44.** [4.7a] 132 orbits **45.** [2.1c] (d) **46.** [5.2a] (b) **47.** [1.2b] (b) **48.** [5.5a] $10\frac{1}{2}$ ft

CHAPTER 6

Margin Exercises, Section 6.1, pp. 342–345

1. $\frac{70}{100}$; $70 \times \frac{1}{100}$; 70×0.01 **2.** $\frac{23.4}{100}$; $23.4 \times \frac{1}{100}$; 23.4×0.01
3. $\frac{100}{100}$; $100 \times \frac{1}{100}$; 100×0.01 **4.** 0.34 **5.** 0.789
6. 0.06625 **7.** 0.18 **8.** 0.0008 **9.** 24% **10.** 347%
11. 100% **12.** 32.1% **13.** 25.3%

Calculator Corner, p. 343

1. 0.14 **2.** 0.00069 **3.** 0.438 **4.** 1.25

Exercise Set 6.1, p. 346

1. $\frac{90}{100}$; $90 \times \frac{1}{100}$; 90×0.01 **3.** $\frac{12.5}{100}$; $12.5 \times \frac{1}{100}$; 12.5×0.01
5. 0.67 **7.** 0.456 **9.** 0.5901 **11.** 0.1 **13.** 0.01
15. 2 **17.** 0.001 **19.** 0.0009 **21.** 0.0018 **23.** 0.2319
25. 0.14875 **27.** 0.565 **29.** 0.09; 0.58 **31.** 0.44
33. 0.36 **35.** 47% **37.** 3% **39.** 870% **41.** 33.4%
43. 75% **45.** 40% **47.** 0.6% **49.** 1.7% **51.** 27.18%
53. 2.39% **55.** 26%; 38% **57.** 17.7% **59.** 21.5%
61. D_W **63.** $33\frac{1}{3}$ **64.** $37\frac{1}{2}$ **65.** $9\frac{3}{8}$ **66.** $18\frac{9}{16}$
67. $5\frac{11}{14}$ **68.** $111\frac{2}{3}$ **69.** $0.\overline{6}$ **70.** $0.\overline{3}$ **71.** $0.8\overline{3}$
72. $1.41\overline{6}$ **73.** $2.\overline{6}$ **74.** 0.9375

Margin Exercises, Section 6.2, pp. 350–352

1. 25% **2.** 62.5%, or $62\frac{1}{2}\%$ **3.** $66.\overline{6}\%$, or $66\frac{2}{3}\%$
4. $83.\overline{3}\%$, or $83\frac{1}{3}\%$ **5.** 57% **6.** 76% **7.** $\frac{3}{5}$ **8.** $\frac{13}{400}$
9. $\frac{2}{3}$
10.

FRACTION NOTATION	$\frac{1}{5}$	$\frac{5}{6}$	$\frac{3}{8}$
DECIMAL NOTATION	0.2	$0.83\overline{3}$	0.375
PERCENT NOTATION	20%	$83.\overline{3}\%$, or $83\frac{1}{3}\%$	$37\frac{1}{2}\%$

Calculator Corner, p. 350

1. 52% **2.** 38.46% **3.** 110.26% **4.** 171.43%
5. 59.62% **6.** 28.31%

Calculator Corner, p. 353

1. 30.54; 1.31% **2.** 32.05; 1.20% **3.** 34.47; 1.19%
4. 26.47; 1.00% **5.** 11.98; 4.32% **6.** 17.52; 0.89%

Exercise Set 6.2, p. 354

1. 41% **3.** 5% **5.** 20% **7.** 30% **9.** 50%
11. 87.5%, or $87\frac{1}{2}\%$ **13.** 80% **15.** $66.\overline{6}\%$, or $66\frac{2}{3}\%$
17. $16.\overline{6}\%$, or $16\frac{2}{3}\%$ **19.** 18.75%, or $18\frac{3}{4}\%$
21. 81.25%, or $81\frac{1}{4}\%$ **23.** 16% **25.** 5% **27.** 34%
29. 40%; 18% **31.** 22% **33.** 5% **35.** 9% **37.** $\frac{17}{20}$
39. $\frac{5}{8}$ **41.** $\frac{1}{3}$ **43.** $\frac{1}{6}$ **45.** $\frac{29}{400}$ **47.** $\frac{1}{125}$ **49.** $\frac{203}{800}$
51. $\frac{176}{225}$ **53.** $\frac{711}{1100}$ **55.** $\frac{3}{2}$ **57.** $\frac{13}{40,000}$ **59.** $\frac{1}{3}$ **61.** $\frac{2}{25}$
63. $\frac{3}{5}$ **65.** $\frac{1}{50}$ **67.** $\frac{7}{20}$ **69.** $\frac{47}{100}$
71.

FRACTION NOTATION	DECIMAL NOTATION	PERCENT NOTATION
$\frac{1}{8}$	0.125	12.5%, or $12\frac{1}{2}\%$
$\frac{1}{6}$	$0.1\overline{6}$	$16.\overline{6}\%$, or $16\frac{2}{3}\%$
$\frac{1}{5}$	0.2	20%
$\frac{1}{4}$	0.25	25%
$\frac{1}{3}$	$0.\overline{3}$	$33.\overline{3}\%$, or $33\frac{1}{3}\%$
$\frac{3}{8}$	0.375	37.5%, or $37\frac{1}{2}\%$
$\frac{2}{5}$	0.4	40%
$\frac{1}{2}$	0.5	50%

73.

FRACTION NOTATION	DECIMAL NOTATION	PERCENT NOTATION
$\frac{1}{2}$	0.5	50%
$\frac{1}{3}$	$0.\overline{3}$	$33.\overline{3}\%$, or $33\frac{1}{3}\%$
$\frac{1}{4}$	0.25	25%
$\frac{1}{6}$	$0.1\overline{6}$	$16.\overline{6}\%$, or $16\frac{2}{3}\%$
$\frac{1}{8}$	0.125	12.5%, or $12\frac{1}{2}\%$
$\frac{3}{4}$	0.75	75%
$\frac{5}{6}$	$0.8\overline{3}$	$83.\overline{3}\%$, or $83\frac{1}{3}\%$
$\frac{3}{8}$	0.375	37.5%, or $37\frac{1}{2}\%$

75. D_W **77.** 70 **78.** 5 **79.** 400 **80.** 18.75 **81.** 23.125 **82.** 25.5 **83.** 4.5 **84.** $8\frac{3}{4}$ **85.** $33\frac{1}{3}$ **86.** $37\frac{1}{2}$ **87.** $83\frac{1}{3}$ **88.** $20\frac{1}{2}$ **89.** $43\frac{1}{8}$ **90.** $62\frac{1}{6}$ **91.** $18\frac{3}{4}$ **92.** $7\frac{4}{9}$ **93.** $\frac{18}{17}$ **94.** $\frac{209}{10}$ **95.** $\frac{203}{2}$ **96.** $\frac{259}{8}$ **97.** $11.\overline{1}\%$ **99.** $257.\overline{46317}\%$ **101.** $0.01\overline{5}$ **103.** $1.04\overline{142857}$

Margin Exercises, Section 6.3, pp. 358–361

1. $12\% \times 50 = a$ **2.** $a = 40\% \times 60$ **3.** $45 = 20\% \times t$ **4.** $120\% \times y = 60$ **5.** $16 = n \times 40$ **6.** $b \times 84 = 10.5$ **7.** 6 **8.** \$35.20 **9.** 225 **10.** \$50 **11.** 40% **12.** 12.5%

Calculator Corner, p. 362

1. 1.2 **2.** \$5.04 **3.** 48.64 **4.** \$22.40 **5.** 0.0112 **6.** \$29.70 **7.** 450 **8.** \$1000 **9.** 2.5% **10.** 12%

Exercise Set 6.3, p. 363

1. $y = 32\% \times 78$ **3.** $89 = a \times 99$ **5.** $13 = 25\% \times y$ **7.** 234.6 **9.** 45 **11.** \$18 **13.** 1.9 **15.** 78% **17.** 200% **19.** 50% **21.** 125% **23.** 40 **25.** \$40 **27.** 88 **29.** 20 **31.** 6.25 **33.** \$846.60 **35.** D_W **37.** $\frac{9}{100}$ **38.** $\frac{179}{100}$ **39.** $\frac{875}{1000}$, or $\frac{7}{8}$ **40.** $\frac{125}{1000}$, or $\frac{1}{8}$ **41.** $\frac{9375}{10{,}000}$, or $\frac{15}{16}$ **42.** $\frac{6875}{10{,}000}$, or $\frac{11}{16}$ **43.** 0.89 **44.** 0.07 **45.** 0.3 **46.** 0.017 **47.** \$800 (can vary); \$843.20 **49.** \$10,000 (can vary); \$10,400 **51.** \$1875

Margin Exercises, Section 6.4, pp. 366–368

1. $\frac{12}{100} = \frac{a}{50}$ **2.** $\frac{40}{100} = \frac{a}{60}$ **3.** $\frac{130}{100} = \frac{a}{72}$ **4.** $\frac{20}{100} = \frac{45}{b}$ **5.** $\frac{120}{100} = \frac{60}{b}$ **6.** $\frac{N}{100} = \frac{16}{40}$ **7.** $\frac{N}{100} = \frac{10.5}{84}$ **8.** \$225 **9.** 35.2 **10.** 6 **11.** 50 **12.** 30% **13.** 12.5%

Exercise Set 6.4, p. 369

1. $\frac{37}{100} = \frac{a}{74}$ **3.** $\frac{N}{100} = \frac{4.3}{5.9}$ **5.** $\frac{25}{100} = \frac{14}{b}$ **7.** 68.4 **9.** 462 **11.** 40 **13.** 2.88 **15.** 25% **17.** 102% **19.** 25% **21.** 93.75% **23.** \$72 **25.** 90 **27.** 88 **29.** 20 **31.** 25 **33.** \$780.20 **35.** D_W **37.** 8 **38.** 4000 **39.** 8 **40.** 2074 **41.** 100 **42.** 15 **43.** $8.0\overline{4}$ **44.** $\frac{3}{16}$, or 0.1875 **45.** $\frac{43}{48}$ qt **46.** $\frac{1}{8}$ T **47.** \$1134 (can vary); \$1118.64

Margin Exercises, Section 6.5, pp. 372–377

1. About 9.5% **2.** 14,560,000 workers **3.** **(a)** \$1475; **(b)** \$38,350 **4.** **(a)** \$9218.75; **(b)** \$27,656.25 **5.** About 2.9% **6.** About 76.6%

Translating for Success, p. 378

1. J **2.** M **3.** N **4.** E **5.** G **6.** H **7.** O **8.** C **9.** D **10.** B

Exercise Set 6.5, p. 379

1. About 13,247 wild horses **3.** 3 years: \$21,080; 5 years: \$17,680 **5.** Overweight: 176.4 million people; obese: 73.5 million people **7.** Acid: 20.4 mL; water: 659.6 mL **9.** About 1808 miles **11.** About 39,867,000 people **13.** 36.4 correct; 3.6 incorrect **15.** 95 items **17.** 25% **19.** 166; 156; 146; 140; 122 **21.** 8% **23.** 20% **25.** About 86.5% **27.** \$30,030 **29.** \$16,174.50; \$12,130.88 **31.** About 27% **33.** 34.375%, or $34\frac{3}{8}\%$ **35.** **(a)** 7.5%; **(b)** 879,675 **37.** 71% **39.** \$1560 **41.** 80% **43.** 98,775; 18.0% **45.** 799,065; 14.8% **47.** 4,550,688; 38.1% **49.** \$36,400 **51.** 40% **53.** D_W **55.** $2.\overline{27}$ **56.** 0.44 **57.** 3.375 **58.** $4.\overline{7}$ **59.** 0.92 **60.** $0.8\overline{3}$ **61.** 0.4375 **62.** 2.317 **63.** 3.4809 **64.** 0.675 **65.** About 5 ft 6 in. **67.** $83\frac{1}{3}\%$

Margin Exercises, Section 6.6, pp. 385–389

1. \$48.50; \$717.45 **2.** \$5.39; \$140.14 **3.** 6% **4.** \$999 **5.** \$5628 **6.** 12.5%, or $12\frac{1}{2}\%$ **7.** \$1675 **8.** \$180; \$360 **9.** 20%

Exercise Set 6.6, p. 390

1. \$19.53 **3.** \$2.65 **5.** \$16.39; \$361.39 **7.** 5% **9.** 4% **11.** \$2000 **13.** \$800 **15.** \$719.86 **17.** 5.6% **19.** \$2700 **21.** 5% **23.** \$980 **25.** \$5880 **27.** 12% **29.** \$420 **31.** \$30; \$270 **33.** \$2.55; \$14.45 **35.** \$125; \$112.50 **37.** 40%; \$360 **39.** \$30; 16.7% **41.** \$549; 36.4% **43.** D_W **45.** D_W **47.** 18 **48.** $\frac{22}{7}$ **49.** 265.625 **50.** 1.113 **51.** $0.\overline{5}$ **52.** $2.\overline{09}$ **53.** $0.91\overline{6}$ **54.** $1.\overline{857142}$ **55.** $2.\overline{142857}$ **56.** $1.58\overline{3}$ **57.** 4,030,000,000,000 **58.** 5,800,000 **59.** 42,700,000 **60.** 6,090,000,000,000 **61.** \$2.69 **63.** He bought the plaques for \$$166\frac{2}{3}$ + \$250, or \$$416\frac{2}{3}$, and sold them for \$400, so he lost money.

Margin Exercises, Section 6.7, pp. 394–397

1. \$301 **2.** \$225.75 **3.** **(a)** \$33.53; **(b)** \$4833.53 **4.** \$2376.20 **5.** \$7690.94

Calculator Corner, p. 397

1. \$16,357.18 **2.** \$12,764.72

Exercise Set 6.7, p. 398

1. \$8 **3.** \$84 **5.** \$113.52 **7.** \$925 **9.** \$671.88 **11.** **(a)** \$147.95; **(b)** \$10,147.95 **13.** **(a)** \$80.14; **(b)** \$6580.14 **15.** **(a)** \$46.03; **(b)** \$5646.03 **17.** \$441 **19.** \$2802.50 **21.** \$7853.38 **23.** \$99,427.40 **25.** \$4243.60 **27.** \$28,225.00 **29.** \$9270.87 **31.** \$129,871.09 **33.** \$4101.01 **35.** \$1324.58 **37.** D_W **39.** reciprocals **40.** divisible by 6 **41.** additive **42.** unit rate **43.** perimeter **44.** divisible by 3 **45.** prime **46.** proportional **47.** 9.38%

Margin Exercises, Section 6.8, pp. 402–406

1. **(a)** \$97; **(b)** interest: \$86.40; amount applied to principal: \$10.60; **(c)** interest: \$55.17; amount applied to principal: \$41.83; **(d)** At 13.6%, the principal was decreased by \$31.23 more than at the 21.3% rate. The interest at 13.6% is \$31.23 less than at 21.3%. **2.** **(a)** Interest: \$91.78; amount applied to principal: \$229.22; **(b)** \$58; **(c)** \$4080 **3.** Interest: \$843.94; amount applied to principal: \$135.74 **4.** **(a)** Interest: \$844.69; amount applied to principal: \$498.64; **(b)** \$88,799.40; **(c)** The Sawyers will pay \$110,885.40 less in interest with the 15-yr loan than with the 30-yr loan. **5.** **(a)** Interest: \$701.25; amount applied to principal: \$548.89; **(b)** \$72,025.20; **(c)** The Sawyers will pay \$16,774.20 less in interest with the 15-yr loan at $5\frac{1}{2}$% than with the 15-yr loan at $6\frac{5}{8}$%.

Exercise Set 6.8, p. 407

1. **(a)** \$98; **(b)** interest: \$86.56; amount applied to principal: \$11.44; **(c)** interest: \$51.20; amount applied to principal: \$46.80; **(d)** At 12.6%, the principal is decreased by \$35.36 more than at the 21.3% rate. The interest at 12.6% is \$35.36 less than at 21.3%. **3.** **(a)** Interest: \$125.14; amount applied to principal: \$312.79; **(b)** \$51.24; **(c)** \$7991.60, \$11,504, \$3512.40 **5.** **(a)** Interest: \$854.17; amount applied to principal: \$155.61; **(b)** \$199,520.80; **(c)** new principal: \$163,844.39; interest: \$853.36; amount applied to principal: \$156.42 **7.** **(a)** Interest: \$854.17; amount applied to principal: \$552; **(b)** \$89,110.60; **(c)** The Martinez family will pay \$110,410.20 less in interest with the 15-yr loan than with the 30-yr loan. **9.** \$99,917.71; \$99,834.94 **11.** \$99,712.04; \$99,422.15 **13.** \$149,882.75; \$149,764.79 **15.** \$199,382.07; \$198,760.41 **17.** **(a)** \$2395, \$21,555; **(b)** \$52.09, \$401.97; **(c)** \$239.88 **19.** **(a)** \$595; \$11,305; **(b)** interest: \$87.61; amount applied to principal: \$273.47; **(c)** \$1693.88 **21.** D_W **23.** D_W **25.** 17.5, or $\frac{35}{2}$ **26.** 28 **27.** 100 **28.** 56.25, or $\frac{225}{4}$ **29.** $\frac{66{,}875}{36}$, or approximately 1857.64 **30.** 787.69 **31.** 71.81 **32.** 17.5, or $\frac{35}{2}$ **33.** 56.25, or $\frac{225}{4}$ **34.** $\frac{66{,}875}{36}$, or approximately 1857.64

Concept Reinforcement, p. 410

1. True **2.** False **3.** True **4.** False

Summary and Review: Chapter 6, p. 410

1. 56% **2.** 1.7% **3.** 37.5% **4.** $33.\overline{3}$%, or $33\frac{1}{3}$% **5.** 0.735 **6.** 0.065 **7.** $\frac{6}{25}$ **8.** $\frac{63}{1000}$ **9.** $30.6 = p \times 90$; 34% **10.** $63 = 84\% \times n$; 75 **11.** $y = 38\frac{1}{2}\% \times 168$; 64.68 **12.** $\frac{24}{100} = \frac{16.8}{b}$; 70 **13.** $\frac{42}{30} = \frac{N}{100}$; 140% **14.** $\frac{10.5}{100} = \frac{a}{84}$; 8.82 **15.** 223 students; 105 students **16.** 42% **17.** 2500 mL **18.** 12% **19.** 93.15 **20.** \$14.40 **21.** 5% **22.** 11% **23.** \$42; \$308 **24.** \$42.70; \$262.30 **25.** \$2940 **26.** Approximately 18.4% **27.** \$36 **28.** **(a)** \$394.52; **(b)** \$24,394.52 **29.** \$121 **30.** \$7727.26 **31.** \$9504.80 **32.** **(a)** \$129; **(b)** interest: \$100.18; amount applied to principal: \$28.82; **(c)** interest: \$70.72; amount applied to principal: \$58.28; **(d)** At 13.2%, the principal is decreased by \$29.46 more than at the 18.7% rate. The interest at 13.2% is \$29.46 less than at 18.7%. **33.** D_W No; the 10% discount was based on the original price rather than on the sale price. **34.** D_W A 40% discount is better. When successive discounts are taken, each is based on the previous discounted price rather than on the original price. A 20% discount followed by a 22% discount is the same as a 37.6% discount off the original price. **35.** 19.5% increase **36.** $66\frac{2}{3}$% **37.** \$168

Test: Chapter 6, p. 413

1. [6.1b] 0.064 **2.** [6.1b] 38% **3.** [6.2a] 137.5% **4.** [6.2b] $\frac{13}{20}$ **5.** [6.3a, b] $a = 40\% \cdot 55$; 22 **6.** [6.4a, b] $\frac{N}{100} = \frac{65}{80}$; 81.25% **7.** [6.5a] 400 passengers; 575 passengers **8.** [6.5a] About 539 at-bats

9. [6.5b] $50.\overline{90}\%$ **10.** [6.5a] 5.5% **11.** [6.6a] \$16.20; \$340.20 **12.** [6.6b] \$630 **13.** [6.6c] \$40; \$160 **14.** [6.7a] \$8.52 **15.** [6.7a] \$5356 **16.** [6.7b] \$1110.39 **17.** [6.7b] \$11,580.07 **18.** [6.5b] Registered nurses: 2.9, 26.1%; post-secondary teachers: 0.6, 37.5%; food preparation and service workers: 2.0, 20%; restaurant servers: 2.1, 0.4 **19.** [6.6c] \$275, about 14.1%
20. [6.8a] \$119,909.14; \$119,817.72
21. [6.6b] \$194,600 **22.** [6.6b], [6.7b] \$2546.16

Cumulative Review/Final Examination: Chapters 1–6, p. 415

1. [6.1b] 0.53 **2.** [6.1b] 37.5% **3.** [6.2a] 112.5%
4. [4.5a] $2.1\overline{6}$ **5.** [5.1a] $\frac{10}{1}$ **6.** [5.2a] $23\frac{1}{3}$ km/h
7. [3.3b] $<$ **8.** [3.3b] $<$ **9.** [4.6a] 296,200
10. [1.4b] 50,000 **11.** [1.9c, d] 13 **12.** [4.4c] 1.5
13. [3.5a] $3\frac{1}{30}$ **14.** [4.2a] 49.74 **15.** [1.2a] 515,150
16. [4.2b] 0.02 **17.** [3.5b] $\frac{2}{3}$ **18.** [3.3a] $\frac{2}{63}$
19. [2.4b] $\frac{1}{6}$ **20.** [1.5a] 853,142,400 **21.** [4.3a] 1.38036
22. [3.6b] $1\frac{1}{2}$ **23.** [4.4a] 12.25 **24.** [3.4b] $123\frac{1}{3}$
25. [1.7b] 95 **26.** [4.2c] 8.13 **27.** [2.7c] 9 **28.** [3.3c] $\frac{1}{12}$
29. [5.3b] 40 **30.** [5.3b] $8\frac{8}{21}$
31. [4.3b], [4.7a], [6.5b] **(a)** \$15,100,000; \$16,400,000; \$9,100,000; \$6,600,000; \$3,300,000; **(b)** \$50.5 million; **(c)** \$10.1 million; **(d)** 8.6%; **(e)** 44.5%

32. [4.7a], [6.5b] **(a)** \$1762.3 billion; **(b)** about 14.9%
33. [4.7a] About 47 messages **34.** [5.4a] About 53 games
35. [4.7a] \$42.50 **36.** [5.2b] 4.995¢/oz **37.** [3.2b] $1\frac{1}{2}$ mi
38. [5.4a] 60 mi **39.** [6.7b] \$12,663.69
40. [3.6c] 5 pieces **41.** [6.5b] Office manager: 1924, 19.4%; Office clerk: 463, 15.3%; Teacher assistant: 1192, 31.5%; Host/hostess: 297, 54 **42.** [6.5b] (b)
43. [4.5a] (e) **44.** [4.7a] (d) **45.** [3.3a] (c)
46. **(a)** [5.4a]

Calories 120; 160		
Calories from fat $6.\overline{6}$; $6.\overline{6}$		
Total fat $0.\overline{6}$ g	1.3%	1.3%
Saturated fat 0 g	0%	0%
Polyunsaturated fat 0 g		
Monounsaturated fat 0 g		
Cholesterol 0 mg	0%	0%
Sodium 280 mg	12%	14%
Potassium $226.\overline{6}$ mg	$6.\overline{6}\%$	$12.\overline{6}\%$
Total carbohydrate: $30.\overline{6}$ g	$10.\overline{6}\%$	$12.\overline{6}\%$
Dietary fiber $6.\overline{6}$ g	$26.\overline{6}\%$	$26.\overline{6}\%$
Soluble fiber $1.\overline{3}$ g		
Insoluble fiber $5.\overline{3}$ g		
Sugars $6.\overline{6}$ g		
Other carbohydrate $17.\overline{3}$ g		
Protein 4 g;		

(b) Complete; **(c)** Wheaties; **(d)** Complete; **(e)** Complete

Glossary

A

Absolute value The distance that a number is from 0 on the number line

Addend In addition, a number being added

Additive identity The number 0

Area The number of square units that fill a plane region

Arithmetic numbers The whole numbers and the positive fractions

Associative law of addition The statement that when three numbers are added, regrouping the addends gives the same sum

Associative law of multiplication The statement that when three numbers are multiplied, regrouping the factors gives the same product

Average A center point of a set of numbers found by adding the numbers and dividing by the number of items of data; also called the arithmetic mean or mean

B

Base In exponential notation, the number being raised to a power

C

Commutative law of addition The statement that when two numbers are added, changing the order in which the numbers are added does not affect the sum

Commutative law of multiplication The statement that when two numbers are multiplied, changing the order in which the numbers are multiplied does not affect the product

Commission A percent of total sales paid to a salesperson

Composite number A natural number, other than 1, that is not prime

Compound interest Interest paid on interest

Cross products Given an equation with a single fraction on each side, the products formed by multiplying the left numerator and the right denominator, and the left denominator and the right numerator

D

Decimal notation A representation of a number containing a decimal point

Denominator The number below the fraction bar in a fraction

Difference The result of subtracting one number from another

Digit A number 0, 1, 2, 3, 4, 5, 6, 7, 8, or 9 that names a place-value location

Discount The amount subtracted from the original price of an item to find the sales price

Distributive law of multiplication over addition The statement that multiplying a factor by the sum of two numbers gives the same result as multiplying the factor by each of the two numbers and then adding

Distributive law of multiplication over subtraction The statement that multiplying a factor by the difference of two numbers gives the same result as multiplying the factor by each of the two numbers and then subtracting

Dividend In division, the number being divided

Divisible The number a is divisible by another number b if there exists a number c such that $a = b \cdot c$.

Divisor In division, the number dividing another number

E

Equation A number sentence that says that the expressions on either side of the equals sign, =, represent the same number

Even number A number that is divisible by 2; that is, it has an even ones digit

Exponential notation A representation of a number using a base raised to a power

Exponent In expressions of the form a^n, the number n is an exponent. For n a natural number, a^n represents n factors of a.

F

Factor *Verb:* to write an equivalent expression that is a product. *Noun:* a multiplier

Factorization A number expressed as a product of natural numbers
Fraction notation A number written using a numerator and a denominator

I

Inequality A mathematical sentence using $<$, $>$, $\leq$, $\geq$, or $\neq$.
Interest A percentage of an amount invested or borrowed

L

Least common denominator (LCD) The least common multiple of the denominators of two or more fractions
Least common multiple (LCM) The smallest number that is a multiple of two or more numbers

M

Marked price The original price of an item
Minuend The number from which another number is being subtracted
Mixed numeral A number represented by a whole number and a fraction less than 1
Multiple of a number A product of the number and some natural number
Multiplicative identity The number 1

N

Natural numbers The counting numbers: 1, 2, 3, 4, 5, ...
Numerator The number above the fraction bar in a fraction

O

Original price The price of an item before a discount is deducted

P

Percent notation A representation of a number as parts per 100
Perimeter The sum of the lengths of the sides of a polygon
Prime factorization A factorization of a composite number as a product of prime numbers
Prime number A natural number that has exactly two different factors: itself and 1
Purchase price The price of an item before sales tax is added

Q

Quotient The result when one number is divided by another

R

Rate A ratio used to compare two different kinds of measure
Ratio The quotient of two quantities
Rational number A number that can be written in the form a/b, where a and b are integers and $b \neq 0$
Reciprocal Two numbers are reciprocals if their product is 1.
Right angle An angle whose measure is $90°$
Rounding Approximating the value of a number; used when estimating

S

Sale price The price of an item after a discount has been deducted
Sales tax A tax added to the purchase price of an item
Simple interest A percentage of an amount P invested or borrowed for t years, computed by calculating principal $\times$ interest rate $\times$ time
Simplify To rewrite an expression in an equivalent, abbreviated form
Solution of an equation A replacement or substitution that makes an equation true
Subtrahend In subtraction, the number being subtracted
Sum The result in addition

T

Terminating decimal A decimal that can be written using a finite number of decimal places
Total price The sum of the purchase price of an item and the sales tax on the item
Triangle A three-sided polygon

U

Unit price The ratio of price to the number of units; also called unit rate
Unit rate The ratio of price to the number of units; also called unit price

V

Variable A letter that represents an unknown number

W

Whole numbers The natural numbers and 0: 0, 1, 2, 3, ...

Photo Credits

xxix, Michael St.–Andre **6,** © Ed Quinn, CORBIS **7,** Tony Stone/Getty **8 (left),** © Jeff Greenberg/The Image Works; **(right),** © CLARO CORTES IV/Reuters/Corbis **36,** © Kin Cheung/Reuters/Corbis **38,** © Michael S. Yamshita/CORBIS **69,** © Pete Saloutos/CORBIS **72,** CORBIS **84,** Tony Stone/Getty **104 (top),** © Keith Dannemiller/CORBIS; **(bottom),** CORBIS **107 (Kidman),** © Rufus F. Folkks/CORBIS; **(Berry),** © Eric Robert/CORBIS SYGMA; **(Dole),** © Reuters/CORBIS; **(Swank),** © Christy Bowe/CORBIS; **(Paltrow),** © Dusko Despotovic/Corbis; **(Roberts),** © Reuters/CORBIS; **(Feinstein),** © Roger Ressmeyer/CORBIS **112,** © Ariel Skelley/Corbis Stock Market **113,** © LWA-JDC/CORBIS **119,** © Laura Dwight/CORBIS **128,** The Image Bank **138,** © Bill Varie/CORBIS **146 (left),** Greg Probst/Stone/Getty; **(right),** © Craig Aurness/CORBIS **163,** photodisc **164,** © Ted Horowitz/CORBIS **171,** © Royalty-Free/Corbis **177,** photodisc **182,** J. P. Jenkins **183,** © Joseph Sohm; ChromoSohm Inc./CORBIS **184,** John Deere **186,** © Anthony Redpath/CORBIS **197 (left),** © Ted Spiegel/CORBIS; **(right),** © Royalty-Free/Corbis **199,** photodisc **202,** © Annie Griffiths Belt/CORBIS **211,** CORBIS **214 (left),** photodisc; **(right),** Karen Bittinger **219 (left),** Plush Studios/Getty; **(right),** © Rick Rickman/NewSport **225 (left),** Digital Vision/Getty; **(right),** © Sion Touhig/Corbis **239,** Bettman/CORBIS **257,** Mark Gibson, Digital Vision/Getty **261,** Chuck Savage/CORBIS **262, 265, 268,** © Reuters/CORBIS **276,** Thinkstock/Getty **285,** © Tom Nebbia/Corbis Stock Market **286,** © Royalty-Free/Corbis **292,** David Muench, CORBIS **293,** © Matt A. Brown/NewSport **294,** apwire photos **296 (left),** © Royalty-Free/Corbis; **(right),** Albert J. Copley, photodisc **297,** © Reuters/CORBIS **300 (top),** Eyewire Collection; **(bottom),** © Reuters/CORBIS **301,** Dick Morton **303 (top),** © JOE GIZA/Reuters/Corbis; **(left),** © Royalty-Free/Corbis; **(right),** © Joe McDonald/CORBIS **304,** Jeremy Woodhouse, Photodisc **306 (top),** © Reuters/CORBIS; **(bottom),** © Bil Janscha/Reuters/Corbis **314,** Brian Spurlock **315,** Buccina Studios, photodisc **318,** Nicholas Devore III/Photographers Aspen **321 (left),** Barry Rosenthal, FPG International; **(right),** Keith Brofsky, photodisc **322 (left),** © Rob Lewine/CORBIS; **(right),** © Kevin Fleming, Corbis **323 (left),** © Royalty-Free/Corbis; **(right),** © Chase Jarvis/CORBIS **333,** Karen Bittinger **334,** Dick Morton **335 (top),** Patrick Clark, photodisc; **(bottom),** Phil Schermeister, Corbis **339,** © Carl & Ann Purcell/CORBIS **342,** photodisc **343,** © Turbo/zefa/CORBIS **347, 348,** photodisc **350,** © Guzelian Asadour/CORBIS **359, 360, 361, 365, 372,** © Royalty-Free/CORBIS **376,** Digital Vision **379,** photodisc **380,** © David H. Wells/CORBIS **391,** Christopher Bissell, Stone/Getty Images **401,** photodisc **403, 404, 413,** © Royalty-Free/CORBIS **415,** EyeWire

Index

A

B

C

Q

R

S

T

U

W

Y

Z

Geometric Formulas

PLANE GEOMETRY

Rectangle
Area: $A = l \cdot w$
Perimeter: $P = 2 \cdot l + 2 \cdot w$

Square
Area: $A = s^2$
Perimeter: $P = 4 \cdot s$

Triangle
Area: $A = \frac{1}{2} \cdot b \cdot h$

Sum of Angle Measures
$A + B + C = 180°$

Right Triangle
Pythagorean Theorem:
$a^2 + b^2 = c^2$

Parallelogram
Area: $A = b \cdot h$

Trapezoid
Area: $A = \frac{1}{2} \cdot h \cdot (a + b)$

Circle
Area: $A = \pi \cdot r^2$
Circumference:
$C = \pi \cdot d = 2 \cdot \pi \cdot r$
($\frac{22}{7}$ and 3.14 are different approximations for π)

SOLID GEOMETRY

Rectangular Solid
Volume: $V = l \cdot w \cdot h$

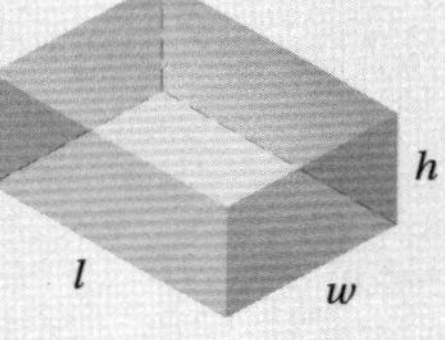
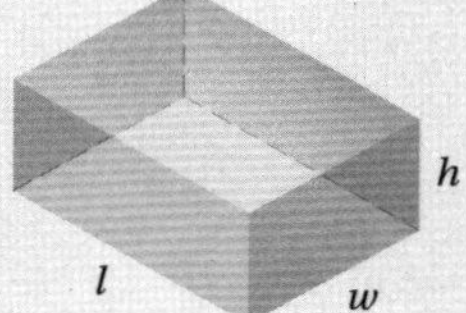

Cube
Volume: $V = s^3$

Right Circular Cylinder
Volume: $V = \pi \cdot r^2 \cdot h$
Surface Area:
$S = 2 \cdot \pi \cdot r \cdot h + 2 \cdot \pi \cdot r^2$

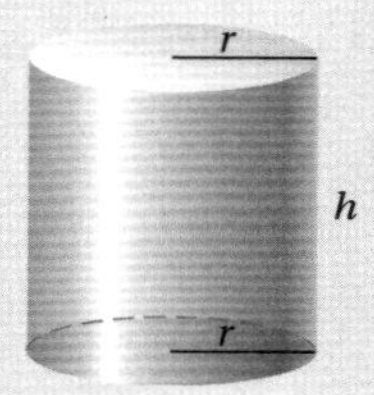

Right Circular Cone
Volume: $V = \frac{1}{3} \cdot \pi \cdot r^2 \cdot h$
Surface Area: $S = \pi \cdot r^2 + \pi \cdot r \cdot s$

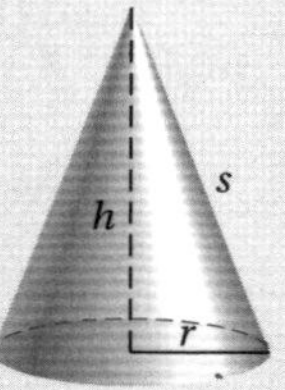

Sphere
Volume: $V = \frac{4}{3} \cdot \pi \cdot r^3$
Surface Area: $S = 4 \cdot \pi \cdot r^2$

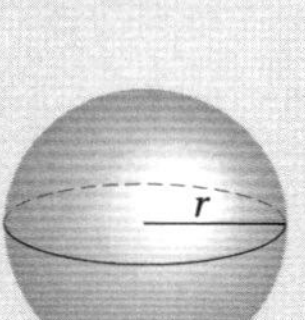

Fraction, Decimal, and Percent Equivalents

Fraction Notation	$\frac{1}{10}$	$\frac{1}{8}$	$\frac{1}{6}$	$\frac{1}{5}$	$\frac{1}{4}$	$\frac{3}{10}$	$\frac{1}{3}$	$\frac{3}{8}$	$\frac{2}{5}$	$\frac{1}{2}$	$\frac{3}{5}$	$\frac{5}{8}$	$\frac{2}{3}$	$\frac{7}{10}$	$\frac{3}{4}$	$\frac{4}{5}$	$\frac{5}{6}$	$\frac{7}{8}$	$\frac{9}{10}$	$\frac{1}{1}$
Decimal Notation	0.1	0.125	$0.16\overline{6}$	0.2	0.25	0.3	$0.33\overline{3}$	0.375	0.4	0.5	0.6	0.625	$0.66\overline{6}$	0.7	0.75	0.8	$0.83\overline{3}$	0.875	0.9	1
Percent Notation	10%	12.5% or $12\frac{1}{2}$%	$16.\overline{6}$% or $16\frac{2}{3}$%	20%	25%	30%	$33.\overline{3}$% or $33\frac{1}{3}$%	37.5% or $37\frac{1}{2}$%	40%	50%	60%	62.5% or $62\frac{1}{2}$%	$66.\overline{6}$% or $66\frac{2}{3}$%	70%	75%	80%	$83.\overline{3}$% or $83\frac{1}{3}$%	87.5% or $87\frac{1}{2}$%	90%	100%

Fundamental Mathematics, Fourth Edition

Marvin L. Bittinger

The Bittinger System for Success—Make It Work for You!

Building on its reputation for accurate content and a unified system of instruction, the fourth edition of Bittinger's *Fundamental Mathematics* integrates success-building study tools, innovative pedagogy, and a comprehensive student-support package.

These resources, available through your bookstore or online at www.aw-bc.com/math, are designed to help you succeed in your course. For a complete supplements list with product descriptions, please refer to the Preface.

Student Supplements:

Student's Solutions Manual	(ISBN: 0-321-29607-9)
Collaborative Learning Activities Manual	(ISBN: 0-321-29604-4)
Digital Video Tutor	(ISBN: 0-321-29605-2)
Math Study Skills for Students Video on CD	(ISBN: 0-321-29745-8)
Work It Out! Chapter Test Video on CD	(ISBN: 0-321-41985-5)
MathXL® Tutorials on CD	(ISBN: 0-321-29747-4)
MathXL® (an access code is required)	www.mathxl.com
InterAct Math Tutorial Web site	www.interactmath.com
Addison-Wesley Math Tutor Center	www.aw-bc.com/tutorcenter

ISBN 0-321-31907-9